COLORIMETRY

The Wiley Bicentennial–Knowledge for Generations

Each generation has its unique needs and aspirations. When Charles Wiley first opened his small printing shop in lower Manhattan in 1807, it was a generation of boundless potential searching for an identity. And we were there, helping to define a new American literary tradition. Over half a century later, in the midst of the Second Industrial Revolution, it was a generation focused on building the future. Once again, we were there, supplying the critical scientific, technical, and engineering knowledge that helped frame the world. Throughout the 20th Century, and into the new millennium, nations began to reach out beyond their own borders and a new international community was born. Wiley was there, expanding its operations around the world to enable a global exchange of ideas, opinions, and know-how.

For 200 years, Wiley has been an integral part of each generation's journey, enabling the flow of information and understanding necessary to meet their needs and fulfill their aspirations. Today, bold new technologies are changing the way we live and learn. Wiley will be there, providing you the must-have knowledge you need to imagine new worlds, new possibilities, and new opportunities.

Generations come and go, but you can always count on Wiley to provide you the knowledge you need, when and where you need it!

William J. Pesce
President and Chief Executive Officer

Peter Booth Wiley
Chairman of the Board

COLORIMETRY

UNDERSTANDING THE CIE SYSTEM

Edited by

JÁNOS SCHANDA
University of Pannonia, Hungary

COMMISSION INTERNATIONALE DE L'ECLAIRAGE
INTERNATIONAL COMMISSION ON ILLUMINATION
INTERNATIONALE BELEUCHTUNGSKOMMISSION

BICENTENNIAL 1807 WILEY 2007

WILEY-INTERSCIENCE
A JOHN WILEY & SONS, INC., PUBLICATION

Published by John Wiley & Sons, Inc., Hoboken, New Jersey
Published simultaneously in Canada

Wiley Bicentennial Logo: Richard J. Pacifico

Library of Congress Cataloging-in-Publication Data:

Colorimetry : understanding the CIE system / edited by János Schanda.
p. cm.
ISBN 978-0-470-04904-4
1. Colorimetry. I. Schanda, János.
QD113.C65 2007
543′.55–dc22 2007026256

CONTENTS

Part II Colorimetric fundamentals

PREFACE

Colorimetry deals with the objective description of the physical correlates of color perception. Color, a fundamental part of visual perception, has fascinated mankind since the early days of human history. Initially, colored samples were used to give an objective description of color, and only at the beginning of the twentieth century, when the objective measurement of colored light became possible, did physicists develop methods for direct comparison of colored lights with reference lights.

In 1931, the International Commission on Illumination (Commission Internationale d'Eclairage: CIE) defined the Standard Observer for Colorimetry and the 1931 CIE System of Colorimetry. The basics of this system are still valid, and many industrial color evaluation methods are based on CIE colorimetry. Nevertheless, the past 75 years have seen a number of further recommendations, and CIE has published from time to time updates of its fundamental colorimetric publication (CIE Publication 15), the latest edition in 2004. Several further CIE publications, and recently also CIE standards, provide the necessary recommendations to perform laboratory and industrial color measurement.

The present book aims to collect all this scattered information under one cover, providing at the same time tutorial help to understand CIE recommendations. The book is divided into three main parts.

Part I, Historic retrospection, starts with an English translation of the 1931 colorimetry resolutions of the CIE. This is followed by reminiscences written 50 years later by Professor W. David Wright. In this essay Prof. Wright summarized his recollections of the turbulent session, when the colorimetric recommendations were drawn up. Professor Wright wrote this article for the Golden Jubilee celebration symposium held in London in 1981. It is a great honor for us that the heirs of Professor Wright and the copyright owner of the material gave their permission to reproduce this very interesting summary of the historic event of the 1931 CIE session.

Part II, Colorimetric fundamentals, discusses the CIE recommendations on colorimetry in a more tutorial form than can be found in CIE publications. This part consists of six chapters. It starts with the recommendations on basic colorimetry, and in this chapter the reader will find all the fundamental definitions and recommendations — but with some explanations that will make their practical use easier to understand. One of the most frequent uses of CIE colorimetry is color-difference calculation, and CIE has struggled with this subject ever since 1931. Chapter 4 provides a short overview of the evolution of CIE color-difference calculation and explains in detail the structure and use of the present recommendation, the CIEDE2000 formula. Two further chapters, belonging to this part of the book, deal with practical colorimetry: determination of tristimulus values from spectral measurements and measurement by the use of tristimulus colorimeters. Questions that are dealt with in fundamental CIE publications using only a few sentences are discussed here in detail and, with additional guidance on the necessary uncertainty analysis, the reader will be able to understand the fundamentals of colorimetric uncertainty analysis, a subject one cannot find explained in other texts. Two further chapters apply the colorimetric knowledge on the evaluation of the colorimetric parameters of computer input and output devices, the problem of the so-called color management and on the colorimetric qualification of lamplight, light-source color rendering, another hot topic of modern colorimetry due to the introduction of solid-state light sources, LEDs, and the difficulties encountered in the description of their color rendering. CIE is still working on this topic; thus no final recommendations will be found in this chapter, but the reader will get a clear picture of where the problems are with the present system and some hints as to how these problems could be overcome.

Part III, Advances in colorimetry, starts with the physiological basis of color-matching functions, showing the limits of classical colorimetry and areas for possible revision. This and the following chapter deal with the open problems of colorimetry, provide an insight into the physiological and psychological problems that colorimetry encounters, and, as far as possible, show how these problems could be solved. In this part of the book, the reader will also find two chapters dealing with the modern approaches of the colorimetric questions: a description of the color appearance models and the explanations as to how the CIECAM02 model, the model recently recommended by the CIE, can be used. The next step however, after understanding the color appearance of simple geometric shapes (color patches), is to understand the appearance of color in images. A chapter is devoted to this question, a research area, where no CIE recommendations are available at this moment, but CIE experts describe possible future developments. The situation is similar with the temporal and spatial problems of colorimetry, a field where, after 75 years of colorimetric calculations, only the very first steps of international consensus are under preparation. The section ends with a look into the future, written by Professor Robert Hunt, the well-known colorimetry expert.

The book contains a number of appendixes. The first two deal with the fundamental issues of uncertainty analysis and its colorimetric application. The reader will also find in this part of the book an outlook to other industries, mainly the paper, pulp, and the textile industries, where CIE recommendations are used in daily practice. Further, appendixes enumerate CIE publications and standards on colorimetry.

The book concentrates on CIE colorimetry and discusses its most recent achievements, and does not deal with other aspects of color; thus, for example, the development of color order systems, color esthetics, and color design principles are not included. We would like to stress, however, that the book is not a CIE publication. It has been written by experts, who participated in the preparation of CIE technical reports and standards, but they give their own interpretation on the content of these CIE recommendations. In the case of the most advanced subjects, they often convey their own evaluation of the subject (but in these instances the authors have tried to distinguish between the generally accepted, CIE-endorsed knowledge and their new ideas that might become part of future CIE publications). Authors come from different disciplines, have their own view on questions of colorimetry, and show the subject of color from their own perspective. This leads to the problem that some items are dealt in more then one chapter, but the intention of the Editor of the book was to permit these overlaps, as they enlighten the subject from the slightly different viewpoints, and the Editor was of the opinion that this will enrich the book further.

The book is intended for all those who have an interest in colorimetry and would like to learn about the most recent achievements in color measurement, including color appearance analysis. It should be useful both as background material for tutorial courses and to help the person in the laboratory or production facility to understand the underlying colorimetric principles of the equipment they use and the measurements they make.

The book provides the necessary information to use the CIE color-related technical reports and standards efficiently. Although the theme of the book is "Understanding the CIE System," we would like to emphasize once again that the content of the single chapters reflects the personal view of the authors and is not a CIE consensus publication.

The material of "Colorimetry: Understanding the CIE System; CIE Colorimetry 1931–2006" has been published to celebrate the 75th anniversary of the CIE Colorimetric System.

On behalf of the collective of the authors,
Ottawa, May 16, 2006.

János Schanda
Editor

Dear Reader,

Authors and Printer made every effort to get the final book in a form to you that is as far as possible free of errors. Unfortunately during multiple editing of the chapters some inconsistencies crept into the texts. We tried to correct most of these, some – hopefully minor – unusual formatting changes have not been corrected, just to keep the changes from the proofs as low as possible. But as you well know, the printer's demon often inserts one error while one corrects an other. The editor and the authors would be most thankful if you would bring to their attention any misprint or error found in the book. Please contact the editor at janos@schanda.hu

We wish you a pleasant reading and successful use of the book.
Veszprém, 2007-05-18.

János Schanda
Editor

CONTRIBUTORS AND REFEREES

The CIE Board of Administration and the Editor of this volume would like to express their sincere thanks to all the Authors and the Reviewers of the single chapters of this book for their help in spreading the information on the CIE recommendations and standards of colorimetry. The following list provides an overview on those who helped to get the scientific-technical content of this book written.

CONTRIBUTORS

Brill, Michael H.: Datacolor, 5 Princess Road, Lawrenceville, NJ 08648, USA. mbrill@datacolor.com

Michael H. Brill is the Principal Color Scientist at Datacolor/ColorVision in Lawrenceville, NJ. Since obtaining his Ph.D. in physics at Syracuse University, he has carried out extensive theoretical research in color in human and computer vision, in geometric/photometric invariance, and in physics-based vision. He is co-inventor of the Emmy-Award-Winning Sarnoff vision model. Dr. Brill is also a Past President of the Inter-Society Color Council, and author of more than 80 refereed technical publications, 10 U.S. patents, numerous national standards, and a SID test pattern. He has chaired or co-chaired four conferences on color technology, vision, and digital display. In addition, he obtained the 1996 Macbeth Award from the ISCC for his work on color constancy. Dr. Brill is a member of the Editorial Board of *Color Research and Application*, and chairs ASTM Subcommittee E12.04 (Color and Appearance Analysis) and also CIE Technical Committee TC1-56 (Improved Color Matching Functions).

Eppeldauer, George P: National Institute of Standards and Technology, 100 Bureau Drive, Mail-Stop 8441, Gaithersburg, Maryland 20899, USA. george.eppeldauer@nist.gov

George Eppeldauer is an electronics engineer in the Optical Technology Division of the National Institute of Standards and Technology (NIST), USA since 1986. He received his Ph.D. at the Technical University of Budapest, Hungary in 1975. His research area is development of transfer and working standard optical radiometers and realization of reference spectral responsivity scales. The radiometer, photometer, and colorimeter standards he developed have been utilized to improve the two SI units, the candela and kelvin, the tristimulus color scale, and the spectral irradiance scale of NIST. In the above field he has more than 130 publications. He received the Outstanding Inventor Gold Medal in 1981 from the Patent Office, the Trade Union Organization, and the Academy of Sciences in Hungary. He received a Research Prize from the Hungarian Academy of Sciences in 1976. He won the Best Paper Award at the NCSLI Conference in 2004. He is chairman of the CIE Technical Committee TC2-48 on Spectral responsivity measurement of detectors, radiometers, and photometers.

Fairchild, Mark D.: Professor & Director, Munsell Color Science Laboratory, Chester F. Carlson Center for Imaging Science, Rochester Institute of Technology, mdf@cis.rit.edu — www.cis.rit.edu/fairchild — www.whyiscolor.org

Mark D. Fairchild is Professor and Director of the Munsell Color Science Laboratory in the Chester F. Carlson Center for Imaging Science at the Rochester Institute of Technology. He received his B.Sc. and M.Sc. degrees in Imaging Science from R.I.T. and Ph.D. in Vision Science from the University of Rochester. He was chair of CIE Technical Committee 1-34 on color appearance models, is currently a member of several other CIE technical committees dealing with color appearance and image technology issues. Mark was presented with the 1995 Bartleson Award by the Colour Group (Great Britain) and the 2002 Macbeth Award by the Inter-Society Color Council for his research work in color appearance and other areas of color science. He is author of over 150 technical publications in color and imaging sciences and the book, Color Appearance Models, 2nd Ed., which serves as a reference to the fundamentals of color appearance and the formulation of specific models. He served as Color Imaging Editor for IS&T's Journal of Imaging Science and Technology for 3 years and was named a Fellow of IS&T (the Society for Imaging Science and Technology) in 2003 for his contributions to digital color imaging. Mark is an active member of IS&T, ISCC, CORM, CIE-USNC, OSA, SID, and ACM-SIGGRAPH.

Gardner, James L.: 5 Adamson Ave, Thornleigh Australia 2120.
jimgardner@optusnet.com.au

Jim Gardner retired from full-time work as leader of the Optical Radionetry Group at the CSIRO National Measurement Laboratory (now National Measurement Institute of Australia) in 2001. Since then he has spent periods as a guest researcher in the national institutes of USA, Korea and England, mostly on propagation of measurement uncertainties from spectral measurements to the various quantities of interest in radiometry, photometry and colorimetry. He is the current Editor for CIE Division 2.

Hirschler, Robert: Attila u. 93-95, 1039 Budapest, Hungary.
robert.hirschler@hunetkft.hu

Dr. Robert Hirschler graduated at the Technical University of Budapest in textile chemistry and worked for 22 years in industry. Since 1989 he's been a consultant in the field of colorimetry and special textile applications, working in SENAI/CETIQT, a textile educational and technology centre in Rio de Janeiro, Brazil. He was the Brazilian representative of CIE Division 1 between 1995 and 2003, he is the chair of CIE TC1-44 and has been active in a number of other technical committees of CIE. He is a member of ASTM E12, CORM and has been active in AIC and ISCC.

Hunt, Robert W.G.: Barrowpoint, 18 Millennium Close, Odstock Road, Salisbury, Wiltshire, SP2 8TB, UK. rwghunt@uwclub.net

After working for 36 years in the Kodak Research Laboratories at Harrow, England, Robert Hunt retired as Assistant Director of Research in 1982. Since then he has worked as a consultant, and is currently a visiting professor of Colour Science at the University of Leeds, England. He was Chairman of the British Colour Group from 1961 to 1963, Chairman of the CIE Colorimetry Committee from 1975 to 1983, and President of the International Colour Association from 1981 to 1985. He has written two books 'The Reproduction of Colour' now in its sixth edition, and 'Measuring Colour' now in its third edition. A recent major interest has been modelling the human system of colour vision so as to be able to predict how colours will appear in different viewing conditions. He gives courses on colour science and imaging in industry and at conferences.

Johnson, Garrett M.: Apple Computer, Inc, 1632 5th St., Santa Monica, CA 90401, USA.HYPERLINK ''mailto:gjohnson@apple.com'' gjohnson@apple.com

Garrett M. Johnson is a Color Scientist at Apple Computer, and an Affiliate Professor with the Munsell Color Science Laboratory, at the Rochester Institute of Technology. He has a Ph.D. in Imaging Science, and M.Sc. in Color Science, both from RIT. His research work focuses on the color imaging chain, and the perceptual appearance of complex images. He is chair of CIE TC8-08, focusing on developing methodologies for testing high-dynamic range rendering algorithms.

Lammers, Johan: Hewlett-Packard, S.L. (ICD), Avda Graells 501, O8190 Sant Cugat del Valles, Barcelona, Spain johan.lammens@hp.com

Johan M. Lammens was born in 1962 in Belgium, where he studied linguistics (BA) and computational linguistics (MA) at the University of Antwerp (1980–1985). He spent a year at the Institute for Perception Research, Technical University of Eindhoven (the Netherlands), working on computer speech synthesis. He worked for two years on computer text-to-speech conversion as a researcher at the University of Utrecht (the Netherlands). From 1988 to 1994 he was at the State University of

New York at Buffalo, where he obtained his MS and PhD degrees in computer science and worked as a teaching, lab, and research assistant. His dissertation research was concerned with a computational model of color perception and color naming, and was partly funded by NASA and Sun Microsystems. At the S. Anna Institute of the University of Pisa (Italy) he held a postdoc position, working on high-speed stereo foveal vision and transputer based motion control (1994–1995). Since 1995 he has been with the R&D department of Hewlett-Packard in Barcelona (Spain), currently in the position of Senior Color Scientist working on color imaging for commercial printing.

Li, Changjun: Department of Colour Science, Leeds University of Leeds, LS2 9JT, UK C.LI@LEEDS.AC.UK

Changjun Li is a Research Fellow at the Department of Colour and Polymer Chemistry, University of Leeds, UK. He received his B.Sc. (1979), M.Sc. (1982), Ph.D (1989) in computational mathematics from Peking University (China), Chinese Academy of Science, and Loughborough University (UK) respectively. He was a Professor of computational mathematics in the Northeastern University of China between 1996 and 2000. Then he joined the Colour & Imaging Institute, University of Derby till 2004. He received the Scientific Progress First Prize awarded by the Educational Committee of Liaoning Province of China in 1998. He is a member of the CIE TC8-01 and a member of the editorial board of International Journal of Computer Mathematics.

Luo M. Ronnier: Department of Colour Science, University of Leeds, Leeds LS2 9JT, UK. m.r.luo@leeds.ac.uk

M. Ronnier Luo is a Professor of Colour and Imaging Science at the Department of Colour Science at the University of Leeds, UK. He was the former Director of the Colour & Imaging Institute at the University of Derby. He received his B.Sc. in Fibre Technology from the National Taiwan University of Science and Technology and his Ph.D. in Colour Physics from the University of Bradford in 1986. He has over 250 publications in the areas of colour measurement, colour difference, colour appearance and colour reproduction. He chairs two Technical Committees of the International Commission on Illumination (CIE): TC 1-52 *Chromatic Adaptation Transforms* and CIE TC 8-2 *Colour Difference Evaluation in Images*. He is a Fellow of Society of Dyers and Colourists and the Society for Imaging Science and Technology. He was also the recipient of the 2003 Royal Photographic Society's Davies Medal and 1994 Bartleson Award for his contribution in the field of colour science.

Martinez-Uriegas, Eugenio: Hewlett-Packard, S.L. (ICD), Avda Graells 501, O8190 Sant Cugat del Valles, Barcelona – Spain.
eugenio.martinez-uriegas@hp.com

In 1973, he obtained the B.Sc. degree in physics from the Universidad Nacional Autonoma de Mexico. In 1979 he received the M.Sc. degree in sensory biophysics from Ohio State University and in 1982 the Ph.D. degree in visual psychophysics, also from Ohio State University.

His professional contributions include basic research on initial processes of human vision with emphasis in design and implementation of psychophysical experiments and mathematical modeling. Related to applied research, his main contributions have been in the areas of analysis and evaluation of digital color imagery, image capture and compression algorithms as well as digital color management.

He has published 36 papers and abstracts in specialized journals and books, and holds three US patents. He is currently the chairman of TC1-60 of CIE Division 1.

Morovič, Jan: Hewlett-Packard, S.L. (ICD) Avda Graells 501, O8190 Sant Cugat del Valles, Barcelona – Spain. jan.morovic@hp.com

Ján Morovic received his Ph.D. in color science from the Colour & Imaging Institute (CII) of the University of Derby (UK) in 1998, where the title of his thesis was To Develop a Universal Gamut Mapping Algorithm. After working as a lecturer in digital color reproduction at the CII he became senior color scientist at Hewlett-Packard in Barcelona, Spain. He is also the chairman of the CIE's Technical Committee 8-03 on Gamut Mapping.

Ohno, Yoshi: National Institute of Standards and Technology, 100 Bureau Drive, Stop 8442, Gaithersburg, MD 20899-8442, USA. ohno@nist.gov

Dr. Yoshi Ohno is the Group Leader for Optical Sensor Group, Optical Technology Division at the National Institute of Standards and Technology (NIST), USA. He joined NIST in 1992 as the project leader for photometry. His recent achievements include development of the Absolute Integrating Sphere Method for realization of the lumen and the Four Colour Method for colorimetry of displays. His researches on photometry of flashing lights and colour rendering of solid-state light sources are also well known. Ohno serves as the Secretary of CIE Division 2 and is active in a number of technical committees including those related to colorimetry in Division 1. He is also active in CORM, IESNA, and ASTM E12. He received his Ph.D. in engineering from Kyoto University, Japan.

Robertson, Alan R.: 2428 Ogilvie Road, Ottawa, Ontario K1J 7N5, Canada. alan.robertson@nrc-cnrc.gc.ca

Alan Robertson obtained his B.Sc. and Ph.D. in Physics from the University of London, England. He worked at Canada's National Research Council (NRC) from 1965 to 2000, first as a research scientist in the Photometry and Radiometry Group specializing in colour science and colour measurement, then as Director of the Chemical and Mechanical Standards Section and finally as Strategic Advisor to the Director General of NRC's Institute for National Measurement Standards. He has served on numerous CIE Technical Committees and was Vice President of the CIE from 1995 to 1999. In 2005, he received the Godlove Award of the Inter-Society Color Council for a lifetime of distinguished service to the color community. Since retirement he has continued his involvement with colour science as a part-time Visiting Worker at NRC.

Sauter, Georg: Physikalisch-Technische Bundesanstalt, AG 4.12 Photometrie, Postfach 3345; D-38023 Braunschweig, Germany. Georg.Sauter@t-online.de

Schanda, János: University of Pannonia, Egyetem u. 10., H-8200 Veszprém, Hungary. Schanda@vision.vein.hu

Dr. János Schanda is a graduated in physics of the Loránd Eötvös University in Budapest. During his technical career he worked at Hungarian Office of Measures, the Research Institute for Technical Physics of the Hungarian Academy of Sciences, and the University of Veszprém (now University of Pannonia) where he heads as professor emeritus "Virtual Environment and Imaging Technologies Laboratory".

He functioned in a number of positions of the CIE. At present he is the Secretary of the Commission, is past Vice President, chaired and chairs several Technical Committees, among others dealing with fundamentals of photometry, colorimetry and colour rendering. At present he is the President of the Hungarian National Committee of the CIE.

Dr. Schanda is member of OSA, of IS&T and of several Hungarian Societies in the fields of light and lighting and optical measurement. He served also on the Board of the International Colour Association (AIC) as its vice-president. He is on the editorial / international advisory board of Color Res. & Appl., Journal of Light & Visual Environment, Light & Engineering and Lighting Research & Technology.

Viénot, Françoise: Muséum national d'histoire naturelle, CRCDG, 36 rue Geoffroy Saint-Hilaire, 75005 Paris France. vienot@mnhn.fr

Professor Françoise Viénot is a physicist from the permanent staff of the Muséum National d'Histoire Naturelle, an institution of the Ministry of Education. There, she has been trained by Yves Le Grand. She has also spent a few months at Professor Billmeyer's Laboratory at Rensselaer Polytechnic Institute, Troy, New York. She has been conducting research and supervising graduate and postgraduate studies at the Muséum National d'Histoire Naturelle in colorimetry, photometry, color vision and the measurement of appearance. She has been responsible for applied research projects on visual metrics. She is teaching Color Vision and Colorimetry at the University in Paris. She is co-author of *Vision et Mesure de la Couleur* by Kowaliski, Viénot and Sève, and coordinator with Roque and Bodo of *Michel-Eugène Chevreul. Un savant, des couleurs!* She is Past President of the Centre National Français de l'Eclairage, chair of TC 1-36 on the "Chromaticity diagram with physiologically significant axes" and has served as Associated Director for Vision for Division 1 of the CIE, and on the editorial board of *Color Research and Application.*

Walraven, Pieter: TNO, Kampweg 5, 3769 DE Soesterberg, The Netherlands. plwalraven@hotmail.com

Pieter L. Walraven has carried out extensive experimental and theoretical research in color vision at the TNO Human Factors Institute in Soesterberg, The Netherlands. He obtained his Ph.D. in physics and mathematics at the University of Utrecht in 1962 with his dissertation "On the mechanisms of colour vision". Together with Dr. Hans Vos he received in 1991 the AIC Judd Award for his many contributions to the understanding of the mechanisms of color vision. Vos and Walraven combined approach gave the mathematical basis to the two-stage color vision model, which brings together the trichromatic theory and the opponent colors theory.

He is author of more than 50 refereed technical publications, chaired the AIC-conference in 1971 on Color Metrics in Driebergen, The Netherlands. He chaired also several CIE committees, and is presently secretary of the CIE-committee TC 1-36 "Fundamental chromaticity diagram with physiological axes". Dr Walraven has been a member of the Board of CIE and of AIC.

Witt, Klaus: Hortensienstr. 25, D-12203 Berlin, Germany. klaus.witt@gmx.de

Klaus Witt received a doctoral degree in physics at the university of Freiburg, Germany. Later he joined the Federal Institute of Materials Research and Testing (BAM) in Berlin, where he became head of the laboratory of colour engineering. His main interests dealt with colour-difference evaluation, colour systematics and colour measurement. He became member of different CIE technical committees of Divisions 1 and 2, chairing one committee on parameters in colour-difference evaluation and serving as reporter on colour-difference evaluation. He retired in 2002 still being interested in further developments in colorimetry.

Zwinkels, Joanne: National Research Council of Canada, Institute for National Measurement Standards, 1200 Montreal Road, Ottawa, Ontario, Canada, K1A 0R6. Joanne.Zwinkels@nrc-cnrc.gc.ca

Dr. Joanne Zwinkels is Group Leader for Photometry and Radiometry at the Institute for National Measurement Standards, National Research Council of Canada (NRC). She joined NRC in 1984 to work in the field of spectrophotometry. Her principal research activities involve the development of new instrumentation, procedures, and reference standards for high-accuracy spectrophotometry, spectrofluorimetry and color and appearance measurements. Her recent accomplishments include the development of a versatile reference goniospectrophotometer for both primary level specular gloss measurements at several standard geometries and total appearance characterization of gonioapparent materials. Zwinkels is the Past President of the Inter-Society Color Council (2004-2006), and currently serves as Vice President of the Canadian National Committee of the CIE, Canadian member for CIE Division 2, and is an active member of CORM and several technical committees in CIE Divisions 1 and 2, including Chair of a technical committee (TC2-25) on fluorescence measurements. She obtained her Ph.D. in Physical Chemistry from the University of Alberta, Canada

REFEREES

Authors and the CIE are indebted to the following persons for reviewing one or more of the chapters and helping with their constructive remarks both the authors and the editor.

Alman, David H. Ph. D.

Dave Alman is a colour scientist with DuPont Performance Coating where he develops colour control systems for coatings for the transportation industries. His research interests include pigment coloration processes, psychophysics of appearance and colour-difference assessment. He was chairman of two CIE Technical Committees that investigated industrial color-difference evaluation.

Gardner, James L: see Contributors

Hunt, Robert WG: see Contributors

Kemenade, Johan:

Dr. Ir. Johan van Kemenade is a senior application scientist with Philips Lighting where he is responsible for specifying and testing new lighting systems for optimal customer satisfaction.

His research interest include human factors linked with lighting such as the perception of lit scenes, description of spots lights, defining colour rendition e.g by the CRV vectors. He participates in some CIE committees in Division 1 and 3.

Luo, M Ronnier: see Contributors

Melgosa, Manuel

Born in 1958. Ph.D. in Physics (Optics) in 1989 (University of Granada, Spain). He has published more than 50 articles on colorimetry in different journals: Color Research and Application, Applied Optics, Journal Optical Society of America A, Journal of Optics, Optik, Die Farbe, Atti della Fondazione Giorgio Ronchi, etc. He served as advisor of CIE TC 1-29 and member of CIE TC 1-47. Currently he is a member of CIE TC 1-63 "Validity of the range of CIEDE2000", and chairman of CIE TC 1-55 "Uniform colour space for industrial colour-difference evaluation". In 1990 he received the Expo-Óptica 90 Award for his Ph.D. work, and in 1997 the VDC-Peter-Abel International Award. Since 1995 he is Representative of the Optical Society of America at the Inter Society Color Council (ISCC). He is the current Secretary of the Study Group "Color Perception in the Elderly" of the International Colour Association (AIC), and a member of ISCC Project 52 "Comparative list of color terms II". Since 2000 Secretary of the Color Committee of the Optical Society of Spain, and Associate Editor of the journal Color Research and Application.

Newman, Todd:

Todd Newman is Director of CIE Division 8, Image Technology. He is also the chair of CIE TC8-10, Office Lighting for Imaging. He was the first chairman of the International Color Consortium. He has worked for Microsoft, Digital Equipment Corporation, Silicon Graphics, Inc.

Ohno, Yoshi: see Contributors

Pointer, Mike

Dr Michael R Pointer is a past principal scientist in the Optical Radiation Measurement team at the National Physical Laboratory in Teddington, UK where his interests included looking for innovative methods to measure correlates of the visual appearance of materials using spectrophotometry and colorimetry, as well as novel measurement techniques.

He has authored over 70 scientific papers, is a Fellow of The Royal Photographic Society and the Institute of Physics. He is active in CIE Division 1 *Vision and Colour*, ASTM Committee E12 *Color and Appearance*, and the Colour Group (Great Britain). He is Associate Editor of the journal *Color Research & Application*, and a Visiting Professor at the University of Leeds. In 2004, his contribution to colour science was recognised by the award of a silver achievement medal by the Society of Dyers and Colourists.

Robertson, Alan R: see Contributors

Sagawa, Ken

Dr. Ken Sagawa is Group leader of Accessible Design Group, Institute for Human Science and Biomedical Engineering, National Institute of Advanced Industrial Science and Technology (AIST).

He graduated from Tokyo Institute of Technology in Japan in 1975. Since 1975 he worked for some National Research Institutes of METI such as Industrial Products Research Institute, National Institute of Bioscience and Human-Technology, and National Institute of Advanced Industrial Science and Technology (AIST). Currently he is working for AIST as a group leader of Accessible Design Group of Human Science and Biomedical Engineering Division in AIST.

His major field is visual psychophysics to be applied for photometry, colorimetry and assessment of visual environment, especially at the mesopic environment. His recent studies have been directed to age-related change of visual functions and the evaluation of visual environment for the older persons and persons with visual impairment.

He has been long involved in CIE as a Chair of some technical committees in Division 1 related to mesopic photometry and to age-related change of vision. For 1995-2003 her served for CIE as a Director of Division 1 "Vision and Colour" and from 2003 as a Vice President. He has also worked for ISO in the field of Ergonomics for accessible design for the older persons and persons with disabilities. He has been a convenor of ISOTC159/WG2 since 2004 and was Japanese representative to ISO COPOLCO (2002–2005).

Tarrant, Arthur

Dr. Arthur Tarrant worked originally at the National Physical Laboratory on spectrophotometry and spectroradiometry. He then went to the University of Surrey, working in spectroscopy, and on the spectral composition of daylight. In later years he became involved with colour in visual displays and the names that people attach to colours. Since his retirement he has worked as a visiting scholar at the University of Derby, and built a novel visual colorimeter. Latterly he has continued that work as a visiting research fellow at the University of Leeds. He has recently been awarded with Honorary Membership of the Colour Group of Great Britain.

1

TRANSLATION OF CIE 1931 RESOLUTIONS ON COLORIMETRY*

Translated by PETER BODROGI

DECISION 1**

It is recommended that in order to obtain a normalized basis to specify colorimetric quantities, regarding its technical and commercial applications, these quantities shall be expressed in relation to a hypothetic observer to be called CIE 1931 reference observer and characterized by its visibility curve, which is the normalized curve adopted by the CIE. For this observer, there shall be a chromatic equivalence between the monochromatic stimulus defined by the wavelength indicated in the fourth column of Table 1.1, and the mixture of three monochromatic stimuli of the following wavelengths: 700.0 nm, 546.1 nm, and 435.8 nm, in the proportions listed in columns 1, 2, and 3 of Table 1.1.

In Table 1.1, the units, in which r, g, and b are expressed, are chosen so that the equal quantity mixture of these three monochromatic stimuli shall be chromatically equivalent to a specific nonmonochromatic stimulus for which the total radiant energy of all wavelengths in a range of two extreme wavelengths situated in the visible spectrum is proportional to the difference between these wavelengths.

Note: The relative luminances of the units of these three monochromatic stimuli are defined for the reference observer in the following ratio: 1:4.5907:0.0601, and the values of the ordinates of the distribution curves of these stimuli that constitute the equal energy spectrum are listed in columns 5, 6, and 7.

*see Commission Internationale de l'Eclairage Huitième Session Cambridge – Septembre 1931. Cambridge at the University Press, 1932. pp. 19–24.

**Germany and France changed their positive vote and voted against.

TABLE 1.1 CIE 1931 reference observer

Trichromatic coefficients (coordinates in the color triangle) (chromaticity coordinates)			Wavelength (nm)	Distribution coefficients for the stimuli of the same energy (color-matching functions)		
r	g	b	λ	$\bar{r}$	$\bar{g}$	$\bar{b}$
0.0272	−0.0115	0.9843	380	0.00003	−0.00001	0.00117
0.0268	−0.0114	0.9846	385	0.00005	−0.00002	0.00189
0.0263	−0.0114	0.9851	390	0.00010	−0.00004	0.00359
0.0256	−0.0113	0.9857	395	0.00017	−0.00007	0.00647
0.0247	−0.0112	0.9865	400	0.00030	−0.00014	0.01214
0.0237	−0.0111	0.9874	405	0.00047	−0.00022	0.01969
0.0225	−0.0109	0.9884	410	0.00084	−0.00041	0.03707
0.0207	−0.0104	0.9897	415	0.00139	−0.00070	0.06637
0.0181	−0.0094	0.9913	420	0.00211	−0.00110	0.11541
0.0142	−0.0076	0.9934	425	0.00266	−0.00143	0.18575
0.0088	−0.0048	0.9960	430	0.00218	−0.00119	0.24769
0.0012	−0.0007	0.9995	435	0.00036	−0.00021	0.29012
−0.0084	0.0048	1.0036	440	−0.00261	0.00149	0.31228
−0.0213	0.0120	1.0093	445	−0.00673	0.00379	0.31860
−0.0390	0.0218	1.0172	450	−0.01213	0.00678	0.31670
−0.0618	0.0345	1.0273	455	−0.01874	0.01046	0.31166
−0.0909	0.0517	1.0392	460	−0.02608	0.01485	0.29821
−0.1281	0.0762	1.0519	465	−0.03324	0.01977	0.27295
−0.1821	0.1175	1.0646	470	−0.03933	0.02538	0.22991
−0.2584	0.1840	1.0744	475	−0.04471	0.03183	0.18592
−0.3667	0.2906	1.0761	480	−0.04939	0.03914	0.14494
−0.5200	0.4568	1.0632	485	−0.05364	0.04713	0.10968
−0.7150	0.6996	1.0154	490	−0.05814	0.05689	0.08257
−0.9459	1.0247	0.9212	495	−0.06414	0.06948	0.06246
−1.1685	1.3905	0.7780	500	−0.07173	0.08536	0.04776
−1.3182	1.7195	0.5987	505	−0.08120	0.10593	0.03688
−1.3371	1.9318	0.4053	510	−0.08901	0.12860	0.02698
−1.2076	1.9699	0.2377	515	−0.09356	0.15262	0.01842
−0.9830	1.8534	0.1296	520	−0.09264	0.17468	0.01221
−0.7386	1.6662	0.0724	525	−0.08473	0.19113	0.00830
−0.5159	1.4761	0.0398	530	−0.07101	0.20317	0.00549
−0.3304	1.3105	0.0199	535	−0.05316	0.21083	0.00320
−0.1707	1.1628	0.0079	540	−0.03152	0.21466	0.00146
−0.0293	1.0282	0.0011	545	−0.00613	0.21487	0.00023
0.0974	0.9051	−0.0025	550	0.02279	0.21178	−0.00058
0.2121	0.7919	−0.0040	555	0.05514	0.20588	−0.00105
0.3164	0.6881	−0.0045	560	0.09060	0.19702	−0.00130
0.4112	0.5932	−0.0044	565	0.12840	0.18522	−0.00138
0.4973	0.5067	−0.0040	570	0.16768	0.17087	−0.00135
0.5751	0.4283	−0.0034	575	0.20715	0.15429	−0.00123
0.6449	0.3579	−0.0028	580	0.24526	0.13610	−0.00108
1	2	3	4	5	6	7

(*continued*)

TABLE 1.1 (*Continued*)

Trichromatic coefficients (coordinates in the color triangle) (chromaticity coordinates)			Wavelength (nm)	Distribution coefficients for the stimuli of the same energy (color-matching functions)		
r	g	b	λ	$\bar{r}$	$\bar{g}$	$\bar{b}$
0.6449	0.3579	−0.0028	580	0.24526	0.13610	−0.00108
0.7071	0.2952	−0.0023	585	0.27989	0.11686	−0.00093
0.7617	0.2402	−0.0019	590	0.30928	0.09754	−0.00079
0.8087	0.1928	−0.0015	595	0.33184	0.07909	−0.00063
0.8475	0.1537	−0.0012	600	0.34429	0.06246	−0.00049
0.8800	0.1209	−0.0009	605	0.34756	0.04776	−0.00038
0.9059	0.0949	−0.0008	610	0.33971	0.03557	−0.00030
0.9265	0.0741	−0.0006	615	0.32265	0.02583	−0.00022
0.9425	0.0580	−0.0005	620	0.29708	0.01828	−0.00015
0.9550	0.0454	−0.0004	625	0.26348	0.01253	−0.00011
0.9649	0.0354	−0.0003	630	0.22677	0.00833	−0.00008
0.9730	0.0272	−0.0002	635	0.19233	0.00537	−0.00005
0.9797	0.0205	−0.0002	640	0.15968	0.00334	−0.00003
0.9850	0.0152	−0.0002	645	0.12905	0.00199	−0.00002
0.9888	0.0113	−0.0001	650	0.10167	0.00116	−0.00001
0.9918	0.0083	−0.0001	655	0.07857	0.00066	−0.00001
0.9940	0.0061	−0.0001	660	0.05932	0.00037	0.00000
0.9954	0.0047	−0.0001	665	0.04366	0.00021	0.00000
0.9966	0.0035	−0.0001	670	0.03149	0.00011	0.00000
0.9975	0.0025	0.0000	675	0.02294	0.00006	0.00000
0.9984	0.0016	0.0000	680	0.01687	0.00003	0.00000
0.9991	0.0009	0.0000	685	0.01187	0.00001	0.00000
0.9996	0.0004	0.0000	690	0.00819	0.00000	0.00000
0.9999	0.0001	0.0000	695	0.00572	0.00000	0.00000
1.0000	0.0000	0.0000	700	0.00410	0.00000	0.00000
1.0000	0.0000	0.0000	705	0.00291	0.00000	0.00000
1.0000	0.0000	0.0000	710	0.00210	0.00000	0.00000
1.0000	0.0000	0.0000	715	0.00148	0.00000	0.00000
1.0000	0.0000	0.0000	720	0.00105	0.00000	0.00000
1.0000	0.0000	0.0000	725	0.00074	0.00000	0.00000
1.0000	0.0000	0.0000	730	0.00052	0.00000	0.00000
1.0000	0.0000	0.0000	735	0.00036	0.00000	0.00000
1.0000	0.0000	0.0000	740	0.00025	0.00000	0.00000
1.0000	0.0000	0.0000	745	0.00017	0.00000	0.00000
1.0000	0.0000	0.0000	750	0.00012	0.00000	0.00000
1.0000	0.0000	0.0000	755	0.00008	0.00000	0.00000
1.0000	0.0000	0.0000	760	0.00006	0.00000	0.00000
1.0000	0.0000	0.0000	765	0.00004	0.00000	0.00000
1.0000	0.0000	0.0000	770	0.00003	0.00000	0.00000
1.0000	0.0000	0.0000	775	0.00001	0.00000	0.00000
1.0000	0.0000	0.0000	780	0.00000	0.00000	0.00000
1	2	3	4	5	6	7

Relative luminance factors: 1:4.5907:0.0601.

DECISION 2†

It is recommended that the following three light sources shall be adopted as the standards of general colorimetry of materials:

1. A gas atmosphere lamp at the color temperature of 2848 K.‡
2. The same lamp used in combination with a filter composed of a 1 cm thick layer of two solutions B_1 and B_2 in a double box of uncolored optical glass.

The two solutions are built up in the following way:

- *Solution B_1*:

Copper sulfate ($SO_4Cu, 5H_2O$)	2.452 g
Mannite ($C_6H_8(OH)_6$)	2.452 g
Pyridine (C_5H_5N)	30.0 cc
Distilled water to be completed to	1000 cc

- *Solution B_2*:

Cobalt and ammonium bisulfate ($SO_4Co_2, SO_4(NH_4)_2, 6H_2O$)	21.71 g
Copper sulfate ($SO_4Cu, 5H_2O$)	16.11 g
Sulfuric acid (density 1.835)	10.0 cc
Distilled water to be completed to	1000 cc

3. The same lamp used in combination with a filter composed of a 1 cm thick layer of two solutions C_1 and C_2 in a double box of uncolored optical glass.

The two solutions are built up in the following way:

- *Solution C_1*:

Copper sulfate ($SO_4Cu, 5H_2O$)	3.412 g
Mannite ($C_6H_8(OH)_6$)	3.412 g
Pyridine (C_5H_5N)	30.0 cc
Distilled water to be completed to	1000 cc

- *Solution C_2*:

Cobalt and ammonium bisulfate ($SO_4Co_2, SO_4(NH_4)_2, 6H_2O$)	30.580 g
Copper sulfate ($SO_4Cu, 5H_2O$)	22.520 g
Sulfuric acid (density 1.835)	10.0 cc
Distilled water to be completed to	1000 cc

It is also recommended that the values of the spectral power distribution that are published in the appendix of this decision shall be used to determine the colorimetric quantities of the spectrophotometric measurements.

†France changed its positive vote and voted against.

‡In the spectral power distribution calculations, the Planckian constant c_2 is set equal to 1.350×10^{-2} m.K.

Note: It is recognized that for certain special applications (e.g., the specification of signaling glasses), other light sources can be prescribed, but in the absence of special conditions, one of the three indicated sources shall be used.

APPENDIX TO DECISION 2

Power distribution in the luminous spectra of the sources A, B, and C.

Source A: The spectral power distribution of this source shall be equal to that of the black body at the temperature of 2848 K, for all colorimetric applications.

Sources B and C: The spectral power distributions of these sources are listed in the table below. This dataset was taken from the spectrophotometric measurement of filter transmissions done by Davis and Gibson (Bureau of Standards).

	Relative energy			Relative energy	
λ (nm)	B	C	λ (nm)	B	C
370	15.2	21.6	560	102.8	105.3
80	22.4	22.0	70	102.6	102.3
90	31.3	47.4	80	101.0	97.8
400	41.3	63.3	90	99.2	93.2
10	52.1	80.6	600	98.0	89.7
20	63.2	98.1	10	98.5	88.4
30	73.1	112.4	20	99.7	88.1
40	80.8	121.5	30	101.0	88.0
450	85.4	124.0	40	102.2	87.8
60	88.3	123.1	650	103.9	88.2
70	92.0	123.8	60	105.0	87.9
80	95.2	123.9	70	104.9	86.3
90	96.5	120.7	80	103.9	84.0
500	94.2	112.1	90	101.6	80.2
10	90.7	102.3	700	99.1	76.3
20	89.5	96.9	10	96.2	72.4
30	92.2	98.0	20	92.9	68.3
40	96.9	102.1			
550	101.0	105.2			

DECISION 3*

In the colorimetric measurement of reflecting materials, except for certain special circumstances that require another method, the incident ray shall go in under 45° to

*France and Germany changed their positive vote and voted against.

TABLE 1.2 CIE 1931 reference observer in the normalized reference system

Trichromatic coefficients (chromaticity coordinates)			Wavelength (nm)	Distribution coefficients for the stimuli of the same energy (color-matching functions)		
x	y	z	λ	$\bar{x}$	$\bar{y}$	$\bar{z}$
0.1741	0.0050	0.8209	380	0.0014	0.0000	0.0065
0.1740	0.0050	0.8210	385	0.0022	0.0001	0.0105
0.1738	0.0049	0.8213	390	0.0042	0.0001	0.0201
0.1736	0.0049	0.8215	395	0.0076	0.0002	0.0362
0.1733	0.0048	0.8219	400	0.0143	0.0004	0.0679
0.1730	0.0048	0.8222	405	0.0232	0.0006	0.1102
0.1726	0.0048	0.8226	410	0.0435	0.0012	0.2074
0.1721	0.0048	0.8231	415	0.0776	0.0022	0.3713
0.1714	0.0051	0.8235	420	0.1344	0.0040	0.6456
0.1703	0.0058	0.8239	425	0.2148	0.0073	1.0391
0.1689	0.0069	0.8242	430	0.2839	0.0116	1.3856
0.1669	0.0086	0.8245	435	0.3285	0.0168	1.6230
0.1644	0.0109	0.8247	440	0.3483	0.0230	1.7471
0.1611	0.0138	0.8251	445	0.3481	0.0298	1.7826
0.1566	0.0177	0.8257	450	0.3362	0.0380	1.7721
0.1510	0.0227	0.8263	455	0.3187	0.0480	1.7441
0.1440	0.0297	0.8263	460	0.2908	0.0600	1.6692
0.1355	0.0399	0.8246	465	0.2511	0.0739	1.5281
0.1241	0.0578	0.8181	470	0.1954	0.0910	1.2876
0.1096	0.0868	0.8036	475	0.1421	0.1126	1.0419
0.0913	0.1327	0.7760	480	0.0956	0.1390	0.8130
0.0687	0.2007	0.7306	485	0.0580	0.1693	0.6162
0.0454	0.2950	0.6596	490	0.0320	0.2080	0.4652
0.0253	0.4127	0.5638	495	0.0147	0.2586	0.3533
0.0082	0.5384	0.4534	500	0.0049	0.3230	0.2720
0.0039	0.6548	0.3413	505	0.0024	0.4073	0.2123
0.0139	0.7502	0.2359	510	0.0093	0.5030	0.1582
0.0389	0.8120	0.1491	515	0.0291	0.6082	0.1117
0.0743	0.8338	0.0919	520	0.0633	0.7100	0.0782
0.1142	0.8262	0.0596	525	0.1096	0.7932	0.0573
0.1547	0.8059	0.0394	530	0.1655	0.8620	0.0422
0.1929	0.7816	0.0255	535	0.2257	0.9149	0.0298
0.2296	0.7543	0.0161	540	0.2904	0.9540	0.0203
0.2658	0.7243	0.0099	545	0.3597	0.9803	0.0134
0.3016	0.6923	0.0061	550	0.4334	0.9950	0.0087
0.3373	0.6589	0.0038	555	0.5121	1.0002	0.0057
0.3731	0.6245	0.0024	560	0.5945	0.9950	0.0039
0.4087	0.5896	0.0017	565	0.6784	0.9786	0.0027
0.4441	0.5547	0.0012	570	0.7621	0.9520	0.0021
0.4788	0.5202	0.0010	575	0.8425	0.9154	0.0018
0.5125	0.4866	0.0009	580	0.9163	0.8700	0.0017
0.5125	0.4866	0.0009	580	0.9163	0.8700	0.0017
1	2	3	4	5	6	7

(*continued*)

TABLE 1.2 (*Continued*)

Trichromatic coefficients (chromaticity coordinates)			Wavelength (nm)	Distribution coefficients for the stimuli of the same energy (color-matching functions)		
x	y	z	λ	$\bar{x}$	$\bar{y}$	$\bar{z}$
0.5448	0.4544	0.0008	585	0.9786	0.8163	0.0014
0.5752	0.4242	0.0006	590	1.0263	0.7570	0.0011
0.6029	0.3965	0.0006	595	1.0567	0.6949	0.0010
0.6270	0.3725	0.0005	600	1.0622	0.6310	0.0008
0.6482	0.3514	0.0004	605	1.0456	0.5668	0.0006
0.6658	0.3340	0.0002	610	1.0026	0.5030	0.0003
0.6801	0.3197	0.0002	615	0.9384	0.4412	0.0002
0.6915	0.3083	0.0002	620	0.8544	0.3810	0.0002
0.7006	0.2993	0.0001	625	0.7514	0.3210	0.0001
0.7079	0.2920	0.0001	630	0.6424	0.2650	0.0000
0.7140	0.2859	0.0001	635	0.5419	0.2170	0.0000
0.7190	0.2809	0.0001	640	0.4479	0.1750	0.0000
0.7230	0.2770	0.0000	645	0.3608	0.1382	0.0000
0.7260	0.2740	0.0000	650	0.2835	0.1070	0.0000
0.7283	0.2717	0.0000	655	0.2187	0.0816	0.0000
0.7300	0.2700	0.0000	660	0.1649	0.0610	0.0000
0.7311	0.2689	0.0000	665	0.1212	0.0446	0.0000
0.7320	0.2680	0.0000	670	0.0874	0.0320	0.0000
0.7327	0.2673	0.0000	675	0.0636	0.0232	0.0000
0.7334	0.2666	0.0000	680	0.0468	0.0170	0.0000
0.7340	0.2660	0.0000	685	0.0329	0.0119	0.0000
0.7344	0.2656	0.0000	690	0.0227	0.0082	0.0000
0.7346	0.2654	0.0000	695	0.0158	0.0057	0.0000
0.7347	0.2653	0.0000	700	0.0114	0.0041	0.0000
0.7347	0.2653	0.0000	705	0.0081	0.0029	0.0000
0.7347	0.2653	0.0000	710	0.0058	0.0021	0.0000
0.7347	0.2653	0.0000	715	0.0041	0.0015	0.0000
0.7347	0.2653	0.0000	720	0.0029	0.0010	0.0000
0.7347	0.2653	0.0000	725	0.0020	0.0007	0.0000
0.7347	0.2653	0.0000	730	0.0014	0.0005	0.0000
0.7347	0.2653	0.0000	735	0.0010	0.0004	0.0000
0.7347	0.2653	0.0000	740	0.0007	0.0003	0.0000
0.7347	0.2653	0.0000	745	0.0005	0.0002	0.0000
0.7347	0.2653	0.0000	750	0.0003	0.0001	0.0000
0.7347	0.2653	0.0000	755	0.0002	0.0001	0.0000
0.7347	0.2653	0.0000	760	0.0002	0.0001	0.0000
0.7347	0.2653	0.0000	765	0.0001	0.0000	0.0000
0.7347	0.2653	0.0000	770	0.0001	0.0000	0.0000
0.7347	0.2653	0.0000	775	0.0000	0.0000	0.0000
0.7347	0.2653	0.0000	780	0.0000	0.0000	0.0000
			Total	21.3713	21.3714	21.3715
1	2	3	4	5	6	7

The luminance factors are 0, 1, and 0, respectively.

the line perpendicular to the surface and the observation direction shall be perpendicular to the surface of the sample.

If the special nature of certain materials or another special aim requires a deviation from these conditions, then the conditions used shall always be published together with the published results.

DECISION 3a†

In the colorimetric measurement of opaque materials, the gloss of the studied material shall be expressed as a function of the gloss of a surface made of magnesium oxide considered under the same illumination and observation conditions.

DECISION 4*

The normalized reference system of colorimetric specifications shall be a system in which the color quality of every stimulus is expressed on three scales by assigning determined values to four stimuli selected in such a way that any two of them cannot be combined to yield a perception that is equivalent to any one of the other two.

DECISION 5*

The four stimuli that define the colorimetric scales will consist of monochromatic radiations of the following wavelengths: 700.0 nm, 546.1 nm, and 435.8 nm, and the standard light source B. These stimuli will be assigned the following values:

$$\begin{aligned} 700.0\,\text{nm} &\equiv 0.73467X + 0.26533Y + 0.00000Z, \\ 546.1\,\text{nm} &\equiv 0.27376X + 0.71741Y + 0.00883Z, \\ 435.8\,\text{nm} &\equiv 0.16658X + 0.00886Y + 0.82456Z, \\ \text{standard light source B} &\equiv 0.34842X + 0.35161Y + 0.29997Z. \end{aligned}$$

Note: The properties of the reference observer, if they are expressed in the above defined system, are given in Table 1.2.

†France changed its positive vote and voted against.

*France and Germany changed its positive vote and voted against.

2

PROFESSOR WRIGHT'S PAPER FROM THE GOLDEN JUBILEE BOOK: THE HISTORICAL AND EXPERIMENTAL BACKGROUND TO THE 1931 CIE SYSTEM OF COLORIMETRY*

W. DAVID WRIGHT
68 Newberries Avenue Radlett Herts. WD7 7EP

The 19th-century concepts of trichromacy led to the pioneer determinations of the colour-mixture curves by Maxwell, König and Abney. Early in the 20th century, American workers made the major contributions to photometry and colorimetry, leading to the Colorimetry Report of the Optical Society of America published in 1922 and to the establishment of the standard V_λ curve by the CIE in 1924 based entirely on studies made in the United States. The Standard Observer for colorimetry, on the other hand, was based solely on studies made in England, but it was a race against time to get the system adopted at the 1931 meeting of the CIE. American views, however, were a determining factor in deciding the form in which the system should be defined.

COLOUR MIXTURE AND MEASUREMENT IN THE 19TH CENTURY

Colour science has excellent credentials as it has claimed the interest of some of the greatest scientists of all time, with Thomas Young, Helmholtz and Maxwell being

*From "Golden Jubilee of Colour in the CIE", The Soc. of Dyers and Colouristst, Bradford, 1981.

Colorimetry: Understanding the CIE System, Edited by János Schanda

outstanding in the 19th century, following on the foundations laid by Isaac Newton some 100 years or more before. A few years ago Paul Sherman made a very interesting historical study of the development of trichromatic concepts in the first half of the 19th century,[1] and showed how the contributions of these scientific giants interlocked with those of lesser figures like Brewster, Herschel, Wollaston and Grassmann. Perhaps, though, the most striking feature of Sherman's thesis was his demonstration of the wide range of Maxwell's contributions to the subject.

We have, for example, Maxwell's disc or top (with which colours could be matched by varying the sector angles of the coloured papers on the rotating top), we have Maxwell's 'spot' (his demonstration of the macular area of the retina in which the yellow macular pigment causes a red patch to be seen when viewing a purple background), we have his 'colour box' (his colour-mixing apparatus using three overlapping spectra), and we have his colour triangle (the equilateral form of the modern chromaticity chart) and the first demonstration of three-colour photography (shown by Ralph Evans to have worked only because of a lucky combination of the reflecting properties of the coloured ribbon he photographed and the ultraviolet sensitivity of his photographic emulsion). Then Maxwell used his colour box to make the first determination of the three-colour mixture curves of the spectrum and he was the first to plot the spectrum locus in the colour triangle. He then went on to measure the two-colour mixture curves of dichromats (observers with one receptor process missing) and showed that the colours they confused were located in the colour triangle on lines radiating from the point in the triangle corresponding to the colour of the missing process. What a series of 'Firsts'! Maxwell also, ahead of Helmholtz, gave careful reasoned support to what is usually referred to as the Young-Helmholtz theory, but which might be more correctly described as the Young-Maxwell theory.

It is, then, legitimate to trace the genealogical thread of trichromacy from Newton through Young to Grassmann (who formulated the laws of colour mixture) and on to Maxwell. After Maxwell came König, who also determined the colour-mixture curves using the colour-mixing apparatus described in Helmholtz's '*Physiological Optics*'. Abney followed with his determination of the mixture curves measured with his own 'colour-patch' apparatus. I would also like to include the name of J. W. Lovibond in the list of these 19th century pioneers, since he described his Tintometer as early as 1887 and blazed the trail for commercial colour measurement, especially as applied to foodstuffs, drink and chemicals. His instrument was, after all, a trichromatic instrument which controlled the red-green-blue mixture by absorption (subtractive mixture) instead of by additive mixture.

Two other 19th-20th century figures who contributed to colour measurement and whom we should mention are A. H. Munsell and W. Ostwald, since they were responsible for two of the most widely used and most highly regarded atlases for identifying and specifying surface colours. To many people, such atlases are still the most meaningful and acceptable method of 'measuring' colour.

AMERICAN CONTRIBUTIONS TO PHOTOMETRY AND COLORIMETRY, 1900–24

There was a great deal of photometric and colorimetric activity in the United States in the first quarter of the 20th century and leading figures in this period included H. E. Ives (United Gas Improvement Company and Bell Telephone), E. P. Hyde (General Electric), L. A. Jones (Kodak), H. P. Gage (Corning Glass), and I. G. Priest, E. C. Crittenden, W. W. Coblentz, W. B. Emerson, K. S. Gibson, E. P. Tyndall, E. A. Weaver (*see* p. 21) and R. Davis (Bureau of Standards). This was a particularly powerful team at the Bureau of Standards, where Priest had been made Chief of the Colorimetry Division as early as 1913, to be joined in 1926 by Deane Judd.

Between 1912 and 1923, Ives carried out extensive studies on heterochromatic photometry and established the conditions under which valid photometric comparisons could be made between lights of different colours, either by flicker photometry or by direct (or step-by-step) brightness matching. The main studies leading up to the 1924 V_λ curve, which was essentially an all-American affair, were carried out at the National Lamp Works of the General Electric Company and at the Bureau of Standards, but at least six different sets of data were included in the proposal submitted by Gibson to the 1924 meeting of the CIE for adoption as the standard visibility curve.[2] Apart from some minor changes, the same curve had been approved a few years earlier by the American Illuminating Engineering Society and by the Optical Society of America.

In 1920 the Optical Society of America appointed a Colorimetry Committee under the Chairmanship of Dr L. T. Troland (a brilliant American psychologist who became President of the OSA in 1922 at the early age of 33), the task of the Committee being to report on the state of the art in colorimetry. This report was published in 1922, as was another massive report which Troland wrote on 'The Present Status of Visual Science'.[3] It was in this report that Troland called for a re-determination of the three colour-excitation curves for the normal eye. I have suggested elsewhere that I regard this as the moment of conception of the 1931 Observer, and since Troland actually wrote his 'Visual Science' report in 1921, this meant that we had a 10-year pre-natal period in which to create and deliver this infant prodigy.

Troland's influence on the development of colorimetry was clearly considerable; and in 1925 he became Research Director of Technicolor while also holding an academic post at Harvard. Sadly, he died in tragic circumstances in 1932.

The OSA Colorimetry Report[4] laid the foundations for modern colorimetry, since it dealt at length with Nomenclature, Standard Psychophysical Data, Physical Standards, and Methods of Colorimetry. The report included a table of Average Normal Visibility Values (the forerunner of the 1924 V_λ curve) and a set of colour-mixture curves known as the OSA Excitation Curves which had been calculated by Weaver from the pioneer results obtained by König and by Abney. In 1923 Ives independently derived an alternative set of colour-mixture curves from König's data alone which became known as the König-Ives colour-mixture curves. This paper of

Ives, together with an earlier one published in 1915,[5] dealt with the transformation of colour-mixture curves from one set of primaries to another. This was an important development as it meant that the set of curves adopted for use in a standard system of colour measurement need not conform to some particular colour vision theory nor be expressed in terms of the instrument primaries of the colorimeter with which the curves were measured.

The Colorimetry Report discussed at some length the question of standard illuminants and described a method due to Priest for obtaining a standard artificial sunlight source using a gas-filled tungsten lamp, together with a pair of crossed Nicol prisms and a $\frac{1}{2}$ mm quartz plate inserted between them. This gave a surprisingly similar energy distribution to that of natural sunlight as determined by Abbot. Another suggestion was to use a No. 78 Wratten blue filter in conjunction with an acetylene lamp. The report had little to say about actual colour measuring instruments, but the equipment available at that time included visual spectrophotometers like the König-Martens instrument, monochromatic colorimeters in which hue and saturation were determined directly by matching the colour under test with a mixture of monochromatic light and white light as in the Nutting instrument, and sub tractive colorimeters like the Lovibond Tintometer and like one developed by L. A. Jones using red, yellow and blue wedges. There were also one or two additive trichromatic colorimeters of no great accuracy or reliability. (See Guild[6] for further information about these instruments.) On the whole, it would seem that the general understanding of the subject was in advance of the instrumental techniques available to put the principles into practice, but in 1924 the lead in photometry and colorimetry was certainly held by the United States.

THE RUN-UP TO THE 1931 OBSERVER: 1924–30

The scene now moves to England where John Guild had taken up the subject of colorimetry at the National Physical Laboratory. He published a number of papers in the Transactions of the Optical Society (London) in 1924-25 and 1925-26, dealing with the transformation of colour-mixture data by algebraic and geometric methods and describing three new colour-measuring instruments, namely his visual spectrophotometer, his vector colorimeter and his trichromatic colorimeter suitable for standardisation work using red, green and blue colour filters as his primaries. His *magnum opus*, however, was his 'Critical Survey of Modern Developments in the Theory and Technique of Colorimetry and Allied Sciences'.[6] This was in effect the English equivalent of Troland's 1922 OSA Colorimetry Report and he echoed Troland's call for new colour-matching data in these words: "There is room, however, for further developments of the practical aspects of the science. . . and most urgently of all, accurate information on the chromatic properties of the average human eye, in order that a 'normal' eye for the purpose of colorimetry may be established by an agreed set of spectral mixture curves in the same way as a 'normal' eye for photometry has been established by an agreed visibility curve (the 1924 V_λ curve)."

Guild presented his 'Survey' at an Optical Convention held at Imperial College in London in 1926 and after the Convention he immediately set to work at the NPL to obtain such a set of spectral mixture curves. He used two Hilger constant deviation monochromators in series to serve as a double monochromator to provide the monochromatic test colour and his own filter trichromatic colorimeter to provide the matching field. Provision was also made for a small amount of one or other of the filter primaries to be transferred to the test colour side of the field to give sufficient desaturation of the spectral colour so that the mixture could be matched by positive amounts of the red-green-blue primaries.

The square field of view in Guild's instrument subtended an angle of approximately 2°, and as Guild explained,[7] this size of field was chosen for the following reasons: "It lies almost entirely within the average 'yellow spot' of the retina (Maxwell's spot), and is of similar dimensions to the field to which the standard visibility data apply. It is of the greatest importance that quantitative work on the properties of the eye should all apply to the same region of the retina, and colour-matching with fields extending beyond the macula lutea may give results which are not at all representative of foveal vision. Experiments also showed that for such small fields the simple two-part division was the most advantageous."

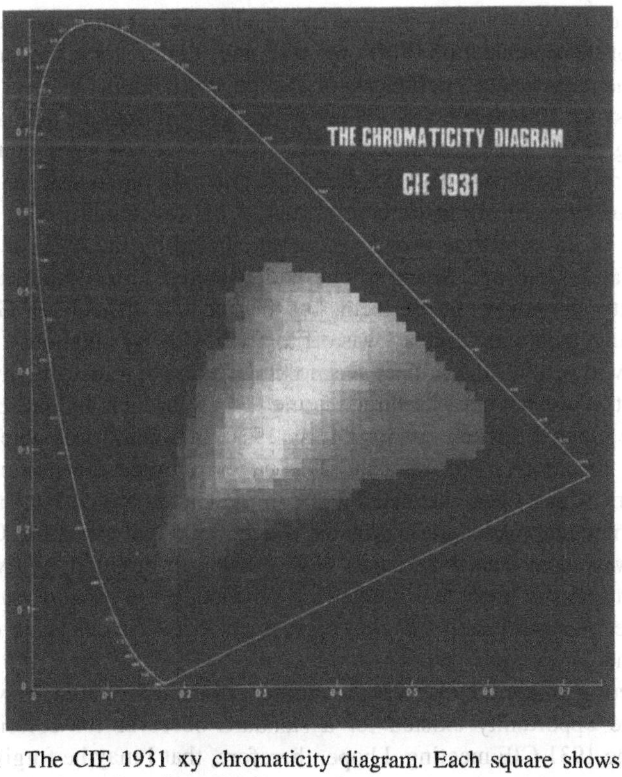

FIGURE 2.1 The CIE 1931 xy chromaticity diagram. Each square shows the brightest surface colour which can be achieved using non-fluorescent dyes or pigments. See color inlet.

Seven observers took part in Guild's investigation and although he did not publish his results until 1931, he made some use of the mixture curves of his average observer in a paper on signal glasses which was given at the CIE meeting held in the United States in 1928.

Thanks to the interest aroused by Guild's paper at the 1926 Optical Convention, the Medical Research Council in England were approached to provide a grant to support further colour vision research at Imperial College where Abney had built and used his colour-patch apparatus. I was the fortunate recipient of the grant and, as a raw research student, I set about building my own design of colorimeter which used spectral primaries but otherwise had the same viewing conditions of a 2° field of view as Guild had used. Here I must point out how much I owed to Guild, partly through his published papers which provided my basic understanding of the subject, partly through the personal advice which he gave me when I went to see him at the NPL, and partly through the great trouble which he took to discuss in detail the papers which in due course I came to present at meetings of our Optical Society. We had hoped that Guild would live to share in this Golden Jubilee meeting, but he died on October 23rd, 1979, shortly before his 90th birthday. He did, however, write to Dr Michael Pointer on June 22nd, 1979, sending his best wishes for a very successful Jubilee meeting and summarising the history of the CIE system as he, not very accurately, remembered it.

This is not the occasion to describe the results which Guild and I obtained, but the timing of their publication is of some relevance to our story. I reported my main results on the trichromatic coefficients of the spectral colours (spectral chromaticity co-ordinates) for 10 observers in a paper to our Optical Society[8] which I presented at a meeting of the Society on March 14th, 1929. I received some quite favourable comments from Guild (two pages of them!), but he took me to task for having given my results in terms of my instrument primaries of wavelengths 650, 530 and 460 nm, instead of the reference primaries recommended by the NPL of wavelengths 700, 546.1 and 435.8 nm. Since no one had bothered to tell me about these primaries, I naturally could not use them, but this made it difficult for Guild to compare his results with mine. Since I was off to a job with Westinghouse in Pittsburgh in the USA within a few days, there was no time for me to transform my data before I left. In fact, it was not until I returned home just a year later and presented another paper to the Optical Society on April 10th, 1930, that Guild was able to make the comparison of our two sets of data. This paper reported the mixture curves of the spectrum[9] which I had calculated while in the United States, but I also included the transformed chromaticity co-ordinates which Guild had asked for. On this occasion I received more than three pages of discussion from Guild, and while he was not very enthusiastic about my method of calculating the mixture curves, he was clearly quite impressed with the close agreement between our basic experimental measurements.

This meeting in April, 1930, was the moment of truth for Guild, in which he now saw that the opportunity existed for a standard observer for colorimetry to be defined at the 1931 CIE meeting. I hope, therefore, that I may be forgiven for quoting some of the comments which Guild made on my results.

He wrote: "I have compared these results with my own figures at various critical parts of the chart (the chromaticity chart) and am pleased to say that no discrepancies exceeding the colour limen in the neighbourhood were found at any of the points checked. This agreement adds enormously to the value to be attached both to Mr Wright's data and to our own, because it must be remembered that the two determinations were made by different groups of observers employing entirely dissimilar apparatus. It shows in a very striking manner that, provided adequate attention is given to the standardisation of the auxiliary apparatus and to such points as field size, these determinations can be carried out with a much higher degree of accuracy and reliability than is still supposed, in many quarters, to be the case, and that a comparatively small group of observers is sufficient to reduce the effects of personal variations of vision to quite small amounts.

It is thus clear that Mr Wright's trichromatic coefficients, which are, so far, the only published set based on modern experimental work, can confidently be regarded as a very close approximation to the properties of an average eye."[9]

This was really an extremely generous comment for Guild to have made to a youngster like myself, especially seeing that his own results had been obtained two or three years before mine, only they had not been published. What surprises me is that Guild had not thought it worth while to transform my data himself immediately after I reported them in 1929. This would have avoided the intense time pressure under which he found himself in 1931. I can only assume that he could not really believe that my results would be any good! Before, though, dealing with the events of 1931, we must retrace our steps and return to the 1924 meeting of the CIE.

In addition to defining the Standard V_λ curve at that meeting, the Commission also decided to set up a Colorimetry Study Committee, and I. G. Priest of the Bureau of Standards and T. Smith, Head of the Optics Department of the National Physical Laboratory, were appointed as its members. No great progress seems to have been made by this Study Committee by the time of the next CIE meeting in 1928, but it did at least lead to a Colorimetry Committee meeting being held then, although neither Smith nor Guild was present. (British colorimetry interests were represented by Dr J. W. T. Walsh and Mr H. Buckley from the NPL, by Dr W. M. Hampton from Chance Bros. and by Professor J. T. MacGregor-Morris of Queen Mary College, London University.) Britain did, however, submit a programme of work in colorimetry for the next three-year period and as a reward for this initiative, they were given the Secretariat responsibilities for the subject. Priest then made the following proposal:

"I think that the first duty of the Secretariat Committee should be to obtain agreement with regard to nomenclature, which is a very complex subject. It is desirable that first of all the two English-speaking nations should agree and in this connection I propose that the Report of the Optical Society of America (their 1922 Colorimetry Report) be taken as a basis and that the British Committee be asked for their comments upon it. I propose that the following additional items of work mentioned in the British report should be undertaken:

(1) Agreement upon a standard of white: This may be approached either on the psycho-physical side or by assuming that it is desirable to adopt an approximation to daylight.

(2) Agreement upon the sensation curves."

Rather independently, the CIE Committee responsible for the specification of coloured glasses for signal lights, a subject in which there was much activity at the 1928 meeting, made the following decision among a number of other recommendations:

> "Il est recommandé d'utiliser les valeurs d'excitation (courbes de sensation) données par le Rapport sur la Colorimétrie d'Optical Society of America publié dans le Journal of the Optical Society, Vol. VI, p. 549, 1922, et cela jusqu 'à ce que des valeurs meilleures aient été approuvées par la C.I.E."

These proposals call for two comments: first, the importance that was still attached in 1928 to the OSA Colorimetry Report published in 1922 and, second, the practical use that was being made of the OSA Excitation Curves worked out by Weaver and included in the 1922 Report. It is no wonder that the Americans hesitated about adopting new colour-mixture data in 1931 unless they could be shown to be significantly superior to their Excitation Curves.

After the 1928 CIE meeting, there was quite a speedy response by the British Committee to the proposal for agreement on a standard white. This took the form of a Memorandum from the NPL dated June 13th, 1929, which opened with the sentence: "The secretariat Committee is giving consideration to the establishment of a standard white for colorimetrie work." It then went on to make definite recommendations about the use of a gas-filled tungsten lamp, how it should be rated, and the use of liquid filters to be used with the tungsten lamp to produce a representative white light. These firm recommendations were then followed by tentative recommendations about the preferred colour temperature of the lamp and the type of liquid filters to be used. Filters made up according to chemical formulae developed a few years earlier at the NPL were already in use in some industrial laboratories, while slightly different formulae had been developed at the Bureau of Standards by Davis and Gibson.[10] The Memorandum was quite a statesmanlike and diplomatic document, but came out clearly in favour of the NPL filters, mainly on account of the inconvenience to existing users if they had to make a change from an illuminant which they had accepted as a standard. The Memorandum concluded: "Since the inconvenience arising from any changes will be more serious the longer a decision is delayed, it is hoped that the presentation of any adverse criticism of these proposals will be regarded as a matter of great urgency and importance."

The first opportunity for this Memorandum to be discussed in the United States was at the Autumn meeting of the Optical Society of America which was held at Cornell University. A special session was arranged for Saturday morning, October 26th, 1929, and I still have a copy of the Minutes of the meeting, since I was one of a dozen or so participants who attended and my views are duly recorded as contributing to the American opinion on the subject! (I had managed to persuade Westinghouse that it would be to their advantage if they allowed me to attend the OSA meeting. It certainly proved to be to my advantage.) The minutes included an extremely lengthy and involved contribution from Priest, but the main points of contention concerned the choice of the Davis-Gibson or the NPL filters and whether the source should be defined by its spectral energy distribution or by the colour temperature of the tungsten lamp and the chemical constitution of the filters. Naturally enough, the Americans preferred the Davis-Gibson filters, and when the final decision was made in 1931, this point was conceded by Guild in view of the very extensive studies that had been made on the Davis-Gibson filters.

The method of definition of the illuminants, however, touched on fundamental concepts on how standards should be defined. Guild was adamant that the filters should be defined in terms of their chemical composition so that the source could be produced without ambiguity, whereas Priest and his colleagues thought that the lamp-plus-filter source should be defined by its energy distribution. Since there were limits to the precision with which the energy distribution could be measured, Guild maintained that a faulty distribution might be adopted as standard and we would then be saddled with a standard source which could not be accurately reproduced in the laboratory. By 1931 Guild had won this argument and the three sources S_A, S_B and S_C were defined by the colour temperature of the tungsten lamp and the chemical formulae for the B and C filters. The energy distributions were, however, included as a supplement to the CIE resolutions.

Guild, I am sure, based his argument on what he believed was correct standardisation practice, but his viewpoint was reinforced by his belief that most practical colorimetry would be carried out by direct visual colour matching in which the sample to be measured had to be illuminated by the standard source. He even went so far as to say in his 1926 'Survey': "The writer's trichromatic colorimeter is, as far as he is aware, the only instrument at present available which offers, in principle, a complete solution to the problems of colorimetry." He never seemed to grasp the future potentiality of the recording spectro-photometer as a tool for accurate colour measurement. Perhaps, though, the CIE should have heeded Guild's warning when the new daylight distribution D_{65} was defined in 1964 by its spectral power distribution. When samples which fluoresce have to be measured, we often need a laboratory source which simulates D_{65}, but none is available which exactly reproduces its spectral distribution. It might have been better to have developed a source that simulated the D_{65} distribution closely enough for most practical purposes and to have adopted that source as the standard, with its energy distribution being given as a supplement to the definition.

THE DRAMA OF 1931

After our Optical Society meeting in April, 1930, in which the close agreement between Guild's results and mine was established, Guild went ahead with the preparation of a paper giving his own results and the mean of our two sets of colour-mixture data. He submitted this paper on 'The Colorimetric Properties of the Spectrum'[7] to the Royal Society in February, 1931, and it was read, before the Society on April 30th, 1931. At the same time he prepared a Memorandum on 'A Normal Eye for Colorimetry' on behalf of the British Secretariat Committee in which, in summarising the results, he stated:

> "The data therefore fulfil the following conditions:
> 1. The unit equations are the mean of those obtained by Wright and the author.
> 2. The spectral distribution curves, when weighted by the luminosity factors of the primaries, summate exactly to the standard 'normal' visibility curve.
> 3. The luminosity factors are in reasonably close agreement with those determined by independent methods in the N.P.L. investigation."

He therefore recommended that a 'normal' eye for technical colorimetric work could now be adopted as a standard, with the same status, and limitations as to permanence, as the International Visibility Curve.

This document was transmitted to the United States in February, 1931, and was read in part at a meeting of the Optical Society of America in New York on February 28th. Copies were also circulated to some five or ten individuals particularly active in the subject, including Ives, Gage and Judd. Their views were collated by Priest with those of his own and a considered reply was sent to Guild in July, 1931, only two months before the CIE was due to meet in Cambridge, England.

This reply from Priest is a very interesting document indeed and is really worthy of more detailed discussion than space permits. So far as the Guild-Wright data were concerned, they were generally accepted as being the best available, and the two industrialists, Ives from Bell Telephone and Gage from Corning Glass, both supported a standard observer based on the new data. Priest, however, discussed at considerable length whether the new data were significantly different or significantly better than the OSA excitation curves which Weaver had calculated back in 1922. The main use to which colorimetry was being put in the United States in 1931 was still in the specification of signal lights and in the colour of signal glasses, and these specifications were generally given in terms of dominant wavelength and saturation. Priest therefore included in his reply a table of dominant wavelengths for four different glasses as derived (a) from direct experiment, and as computed (b) from the OSA excitations, (c) from Guild's unpublished excitations, and (d) from another set of unpublished excitations due to Priest and Judd. (I have been unable to trace any information about these Priest-Judd curves.) Priest then concludes: "These comparisons cannot be regarded as demonstrating any peculiar superiority of Mr Guild's data over the OSA data for the purpose in standardising the colors of signal glasses." And at another point in his reply, he states: "But we are of the opinion that, to all practical intents and purposes, the OSA data are adequate for the limited purposes for which we use them seriously." This last statement reveals a surprising denigration of the importance of colorimetry, especially by someone who was Chief of the Colorimetry Division at the Bureau of Standards. Judd's view was that the OSA excitation curves did not represent a sufficiently abnormal eye to warrant their rejection and that the most compelling reason for urging adoption of Guild's curves was that they incorporated the standard visibility function.

One feature that intrigues me about these American comments on the British proposal is the almost patriotic fervour with which they espoused the OSA excitation curves. These curves were admittedly calculated by Weaver, but the experimental data on which they were based were provided by König in Germany and by Abney in England!

One of the complaints that Priest made in his comments was the lack of adequate time or opportunity for critical consideration and discussion of the data. One can sympathise with him about this, because although Guild's Memorandum was sent in February, 1931, the full account of Guild's experimental work and the derivation of our mean data were contained in Guild's Royal Society paper and Priest did not receive a copy of this until July 18th, 1931. There is, however, a certain irony in his statement: "The adoption of standard visibility was arrived at after the results of much more extensive and diverse experimental work were available, and after these results had received much more extensive, critical and mature consideration." The irony is that in spite of this mature consideration, the V_λ curve has proved to be significantly in error at the violet end of the spectrum because insufficient weight was given to the best experimental data at these wavelengths (data that were actually obtained at the Bureau of Standards), whereas our colour-mixture data have not, so far as I am aware, been shown to be incorrect in any respect for the viewing conditions for which they were measured.

The one criticism of Guild's Memorandum that ran right through Priest's document was dissatisfaction with his choice of primaries – the spectral stimuli at wavelengths of 700.0, 546.1 and 435.8 nm. The use of such real physical primaries meant, as was well-known, that negative quantities inevitably appear in the colour-mixture curves, and these were regarded by all the Americans as quite unacceptable in a colour-measuring system for use in commerce and industry. Guild's choice of these primaries to define the colour-matching characteristics of a standard observer was based, once again, on his underlying philosophy about standards, namely that they should be unambiguous and physically realisable with certainty. Thus, 700.0 nm is a wavelength at the red end of the spectrum in a region where there is very little variation of hue with wavelength, while 546.1 and 435.8 nm are prominent green and violet lines in the mercury spectrum.

While Guild, therefore, pressed for the standard observer to be defined in terms of these primaries, he left himself open in his Memorandum for the data to be transformed to a different set of primaries giving an all-positive system, if this was thought to be more convenient for practical commercial colorimetry. This is evident from the final paragraph of his Memorandum:

> "The proposal to standardise the data of Table IV does not, therefore, necessarily carry with it the proposal to adopt either the N.P.L. system of reference primaries, or the N.P.L. standard white. The question whether the visual relations embodied in the

> data are suitable to represent a standard normal eye for technical colorimetry, may be, and ought to be, considered separately from those questions which are relevant to the choice of suitable reference standards."

This was not, perhaps, as clearly worded as it might have been, otherwise Priest would have realised that Guild would not oppose an all-positive system for actual colour specification.

When Priest arrived in England in September, 1931, he almost certainly had no intention of agreeing to the adoption by the CIE of new standard observer data and a CIE system of colorimetry at the forthcoming meeting. He came armed with all the objections discussed in his document, but after an exhausting week at the NPL prior to the CIE meeting, his objections were met one by one as Smith and Guild transformed the data to meet Priest's criticisms. So when we assembled at 9.30 am on Thursday, September 18th, at Trinity College, Cambridge, for the fateful meeting of the CIE Colorimetry Committee, we were presented with fresh resolutions which were, like newly baked bread, still warm from the heated discussions in which Smith, Guild and Priest had indulged the previous week. They were, though, resolutions which, with one exception, could be endorsed by both the British and American representatives.

There were 21 delegates present from 6 countries, namely France, Germany, Great Britain, Japan, Netherlands, and the United States. Agenda papers had been circulated in advance but these were now out of date, thanks to the Priest-Smith-Guild discussions. Mr Guild expressed his regret that so little notice had been given of the revised resolutions that were tabled and explained that the British Secretariat Committee would have hesitated to present them were it not for the fact that the Commission allowed a period of four months in which the resolutions could be considered by National Committees. This interlude for second thoughts nearly wrecked Guild's hopes for the new system, since although all the resolutions were approved at the meeting itself, four of the five by a unanimous vote, France subsequently reversed all of its votes and Germany reversed four of them. This presumably still left a sufficient majority of votes in favour, since in due course all the resolutions were formally approved by the CIE.

The first of the five resolutions defined the colour-matching functions of the standard observer in terms of Guild's spectral primaries. The second defined the three standard illuminants, S_A, S_B and S_C. The third defined the illuminating and viewing conditions for measuring surface colours, namely illumination at 45° to the normal and viewing normal to the surface. (This was the resolution which America did not accept, since it differed from the normal illumination and diffuse collection used in the General Electric (Hardy) spectrophotometer.) The fourth resolution was a rather obscure statement laying down the principle on which the reference primaries, X, Y and Z should be chosen. My recollection is that no one at the meeting understood this resolution and it was certainly accepted without discussion. The fifth resolution consisted of colour matching equations defining X, Y and Z in terms of the standard observer system and the tables of trichromatic coefficients and spectral mixture curves in terms of these primaries.

The particular choice of reference primaries X, Y and Z satisfied the American demand for an all-positive system and also took advantage of a device originally suggested by Schrödinger and elaborated in a very impressive paper by Judd[11] in which X and Z were located on the alychne, the locus in the chromaticity chart of colours of zero luminance. This meant that all the luminance information was given by the Y tristimulus value and that the $\bar{y}_\lambda$ colour-matching function was identical with the V_λ curve. In the collection which I still have of the papers tabled at the meeting, I see that we were offered two alternative systems to choose from, Alychne System 1 and Alychne System 2. System 2 was the one that was adopted, the main difference being in the Z trichromatic coefficients and tristimulus values. I cannot recall any reference to System 1 and there is no mention of it in the Minutes of the meeting.

Looking back on all the activities leading up to the 1931 decisions, I keep wondering why it was so much an Anglo-American concern. I assume that Guild's Memorandum on 'A Normal Eye for Colorimetry' was circulated to other member countries of the CIE, but there is no evidence of any comments or criticisms having been received other than those from the United States. And at the Cambridge meeting itself, only Professor L. S. Ornstein (Netherlands), Professor M. Pirani (Germany) and Mr R. Tajima (Japan) made any significant contributions to the discussion, apart from the American and British delegates. In the aftermath of the Great War and the devastation in Europe, colorimetry cannot have had a very high priority in European countries, and perhaps this helps to explain why France and Germany reversed their votes. They may well have felt they were being rushed into making decisions in a subject in which they were only just beginning to gain any practical experience of their own. They needed more time to think.

POSTSCRIPT TO 1931

Once the CIE System was established, the Americans threw off all their reservations and abandoned the OSA Excitation Curves once and for all. Unhappily, Priest was taken ill soon after returning from the CIE meetings and died in 1932. His mantle fell on Deane Judd, who prepared a report on the CIE recommendations and presented them in a paper to the Optical Society of America in a form designed to suit American interests in colorimetry.[12] Hardy's 'Handbook of Colorimetry' published in 1936 made the CIE system even more readily available and acceptable to a wide range of colour technologists. In England, Smith and Guild published a paper in the Transactions of our Optical Society on the CIE standards and their use, but hardly in a form to appeal to the non-mathematical reader.[13] (Guild once confessed to me that he was one of the many readers who did not understand the paper!)

The first request that I received to specify a colour on the CIE system was from our Ministry of Agriculture and Fisheries, who wished to define the colour of forced rhubarb in connection with a National Mark scheme they were introducing. This showed surprising enterprise on the part of our civil servants and it also provoked an amusing article in our humorous magazine, *Punch*.[14] More seriously, the acceptance of the CIE system was shown most immediately in its application to coloured light signals. As John Holmes will be describing, recommendations were made at the CIE meeting in Berlin in 1935 defining the colour limits on the 1931 chromaticity chart for road, rail and aviation signal colours. I could, but I must not, trespass on other papers in this Golden Jubilee to provide further vivid evidence of the flood of activity which 1931 unleashed in the colour industries and in colour technology.

I regard Troland, Ives, Priest, Guild and Judd as the main architects of the 1931 CIE System and I would like to suggest that we think of this Golden Jubilee as a tribute to the vision and initiative which they showed 50 and more years ago.

NOTE ADDED IN PROOF

My suggestion on p. 11 that E. A. Weaver worked at the Bureau of Standards is almost certainly incorrect. I am indebted to Mr R. S. Hunter for the information that a Mr Eastman A. Weaver joined the Optical Society of America on 10 April 1920 and that he had been recommended for membership by Professor L. T. Troland. Since his home was in Boston, Massachusetts, and as Troland indicated in his 'Visual Science' report that he and Weaver were planning to make a re-determination of the three-colour excitation functions, it seems probable that Weaver was either a graduate student of Troland's or a member of his staff at Harvard. As we have seen, the OSA Excitation Curves were widely accepted both nationally and internationally in the 1920s, yet so far as I know their derivation by Weaver was never published other than in the 1922 OSA Colorimetry Report. Weaver appears to have been something of a mystery man, but that is no reason why we should not acknowledge his very significant contribution to the early development of colorimetry.

REFERENCES

[1] Sherman, P. D., Problems in the theory and perception of colour: 1800–1860, Univ. London PhD Thesis (1971). (To be published by Adam Hilger Ltd under the title 'Colour Vision in the Nineteenth Century; The Young–Helmholtz–Maxwell Theory').

[2] Gibson, K. S., The relative visibility function, *Recueil des Travaux, 6th Session (1924) CIE*, Cambridge Univ. Press, pp. 232–238 (1926).

[3] Troland, L. T., The present status of visual science, *Bull. National Research Council*, **5**, Pt. 2, No. 27, 1–120 (1922).

[4] Troland, L. T., Report of committee on colorimetry for 1920–21, *J. Opt. Soc. Am.*, **6**, 527–595 (1922).

[5] Ives, H. E., The transformation of color-mixture equations from one system to another, *J. Frank. Inst.*, **180**, 673–701 (1915); **195**, 23–44 (1923).

[6] Guild, J., A critical survey of modern developments in the theory and technique of colorimetry and allied sciences, *Proc. Optical Convention*, Pt. 1, 61–146 (Optical Convention, London, 1926).

[7] Guild, J., The colorimetric properties of the spectrum, *Phil. Trans. Roy. Soc. Lond., Ser. A*, **230**, 149–187 (1931).

[8] Wright, W. D., A re-determination of the trichromatic coefficients of the spectral colours, *Trans. Opt. Soc. Lond.*, **30**, 141–164 (1928–29).

[9] Wright, W. D., A re-determination of the mixture curves of the spectrum, *ibid.*, **31**, 201–218 (1929–30).

[10] Davis, R. and Gibson, K. S., Filters for the reproduction of sunlight and daylight and the determination of color temperature, *Misc. Pub. Bur. Stand.*, No. 114, 1–165 (1931).

[11] Judd, D. B., Reduction of data on mixture of color stimuli, *J. Res. Nat. Bur. Stand. (U.S.)*, **4**, 515–548 (1930).

[12] Judd, D. B., The 1931 ICI standard observer and co-ordinate system for colorimetry, *J. Opt. Soc. Am.*, **23**, 359–374 (1933).

[13] Smith, T. and Guild, J., The CIE colorimetric standards and their use, *Trans. Opt. Soc. Lond.*, **33**, 73–134 (1931–32).

[14] Rhubarbarisms, H.F.E., *Punch*, **189**, 646 (11 Dec. 1935).

3

CIE COLORIMETRY

JÁNOS SCHANDA
University of Pannonia, Egyetem u. 10., H-8200 Veszprém, Hungary

INTRODUCTION

As can be seen from the first two chapters of this book, CIE colorimetry looks back over an evolution of 75 years.[1] During this time colorimetry progressed considerably. In this chapter we will summarize the present day knowledge of color stimulus metrics. In this respect first we have to stress that color is a perception, and as such it is not accessible to engineering measurement. Metrology can access only the stimulus that will have as a consequence the perception. Thus CIE colorimetry is the metric of the psychophysical color stimulus.

In psychophysics one often distinguishes between class A and class B observations, where class A relates to observations where two stimuli cause indistinguishable perceptions. Color matches are class A observations: If two stimuli with unequal physical characteristics produce—under otherwise similar exterior circumstances—the same sensation, we regard them to be equivalent. CIE colorimetry in its very fundamental form relates to such phenomena. In some cases, where we try to describe nonequality of the perceptions, such as when we try to describe color difference evaluation, we take colorimetry to its limits. More advanced descriptions of psychophysical phenomena, for example, the determination of the brightness of different colored lights, is beyond the realms of basic colorimetry, and the determination of its physical correlates is part of *advanced colorimetry.* Items belonging to this later group are, for example, color appearance models. Such phenomena are called class B observations and will be dealt with in later chapters of this book.

Colorimetry: Understanding the CIE System, Edited by János Schanda

Light sensation is produced by visible radiation, electromagnetic radiation falling within the wavelength limits of 380 nm and 780 nm.* Radiation from the short wavelength region of this radiation produces usually the sensation of blue light, radiation with wavelengths between 520 nm and 550 nm are seen as green light, and above about 650 nm we perceive the light usually to be of red color. These limits are not well defined, and the actual perception depends strongly on the adaptation state of the eye and on light stimuli surrounding the test object. The modification by the latter is called simultaneous contrast, and its effects will be discussed in the chapter regarding color appearance models.

There are two fundamental methods of producing color stimuli: additive and subtractive color mixing. In additive color mixing lights are mixed, as in color TV displays, where the color sensation in our eye is produced by the additive mixture of tiny red, green, and blue lights, where the single spots are so near to each other that our eye is unable to resolve them spatially and we see the mixture of the lights. By changing the intensity of the single spots, different mixed colors can be produced. In subtractive color mixing, colorants remove some part of the visible spectrum. Superposing several colorants of different concentrations on each other will change the color of the transmitted light.

Basic colorimetry, the description of the results of color matching experiments, is built on additive color mixing because the laws of additive color mixing are simpler than those of subtractive color mixing. The four basic empirical laws of additive color mixing were formulated in 1853 by H. G. Grassmann[2]:

1. Every impression of color may be analyzed into three mathematically determinable elements, the hue, the brightness of color, and the brightness of the intermixed white.
2. In the second place we assume that if one of two mingling lights is continuously altered (while the other remains unchanged), the impression of the mixed light is also continuously changed.
3. Two colors, both of which have the same hue and the same proportion of intermixed white, also give identical mixed colors, no matter what homogeneous colors they may be composed of.
4. The total intensity† of any mixture is the sum of the intensities of the lights mixed.

*These wavelength limits are mean values of many observers and have been standardized by the CIE. There are observers, who also see in the near ultraviolet (UV) and/or near infrared (IR) region of the spectrum, up to about 360 nm in the UV and 850 nm in the IR.

†In the original text the term "intensity" was used, in the updated translation at most places it was substituted by the terms "brightness" or "illumination," corresponding better to modern terminology, but in the last statement the original term was retained with the meaning something like "visual luminance."

In modern textbooks one finds only the first three (the accurate) empirical laws of *color-matching* properties of additive mixtures of color stimuli in the following form[3]:

1. To specify a color match, three independent variables are necessary and sufficient.
2. For an additive mixture of color stimuli, only their tristimulus values are relevant, not their spectral compositions.
3. In additive mixtures of color stimuli, if one or more components of the mixture are gradually changed, the resulting tristimulus values also change gradually.

CIE colorimetry builds on these empirical laws that hold reasonably well as long as the observation conditions (e.g., size of stimuli, presentation on the retina: foveal or parafoveal, etc.), the previous exposure of the observer's eye, and the person who makes the matching are kept the same. Therefore the observation conditions have been standardized: foveal vision, 2° or 10° field size, dark surrounding; as previous exposure a sufficiently long dark adaptation is supposed and the standardized color-matching functions (CMFs) have been determined by averaging the results of a large number of observers. Further questions relating to the validity of Grassmann's laws will be discussed in Chapter 10.

In the following sections we will discuss the recommendations of CIE colorimetry, based on the most recent publication on colorimetry[4]; we will, however, use a somewhat different approach to the recommendations, as this corresponds better to a general understanding of the subject.

CIE STANDARD COLORIMETRIC OBSERVERS

According to Grassman's laws a color stimulus can be matched by the additive mixture of three properly selected stimuli (properly selected includes independent, i.e., none of the stimuli can be matched by the additive mixture of the other two stimuli). Figure 3.1 shows the basic experiment of obtaining a color match. The test stimulus is projected on one side of a bipartite field, the additive mixture of the three matching stimuli (it is practical to use monochromatic red, green, and blue lights, see later) is projected onto the other side of the field. By using adjustable light attenuators, the light flux of the three matching stimuli are adjusted to obtain a color appearance match between the two fields. When this situation is reached the test stimulus can be characterized by the three luminance values of the matching stimuli reaching the eye of the observer.

The spectral power distributions (SPDs) of the test stimulus and of the additive mixture of the three matching stimuli are usually different. In such cases we speak about metameric colors: They look alike to the human observer (having equal tristimulus values, see later), but their SPD is different. Metamerism is fundamental in colorimetry.

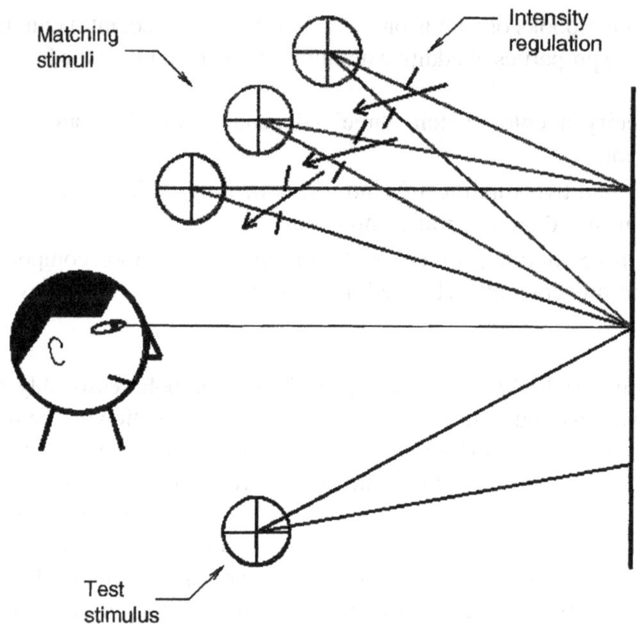

FIGURE 3.1 Basic experiment of color matching. See color insert.

To obtain a colorimetric system one has to define the matching stimuli, specifying both their spectral composition and the units in which their amounts are measured. If this is done one can describe a color match in the following form:

$$[\mathrm{C}] \equiv R[\mathrm{R}] + G[\mathrm{G}] + B[\mathrm{B}] \tag{3.1}$$

where [C] is the unknown stimulus; "≡" reads as "matches"; [R], [G], [B] are the units of the matching stimuli and R, G, B represent the amounts to be used, expressed in the adopted units, of the matching stimuli to reach a match.

As a next step one has to determine for every monochromatic constituent of the equienergy spectrum (the spectrum having equal power per small constant wavelength interval throughout the visible spectrum) the amounts of the three matching stimuli needed to achieve a match. The wavelength dependent amounts needed for the above color match of the monochromatic test stimuli are called CMFs and are written in the following form: $\bar{r}(\lambda), \bar{g}(\lambda), \bar{b}(\lambda)$. Because of the additivity and multiplicativity of color stimuli, for a nonmonochromatic test color stimulus, $P(\lambda)$, the amounts of the matching stimuli needed for a match can be determined by adding the amounts needed to match the monochromatic components of the test stimulus (for a detailed analysis see, e.g., Ref. 5):

$$[\mathrm{C}] = \int_{380\,\mathrm{nm}}^{780\,\mathrm{nm}} \bar{r}(\lambda)P(\lambda)\mathrm{d}\lambda \cdot [\mathrm{R}] + \int_{380\,\mathrm{nm}}^{780\,\mathrm{nm}} \bar{g}(\lambda)P(\lambda)\mathrm{d}\lambda \cdot [\mathrm{G}] + \int_{380\,\mathrm{nm}}^{780\,\mathrm{nm}} \bar{b}(\lambda)P(\lambda)\mathrm{d}\lambda \cdot [\mathrm{B}] \tag{3.2}$$

The $\int_{380\,\text{nm}}^{780\,\text{nm}} \bar{r}(\lambda)P(\lambda)\text{d}\lambda$, $\int_{380\,\text{nm}}^{780\,\text{nm}} \bar{g}(\lambda)P(\lambda)\text{d}\lambda$, $\int_{380\,\text{nm}}^{780\,\text{nm}} \bar{b}(\lambda)P(\lambda)d\lambda$ integrals are called tristimulus values and can serve as the descriptors of the color stimulus and according to Equation (3.1) the symbols R, G, B are used.

The CIE 1931 Standard Colorimetric Observer

To be able to repeat an additive color match precisely the observation conditions have to be standardized. For the CIE 1931 standard colorimetric observer a 2° foveal field of observation and a dark surround was chosen (the 2° field is well within Maxwell's spot, i.e., where the macula lutea has an almost constant density).

Determination of the $\bar{r}(\lambda), \bar{g}(\lambda), \bar{b}(\lambda)$ *Color-Matching Functions*

The CIE 1931 standard colorimetric observer was derived from the results of two experimental investigations, conducted by W. D. Wright[6,7] and J. Guild[8] (for historic details see Chapter 2, a facsimile reproduction of the authentic work written by Dr. Wright for the Golden Jubilee of the CIE system of colorimetry). The two investigations used different *matching stimuli* (called also "primaries"), but when transforming the results to a common system the agreement was surprisingly good, despite the fact that the number of observers was only seven in Guild's work and only 10 in Wright's.

To be able to define a standard observer the spectral compositions and the luminances of the primaries have to be specified. Single wavelengths were used: 700 nm for the red, 546.1 nm for the green, and 435.8 nm for the blue primary. The "unit intensity" of the primaries was defined by stating their luminances. The requirement was that for an equienergy spectrum the addition of the unit amounts of the three primaries should give a color match. If 1 cd/m^2 of red light was used, then 4.5907 cd/m^2 of green and 0.0601 cd/m^2 blue light was needed to match the color of an equienergy spectrum.

Performing color matches using these matching stimuli one gets the CMFs depicted in Figure 3.2. The negative lobes in these curves refer to the fact that in some parts of the spectrum a match can be obtained only if one of the matching stimuli is added to the test stimulus.

As mentioned the units of the three primaries have been defined by their luminances and thus the luminance of a color stimulus with the tristimulus values of R, G, B will be

$$L = 1.0000\,R + 4.5907\,G + 0.0601\,B \tag{3.3}$$

But the units used are very often only defined as relative luminances, so that L is in these cases only a relative luminance.

Derivation of the CIE XYZ Trichromatic System from the CIE RGB Trichromatic System

In many colorimetric calculations—especially at the time of standardizing the trichromatic system, when no computers were available—the negative lobes in the

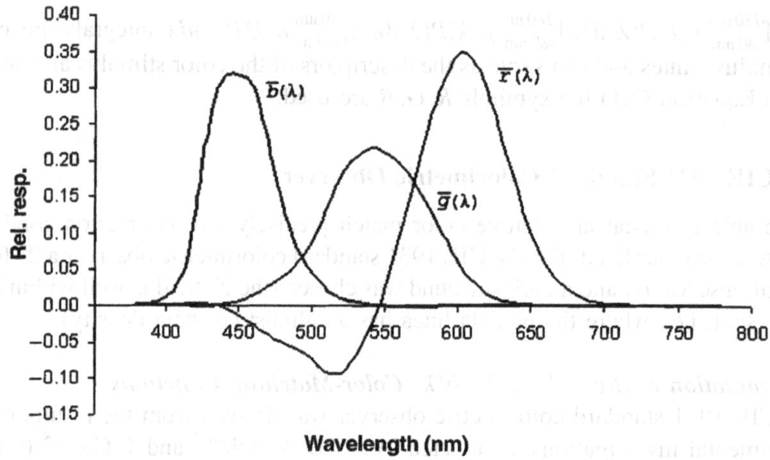

FIGURE 3.2 $\bar{r}(\lambda), \bar{g}(\lambda), \bar{b}(\lambda)$ CMFs of the CIE 1931 standard colorimetric observer.

CMFs made calculations more difficult, therefore in 1931 the CIE decided to transform from the real [R], [G], [B] primaries to a set of imaginary primaries [X], [Y], [Z], where the CMFs have no negative lobes. Further requirements were that the tristimulus values of an equienergy stimulus should be equal ($X = Y = Z$), that one of the tristimulus values should provide photometric quantities, and that the volume of the tetrahedron set by the new primaries should be as small as possible.

Based on above requirements one gets the following matrix transformation between the R, G, B and the new X, Y, Z tristimulus values:

$$\begin{vmatrix} X \\ Y \\ Z \end{vmatrix} = \begin{vmatrix} 2.768\,892 & 1.751\,748 & 1.130\,160 \\ 1.000\,000 & 4.590\,700 & 0.060\,100 \\ 0 & 0.056\,508 & 5.594\,292 \end{vmatrix} \bullet \begin{vmatrix} R \\ G \\ B \end{vmatrix} \tag{3.4}$$

As can be seen the Y tristimulus value will add up to a (relative) photometric quantity as defined in Equation (3.3). The CMFs are the tristimulus values of monochromatic radiations, thus the $\bar{x}(\lambda)$, $\bar{y}(\lambda)$, $\bar{z}(\lambda)$ functions can be calculated from the $\bar{r}(\lambda)$, $\bar{g}(\lambda)$, $\bar{b}(\lambda)$ CMFs using the above equation.

Figure 3.3 shows the CMFs of the CIE 1931 standard colorimetric observer. This observer should be used if the fields to be matched subtend between about 1° and about 4° at the eye of the observer. In technical applications this observer is often written as 2°-standard colorimetric observer. (A 2° visual field represents a diameter of about 17 mm at a viewing distance of 0.5 m.)

Values of the CIE 1931 standard colorimetric observer have been standardized.[9,10] The color-matching functions are given in the standard as values from 360 nm to 830 nm at 1 nm intervals with seven significant figures. For almost all practical applications an abridged and coarser set of data is adequate. CIE Publication 15[4] states "In the case where more coarsely sampled data will

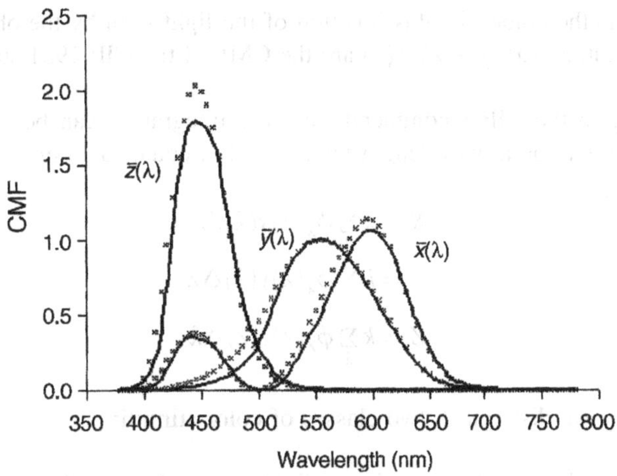

FIGURE 3.3 The $\bar{x}(\lambda)$, $\bar{y}(\lambda)$, $\bar{z}(\lambda)$ CMFs of the CIE 1931 standard (2°) colorimetric observer and, shown by ...x..., the $\bar{x}_{10}(\lambda), \bar{y}_{10}(\lambda), \bar{z}_{10}(\lambda)$ CMFs of the CIE 1964 standard observer (see later).

produce no significant calculation error selected values taken from the standard at 5 nm intervals, rounded to six decimal places, and reproduced in the above publication both in printed and electronic form will be sufficient." Further on the CIE publication states

> The color-matching functions $\bar{x}(\lambda)$, $\bar{y}(\lambda)$, $\bar{z}(\lambda)$ of current practice agree closely with those defined originally in 1931. Three minor changes have been introduced: at $\lambda = 775\,\text{nm}$ the new value of $\bar{x}(\lambda)$ is 0.000 059 instead of 0.0000; at $\lambda = 555\,\text{nm}\,\bar{y}(\lambda)$ is 1.0000 instead of 1.0002; and at $\lambda = 740\,\text{nm}\,\bar{y}(\lambda)$ is 0.000 249 instead of 0.0003. These changes are considered insignificant in most colorimetric computations. From these corrected tables the CIE standard colorimetric observer[9] was determined.
>
> If the color-matching functions taken at 5 nm intervals are not sufficient, the values given in the CIE Standard[9] should be used. For interpolation at wavelength intervals smaller than 1 nm a linear interpolation should be used.

Tristimulus Values and Chromaticity Coordinates

As mentioned in connection with Equation (3.2) the amounts of the primaries to achieve a match are called tristimulus values. In the case of the CIE-XYZ trichromatic system the tristimulus values are defined as

$$X = k\int_{380\,\text{nm}}^{780\,\text{nm}} \phi_\lambda(\lambda)\bar{x}(\lambda)\mathrm{d}\lambda,\ Y = k\int_{380\,\text{nm}}^{780\,\text{nm}} \phi_\lambda(\lambda)\bar{y}(\lambda)\mathrm{d}\lambda,\ Z = k\int_{380\,\text{nm}}^{780\,\text{nm}} \phi_\lambda(\lambda)\bar{z}(\lambda)\mathrm{d}\lambda \tag{3.5}$$

where $\phi(\lambda)$ is the color stimulus function of the light seen by the observer, k is a constant, and $\bar{x}(\lambda)$, $\bar{y}(\lambda)$, $\bar{z}(\lambda)$ are the CMF of the CIE 1931 standard observer.

According to the CIE recommendation, the integration can be carried out by numerical summation at wavelength intervals, $\Delta\lambda$, equal to 1 nm:

$$\begin{aligned} X &= k\sum_{\lambda}\phi_{\lambda}(\lambda)\bar{x}(\lambda)\Delta\lambda \\ Y &= k\sum_{\lambda}\phi_{\lambda}(\lambda)\bar{y}(\lambda)\Delta\lambda \\ Z &= k\sum_{\lambda}\phi_{\lambda}(\lambda)\bar{z}(\lambda)\Delta\lambda \end{aligned} \tag{3.6}$$

In colorimetry we distinguish two classes of color stimuli:

- Those reaching us from a primary light source directly, for example, from a lamp, a color monitor, direct sunlight, among others;
- Those reaching us from a reflecting or transmitting material, that is, it is the reflected/transmitted light, where the nonluminous material object changes the spectral distribution of the light by (selective) absorption. We are often interested in the colorimetric characteristics of such materials, for example, the color of a painted surface, the transmission of a tinted glass, among others. For the two classes the k constant in Equation (3.5) has been defined in a different form:

Tristimulus Values of Self-Luminous Objects. The Y tristimulus value is proportional to a photometric quantity because $\bar{y}(\lambda) = V(\lambda)$. A photometric quantity can be calculated from the corresponding radiometric one

$$\phi_{\mathrm{v}} = K_{\mathrm{m}} \int_{0}^{\infty} \phi_{\mathrm{e},\lambda} \cdot V(\lambda)\mathrm{d}\lambda \tag{3.7}$$

where K_{m} is the maximum value of the luminous efficacy of radiation, $K_{\mathrm{m}} = 683\,\mathrm{lm/W}$, $\phi^{\ddagger}_{\mathrm{e},\lambda}$ is a radiometric quantity, for example, spectral radiance, $V(\lambda)$ is the spectral luminous efficiency function, and ϕ_{v} is the corresponding photometric quantity, for example, the luminance.

Based on the above considerations, if a spectral radiance quantity is inserted in Equation (3.5) or (3.6), and k is set equal to K_{m}, we get Y in photometric units. In this case the same k has to be used also for determining the X and Z tristimulus values. In some cases it is convenient to set $Y = 100$ and scale X and Z accordingly.

‡$\phi_{\mathrm{e},\lambda}$ stands for a radiometric quantity, and this is shown by the index e. The index λ shows that the function is a spectral distribution, that is, $\phi_{\mathrm{e},\lambda} = \mathrm{d}\phi_{\mathrm{e}}(\lambda)/\mathrm{d}\lambda$.

Tristimulus Values of Non-Self-Luminous Objects. In colorimetry reflecting and transmitting objects are called secondary light sources or non-self-luminous objects. If the light of a source falls on a reflecting or transmitting material, part of this light is reflected/transmitted and this light reaches the eye of the observer, thus this reflected/transmitted part is the stimulus that has to be inserted into the Equations (3.5) or (3.6) as ϕ_λ. The spectral reflection of the surface is described, for example, by the spectral reflectance factor $R(\lambda)$ and the spectral transmission is described, for example, by the spectral transmittance factor $T(\lambda)$ (for more detail on the properties of reflection and transmission see Section "Standards and recommendations for measuring reflecting/transmitting materials"). When we are interested in the colorimetric properties of the reflecting/transmitting materials it is enough to know the relative spectral properties of the source illuminating the samples.

Based on the above considerations, the relative color stimulus function, $\phi(\lambda)$, for reflecting or transmitting objects is given by

$$\phi(\lambda) = R(\lambda) \cdot S(\lambda) \quad \text{or} \quad \phi(\lambda) = T(\lambda) \cdot S(\lambda) \tag{3.8}$$

where $R(\lambda)$ is the spectral reflectance factor, $T(\lambda)$ is the spectral transmittance factor, of the object color (preferably evaluated for one of the geometric conditions given in Section "Measuring geometries"), and $S(\lambda)$ is the relative SPD of the illuminant (which, whenever possible, should be one of the CIE standard illuminants; see Section "CIE illuminants and sources").

Similar equations can be written for other quantities related to reflection and transmission (see Section "Quantities to describe reflection and transmission").

In this case the constant k is chosen so that $Y = 100$ for objects for which $R(\lambda)$, or $T(\lambda) = 1$ for all wavelengths, and hence

$$k = \frac{100}{\sum_\lambda S(\lambda) \cdot \bar{y}(\lambda) \mathrm{d}(\lambda)} \tag{3.9}$$

Chromaticity Coordinates and Chromaticity Diagram. A color stimulus can be completely described by the three tristimulus values,[§] but this is not a very easily conceivable description. It is hard to imagine a stimulus if only its tristimulus values are given, and frequently we are not interested in the absolute values of the tristimulus values. In such cases the chromaticity coordinates can be used.

[§]Do not forget that this is the description of the stimulus, that is, two stimuli with equal tristimulus values will be undistinguishable for the observer if seen under the same exterior circumstances, but this is not a description of the color perception, see Chapter 11 on CIE color appearance models.

Chromaticity coordinates are defined as

$$x = \frac{X}{X+Y+Z}$$
$$y = \frac{Y}{X+Y+Z} \qquad (3.10)$$
$$z = \frac{Z}{X+Y+Z}$$

where $x + y + z = 1$, thus it is enough to describe the chromaticity with two numbers, usually x and y. One should not forget, however, that a color stimulus can be described only with three characteristic quantities, thus if x, y are used one has to quote Y as well.

Plotting the x, y chromaticity coordinates in a rectangular coordinate system, we get the diagram seen in Figure 3.4. Here the chromaticity of the equienergy

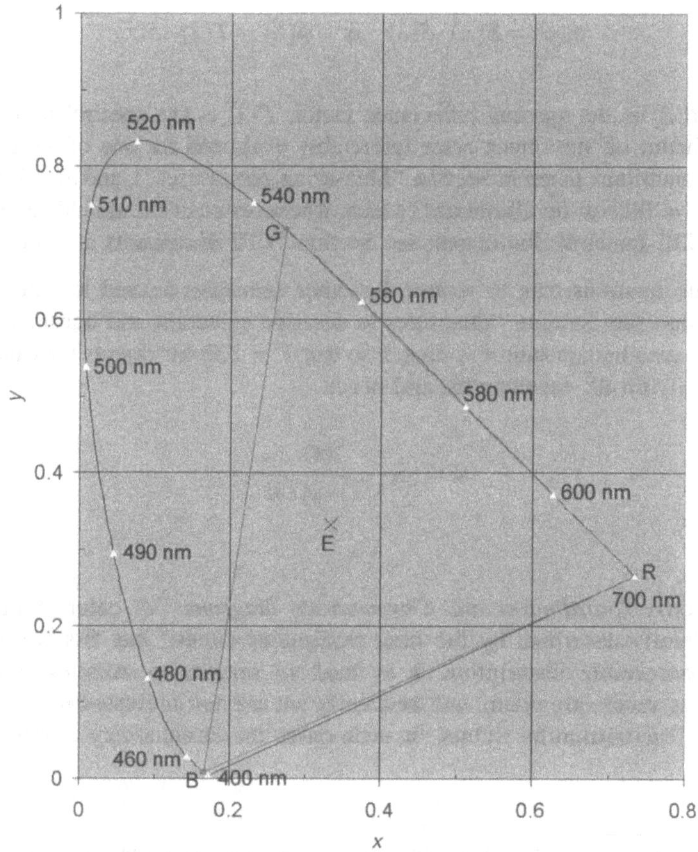

FIGURE 3.4 x, y chromaticity diagram of the CIE 1931 trichromatic system. The triangle shown refers to the R, G, B primaries used to define the CIE trichromatic system.

spectrum (labeled by E) and of monochromatic radiations have been plotted. The wavelengths of some of the monochromatic stimuli are shown. The straight line between the chromaticity of the 380 nm and 780 nm wavelength monochromatic stimuli is called the purple boundary.

Chromaticities within the diagram can be obtained by mixing two stimuli, for example, two monochromatic stimuli, as in case of the purple line, where a chromaticity along this line can be obtained by mixing the 380 nm and the 780 nm monochromatic stimuli. The R, G, B primaries, used to define the CIE 1931 RGB trichromatic system are also shown. From this figure it becomes obvious that, for example, a 520 nm monochromatic stimulus cannot be matched by the simple additive superposition of the three primaries. Only chromaticities within the triangle defined by the RGB primaries can be matched by simple additive mixture. For every chromaticity outside the triangle the additive mixture of two primaries can be matched with the additive mixture of the third primary and the unknown stimulus, for example, matching the additive mixture of given amounts of the G and B primaries with the additive mixture of the R primary and the 520 nm monochromatic stimulus.

CIE 1964 Standard Colorimetric Observer

The CIE 1931 trichromatic system is recommended only for small, 1°–4° size, stimuli. We need, however, the description of larger stimuli as well, where the stimulus falls on a larger area of the retina than the one covered by the macula lutea, or where we see the stimulus partly parafoveally. For that purpose the CIE standardized a large field colorimetric system,[11] based on the visual observations conducted on a 10° visual field. A 10° visual field represents a diameter of about 90 mm at a viewing distance of 0.5 m.

CIE committee W-1.3.1 recommended in 1959[12] the adoption of a 10° colorimetric observer, based on the works of Stiles and Burch (see Ref. 13) and Speranskaya.[14] In these investigations different sets of monochromatic primary stimuli were used, the CMFs were obtained directly from the observations, and no appeal to heterochromatic brightness measurements or to any luminous efficiency function was required. Stiles and Burch made flicker photometric comparisons for a red, green, and blue stimulus, thus it became possible to develop a large field photometric system as well.[15] The two sets of CMFs were derived differently, with different numbers of observers, at different luminance levels (and thus with different contribution of rod-vision, the so-called rod-intrusion, that produces deviations from additivity, see Refs. 11,16). The data were transformed to monochromatic primaries of R (645.2 nm), G (526.3 nm), and B (444.4 nm). The original data were taken at an equal wave number scale, and this had to be transformed to a wavelength scale and corrected for rod-intrusion, and harmonization between the two sets of data had to be done. Finally Judd came up with a transformation from the RGB-CMFs to an XYZ system that resembled the CIE 1931 XYZ system.[17] For more detail see Appendix B of the CIE recommendation.[4]

The CMFs of the 10°-system are distinguished from the 2°-system by a 10 in the subscript. The official transformation from the visually determined RGB CMFs to the 10° XYZ functions is

$$\begin{vmatrix} \bar{x}_{10}(\lambda) \\ \bar{y}_{10}(\lambda) \\ \bar{z}_{10}(\lambda) \end{vmatrix} = \begin{vmatrix} 0.341\,080 & 0.189\,145 & 0.387\,529 \\ 0.139\,058 & 0.837\,460 & 0.073\,160 \\ 0 & 0.039\,553 & 1.026\,200 \end{vmatrix} \bullet \begin{vmatrix} \bar{r}_{10}(\lambda) \\ \bar{g}_{10}(\lambda) \\ \bar{b}_{10}(\lambda) \end{vmatrix} \qquad (3.11)$$

These color-matching functions are given in the standard[9,10] as values from 360 nm to 830 nm at 1 nm intervals with six significant figures, and they define the CIE 1964 standard colorimetric observer. The CIE recommendation[4] states "In the case where more coarsely sampled data will produce no significant calculation error selected values taken from the standard at 5 nm intervals, rounded to six decimal places, and reproduced in the above publication both in printed and electronic form will be sufficient. For values between the 1 nm intervals, linear interpolation should be used." Figure 3.3 shows the CMFs as small crosses. The tristimulus values are calculated similar to Equation (3.5) as

$$X_{10} = k_{10} \int_{380\,\text{nm}}^{780\,\text{nm}} \phi_\lambda(\lambda) \cdot \bar{x}_{10}(\lambda) d\lambda, \quad Y_{10} = k_{10} \int_{380\,\text{nm}}^{780\,\text{nm}} \phi_\lambda(\lambda) \cdot \bar{y}_{10}(\lambda) d\lambda,$$
$$Z_{10} = k_{10} \int_{380\,\text{nm}}^{780\,\text{nm}} \phi_\lambda(\lambda) \cdot \bar{z}_{10}(\lambda) d\lambda \qquad (3.12)$$

and similar to Equation (3.6).

***k_{10}* in the Tristimulus Values of Self-Luminous Objects for the 10° Observer**
At the time, when the latest official CIE colorimetric document was produced, the CIE 10° photometric observer was not yet accepted, and thus there is no official recommendation for k_{10} for self-luminous objects. Since publishing the CIE recommendation on colorimetry,[4] the CIE accepted a $V_{10}(\lambda)$ function and defined a constant $K_{m,10}$ for practical purposes as 683.6 lm/W[1]. Thus now one can build up a colorimetric system for self-luminous objects using $k_{10} = 683.6$.

***k_{10}* in the Tristimulus Values of Non-Self-Luminous Objects for the 10° Observer**
In the case of non-self-luminous objects, that is, in the case of reflecting or transmitting samples, the equation for k_{10} is, by analogy to Equation (3.9)

$$k_{10} = \frac{100}{\sum_\lambda S(\lambda) \cdot \bar{y}_{10}(\lambda) d(\lambda)} \qquad (3.13)$$

Chromaticity Coordinates for the 10° Observer

Also the definition of the chromaticity coordinates is similar to that shown in Equation (3.10):

$$x_{10} = \frac{X_{10}}{X_{10} + Y_{10} + Z_{10}}, \quad y_{10} = \frac{Y_{10}}{X_{10} + Y_{10} + Z_{10}}, \quad z_{10} = \frac{Z_{10}}{X_{10} + Y_{10} + Z_{10}} \tag{3.14}$$

The chromaticity diagram of the 10° observer is quite similar to the diagram shown in Figure 3.4.

Notes on the Use of the CIE 1964 Standard Colorimetric Observer

There are several caveats that have to be taken into consideration when using the CIE 1964 trichromatic system. In principle the precision of the system is higher than that of the CIE 1931 trichromatic system,[18] as it has been determined with a higher number of observers (49 in case of Stiles and Burch and 18 + 9 in case of Speranskaya), but with the larger stimulus area rod-intrusion had to be considered as well. In the case of the Speranskaya experiments the luminance level was partly quite low, and thus the rod-correction had to be large. Also in the case of using the 10° observer without any rod-correction the luminance levels have to be high enough. While in the case of a 2° field one can calculate with photopic adaptation down to about 10 cd/m^2, this is not the case for the larger field size. The CIE recommendation[4] states the following:

> "The large-field color matching data as defined by the CIE 1964 standard colorimetric observer are intended to apply to matches where the luminance and the relative spectral power distributions of the matched stimuli are such that no participation of the rod receptors of the visual mechanism is to be expected. This condition of observation is important as 'rod intrusion' may upset the predictions of the standard observer. For daylight, possible participation of rod vision in color matches is likely to diminish progressively above about 10 cd·m^{-2} and be entirely absent at about 200 cd·m^{-2}."

A comment details this further:

> "For daylight illuminant D65 2.464 scotopic trolands corresponds to 1 photopic troland[19]. Rod saturation in 9° extrafoveal vision occurs at about 2000–5000 scotopic trolands[20]. Thus rod saturation would occur at a photopic light level between 812 troland and 2,029 troland. Working from the Table in Le Grand[21] that takes into account variation of pupil size with light level and the Stiles-Crawford effect, this would correspond to 130-380 cd/m^2 (kind contribution by J. Pokorny)."

CIE ILLUMINANTS AND SOURCES

In Equation (3.8) we have seen that when we want to describe the colorimetric characteristics of a reflecting or transmitting material, we have to irradiate it, and the relative SPD, symbol: $S(\lambda)$, is part of the relative color stimulus function. The radiation $S(\lambda)$ is modified by the reflectance (or transmittance) of the material. Thus to be able to reproduce colorimetric measurements, the SPD of the irradiating source has to be reproduced too. The CIE has standardized a few SPDs and recommends that these should be used whenever possible when colorimetric characterization of materials is made.

A further distinction is the following: For calculations only the relative SPD is needed, such theoretical sources are called illuminants. There are two standard illuminants: CIE standard illuminant A and D65, and several secondary illuminants. Practical realizations of a CIE illuminant are called CIE sources. Often an illuminant cannot be reproduced accurately, in such cases we speak about a simulator (characterization of simulators will be discussed in a later section).

In 1931 the CIE decided to introduce three standard illuminants, termed illuminants A, B, and C. They were chosen in such a form that illuminant A should resemble the SPD of an average incandescent light, and it was thought that direct sunlight might be a good second choice (illuminant B) and average daylight (illuminant C) as a further choice. During the years it turned out that illuminant B was very seldom used and was soon dropped. Illuminant C is still in use in some industries, but in 1964 the CIE recommended a new set of daylight illuminants,[22] where the SPD was also defined in the ultraviolet (UV) part of the spectrum.†† One phase of daylight was selected as the most representative and is now known as CIE standard illuminant D65. It is usual to term one further illuminant with a letter: Illuminant E has an SPD independent of wavelength, and it represents the equienergy spectrum.

CIE Standard Illuminant A and Planckian Radiators

A CIE draft standard[23] states

> "CIE standard illuminant A is intended to represent typical, domestic, tungsten-filament lighting. Its relative spectral power distribution is that of a Planckian radiator at a temperature of approximately 2 856 K. CIE standard illuminant A should be used in all applications of colorimetry involving the use of incandescent lighting, unless there are specific reasons for using a different illuminant.

The radiation of a coiled tungsten filament incandescent lamp can be well approximated by blackbody radiation, one of the few radiations whose SPD can be described using fundamental physical laws and constants. Planck's radiation law describes the spectral concentration of radiant exitance, M_e, in W/m^3 (power per area of source per wavelength interval), as a function of wavelengths, λ, in meters, and temperature, T, in Kelvins, by the equation

$$M_{e,\lambda}(\lambda, T) = \frac{c_1}{\lambda^5 \cdot \left[\exp\left(\frac{c_2}{\lambda \cdot T}\right) - 1\right]}, \text{ units W/m}^3 \qquad (3.15)$$

††This became important because in the 1950s and 1960s optical brighteners and fluorescent pigments became popular, and to measure their color correctly one has to include the UV excitation too.

where $c_1 = 2\pi \cdot h \cdot c^2 = 3.74183 \times 10^{-16}\,\mathrm{W/m^2}$, $c_2 = h \cdot c/k = 1.4388 \times 10^{-2}\,\mathrm{m \cdot K}$, c is the speed of light in vacuum, h is Planck's constant, and k is the Boltzmann constant.

With the above values of c_1 and c_2 CIE standard illuminant A has a temperature of 2856 K. At the time of standardizing illuminant A the values of the constants were different, the value of c_2 was $1.435 \times 10^{-2}\,\mathrm{m \cdot K}$ (c_1 is not important if only the relative spectral distribution is of concern), and for the same power distribution the temperature was 2848 K. In 2004, the CIE decided to state in the future not the temperature of the blackbody radiation in the definition of illuminant A, but define it directly so that if in the future the values of the constants might be changed, no change in the temperature of the blackbody should be needed. Thus in the new edition of the fundamental publication on colorimetry[4] the following equation defines CIE standard illuminant A:

$$S_\mathrm{A}(\lambda) = 100\left(\frac{560}{\lambda}\right)^5 \times \frac{\exp\dfrac{1.435 \times 10^7}{2\,848 \times 560} - 1}{\exp\dfrac{1.435 \times 10^7}{2\,848\,\lambda} - 1} \tag{3.16}$$

where λ is the wavelength in nanometers in standard air.[††]

The relative SPD that this equation defines is the same as that defined in 1931. The wavelength range for this illuminant is 380 nm – 830 nm, and the equation should be used to calculate the relative power at any wavelength between these limits. This SPD is normalized to the value 100 (exactly) at the wavelength 560 nm (exactly), and the wavelength dependence is shown in Figure 3.5).

The tristimulus values and chromaticity coordinates of CIE standard illuminant A are

$$X = 109.85; \quad Y = 100.00; \quad Z = 35.58; \quad x = 0.447\,58; \quad y = 0.407\,45$$

As mentioned, blackbody radiation belongs to the very few radiations whose SPD can be calculated from basic physical laws, and the SPDs of many practical radiators come close to the SPD of a blackbody radiation. Furthermore, blackbody radiation if seen without an external reference is seen in a wide range of temperatures as "white." For all these reasons the chromaticities of blackbodies of different temperatures, called the Planckian locus, in the chromaticity diagram is a curve one often refers to in colorimetry. Figure 3.6 shows the Planckian locus, on which we have inserted temperature values for some points; the chromaticity of CIE standard illuminant A is also shown (for further items shown see next section).

[††]Despite the fact that Equation (3.16) is based on Planck's equation for a vacuum, the wavelengths are to be taken as being in standard air (dry air at 15°C and 10,325 Pa, containing 0.03% by volume of carbon dioxide). This makes CIE standard illuminant A compatible with other CIE colorimetric and photometric data.

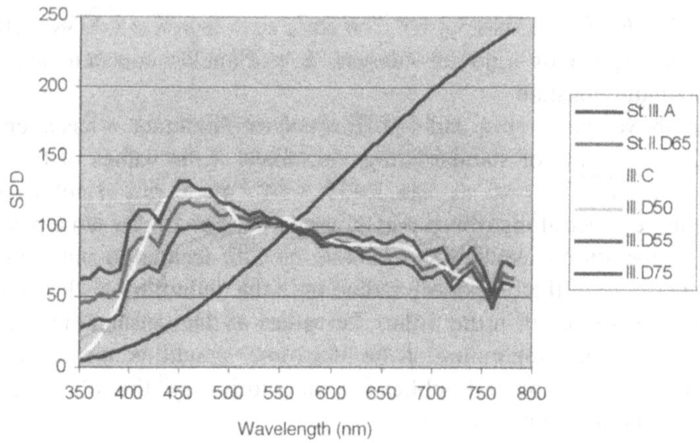

FIGURE 3.5 Relative spectral power distribution of the CIE standard illuminants and a further three daylight illuminants and illuminant C. See color insert.

Daylight Illuminants

For the human visual organ the "natural" illumination is daylight, thus it seems obvious that if further illuminants should be selected these should be phases of daylight. The SPD of daylight (sunlight + scattered skylight, influenced also by the

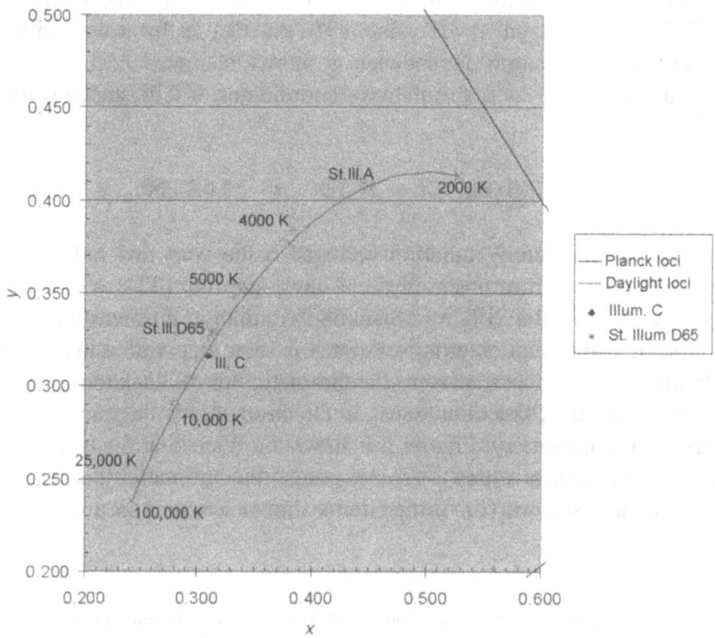

FIGURE 3.6 Part of the CIE 1931 chromaticity diagram with Planckian locus, the daylight locus, and some color temperature and illuminant points.

light of clouds) is variable, both during the day, but also depending on the season and the weather conditions. The CIE accepted a recommendation by Judd and coworkers[24] to describe phases of daylight.[25] These authors found that although daylight is highly variable the chromaticities of different phases of daylight fall on a curve more or less parallel to the Planckian locus on the chromaticity diagram (see Figure 3.6). But not only are the chromaticities of phases of daylight located on a simple curve in the x, y diagram, but even their SPDs can be described using only three basic functions.

To calculate a daylight phase, first its chromaticity (characterized by its correlated color temperature (CCT) a chromaticity that is not too different from a chromaticity on the Planckian locus, see later) has to be determined. Phases of daylight have been determined for the CCT range between 4000 K and 25,000 K. For the determination of the x chromaticity coordinates, this range of temperatures has been subdivided into two sections. For CCTs between 4000 K and 7000 K the following equation has to be used:

$$x_{\mathrm{D}} = \frac{-4.6070 \times 10^9}{(T_{\mathrm{cp}})^3} + \frac{2.9678 \times 10^6}{(T_{\mathrm{cp}})^2} + \frac{0.09911 \times 10^3}{(T_{\mathrm{cp}})} + 0.244063 \qquad (3.17)$$

for CCTs greater than 7000 K to approximately 25,000 K the equation is

$$x_{\mathrm{D}} = \frac{-2.0064 \times 10^9}{(T_{\mathrm{cp}})^3} + \frac{1.9018 \times 10^6}{(T_{\mathrm{cp}})^2} + \frac{0.24748 \times 10^3}{(T_{\mathrm{cp}})} + 0.237040^{\S\S} \qquad (3.18)$$

where T_{cp} is the CCT of the phase of daylight.

With the help of x_{D} the corresponding y_{D} can be calculated:

$$y_{\mathrm{D}} = -3.000\, x_{\mathrm{D}}^2 + 2.870\, x_{\mathrm{D}} - 0.275 \qquad (3.19)$$

The prescription to determine the relative SPD of a phase of daylight is a little bit cumbersome, but the CIE decided at the last update of the system to stay with the following method. In 1964, when Judd and coworkers suggested their method[24] it was reasonable to define the characteristic vectors only at 10 nm intervals and suggest linear interpolation if the values of the phase of daylight are needed at non-full 10 nm wavelengths, but this now causes some inconsistency in colorimetric functions, as only the daylight spectra are not smooth curves, but break at every full 10 nm value. Also the rounding of values during the calculations has to be made exactly as prescribed here below to get to the internationally agreed values. Steps of the calculation are as follows.

§§Constants in the original equations differ from those shown here because the correlated color temperature assigned to a given chromaticity changed due to the change in the c_2 constant of Planck's equation, as discussed in the previous section.

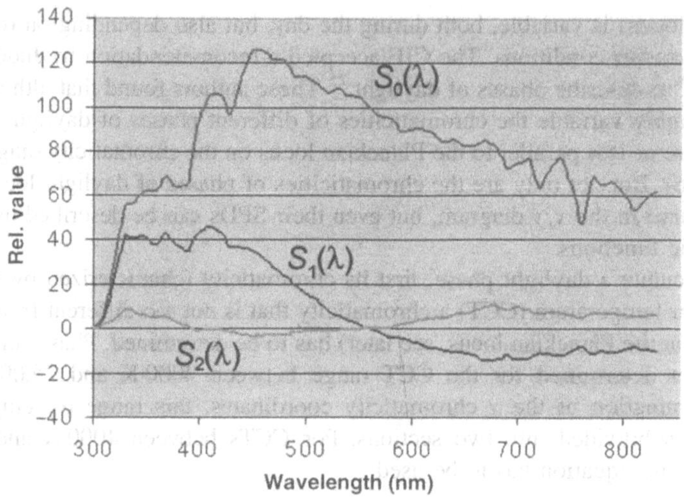

FIGURE 3.7 Characteristic vectors used to reconstitute phases of daylight.

Based on the x_D, y_D chromaticity coordinates of the selected daylight phase one has to calculate the M_1 and M_2 factors:

$$M_1 = \frac{-1.3515 - 1.7703\,x_D + 5.9114\,y_D}{0.0241 + 0.2562\,x_D - 0.7341\,y_D}$$
$$M_2 = \frac{0.0300 - 31.4424\,x_D + 30.0717\,y_D}{0.0241 + 0.2562\,x_D - 0.7341\,y_D} \quad (3.20)$$

The values of M_1 and M_2 have to be rounded to three decimal places, with the rounded values the spectral distribution of the daylight phase is calculated using the following equation for every 10 nm between 300 nm and 830 nm:

$$S(\lambda) = S_0(\lambda) + M_1 S_1(\lambda) + M_2 S_2(\lambda) \quad (3.21)$$

The $S_0(\lambda)$, $S_1(\lambda)$, and $S_2(\lambda)$ characteristic vectors are shown in Figure 3.7.

The daylight SPD for other wavelengths between 300 nm and 830 nm can be found by linear interpolation.***

CIE Standard Illuminant D65

The CIE selected from the infinite number of daylight phases one to substitute illuminant C, mainly because illuminant C was defined only in the visible part of the spectrum. In the 1960s the evaluation of optical brighteners and of fluorescent pig-

***An alternative method that provides smooth $S(\lambda)$ functions can be found in Appendix C of the CIE Publication 15:2004,[4] recommended for evaluation.

ments became important, and to evaluate these correctly the UV excited luminescence has to be considered as well. The D illuminants—as mentioned in the previous section—were defined from 300 nm on, a wavelength range where real daylight differs from zero (below 300 nm atmospheric absorption becomes considerable, and practically no shorter wavelength radiation reaches the earth surface). A further problem with illuminant C was that it was slightly purplish, at least compared to phases of daylight.

The first CIE recommendations[11] thought to supplement the CIE standard A, B, and C sources with four further ones at 3900 K, 5500 K, 6500 K, and 7500 K, but at its next session[25] it recommended that the CIE should retain only two standard illuminants: CIE standard illuminant A and CIE standard illuminant D65. The former had an SPD unchanged from its 1931 values; the SPD of the latter was defined using the method described in the previous section. The CIE recommended also that if other daylight phases should be necessary one with a CCT of 5500 K, or 7500 K should be used.

A CIE standard[23] states

> "CIE standard illuminant D65 is intended to represent average daylight and has a correlated color temperature of approximately 6500 K. CIE standard illuminant D65 should be used in all colorimetric calculations requiring representative daylight, unless there are specific reasons for using a different illuminant. Variations in the relative spectral power distribution of daylight are known to occur, particularly in the ultraviolet spectral region, as a function of season, time of day, and geographic location. However, CIE standard illuminant D65 should be used pending the availability of additional information on these variations."

In 1967 the International Practical Temperature Scale, 1948, amended 1960 was in use. With that temperature scale c_2 was 1.438×10^{-2} m · K. In 1968, the International Practical Temperature Scale changed the value of c_2 to $1.438\,8 \times 10^{-2}$ m · K. Because of this fact the CCT of a daylight phase of T K on the 1948/1960 scale changed to $1.4388/1.4380 \times T$, thus D65 with its "nominal CCT" has now a CCT of approximately 6504 K, and this temperature has to be set into the Equations (3.17) and (3.18) to get to the SPD as defined in 1967.

CIE Illuminants

As mentioned, at present two illuminants are called CIE standard illuminant: illuminant A and D65. While CIE standard illuminant A is defined by Equation (3.16), CIE standard illuminant D65 is defined by the table published in the CIE standard.[23,26] Among the daylight phases the CIE selected D50, D55 and D75 as preferred daylight illuminants if D65 is not applicable. The graphic arts industry selected D50 as its reference illuminant, as it is somewhere between average daylight and incandescent light. Other technologies might use different reference illuminants.

Because of the change of the c_2 constant in Planck's equation to get the original SPDs for the daylight illuminants, the CCT for D50 is not the originally selected

TABLE 3.1 Nominal and actual CCT of CIE daylight illuminants

Illuminant	Nominal CCT, K	Actual CCT (approximate), K
CIE illuminant D_{50}	5000	5003
CIE illuminant D_{55}	5500	5503
CIE standard illuminant D_{65}	6500	6504
CIE illuminant D_{75}	7500	7504

"nominal" 5000 K (and similarly for the other daylight illuminants), but the nominal CCT has to be multiplied by 1.4388/1.4380. For the four preferred daylight illuminants. Table 3.1 shows the corresponding actual CCTs.

Figure 3.5 shows the SPD of the two CIE standard illuminants, as well as of the further three CIE daylight illuminants and illuminant C (not a CIE standard illuminant anymore). Tristimulus values and chromaticity coordinates for the six illuminants are reproduced in Appendix A of this chapter.

The CIE published the SPDs[4] of a number of further illuminants, representative of fluorescent lamps and high pressure discharge lamps. The use of these SPDs is recommended if SPDs of such lamps of different CCTs are needed for testing. Appendix B of this chapter shows the colorimetric characteristics of these fluorescent and high pressure discharge lamps. FL 1–6 are standard (traditional), FL 7–9 are broad-band, and FL 10–12 are narrow band, fluorescent lamps. From this group FL 2, FL 7, and FL 11 should take priority over others when a few typical fluorescent lamp illuminants are to be selected. FL 3.4–6 are delux lamp spectra, and FL 3.7–11 are representative three band lamp spectra. FL 3.12–14 represent multiband fluorescent lamps, whereas FL 3.15 is a D65 simulator fluorescent lamp. HP 1 is a standard high pressure sodium lamp, whereas HP 2 is a color enhanced high pressure sodium lamp. HP 3–5 are typical high pressure metal halide lamps. Spectra of these lamps can be found in the CIE publication 15:2004.[4]

CIE Sources and Simulators for Colorimetry

In 1931 the CIE defined sources to represent Illuminant A, B, and C. These were artificial sources to be used whenever the illuminants had to be realized.

Source A

The definition of CIE source A is still a CIE source, and according to the new CIE draft standard[23] its definition is the following:

"CIE standard illuminant A can be realized by CIE source A, defined as a gas-filled, tungsten-filament lamp operating at a correlated color temperature

$$T = \frac{2848c_2}{14350}\text{K}'' \tag{3.22}$$

This complicated temperature definition is caused by the fact that in 1931 the value of c_2 was 14,350 μm·K, and thus the temperature was 2848 K, but now we use the International Temperature Scale of 1990 (ITS-90), with $c_2 = 14{,}388$ μm · K, and thus the temperature of the lamp has to be set to 2856 K.

If also the UV content of the radiation is of importance, a lamp with a quartz envelop or window should be used.

The spectral emissivity of tungsten varies with wavelength, and, as a result, light from an incandescent tungsten lamp is somewhat greener than that of a Planckian radiator at the same CCT. The difference for a coiled filament lamp, due to the inter-reflection between the coils is in the visible part of the spectrum well below 5%. Further details on the use of incandescent standard lamps will be found in Chapter 5 "Spectral color measurement" (see also Ref. 27).

Sources B and C

As the CIE illuminants B and C have been deprecated the sources B and C are not CIE sources anymore. The original realization of these sources was based on the source A and some precisely defined liquid filters (see CIE TC Report on colorimetry[4]).

Source D65

The CIE publication on colorimetry[4] states "At present no artificial source is recommended to realize CIE standard illuminant D65 or any other illuminant D of different CCT. It is hoped that new developments in light sources and filters will eventually offer sufficient basis for a CIE recommendation."

Meanwhile the CIE has agreed on a formula to describe the quality of a daylight simulator for colorimetry.[28,29] With the help of this formula, daylight simulators of 5000 K, 5500 K, 6500 K, and 7500 K nominal CCT (D50, D55, D65 and D75 simulators) can be graded both for their visible range and ultraviolet range suitability as substitutions of an illuminant for visual color evaluation. The formula is based on metameric samples: Five sets of metameric data are used to evaluate the colorimetric suitability of the test source for the visible wavelength range. Figure 3.8 shows as an example the five metameric sample pairs for a D65 simulator evaluation. If the simulator is a good reproduction of the D65 SPD, the color difference of the two samples of a pair will be negligible. An ultraviolet range metamerism index is employed with a second set of metameric samples to evaluate the suitability of the test source in relation to ultraviolet-excited luminescent colors. The metameric sample pairs for this assessment are comprised of a luminescent and a nonluminescent sample, which are spectrally identical matches for the daylight illuminant. The calculation is performed using the CIE 1964 standard colorimetric observer. The color differences can be calculated either in the CIELAB or CIELUV color space (see later).

Separate metamerism indices are calculated by averaging the color differences of the metameric pairs for the visible and for the ultraviolet part of the spectrum. In case of using the CIELAB or CIELUV color difference formulas the five grades a daylight simulator might have for the visible and for the UV spectrum are given in

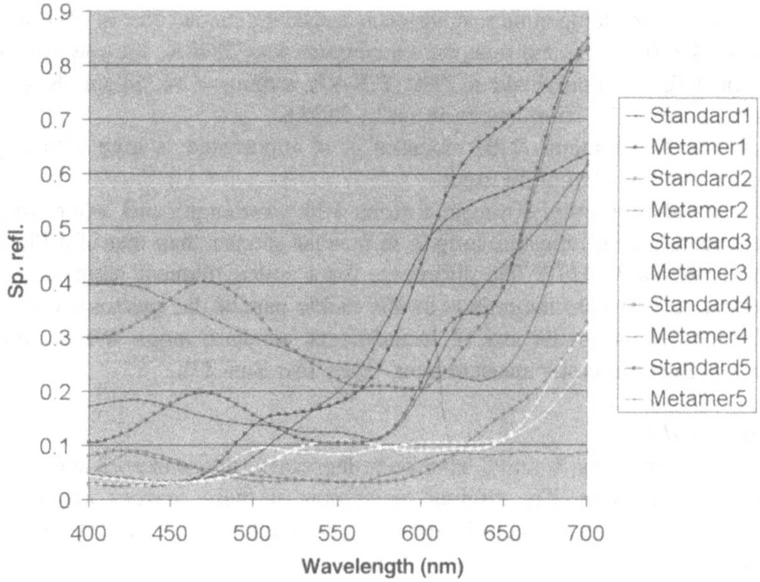

FIGURE 3.8 Metameric sample pairs for D65 simulator evaluation in the visible spectral range.

Table 3.2. The visible and the UV are categorized independently, first the result for the visible category then for the UV category have to be stated, with the indication of the color difference formula used. According to the CIE TC Report[28] daylight simulators of category BC(CIELAB) are found to be useful for many applications.

The above daylight simulator categorization is mainly used in visual inspections of materials. In instrumental measurements, if only the transmitted or reflected radiation needs to be spectrally analyzed, the irradiation is irrelevant. But in the case of measuring fluorescent materials it becomes important that the irradiation should contain the proper amount of UV radiation. This is often secured by inserting adjustable UV absorbing filters into the irradiation beam and by measuring a sample with known fluorescence adjusting the filter so that the proper color is measured.

TABLE 3.2 Daylight simulator categories

Category	CIELAB	CIELUV
A	<0.25	<0.32
B	0.25–0.5	0.32–0.65
C	0.5–1.0	0.65–1.3
D	1.0–2.0	1.3–2.6
E	>2.0	>2.6

STANDARDS AND RECOMMENDATIONS FOR MEASURING REFLECTING/TRANSMITTING MATERIALS

To be able to determine the colorimetric characteristics of materials, there have to be defined not only a standard observer and a standard illuminant but also the measuring geometry and a reference standard.

One can illuminate a material sample by diffuse or collimated light; illuminate it perpendicularly or from a given angle. Many materials will change their colorimetric properties with such changes. Therefore to be able to get to results that can be communicated from one place to another, or to be able to agree on requirements and check their fulfillment, the measuring geometry also has to be standardized.

This again does not mean that what we measure replicates exactly what we see, instrumental color measurement can only correlate with the visual impression, but will never be able to deliver numbers that exactly correspond to a visual sensation.

Nevertheless the CIE tried to standardize such measuring geometries as they come near to how we see materials, and thus colorimetric results can guide in practical decisions well.

Terms Used in Conjunction With Transmission and Reflection Measurement

Terms used to describe the optical properties of materials are often used loosely, and first we would like to recapitulate these.[†††]

Phenomena

The phenomenon that material samples return part of the impinging radiation from their surface or within the medium is called reflection. Reflection can be regular, also called specular reflection, when the laws of geometric optics are followed, or diffuse reflection, when on a macroscopic scale no regular reflection is found, or mixed reflection, when part of the reflected radiation is regular and part of it is diffuse.

For many materials not all of the impinging radiation is reflected, part of it is absorbed (transformed into other forms of energy e.g., heat, or just a change of wavelengths occurs and part of the radiation is reemitted in the form of luminescence) and part of it is transmitted. For the transmitted part similar expressions are used to describe the phenomena: transmission, regular, or direct transmission, diffuse transmission, mixed transmission.

Special (idealized) forms of reflection and transmission are isotropic diffuse reflection/transmission: Diffuse reflection/transmission in which the spatial distribution of the reflected/transmitted radiation is such that the radiance is the same

†††The International Lighting Vocabulary describes terms both as radiant and luminous terms, we will here restrict ourselves to radiant terms, which can be defined for monochromatic components. In that case one uses the "spectral" adjective.

in all directions in the hemisphere into which the radiation is reflected/transmitted. For a surface showing isotropic reflection Lambert's law holds

$$I(\theta) = I_{\mathrm{n}}\cos(\theta) \tag{3.23}$$

where $I(\theta)$ is the radiant intensity at the angle θ from the normal of the surface and I_{n} is the radiant intensity in the direction of the surface normal. A Lambertian surface directs the radiation coming from the surface according to Lambert's law. The numeric value of the radiance of a Lambertian surface, irradiated by E (W/m^2) and having a reflectance (see below for the definition of reflectance) of ρ will be in all directions of the hemisphere:

$$L = \frac{E \cdot \rho}{\pi} \tag{3.24}$$

The material sample that shows diffuse reflection/transmission is called a diffuser. A special form of such a material is a perfect reflecting diffuser, an ideal isotropic diffuser with a reflectance equal to 1 (a similar definition holds for the perfect transmitting diffuser). For defining the colorimetric standard of reflectance this idealized diffuser will be of special importance.

Quantities to Describe Reflection and Transmission

Reflectance/transmittance is the ratio of the reflected/transmitted radiant flux to the incident flux (measured in watts). Similarly to the phenomena one distinguishes between regular and diffuse reflectance/transmittance. The usual symbol is the Greek letter ρ for reflectance and τ for transmittance.

Reflectance factor is a quantity that has to be distinguished from reflectance. It is defined as the ratio of the radiant flux reflected in the direction delimited by a given cone (see Figure 3.9), to the reflected radiant flux reflected in the same directions by a perfect reflecting diffuser identically irradiated. Its usual symbol is R.

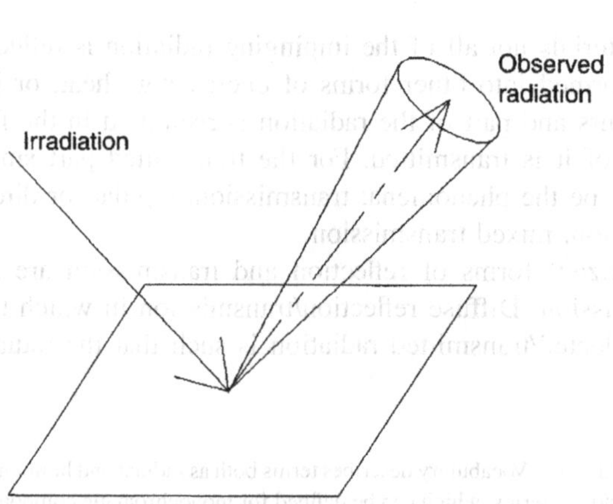

FIGURE 3.9 Schematic representation of the principles of the reflectance factor.

The angular distribution of the irradiation and the cone of observation have to be defined to be able to compare results taken with different instruments. The reflectance factor will approximate the reflectance as the observation cone approaches 2π sr (steradian). On the contrary for regularly reflecting surfaces that are irradiated by a beam of small solid angle, the reflectance factor may be much larger than 1 if the cone includes the mirror image of the source.

If the cone delimiting the reflected radiant flux in the definition of the reflectance factor is negligibly small we get the radiance factor, where the reflected radiance of the surface is compared to the radiance of the equally irradiated perfect reflecting diffuser. It is usual to use the symbol β to describe the radiance factor.

Above are the most important quantities used in describing the colorimetric characteristics of materials, for further details please consult the glossary or the relevant CIE publications.[30,31]

Measuring Geometries

In the visual evaluation of the color of an object one usually illuminates the target from different directions, using collimated and diffuse illumination and views it from different directions, and from the impression thus obtained one gets a mental picture of the color of the sample. Under different illuminating and viewing directions the impression might be different. To simulate this human process in instrumental color stimulus measurement one uses different measuring geometries to determine the spectral (or tristimulus) reflectance or reflectance factor (or transmittance, transmittance factor) of the material sample. To be able to compare the measured results obtained by different equipments the measuring geometries of these have to comply with some standards. The CIE has tightened the standard specifications since the first recommendations were published in 1931, but today one still finds considerable differences between instruments of different manufacturers. Therefore, it is important that the user should understand the critical parameters of an instrument. To enable this the CIE has worked out a terminology that makes the definition of the measurement geometry easier. Manufacturers are urged to follow this new (first published in 2004) terminology, as this will make their communications with their customers more efficient.

In both instrumental color stimulus measurement and visual color inspection of material samples, one has to irradiate the sample and collect the reflected or transmitted radiation for evaluation. First, to distinguish between the instrumental measurement of a color stimulus and the visual evaluation of the color perception one should distinguish between irradiation of a sample in an instrument and the illumination of the sample for visual inspection. Then, in the measuring instrument, one collects the reflected radiation, this is often called the efflux beam, while in human observation the term viewing beam is used. In the following we will deal with the instrumental measurement geometry of reflecting materials, followed by a short description of transmitting materials.

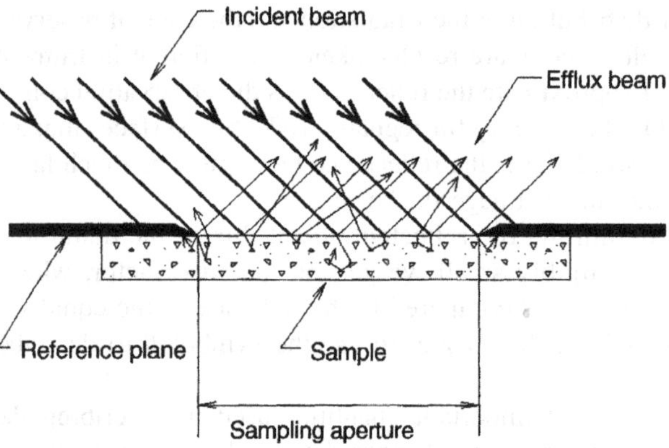

FIGURE 3.10 Reference plane and sampling aperture.

The Sample Plane and Influx Geometry

In a colorimetric measurement of reflecting samples one has to define the plane in which the measurement takes place. As shown in Figure 3.10 the measuring instrument usually has an opening, the sampling aperture, to which the sample can be attached in the reference plane. As seen in the example of the figure, there is often not only reflection from the surface of the sample but also penetration by the radiation into the sample where it is scattered, and some—diffused—radiation reaches the surface again and participates in the efflux beam.

If the sample aperture is overfilled with the incident beam, as shown in the figure, some of the scattered radiation reaches the surface somewhere beyond the sampling aperture and will not reach the measuring detector. For such samples an incident beam is needed that under fills the sampling aperture. For the geometric arrangement of the incident beam the term irradiation (or influx, or illumination[‡‡‡] or incidence) geometry is used, which describes the angular distribution of irradiance at the center of the sampling aperture.

The observed color will change for most samples with the type of illumination. To cope with this visual observation the CIE standardized for the instrumental measurement of color stimuli two groups of influx geometries: diffuse and directional. One of these geometries has to be selected together with an efflux geometry according to the type of the sample.

In the case of diffuse influx geometry the sample has to be irradiated from the upper hemisphere with angle independent constant radiance. This is done in practice by using an integrating sphere as irradiator.

The directional influx geometry has four subclasses: 45° directional and annular, 0° and 8°.

[‡‡‡]As mentioned the term "illumination" should be used for arrangements where the inspection is done visually.

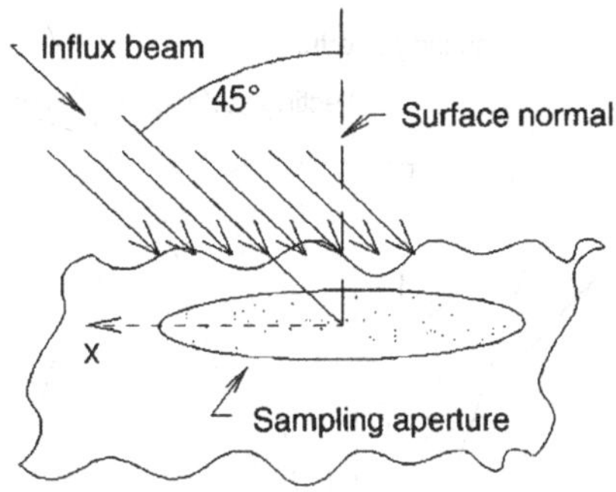

FIGURE 3.11 Schematic view of the 45° directional geometry.

In case of the *forty-five degree directional geometry* the sample is irradiated under 45° (within a cone of ±5°) to the normal of the sample, at one azimuth angle, shown in Figure 3.11 as direction *x*. With this geometry the structure of the sample will highly influence the reading of the reflected radiation. Symbol: 45°x.

If one is not interested in studying the influence of the structure of the sample but would like to average out its influence, it might be more advantageous to irradiate the sample from a 45° annulus, called *forty-five degree annular geometry*, symbol 45°a. This geometry may be achieved by the use of a small source and an elliptic ring reflector, or other aspheric optics. This geometry is sometimes approximated by the use of a number of light sources in a ring or a number of fiber bundles illuminated by a single source and terminated in a ring. Such an approximation to annular geometry is called *circumferential geometry*, symbol 45°c.

One is often interested in the color characteristics of materials if irradiated perpendicularly, for this the CIE standardized two geometries, the *zero degree directional geometry* (symbol 0°), where the reflecting material is irradiated at the normal and the *eight degree geometry* (symbol 8°) where irradiation is 8° to the normal of the sample, at one azimuth angle (as 8° is so near to normal it is usually not necessary to state the azimuthal direction). As we will see when we also discuss the efflux geometries, this arrangement has the advantage that the result will be nearly the same as with zero degree geometry, but it permits differentiation between specular component included and excluded measurements.

The laws of geometric optics say that the direction of light propagation can be reversed. Thus all the above influx geometries could be imagined also as efflux geometries, and combined with one or the other influx geometry provides a complete geometry for the measurement of the reflection properties of the material. The CIE recommended several configurations that will now be discussed:

Diffuse Geometries

(a) Diffuse: eight-degree geometry, specular component included (di:8°) and excluded (de:8°): Figure 3.12 shows the schematic geometry of the diffuse: 8° geometry. The sample is irradiated by the radiation of the irradiator

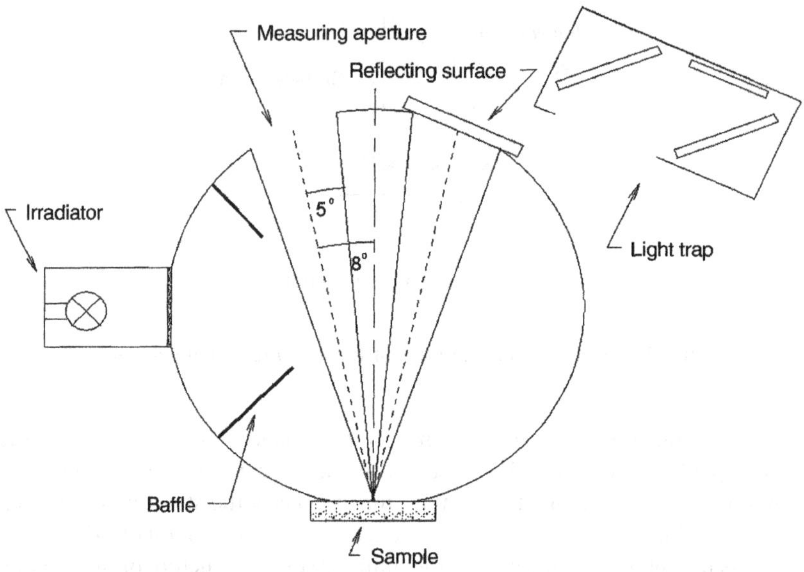

FIGURE 3.12 Schematic view of a diffuse 8° geometry.

diffusely reflected from the interior of the integrating sphere. The baffles prevent direct irradiation from the irradiator to reach the sample or the measuring (efflux) aperture. If one wishes to include also the specularly reflected light in the measurement, then a reflecting surface with the same reflectance as the sphere wall has to be placed in the mirror image of the measuring aperture (as shown in the figure), so that the radiation reaching the sample from this surface-element of the sphere will also contribute to the measured signal. If one wishes to exclude this radiation then a light trap, for example, consisting of a series of black glass samples arranged in a form that no radiation can be reflected back into the sphere, has to be placed on to the opening instead of the reflecting surface. This device is often called a gloss trap. The geometric layout should be built in such a form that there is no radiation reflected in the direction of the receiver by a plane first-surface mirror at the sampling aperture and that there are no rays specularly reflected within 1° of such rays, as a precaution against instrumental scattering of stray light or misalignment. With this arrangement the sample is overfilled. As shown in Figure 3.12 the measuring aperture is centered 8° from the normal of the sample. The geometry has to fulfill the condition that radiation reflected at the sampling aperture be evaluated uniformly at all directions within 5° of the axis of the collection (efflux) beam.[§§§] Further details and practical hints can be found in the relevant CIE publication.[32]

[§§§]At the time of writing this chapter of the book a CIE technical committee (TC 2-39) was still working on some more tight tolerances for colorimetry. Interested parties should check the CIE WEB site for new recommendations.

(b) Eight degree: diffuse geometry, specular component included (8°:di) and excluded (8°:de): The layout of the arrangement is very similar to that shown in Figure 3.12, except that the irradiation is now with a collimating beam at the port shown as measuring aperture, and the detection of the radiation is performed by a detector system at the port where in Figure 3.12 the irradiator is shown. With this arrangement the sampling aperture can and should be underfilled.

(c) Alternative diffuse geometries: (d:0°) and (0°:d): Instead of setting the efflux aperture at 8° from the normal of the sample, one can set it exactly at 0° (or irradiate the sample with a perpendicular ray of light). In this case one measures theoretically in a specular excluded geometry, only it is much more difficult to avoid some reflected light reentering the sphere because some optics have to be used to collimate or collect the radiation and, from its surfaces, reflected light can be redirected into the sphere.

(d) Diffuse/diffuse geometry (d:d): The CIE Technical Report[4] recommends that for this measurement the influx geometry should correspond to the prescription of the di:8° geometry and flux reflected at the sampling aperture should be collected at all angles in the hemisphere bounded by the reference plane.

Integrating spheres are very convenient equipments to produce diffuse radiation or to average the radiation emitted into the hemisphere. But their construction is tricky, the proper size and position of the baffles are critical, the apertures for infux and efflux, as well as of the light trap should be as small as possible, and their total area should be less then 10% of the area of the inner surface of the sphere. The surface of the inner coating of the sphere is critical too: highest diffuse reflectance is desirable, but at the same time the spectral characteristics (diffuse spectral reflectance of the sphere coating: $\rho(\lambda)$) should be flat (i.e., wavelength independent) because the sphere factor that couples the emitted spectral radiation of the sample to the spectral radiant flux leaving the efflux port of the sphere

$$\frac{\rho(\lambda)}{1-\rho(\lambda)} \tag{3.25}$$

magnifies spectral differences.[33] Even the best high reflectance paints have 1–2% drop in reflectance from the long wavelength edge of the visible spectrum to the shortest wavelength. If the reflectance is high (>97%) this will produce a considerable selectivity in the sphere throughput. In case of tristimulus colorimetry this has to be taken into consideration in designing the filters, otherwise the colorimeter will show erroneous results.

Because the test sample is part of the inner surface of the sphere, its reflectance will also influence the diffuse reflectance on the sphere wall. Thus if a simple substitution method is used, measuring a sample and a standard one after the other, the efflux will not be linearly proportional to the reflecting properties of the sample (see, e.g. Ref. 4). Modern instruments usually avoid this problem by using a double beam arrangement, where sample and standard are both continuously attached to the sphere, thus in measuring the efflux of the standard the sphere efficiency is also influenced by the sample reflection properties.

Finally one has to be aware of the fact that influence of the sample reflectance cannot be avoided in case of fluorescent samples, the ultraviolet absorption of the sample will influence the influx spectral distribution, and this can lead to considerable measurement errors[34–37]. Thus the CIE recommends[4] the use of directional geometries to measure fluorescent samples.

Directional Geometries

In connection with the input plane and influx geometry we have already mentioned the 45° influx direction. Many visual inspections come near to a geometry where either the illumination or the viewing is at angles near to 45°, and the corresponding viewing or illumination is at angles near to perpendicular. This has been taken as example by the CIE when it defined the following geometries:

(a) Forty-five degree directional/normal geometry (45°x:0°): This geometry is fulfilled if the sampling aperture is irradiated uniformly at an angle of 45° from the normal of the sample at an azimuthal angle of *x*, with a cone of light having a half angle of 5°. The collection of the radiation is in a cone with its axis on the normal to the sampling aperture, apex at the center of the sampling aperture and a half angle of 5°. Figure 3.13 shows a schematic view of the 45°x:0° measuring geometry, where we depicted the system to show overfilling of the sample area by the influx beam.

(b) Forty-five degree annular/normal geometry (45°a:0°): This geometry differs from the previously described 45°x:0° geometry in that the irradiation takes place from all directions between two right circular cones with their axes on

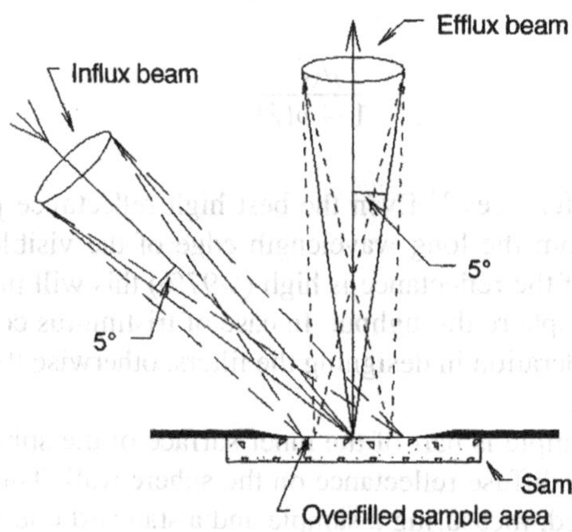

FIGURE 3.13 Schematic view of a 45°x:0° geometry. The *x*-direction is not shown.

the normal to the sampling aperture and apices at the center of the sampling aperture, the smaller cone having a half angle of 40° and the larger of 50°. If the irradiating geometry is approximated by the use of a number of light sources in a ring or a number of fiber bundles irradiated by a single source and terminated in a ring, one gets the circumferential/normal geometry (45°c:0°).

(c) Normal/forty-five degree directional geometry (0°:45°x): This geometry is similar to the 45°x:0° geometry, with the light path reversed, so the sampling aperture is irradiated normally and reflected radiation is collected at one azimuth angle at 45° to the normal. The angular and spatial conditions should be the same as in the case of the 45°x:0° geometry.

(d) Normal/forty-five degree annular geometry (0°:45°a): Here again the angular and spatial conditions for 45°a:0° geometry should be met, with the light path reversed, so the sampling aperture is irradiated normally and reflected radiation is collected within an annulus centered at 45° to the normal.

Quantities Using Different Measuring Geometries

In Section "Quantities to describe reflection and transmission" we have discussed the difference between reflectance and reflectance factor. Among the above described diffuse and directional geometries the di:8°, de:8°, d:0°, 45°x:0°, 45°a:0°, 0°:45°x, and 0°:45°a give values of reflectance factor, $R(\lambda)$. The 8°:di and d:d geometries deliver reflectance.

One should realize in this respect that the 8°:de and 0°:d geometries do not provide standardized quantities because in the definition of reflectance the total reflected radiation within the hemisphere has to be included, and these two geometries exclude the specular component.

Nonstandard Geometries

There are materials (metallic, pearlescent, interference, luster pigments, etc.) whose color changes very strongly with the direction of illumination and viewing. For the characterization of such samples the above discussed geometries do not give an adequate description. Investigations are still continuing to standardize measuring geometries for the determination of the color stimulus of such samples. Measurements have to be performed under several influx and efflux geometries. Table 3.3 shows angles used for such measurements and compares the different angle descriptions used. For further details see, for example, Ref. 38.

Recommended Geometry for Transmission Measurements

The CIE recommended six standard measuring geometries to be used to determine the color stimuli of transmitting materials[4]:

(a) *Normal/normal geometry (0°:0°)*: the irradiating (influx) and measuring (efflux) geometry be of identical right-circular conic form, with their axes on the normal to the center of the sampling aperture, and half-angle of 5°, that

TABLE 3.3 The designation of irradiation (influx) and measuring (efflux) angles for multiangle spectral measurements[38]

Angle designated with respect to the aspecular[a] angle	Angle designated with respect to the normal of the surface to be measured	Angle designated with respect to the surface to be measured
Incident—influx	Incident—influx	Incident—influx
90°	45°	45°
Efflux—detection	Efflux—detection	Efflux—detection
15°	−30°	120°
25°	−20°	110°
45°	0°	90°
75°	30°	60°
110°	65°	25°

[a]The aspecular angle is the angle measured from the specular direction, if it shows towards the incident beam, it is a negative angle, if it is toward the surface of the sample it is a positive angle.

the surface and angular irradiation of the sampling aperture be uniform, and that the surface and angular responsivity of the receiver be uniform.**** An important construction constraint is that the irradiating (influx) and collecting (efflux) beams shall be equal whether there is a sample in place or not. This becomes critical for thick samples because for the empty system the apex of the influx and efflux cone have to be copunctual, but if a material sample is in the beam, the two apexes have to be displaced.

(b) *Diffuse/normal geometry, regular component included (di:0°)*: the sampling aperture be uniformly irradiated from all directions in the hemisphere bounded by the first reference plane and that the measuring (efflux) beam be as specified for 0°:0° geometry.

(c) *Diffuse/normal geometry, regular component excluded (de:0°)*: the geometry be that specified for di:0° except that, with the sampling aperture open (i.e. no sample in place), there be no rays directed toward the receiver and no rays within 1° of such rays, as measured at the center of the sampling aperture.

(d) *Normal/diffuse geometry, regular component included (0°:di)*: the geometry be the reverse of that specified for di:0° geometry.

(e) *Normal/diffuse geometry, regular component excluded (0°:de)*: the geometry be the reverse of that specified for de:0° geometry.

(f) *Diffuse/diffuse geometry (d:d)*: the CIE recommended that the sampling aperture be uniformly irradiated at all angles within the hemisphere bounded by the first reference plane and that the transmitted flux be uniformly evaluated at all directions within the hemisphere bounded by the second reference plane.

****At the time of writing this chapter of the book a CIE technical committee (TC 2-39) was still working on the some more tight tolerances for colorimetry. Interested parties should check the CIE WEB site for new recommendations.

The two geometries, where the diffuse, regular component is excluded [item (c) and (e)] provide the equivalent of reflectance factor in transmission that is called *transmittance factor* in CIE 15:2004[4] and is called in some publications *diffuse transmittance.* All other geometries measure transmittance.

Specifications of the construction of the integrating sphere are similar to those given for reflection measurement. Also with transmission measurement multiple reflections between sample and instrument optical surfaces have to be avoided; such reflections can be eliminated, for example, by slightly tilting the sample.

Further recommendations and practical hints can be found in the CIE publication No. 130.[32]

Standards of Reflectance

In the equations for the colorimetric quantities of materials (see Equations (3.5), (3.6), and (3.8) the reflectance (reflectance factor) and transmittance (transmittance factor) have to be inserted. As seen in Section "Quantities to describe reflection and transmission" to determine quantities describing reflection, the efflux radiant power has to be measured once with the sample and then with the perfect reflecting diffuser (a theoretical construct, no material standard realizing it) in the sample plane. The total reflected power of the perfect reflecting diffuser is equal to the input power, thus if radiation from this reflector is measured in the proper geometric setting we obtain the total input power. In the case of transmittance the situation is simpler, as the transmitted flux of the sample has to be compared to the flux "transmitted" through the system without any substance in the measuring plane (as mentioned in connection with the 0°:0° geometry, the thickness of the transmitting sample has to be taken into consideration).

In CIE terminology the "perfect reflecting diffuser" is the reference standard for reflectance/reflectance factor (see Ref. 4). In practice this means that material standards used to calibrate a colorimeter or spectrophotometer have to be calibrated against the perfect reflecting diffuser. During the twentieth century several methods were developed to perform this measurement, and a number of materials were tested on their suitability as reflectance standards. The CIE publication No. 44[39] provides an overview of the different methods of calibration together with a detailed list for further reading.

Properties of materials that can be used as secondary standards of reflectance are summarized in the CIE publication 46.[40] Until the 1959 CIE Session[12] the colorimetric measurements of materials were referred not to the perfect reflecting diffuser but to smoked magnesium oxide, and its reflectance value was taken to be 100% for all wavelengths. Data of several secondary standards are still referred to this value, for example, the colorimetric values of Munsell samples.[41] The CIE report summarizes properties and reflection values of

- smoked magnesium oxide, pressed powder of magnesium oxide, and pressed powder of barium sulfate; these samples resemble more or less a Lambertian surface and thus can be used to transfer reflectance and reflectance factor data

very fragile surface and have limited stability.
between different instruments. Their drawback is that they usually have a very fragile surface and have limited stability.

- glasses, tiles, and plastics; these materials are usually more stable, but their reflectance characteristic deviate more from the ideal Lambertian distribution.

At the time of writing the CIE Report 46, the use of PTFE as reflectance standard was still in its infancy. Since then two forms of this standard have evolved: pressed and sintered PTFE samples. The pressed samples are somewhat less stable, but it is possible to prepare highly reproducible standards in the laboratory if following the steps recommended in CIE publication 135/6.[42] Sintered PTFE standards may have a slightly lower reflectance, but are much more stable. They are well suited to keep the reflectance scale in the laboratory and to transfer it from one instrument to another.

UNIFORM CHROMATICITY DIAGRAM AND UNIFORM COLOR SPACES

CIE Y, x, y space is quite well suited to describe color stimuli. The practical use of colorimetry, however, very often requires information about whether two samples will be indistinguishable by visual observation or not. David MacAdam showed in his 1942 paper[43] that the chromaticity difference that corresponds to a just noticeable color difference will be different in different areas of the x, y chromaticity diagram, and also at one point in the diagram equal chromaticity differences in different directions represent visual color differences of different magnitudes.

Figure 3.14 shows 10 times just noticeable chromaticity differences in the CIE x, y chromaticity diagram.

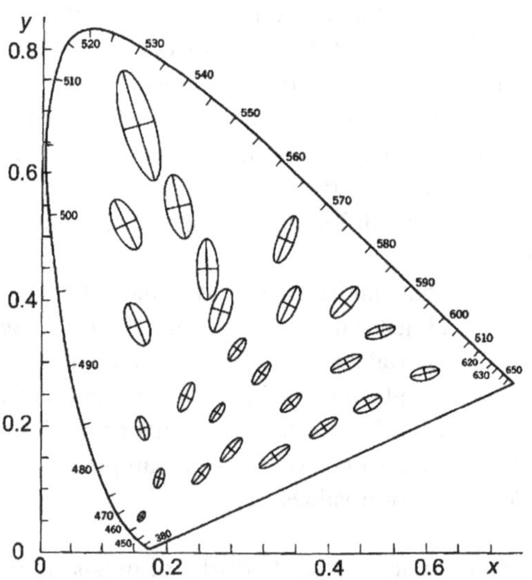

FIGURE 3.14 Ten times just noticeable chromaticity differences according to MacAdam's determination.

Uniform Chromaticity Diagram, CIE 1976 UCS Diagram

Many attempts were made to transform the x, y diagram in such a form that the MacAdam ellipses become circles, but no perfect transformation is available. The CIE recommended a uniform chromaticity scale diagram in 1959,[44††††] based on the MacAdam uniform-chromaticity scale diagram of 1937[45] and amended its recommendation in 1976.[46]

The present recommendation of the *CIE uniform chromaticity scale diagram* (*USC diagram*) is the following:

> "The use of the following chromaticity diagram is recommended whenever a projective transformation of the (x, y)-diagram yielding color spacing perceptually more uniform than that of the (x, y)-diagram is desired. The chromaticity diagram is produced by plotting
>
> $$u' = 4X/(X + 15Y + 3Z) \tag{3.26}$$
>
> as abscissa and
>
> $$v' = 9Y/(X + 15Y + 3Z) \tag{3.27}$$
>
> as ordinate, in which X, Y, Z are tristimulus values. The third chromaticity coordinate w' is equal to $(1 - u' - v')$."

The two samples have to have negligible luminance difference, and CIE Publication 15:2004[4] defines this as "e.g., $\Delta Y < 0.5$. A note to the above definition states:

> "This diagram is intended to apply to comparisons of differences between object colors of the same size and shape, viewed in identical white to middle-grey surroundings, by an observer photopically adapted to a field of chromaticity not too different from that of average daylight."

The same chromaticity diagram is produced by transforming x, y values into u', v' values using the following formulas:

$$u' = 4x/(-2x + 12y + 3), \quad v' = 9y/(-2x + 12y + 3) \tag{3.28}$$

If the angle subtended at the eye of the observer by the pairs of specimens is between 1° and 4° the CIE 1931 standard colorimetric observer should be

††††This recommendation shows up in the first edition of CIE Colorimetry as "approved officially by the CIE in 1960 and the diagram is now known as the CIE 1960 UCS diagram."

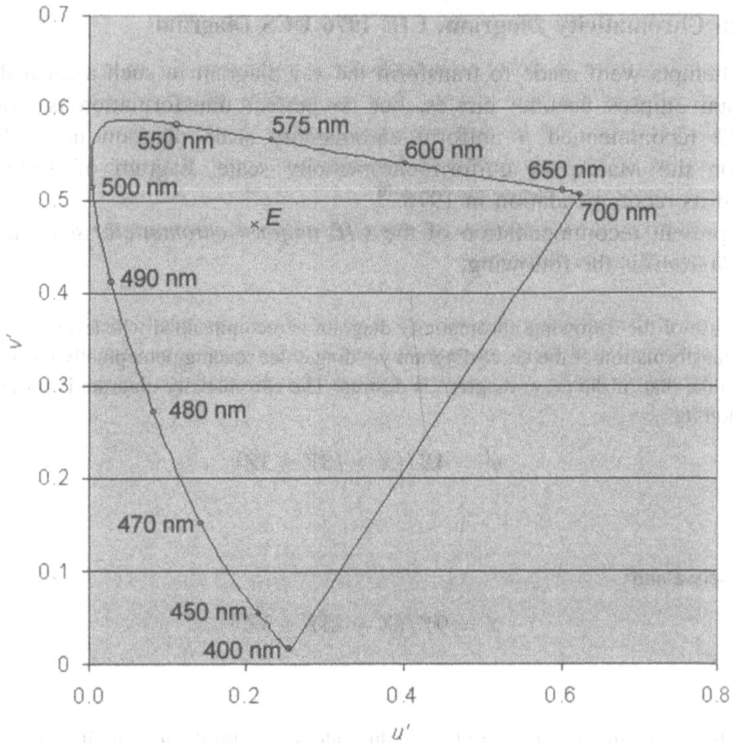

FIGURE 3.15 CIE 1976 u', v' UCS diagram, with spectrum locus and illuminant E.

used to calculate the X, Y, Z (or x, y) values, otherwise the CIE 1964 standard colorimetric observer should be used. Figure 3.15 shows the CIE 1976 u', v' UCS diagram, showing the spectrum locus with some wavelength data and the chromaticity of the equienergy spectrum (E).

Although the CIE recommendation explicitly mentions "object colors," the CIE UCS diagram is often used to describe light source chromaticity and define permissible chromaticity difference between a standard and a test source. Such data should be interpreted with caution, as the UCS diagram was never tested rigorously for this case.

CIE 1976 Uniform Color Spaces

Colour stimuli are three dimensional, and the request to extend the UCS into a three dimensional space was already expressed at the time the 1960 UCS diagram was accepted. The CIE made this extension in what came to be known as the CIE 1964 Uniform Color Space,[47] with the coordinates U^*, V^*, W^*. But this color space was soon surpassed by two new recommendations, which will now be described.

At the 1975 CIE Session, in London, the CIE Technical Committee responsible for colorimetry not only agreed on the new CIE UCS diagram, but made two recommendations for uniform color spaces[46]: CIELAB and CIELUV. Both spaces can be used with the CIE 1931 standard colorimetric observer if the samples are seen within a visual angle between 1° and 4°. For samples seen under a larger visual angle the CIE 1964 standard colorimetric observer should be used.

Both spaces are intended to apply to comparisons of differences between object colors of the same size and shape, viewed in identical white to middle-gray surroundings, by an observer photopically adapted to a field of chromaticity not too different from that of average daylight. Nevertheless in practice both of them have been applied to pseudo object colors (e.g., colors seen on a computer display or a projector screen), under non-daylight adaptation conditions. Such data have to be handled with care, as the spaces have not been tested for these conditions.

CIE 1976 ($L^*a^*b^*$) Color Space, CIELAB Color Space

This color space is defined by the following equations:

$$L^* = 116f(Y/Y_n) - 16 \tag{3.29}$$

$$a^* = 500\,[f(X/X_n) - f(Y/Y_n)] \tag{3.30}$$

$$b^* = 200\,[f(Y/Y_n) - f(Z/Z_n)] \tag{3.31}$$

where
$$f(X/X_n) = (X/X_n)^{1/3} \quad \text{if} \quad (X/X_n) > (24/116)^{3\ddagger\ddagger\ddagger\ddagger} \tag{3.32}$$

$$f(X/X_n) = (841/108)(X/X_n) + 16/116 \quad \text{if} \quad (X/X_n) \leq (24/116)^3 \tag{3.33}$$

and
$$f(Y/Y_n) = (Y/Y_n)^{1/3} \quad \text{if} \quad (Y/Y_n) > (24/116)^3 \tag{3.34}$$

$$f(Y/Y_n) = (841/108)(Y/Y_n) + 16/116 \quad \text{if} \quad (Y/Y_n) \leq (24/116)^3 \tag{3.35}$$

‡‡‡‡As 24/116 is not a simple ratio, in some publications the 6/29 ratio is used, in others the approximate value of 0.008856 (used in earlier editions of CIE 15). Similarly some authors prefer to use instead of 841/108 the expression $(1/3)*(29/6)^2$ or the approximate value of 7.787, or instead of 16/116 the ratio 4/29.

and
$$f(Z/Z_n) = (Z/Z_n)^{1/3} \quad \text{if} \quad (Z/Z_n) > (24/116)^3 \tag{3.36}$$

$$f(Z/Z_n) = (841/108)(Z/Z_n) + 16/116 \quad \text{if} \quad (Z/Z_n) \leq (24/116)^3 \tag{3.37}$$

where X, Y, Z are the tristimulus values of the test object color stimulus considered and X_n, Y_n, Z_n are the tristimulus values of a specified white object color stimulus. In most cases, the specified white object color stimulus should be light reflected from a perfect reflecting diffuser illuminated by the same light source as the test object. In this case, X_n, Y_n, Z_n are the tristimulus values of the light source with Y_n equal to 100. For simulated object colors, the specified white stimulus normally chosen is one that has the appearance of a perfect reflecting diffuser, again normalized by a common factor so that Y_n is equal to 100.

This space is a simplified version of the Adams–Nickerson space.[48,49] The original definitions[46] were modified shortly after their publication by the introduction of the linear parts at low tristimulus values[50] (this version was published in the second edition of CIE Colorimetry[51]), using decimal approximations in the constants of the equations. The CIE Technical Committee TC 1–48 decided at its meeting in Veszprém in 2002 to suggest to the CIE the use of constants expressed as integer ratios, a modification that secured continuity at the breaks between the cube root and linear parts of the equations and prevented the accumulation of rounding errors in repetitive forward and reverse calculations. The 2004 edition of the CIE publication 15[4] contains these new equations that are presently in force.

In this color space the positive a^* axis points approximately in the direction of red color stimuli, the negative axis approximately in the direction of green stimuli; positive b^* points approximately in the direction of yellow stimuli; negative b^* approximately in the direction of blue stimuli. L^* is coupled to the luminance of the stimulus, thus it is a crude correlate of lightness. Thus one can construct approximate correlates of the perceived attributes lightness, chroma, and hue in the following form:

$$\textit{CIE 1976 lightness:} \quad L^* \text{as defined in Equation (3.29)} \tag{3.38}$$

$$\textit{CIE 1976 a,b (CIELAB) chroma:} \quad C^*_{ab} = (a^{*2} + b^{*2})^{1/2} \tag{3.39}$$

$$\textit{CIE 1976 a,b (CIELAB) hue angle:} \quad h_{ab} = \arctan\,(b^*/a^*) \tag{3.40}$$

CIELAB hue angle, h_{ab}, shall lie between 0° and 90° if a^* and b^* are both positive, between 90° and 180° if b^* is positive and a^* is negative, between 180° and 270° if b^* and a^* are both negative, and between 270° and 360° if b^* is negative and a^* is positive.

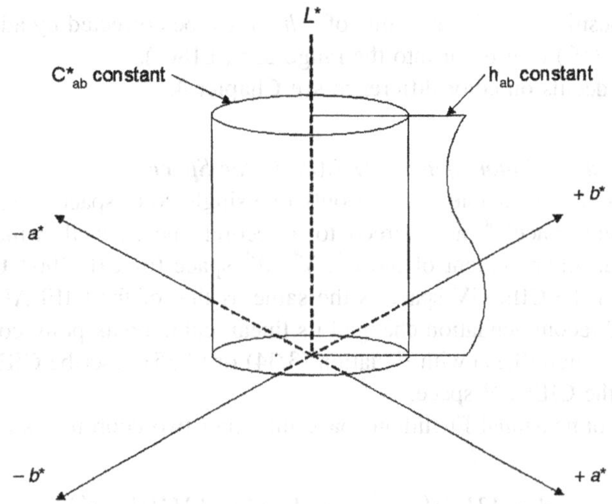

FIGURE 3.16 A three dimensional representation of the CIELAB space showing a cylinder of constant chroma, C_{ab}^{*}, and a plane of constant hue angle, h_{ab}.

McLaren pointed out that hue angles calculated using this equation can lead to anomalous values if any tristimulus ratio is below the critical figure (24/116).[3] This is unlikely for surface colors, but might occur for transparent object colors of low luminance factor lying close to the spectrum locus or purple line. McLaren proposed to use a modified Judd polynomial function in such cases.[52]

Figure 3.16 shows the structure of the CIELAB color space with a surface of constant CIELAB chroma and CIELAB hue angle.

We would like to call attention to the fact that CIE 1976 lightness (Equation (3.29)) is calculated using luminance/luminance factor only, thus it does not take into consideration the lightness—luminance discrepancy of chromatic colors.[53]

Euclidean distances in CIELAB color space can be used to represent approximately the perceived magnitude of color differences between object color stimuli of the same size and shape, viewed in identical white to middle-gray surroundings, by an observer photopically adapted to a field of chromaticity not too different from that of average daylight. Two equivalent equations describing *CIELAB color difference* are

$$\Delta E_{ab}^{*} = [(\Delta L^{*})^2 + (\Delta a^{*})^2 + (\Delta b^{*})^2]^{1/2} \tag{3.41}$$

or

$$\Delta E_{ab}^{*} = [(\Delta L^{*})^2 + (\Delta C_{ab}^{*})^2 + (\Delta H_{ab}^{*})^2]^{1/2} \tag{3.42}$$

where

$$\Delta H_{ab}^{*} = 2(C_{ab,1}^{*} \cdot C_{ab,0}^{*})^{1/2} \cdot \sin(\Delta h_{ab}/2)$$

and if X indicates L^{*} or a^{*} or b^{*}, $\Delta X = X_1 - X_0$, the indices 1 and 0 referring to test and standard, respectively, and Δh_{ab} is in radians. If the line joining the two colors crosses the positive a^* axis, the value of Δh_{ab} must be corrected by adding or subtracting 2π (360°) to bring it into the range $\pm\pi$ (±180°).

For more details on color difference see Chapter 4.

CIE 1976 (L*u*v*) Color Space, CIELUV Color Space

In 1976, the CIE was unable to select only one single color space as representative "uniform color space"[46] and agreed to a second one as well, which could be regarded as an improvement of the U^*, V^*, W^* space (of CIE 1964 UCS[22]). The L^* function of the CIELUV space is the same as that of the CIELAB space, and the CIE 2002 recommendation changed its linear–cubic break point constant similarly, thus Equation (3.29) with Equation (3.34) or (3.35) describe CIE 1976 lightness also in the CIELUV space.

In a three-dimensional Euclidian space the other two coordinates are

$$u^* = 13L^*(u' - u'_n) \quad \text{and} \quad v^* = 13L^*(v' - v'_n) \tag{3.43}$$

where u', v' are the CIE 1976 UCS coordinates of the test stimulus, and u'_n, v'_n are those of a specified white object color stimulus.

In the CIELUV space not only correlates of chroma and hue can be defined but also a correlate of saturation can be defined in the u', v' diagram:

$$\textit{CIE 1976 u,v (CIELUV) saturation}: \; s_{uv} = 13[(u' - u'_n)^2 + (v' - v'_n)^2]^{1/2} \tag{3.44}$$

$$\textit{CIE 1976 u,v (CIELUV) chroma}: \quad C^*_{uv} = (u^{*2} + v^{*2})^{1/2} = L^* \cdot s_{uv} \tag{3.45}$$

$$\textit{CIE 1976, u,v (CIELUV) hue angle}: h_{uv} = \arctan(v^*/u^*) \tag{3.46}$$

CIELUV hue angle, h_{uv}, shall lie between 0° and 90° if u^* and v^* are both positive, between 90° and 180° if v^* is positive and u^* is negative, between 180° and 270° if v^* and u^* are both negative, and between 270° and 360° if v^* is negative and u^* is positive.

CIELUV color difference, ΔE^*_{uv}, between two color stimuli is calculated as the Euclidean distance between the points representing them in the space:

$$\Delta E^*_{uv} = [(\Delta L^*)^2 + (\Delta u^*)^2 + (\Delta v^*)^2]^{1/2} \tag{3.47}$$

Similar to CIELAB hue difference the CIELUV hue difference is defined as

$$\Delta H^*_{uv} = 2(C^*_{uv,1} \cdot C^*_{uv,0})^{1/2} \cdot \sin(\Delta h_{uv}/2) \tag{3.48}$$

where 1 and 0 refer to the two samples between which the color difference is to be calculated and $\Delta h_{uv} = h_{uv,1} - h_{uv,0}$ measured in radians. Similar to the CIELAB hue angle calculation, if the line joining the two colors crosses the positive u^* axis, the value of Δh_{uv} must be corrected by adding or subtracting 360° to bring it into the range ±180°.

The CIE has not officially withdrawn these CIELAB and CIELUV color difference formulas, but for small color differences methods described in Chapter 4 for color differences and for large color differences, the methods based on color appearance models (see Chapter 11) provide much better agreement with visual data.

DESCRIPTORS OF CHROMATICITY

Color information is required in a number of technical subjects, where no deep colorimetric knowledge is available. Thus, for example, for saturated lights, the wavelengths of spectral lights of similar chromaticity as that of the test light can give a shorthand description of the correlate of hue. In illuminating engineering, despite the fact that the whitish light emitted by different light sources differs in color in various respects, one would like to describe this difference with a singe number. Comparing the color stimulus of the source with that of a Planckian radiator provides an opportunity to give single number information of the chromaticity of the light. The most important descriptors will now be introduced.

Dominant/Complementary Wavelength and Purity

If one has the opportunity to look for a few times through a monochromator one easily gets a feeling for what the name of a color of a given wavelength is (naturally for the given adaptation condition). Based on this experience one can start to describe a color by the wavelength of the monochromatic radiation to which it seemed to be similar.

The CIE definition of *dominant wavelength* is "Wavelength of the monochromatic stimulus that, when additively mixed in suitable proportions with the specified achromatic stimulus, matches the color stimulus considered."

In Figure 3.17 we explain the definition of the dominant wavelength in the CIE 1931 chromaticity diagram: Point *E* is the equienergy point as the "specified achromatic stimulus" (it could be just as well be CIE standard illuminant A, or D65, or any other white stimulus). Let us determine the dominant wavelength of the stimulus *F*. If we mix additively a monochromatic stimulus that can be characterized by the chromaticity *G*, in our example a monochromatic radiation of the wavelength 500 nm, with the achromatic stimulus *E*, we can reach the point *F*. Thus chromaticity *G* is the dominant wavelength of chromaticity *F*.

For a stimulus that lies within the triangle determined by the achromatic stimulus and the long wavelength and short wavelength endpoints of the visible spectrum, as in our example chromaticity *H*, the chromaticity *I* is not a

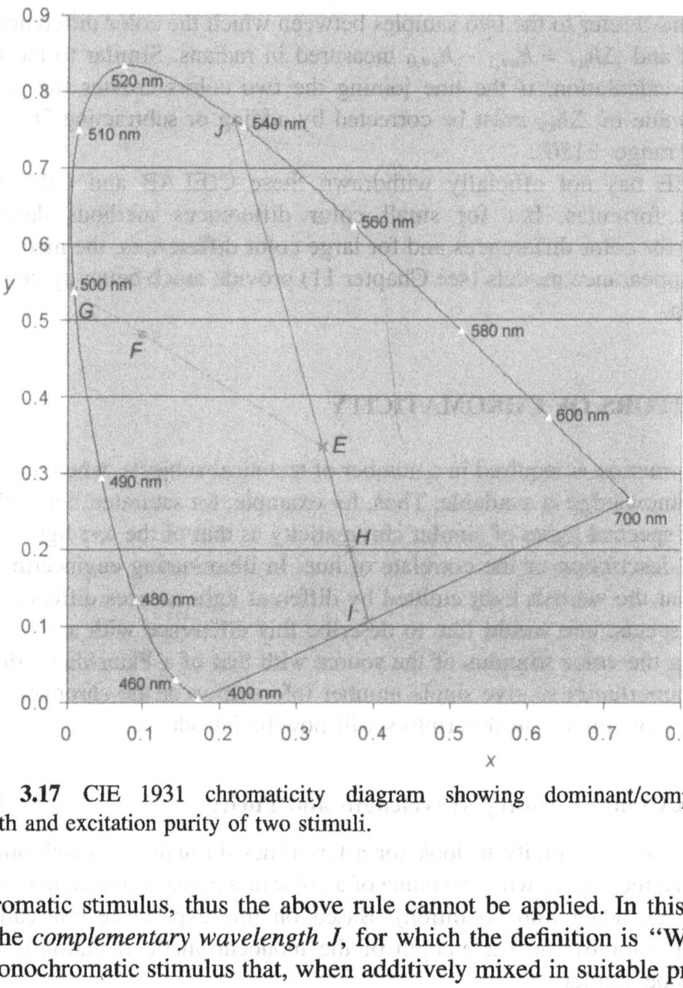

FIGURE 3.17 CIE 1931 chromaticity diagram showing dominant/complementary wavelength and excitation purity of two stimuli.

monochromatic stimulus, thus the above rule cannot be applied. In this case one defines the *complementary wavelength J*, for which the definition is "Wavelength of the monochromatic stimulus that, when additively mixed in suitable proportions with the color stimulus considered, matches the specified achromatic stimulus."

In both cases one gets a complete definition of the chromaticity only if one defines a second characteristic value. This is the *excitation purity* (p_e): the ratio *EF/EG* (or *EH/EI*) of the two collinear distances shown in Figure 3.17 (one could also define it in the CIE 1964 10° chromaticity diagram). From the two-dimensional rectilinear character of the CIE x, y chromaticity diagram it is obvious that the above definition is equivalent to the following two definitions:

$$p_e = \frac{y_F - y_E}{y_G - y_E} \quad \text{or} \quad p_e = \frac{x_F - x_E}{x_G - x_E} \tag{3.49}$$

and similar equation can be written for the stimulus H. From the above two forms (difference in x coordinates or y coordinates) the one to be used is preferably that which has the greater value in the numerator to obtain greater precision.

The CIE defined also *colorimetric purity* (p_c) based on the following relation:

$$p_c = L_d/(L_d + L_a) \tag{3.50}$$

where L_d and L_a are the luminances of the monochromatic stimulus and of the specified achromatic stimulus respectively, that match the color stimulus considered in an additive mixture (for more detail see Ref. 4).

Correlated Color Temperature

Sources look "white" if seen without another light to which one can compare its color and if its color is similar to the color of a Planckian radiator (blackbody) roughly between 2700 and 10,000 K. If the chromaticity lies exactly on the Planckian locus (see Section "CIE standard illuminant A and Planckian radiators") the temperature of the Planckian radiator is the *color temperature* of the test source. However, if the chromaticity of the test source deviates slightly from a point on the Planckian locus, one can still compare its chromaticity to a Planckian radiator, but then one uses the expression CCT (*correlated color temperature*). The CIE definition is

> The correlated color temperature is the temperature of a Planckian radiator having the chromaticity nearest the chromaticity associated with the given spectral distribution on a diagram where the (CIE 1931 standard observer based) *u'*, 2/3*v'* coordinates of the Planckian locus[§§§§] and the test stimulus are depicted.

This means that if the chromaticity of the test source has been determined, one has to find a Planckian radiator, whose chromaticity is least different from the chromaticity of the test source on the u', 2/3v' diagram. Test sources with equal CCT lie on straight lines perpendicular to the Planckian locus on the u', 2/3v' diagram. These are called isotemperature lines.

Naturally if the chromaticity gets further away from the Planckian locus, the test source starts to develop a greenish or purplish tint. Therefore the CIE recommendation restricts the area in the chromaticity diagram where CCT has practical meaning:

> The concept of correlated color temperature should not be used if the chromaticity of the test source differs more than $\Delta C = [(u'_t - u'_P)^2 + \frac{4}{9}(v'_t - v'_P)^2]^{1/2} = 5 \times 10^{-2}$ from the Planckian radiator, where u'_t, v'_t refer to the test source, u'_P, v'_P to the Planckian radiator.

[§§§§]In calculating the chromaticity coordinates of the Planckian radiator the c_2 value according to ITC-90 has to be used ($c_2 = 1.4388$) in Planck's equation for standard air but assuming $n = 1$.

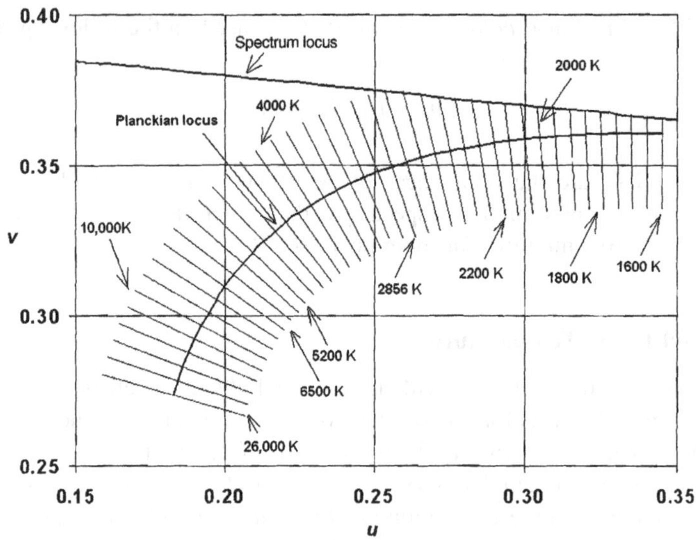

FIGURE 3.18 u, v (i.e., u′, 2/3v′ chromaticity diagram, with Planckian locus and a few isotemperature lines.

The u', $2/3v'$ coordinate system is actually identical with the CIE 1960 UCS system. After 1976, when the new u', v' chromaticity diagram was introduced, and color spaces based on new concepts came into use, the question arose, whether, using one of these, the concept of CCT could be better described than using the CIE 1960 UCS diagram. Experimental investigations have shown, however, that the isotemperature lines in the u′, 2/3v′ diagram describe the smallest visual chromaticity distance between test source and Planckian radiator quite well (Figure 3.18).[54]

Whiteness

CCT describes the color of whitish lights reasonably well. We have a similar problem with white objects: White paper samples of different origin or different quality—when seen without a reference—will look white. As soon as we place the paper samples side by side, we can set up a subjective whiteness scale. Whiteness specification has a long history (see., e.g., Refs. 55–58), and many whiteness formulas have been developed. Slightly bluish whites are regarded as "whiter" than the perfect reflecting diffuser, and, with the invention of optical brighteners that transform part of the UV radiation into bluish or greenish light, reflectances higher than 100% could be achieved. Figure 3.19 shows the spectral reflectances of samples with different amounts of optical brighteners.

With increasing blue emission the perceived whiteness increases, but with too much blue emission it starts to drop. Many formulas were in use in the 1950s

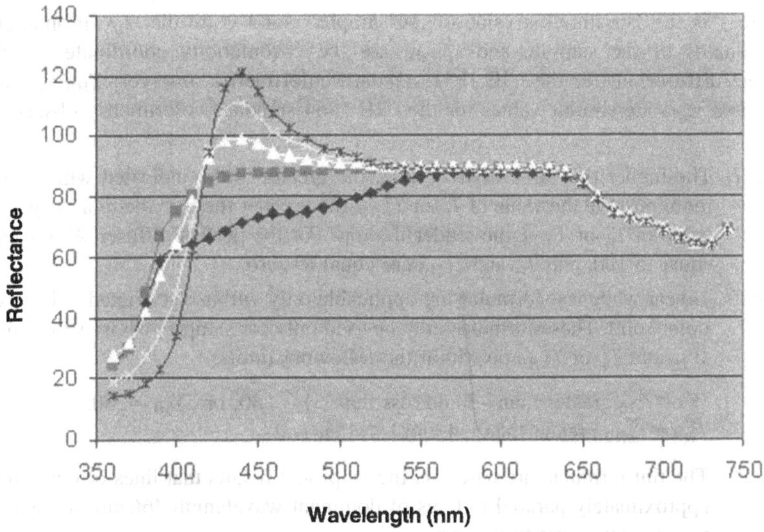

FIGURE 3.19 Reflectances of white samples with different amounts of optical brighteners.

and 1960s. The CIE Colorimetry committee adopted and published in 1986 with the second edition of publication 15, a compromise formula, which has been widely accepted. Actually there are two formulas, one describes the whiteness scale, the other the tint, as samples might show a greenish or reddish tint of white. The CIE recommendation states the following[4]:

> "To promote uniformity of practice in the evaluation of whiteness of surface colors, it is recommended that the formulae for whiteness, W or W_{10}, and for tint, T_w or $T_{w,10}$, given below, be used for comparisons of the whiteness of samples evaluated for CIE standard illuminant D65. The application of the formulae is restricted to samples that are called 'white' commercially, that do not differ much in color and fluorescence, and that are measured on the same instrument at nearly the same time. Within these restrictions, the formulae provide relative, but not absolute, evaluations of whiteness, that are adequate for commercial use, when employing measuring instruments having suitable modern and commercially available facilities.

$$
\begin{aligned}
W &= Y + 800(x_n - x) + 1700(y_n - y) \\
W_{10} &= Y_{10} + 800(x_{n,10} - x_{10}) + 1700(y_{n,10} - y_{10}) \\
T_w &= 1000(x_n - x) - 650(y_n - y) \\
T_{w,10} &= 900(x_{n,10} - x_{10}) - 650(y_{n,10} - y_{10})
\end{aligned}
\tag{3.51}
$$

where Y is the Y-tristimulus value of the sample, x and y are the x, y chromaticity coordinates of the sample, and x_n, y_n are the chromaticity coordinates of the perfect diffuser, all for the CIE 1931 standard colorimetric observer; Y_{10}, x_{10}, y_{10}, $x_{n,10}$ and $y_{n,10}$ are similar values for the CIE 1964 standard colorimetric observer.

Note 1: The higher the value of W or W_{10}, the greater is the indicated whiteness. The more positive the value of T_w or $T_{w,10}$, the greener the tint; the more negative the value of T_w or $T_{w,10}$, the redder the tint. For the perfect diffuser W and W_{10} are equal to 100, and T_w and $T_{w,10}$ are equal to zero.

Note 2: Linear whiteness formulae are applicable only within a restricted volume of the color solid. These formulae may be used only for samples whose values of W or W_{10} and T_w or $T_{w,10}$ lie within the following limits:

W or W_{10} greater than 40 and less than $5Y - 280$, or $5Y_{10} - 280$;
T_w or $T_{w,10}$ greater than -4 and less than $+2$.

Note 3: The tint formulae are based on the empirical results that lines of equal tint run approximately parallel to lines of dominant wavelength 466 nm in the x, y and x_{10}, y_{10} chromaticity diagrams.

Note 4: Equal differences in W or W_{10} do not always represent equal perceptual differences in whiteness, nor do equal differences in T_w or $T_{w,10}$ always represent equal perceptual differences in greenishness or reddishness of whites. Measures of whiteness and tint that correlate uniformly with these perceptual attributes would require more complicated formulae, which is beyond present knowledge."

METAMERISM

Spectrally different color stimuli can have the same tristimulus values in a specified colorimetric system, and this is called metamerism and the color stimuli with the same tristimulus values are *metameric* color stimuli or *metamers*. This is the consequence of moving from a 401 element spectral space (if we consider the spectra between 380 nm and 780 nm at every nanometer) to a three-dimensional space. The official definition is

"Two specimens having identical tristimulus values for a given reference illuminant and reference observer are metameric if their spectral radiance distributions differ within the visible spectrum."

Because the tristimulus values will, in general, no longer be identical if a change is made either to the illuminant or to the observer, a distinction is made between illuminant and observer metamerism. The break down of metamerism can be very important in different industrial applications. Thus, for example, in the automotive industry interior parts of a car are often prepared from different materials, but the color of these parts should match, even under different illuminants, and for observers of slightly different spectral sensitivity (CMFs). To describe how well such a match holds if from a reference illuminant (e.g., CIE standard illuminant D65) the illumination is changed to a test illuminant, the CIE introduced the *special metamerism index: change in illuminant*. To describe the change in tristimulus values if the CIE 1931 (or 1964) standard observer is changed to a test observer the CIE *special metamerism index: change in observer* is used.

Special Metamerism Index: Change in Illuminant

Figure 3.20 shows the spectral radiance distribution of two samples that are metameric for CIE 1964 standard observer under CIE standard illuminant D65, but have different tristimulus values under CIE standard illuminant A. Table 3.4 shows the two sets of tristimulus values.

The CIE method to calculate the special metamerism index: change in illuminant ($M_{X,ilm}$, where X stands for the test illuminant) contains the following steps[4]:

- Determine the tristimulus values under the reference illuminant (preferably CIE st. ill. D65).
- Determine the tristimulus values under the test illuminant. The information on the test illuminant has to be set in the subscript of the metamerism index (X in the above example). It is recommended that either CIE standard illuminant A or one of the sources enumerated in the CIE Publication 15:2004 be used, see Section CIE illuminants.

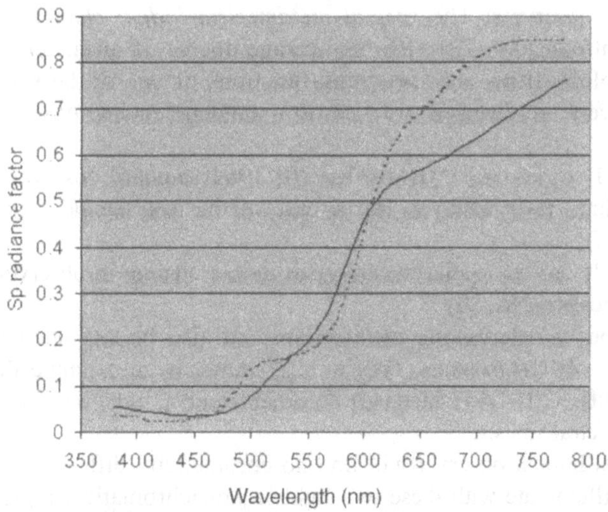

FIGURE 3.20 Spectral radiance distributions of two metameric samples with equal tristimulus values under CIE D65 illumination, but different tristimulus values under CIE illuminant A.

TABLE 3.4 Tristimulus values of the two samples, whose spectral radiance distribution is shown in Figure 3.20

	Sample No. 1			Sample No. 2		
Illuminant	X_{10}	Y_{10}	Z_{10}	X_{10}	Y_{10}	Z_{10}
CIE st. ill. D65	30.29	24.41	4.33	30.29	24.41	4.33
CIE st. ill. A	44.72	30.64	1.52	46.08	30.66	1.66

- CIELAB color difference should be used to calculate the color difference between the two sets of tristimulus values (if a different color difference formula is used, this should be noted).

The special metamerism index: change in illuminant is defined as

$$M_{\text{Xilm}} = \Delta E^*_{\text{ab}}$$

For the samples in Table 3.4 $M_{\text{Ailm}} = 4.12$.

Special Metamerism Index: Change in Observer

The CIE 1931 and 1964 standard colorimetric observers represent the color vision properties of the average population reasonably well. Nevertheless it is well known that individual deviations in the color-matching functions occur among color normal observers. The *special metamerism index: change in observer*[59] (M_{obs}) was introduced to describe the average degree of mismatch found among metameric colors if the color-matching functions of one of the standard colorimetric observers are changed to those of a standard deviate observer of normal color vision.

Figure 3.21 shows the CMFs of the CIE 1964 standard observer and the first standard deviate (test) observer. In the case of the test sample 1 of Figure 3.21 the M_{obs} is 2.7.

More details on the special metamerism index: change in observer is found in the CIE Publication No. 80.[59]

The method of determining metamerism can also be expanded to determine characteristics of light sources. Figure 3.22 shows as an example the spectra of two illuminants: CIE D65 standard illuminant and a light that is composed of only three spectral lines.

The normal human observer will find no chromaticity difference for these two lights. If we illuminate with these two lights a nonachromatic sample, most probably a color difference will be observed. Figure 3.23 shows as an example the chromaticities of a light yellowish sample. With more saturated colors the difference can be much larger.

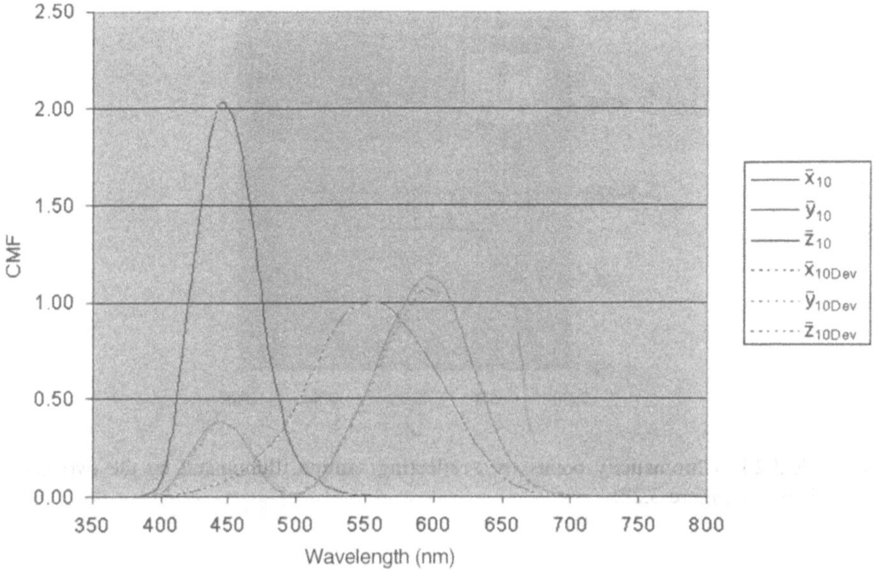

FIGURE 3.21 CIE 1964 standard observer and first deviate observer color-matching functions.

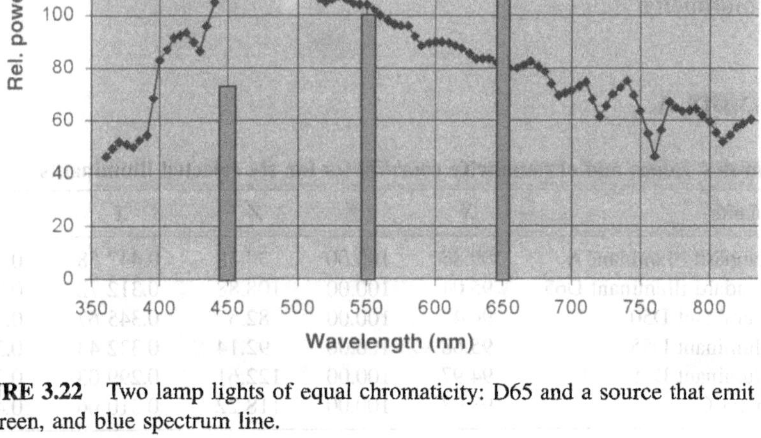

FIGURE 3.22 Two lamp lights of equal chromaticity: D65 and a source that emit only a red, green, and blue spectrum line.

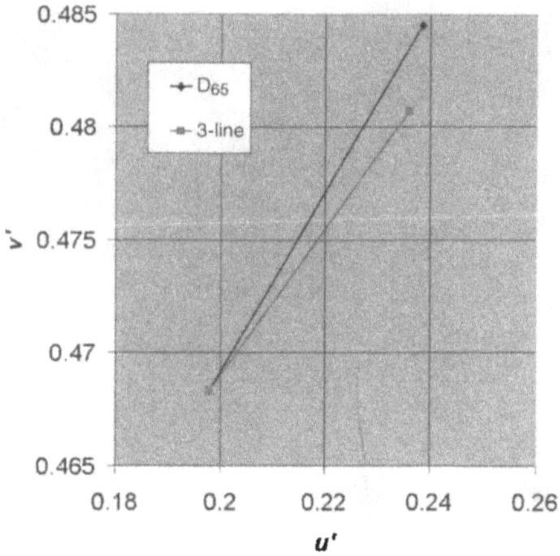

FIGURE 3.23 Chromaticity points of a reflecting sample illuminated by the two lamp lights shown in Figure 3.22.

SUMMARY

In this chapter we have introduced the fundamentals of CIE colorimetry, mainly based on the CIE Publication 15:2004[4] Colorimetry. The CIE standard colorimetric observers and illuminants were introduced, together with the fundamentals of the recommended measuring geometries and the basic equations for determining color differences and other colorimetric descriptors. The task of this chapter was not to substitute the CIE publication but to make the reading of the publication easier by showing some examples. For a thorough understanding of CIE colorimetry the reader has to be directed to the above publication and the further CIE standards and publications dealing with more particular questions of colorimetry; see a list of CIE publications dealing with colorimetry at the end of the book. The subsequent chapters will also provide further insight into colorimetry.

APPENDIX A

Tristimulus values and chromaticity coordinates for six selected illuminants

Illuminants	X	Y	Z	x	y
CIE standard illuminant A	109.85	100.00	35.58	0.447 58	0.407 45
CIE standard illuminant D65	95.04	100.00	108.88	0.312 72	0.329 03
CIE illuminant D50	96.4	100.00	82.5	0.345 67	0.358 5
CIE illuminant D55	95.68	100.00	92.14	0.332 43	0.347 44
CIE illuminant D75	94.97	100.00	122.61	0.299 03	0.314 8
Illuminant C	98.07	100.00	118.22	0.310 06	0.460 89

APPENDIX B

Colorimetric characteristics of the representative fluorescent and high pressure discharge lamps

Lamp	Chromaticity coordinates x	Chromaticity coordinates y	Correlated color temperature (T_{cp}) (kelvins)	General color-rendering index (R_a)
FL 1	0.3131	0.3371	6430	76
FL 2	0.3721	0.3751	4230	64
FL 3	0.4091	0.3941	3450	57
FL 4	0.4402	0.4031	2940	51
FL 5	0.3138	0.3452	6350	72
FL 6	0.3779	0.3882	4150	59
FL 7	0.3129	0.3292	6500	90
FL 8	0.3458	0.3586	5000	95
FL 9	0.3741	0.3727	4150	90
FL 10	0.3458	0.3588	5000	81
FL 11	0.3805	0.3769	4000	83
FL 12	0.4370	0.4042	3000	83
FL 3.1	0.4407	0.4033	2932	51
FL 3.2	0.3808	0.3734	3965	70
FL 3.3	0.3153	0.3439	6280	72
FL 3.4	0.4429	0.4043	2904	87
FL 3.5	0.3749	0.3672	4086	95
FL 3.6	0.3488	0.3600	4894	96
FL 3.7	0.4384	0.4045	2979	82
FL 3.8	0.3820	0.3832	4006	79
FL 3.9	0.3499	0.3591	4853	79
FL 3.10	0.3455	0.3560	5000	88
FL 3.11	0.3245	0.3434	5854	78
FL 3.12	0.4377	0.4037	2984	93
FL 3.13	0.3830	0.3724	3896	96
FL 3.14	0.3447	0.3288	6509	98
HP1	0.5330	0.4150	1959	8
HP2	0.4778	0.4158	2506	83
HP3	0.4302	0.4075	3144	83
HP4	0.3812	0.3797	4002	74
HP5	0.3776	0.3713	4039	87

REFERENCES

1. CIE (1932) *Colorimétrie, Resolutions 1–4*, in: Recueil des travaux et compte rendu des séances, Hutième Session Cambridge – Septembre 1931, Bureau Central de la Commission, The National Physical Laboratory Teddington, Cambridge at the University Press, pp. 19–29.
2. Grassmann HG (1853) Zur Theorie der Farbenmischung/Theory of compound colours, Original published in *Poggendorf' Ann. Phys.*, **89**, 69 original translation in English in *Phil. Mag.*, 4(7), 254–264, 1854; present formulation from *Sources of Color Science* 53–60 MIT Press, 1970, pp. 53–60; see Selected papers in Colorimetry – Fundamentals (Ed., D.L. MacAdam), SPIE Milestone Series MS 77, 1993, pp. 10–13.
3. Hunt RWG (1998) *Measuring Colour*, 3rd ed., Fountain Press, England.
4. CIE Technical Report (2004) *Colorimetry*, 3rd ed., Publication 15:2004, CIE Central Bureau, Vienna.
5. Schanda JD (1997) Colorimetry, in: *Handbook of Applied Photometry* (Ed., C. DeCusatis), AIP Press, Woodbury, NY, 1997, pp. 327–412.
6. Wright WD (1928–1929) A re-determination of the trichromatic coefficients of the spectral colours. *Trans. Opt. Soc. Lond.*, **30**, 141–164.
7. Wright WD (1929–1930) A re-determination of the mixture curves of the spectrum. *Trans. Opt. Soc. Lond.*, **31**, 201–218.
8. Guild J (1931) The colorimetric properties of the spectrum. *Phil. Trans. R. Soc. Lond. Ser. A*, **230**, 149–187.
9. ISO/CIE 10527(E) *joint ISO/CIE standard: Colorimetric observers*, 1991 (S002, 1986).
10. CIE Standard S 014-1/E:2006: *Colorimetry—Part 1: CIE standard colorimetric observers.*
11. CIE (1964) *Proceedings of the CIE 15th Session*, Vienna, Vol. A (**CIE 11A**) 1963, p. 35, Bureau Central de la CIE, Paris, 1964.
12. CIE Committee W-1.3.1. (1959) Colorimetry, report by DB Judd, in: *Proceedings of the CIE Session* Bruxelles, pp. 91–107.
13. Stiles WS, Burch JM (1959) N.P.L. color-matching investigation: Final report (1958). *Optica Acta*, **6**, 1–26.
14. Speranskaya NI (1959) Determination of spectral color co-ordinates for twenty-seven normal observers. *Optics Spectrosc.*, **7**, 424–428.
15. CIE Technical Report (2005) *CIE 10 degree photopic photometric observer*, CIE Publication **165**:2005, CIE Central Bureau, Vienna.
16. Schanda J, Morren L, Rea M, Ronchi L, Walraven P (2002) Does lighting need more photopic luminous efficiency functions? *Lighting Res. Technol.*, **34**, 169–178.
17. CIE (1993) Judd's method for calculating the tristimulus values of the CIE 10° observer, in: *CIE Proceedings of the CIE Symposium on advanced colorimetry*, CIE Publication **x007**, CIE Central Bureau, Vienna, pp. 107–114.
18. Wyszecki G, Stiles WS (1982) *Color Science, Concepts and Methods, Quantitative Data and Formulae*, 2nd ed., Wiley, New York, pp. 132–133, 140–143.
19. Wyszecki G, Stiles WS (1982) *Color Science, Concepts and Methods, Quantitative Data and Formulae*, 2nd ed., Wiley New York, p. 104.
20. Wyszecki G, Stiles WS (1982) *Color Science, Concepts and Methods, Quantitative Data and Formulae*, 2nd ed., Wiley New York, p. 547.
21. Le Grand Y (1968) *Light, Color and Vision*, 2nd ed., Chapman and Hall, London, p. 106.

22. CIE (1964) Official recommendations, Committee E-1.3.1—Colorimetry, in: *Proceedings of the 15th Session*, Vienna, 1963, Vol. **A**, CIE Publication **11 A**, p. 35.
23. CIE Standard S 014-2E:2006: *Colorimetry—Part 2: CIE Standard Illuminants.*
24. Judd DB, MacAdam DL, Wyszecki G (with the collaboration of Budde HW, Condit HR, Henderson ST, Simonds JL) (1964) Spectral distribution of typical daylight as function of correlated color temperature. *J. Opt. Soc. Amer.*, **54**, 1031–1040.
25. CIE (1967) Recommendations on standard illuminants for colorimetry, in: Proceedings of the CIE Washington Session, Vol. A, CIE Publ. **14**, pp. 95–97.
26. ISO 10526/CIE S005/E (1999) *Joint ISO/CIE standard: CIE standard illuminants for colorimetry.*
27. CIE Technical Report (2002) *The use of tungsten filament lamps as secondary standard sources*, CIE **149**:2002.
28. CIE Technical Report (1999) *A method for assessing the quality of daylight simulators for colorimetry* (with supplement 1-1999), CIE **51.2**–1999.
29. CIE Standard (2004) *Standard method of assessing the spectral quality of daylight simulators for visual appraisal and measurement of color*, CIE S 012/E:2004.
30. CIE (1987) *International Lighting Vocabulary*, CIE **17.4**.
31. CIE Technila Report (1977) *Radiometric and photometric characteristics of materials and their measurement*, CIE Publication **38**-1977.
32. CIE Technical Report (1998) *Practical methods for the measurement of reflectance and transmittance*, CIE Publication **130**-1998.
33. CIE Technical Report (1989) *The measurement of luminous flux*, CIE Publication **84**-1989.
34. Gundlach D, Mallwitz E (1976) Fragen der Probenbeleuchtung und Meβgeometrie in der Farbmessung. *Die Farbe*, **25**, 113–130.
35. Alman DH, Billmeyer FW Jr (1976) Integrating sphere errors in the colorimetry of fluorescent materials. *Color Res. Appl.*, **1**, 141–145.
36. CIE Technical Report (1988) *Intercomparison on measurement of (total) spectral radiance factor of luminescent specimens*, CIE Publication **76**–1988.
37. CIE Technical Report (2004) *The effects of fluorescence in the characterization of imaging media*, CIE Publication **163**:2004.
38. Pointer MR (2003) *Measuring visual appearance—A framework for the future*, NPL Report: COAM 19.
39. CIE Technical Report (1979) *Absolute methods for reflection measurements*, CIE Publication **44**-1979.
40. CIE Technical Report (1979) *A review of publications on properties and reflection values of material reflection standards*, CIE Publication **46**–1979.
41. Newhall SM, Nickerson D, Judd DB (1943) Final report of the O.S.A Subcommittee on the spacing of the Munsell Colors. *J. Opt. Soc. Am.*, **43**, 385–418.
42. Barnes PY, Hsia JJ (1999) *45°/0° spectral reflectance factors of pressed polytetrafluoroethylen (PRFE powder)*, CIE Publication **135**/6-1999, pp. 50–64.
43. MacAdam DL (1942) Visual sensitivities to color differences in daylight. *J. Opt. Soc. A.*, **32**, 247–274.

44. CIE (1959) Discussion du rapport du comité W-1.3.1 Séance du 16 Juin 1959. Official recommendations No. 4, in: *Proceedings of the CIE Session Bruxelles*, 1959, pp. 107–109.
45. MacAdam DL (1937) Projective transformation of I.C.I. color specifications, *J. Opt. Soc. Am.*, **27**, 294–299.
46. CIE (1975) Progress report of CIE TC – 1.3 Colorimetry, in: Compte Rendu 18e Session, Londres 1975, CIE Publication **36** pp. 161–172.
47. CIE (1964) Official recommendations, Committee E-1.3.1–Colorimetry, in: *Proceedings of the 15th Session*, Vienna, 1963, Vol. **A**, CIE Publication **11 A**, pp. 35, 112–114.
48. Adams EQ (1942) X-Z planes in the 1931 ICI system of colorimetry. *J. Opt. Soc. Am.*, **32**, 168–173.
49. Nickerson D (1950) Munsell renotations used to study color space of Hunter and Adams. *J. Opt. Soc. Am.*, **40**, 85–88.
50. Pauli H (1976) Proposed extension of the CIE recommendation on uniform color spaces, color difference equations, and metric color terms. *J. Opt. Soc. Am.*, **66**, 866–867.
51. CIE Technical Report (1986) *Colorimetry*, CIE Publication **15.2**-1986.
52. McLaren K (1980) CIELAB hue-angle anomalies at low tristimulus ratios. *Color Res. Appl.*, **5**, 139–143.
53. CIE Technical Report (1988) *Brightness-luminance relations: Classified bibliography*. CIE Publication **78**-1988.
54. Borbély Á, Sámson Á, Schanda J (2001) The concept of correlated color temperature revisited. *Color Res. Appl.*, **26**(6), 450–457.
55. McAdam DL (1934) The specification of whiteness. *J.Opt. Soc. Am.*, **24**, 188–191.
56. Judd DB (1941) Whiteness of light surface-colors. *J Opt. Soc. Am.*, **31**, 462–463.
57. Grum F, Witzel RF (1974) Evaluation of whiteness. *J. Opt. Soc. Am.*, **64**, 210–215.
58. Ganz E (1979) Whiteness formulas: a selection. *Appl. Opt.*, **18**, 1073–1078.
59. CIE Technical Report (1989) *Special metamerism index: Observer metamerism*, CIE Publication **80**-1989.

4

CIE COLOR DIFFERENCE METRICS

KLAUS WITT
Hortensienstr. 25, D-12203, Germany

INTRODUCTION

Since ancient times colorists have tried to reproduce their coloration of materials using their eyes to control the accuracy of matching. With the development of optical measurement techniques in the nineteenth century and the definition of a CIE standard colorimetric observer in 1931, observer judgments on color identity could, hopefully, be replaced by the use of colorimetry. However, exact identity of coloration is a very rare event and, hence, colorimetric values of a matching pair of specimens differ to some extent (including statistical uncertainty of the measurements). The question is what do the measured differences tell us about the magnitude of the perceived color differences? How sensitive is the human eye in relation to the optical apparatus? If the apparatus is the more sensitive, there should be an upper limit of the measured color differences, which are visually equal. Such a limit is termed as a threshold of the perceived color difference. Subthreshold color differences of matched specimens are visually identical. However, though colorists try to attain subthreshold matching, in many cases technical problems provoke beyond threshold results. Now the customer is asked whether he accepts, and the discussion about an acceptable magnitude of the color difference starts. What are the scales of the beyond threshold color differences and how can we assign measured differences to them? To answer this question, psychophysical experimentation is needed, which means to correlate results of the observer tasks with the instrument readings. This is central to the development of color difference evaluation.

Colorimetry: Understanding the CIE System, Edited by János Schanda

Ongoing research since the 1940s has sought color difference solutions for a range of industrial applications. CIE formed technical committees and published technical reports to guide colorists as the improvements in color difference evaluation developed. This chapter surveys these developments over a span of more than 60 years.

MACADAM'S EXPERIMENTS ON VARIABLE STIMULI

In the early 1940s, MacAdam performed a famous experiment.[1] He constructed an optical apparatus with a bipartite visual field, presented in the first half field one of 25 different colors at constant luminance as standard, and allowed in the second half field an adjustment of the same color by changing filter combinations in the second light beam. Two observers, he himself and P. G. Nutting, repeatedly tried to set the correct test color in the variable half field. The colorimetric measures of the final settings of the test colors plotted in an x,y-chromaticity diagram scattered around that of the standard color producing two-dimensional scattergrams, which could be represented by mean mathematical ellipses. The area inside such an ellipse includes all colors visually identical with the standard. However, the threshold of discriminability is not the ellipse boundary directly but is thought to be two to three times larger. Nevertheless, such an ellipse is taken as a basis for describing near-threshold color differences. The remarkable finding for all of the 25 standard colors is the great variation of eccentricity, orientation, and size of ellipses within the chromaticity diagram, see Figure 4.1 for observer P. G. Nutting, with smallest ellipses in the area of blue colors and largest in that of the green colors. This indicates that the x,y-chromaticity diagram is not a good basis for describing near-threshold color differences. MacAdam tried to extend ellipse constants from the 25 colors to the complete chromaticity chart[2] to allow for an unlimited use of his data for color difference evaluation. Later Simon and Goodwin published graphical material to allow for quick calculation of color differences.[3]

As color is represented in a three-dimensional space, MacAdam's two-dimensional experiments needed an extension for the third dimension, lightness. He started this work together with Brown[4] who later found multivariate normality in the color-matching process[5] defined by ellipsoids to represent three-dimensional standard deviations for the variability among 12 observers.[6]

Before MacAdam's experiments, Judd had found that certain equistepped color scales could be better represented in a projectively transformed chromaticity chart (proposed by MacAdam) termed uniform chromaticity chart (UCS),[7] which later was adopted by CIE as the 1960 CIE-UCS diagram (now termed CIE 1960 UCS diagram). Plotting the MacAdam ellipses in such a diagram revealed ellipses with few eccentricities though not circles, see Figure 4.2, expected if the CIE 1960 UCS diagram had ideal form for equal color difference spacing. Nevertheless, Wyszecki took the result as a good compromise and proposed to start with this diagram for the definition of a color difference formula,[8] which CIE quickly

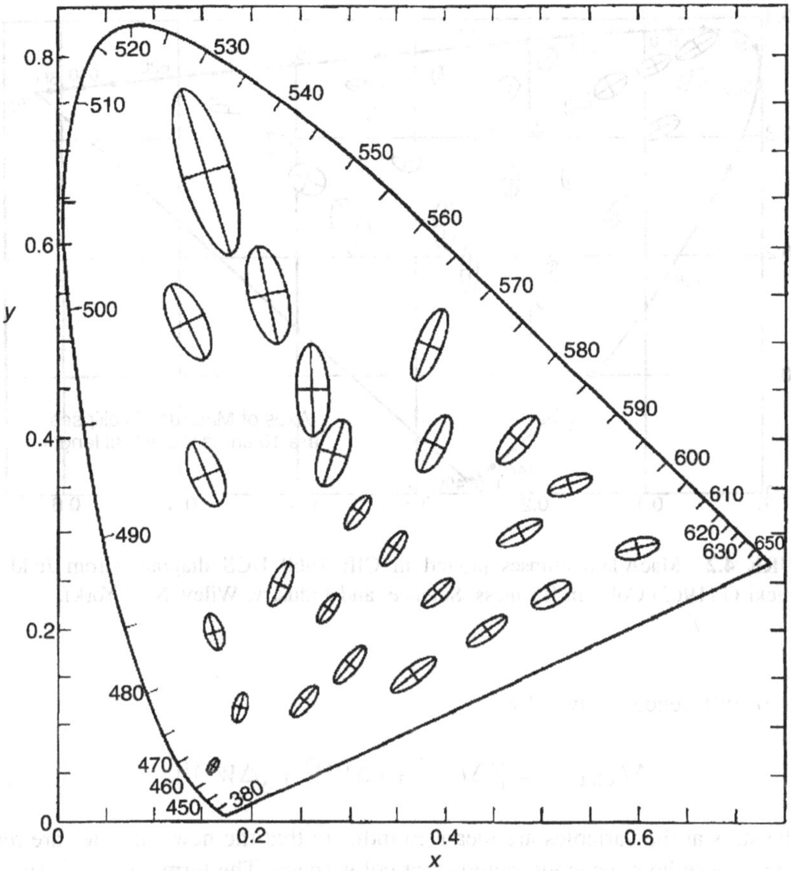

FIGURE 4.1 MacAdam ellipses 10-times enlarged (from Judd DB, Wyszecki G (1963) Color in Business, Science, and Industry, Wiley, New York).

accepted in 1964. This formula has Euclidean form and calculates the color difference as the distance between two points in a three-dimensional color space. The formula is built using a projective transformation as follows:

$$u = 4x/(-2x + 12y + 3), \quad v = 6y/(-2x + 12y + 3) \tag{4.1}$$

where x,y are the chromaticity coordinates of the object.

Coordinates of the color space are described as

$$U^* = 13W^*(u - u_0), \; V^* = 13W^*(v - v_0), \; W^* = 25Y^{1/3} - 17 \tag{4.2}$$

u_0, v_0 are the values for the light source and Y is the luminous tristimulus value of the object.

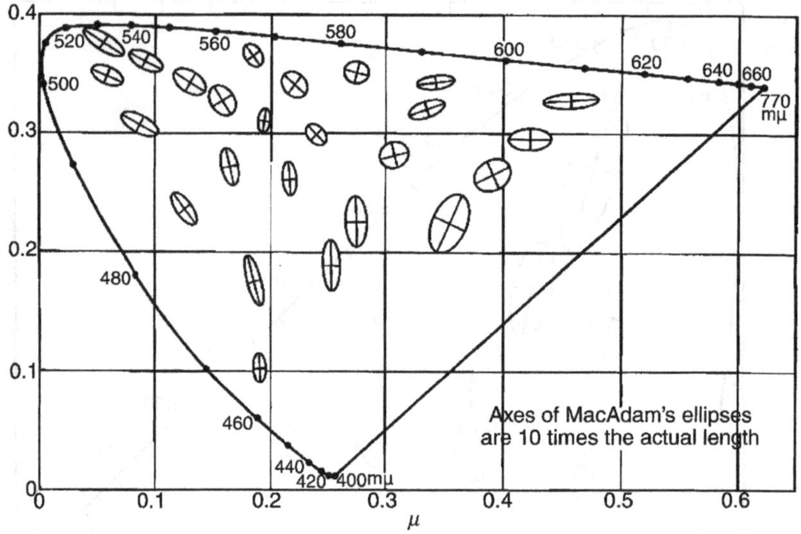

FIGURE 4.2 MacAdam ellipses plotted in CIE 1960 UCS diagram (from Judd DB, Wyszecki G (1963) Color in Business, Science, and Industry, Wiley, New York).

Color difference is given by

$$\Delta E_{\mathrm{CIEUVW}} = [(\Delta U^*)^2 + (\Delta V^*)^2 + (\Delta W^*)^2]^{1/2} \tag{4.3}$$

The stars at the variables are meant to indicate that the new variables are representing a more homogeneous, equidistant color space. The terms $(u - u_0), (v - v_0)$ shift the zero of the CIE 1960 UCS diagram to the value of the light source as the neutral point. They are a measure of color saturation, whereas their multiplication by the lightness variable W^* produce color chroma variables U^* and V^*.

ADAMS' AND NICKERSON'S CONTRIBUTION TO COLOR DIFFERENCE EVALUATION

After MacAdam's experiments on subthreshold of color differences, a second approach to color difference research was the idea to use equispacing of colors in the Munsell color order system as a reference. This requires measurement of the perceived color differences according to an accepted scaling. The Munsell lightness value scale, V, was such an accepted uniform scaling of lightness. Its relation to the Y tristimulus value was given in the form of a fifth-order polynomial function, with reference to the perfect reflecting diffuser:

$$Y = 1.1913V_Y - 0.22533V_Y^2 + 0.23352V_Y^3 - 0.020484V_Y^4 + 0.0008194V_Y^5 \tag{4.4}$$

Adams[9] applied this function to all three tristimulus values and developed a new chromatic-value system:

$$Q_1 = V_X - V_Y, Q_2 = 0.4\,(V_Y - V_Z), Q_3 = V_Y \tag{4.5}$$

Nickerson[10] used these coordinates to construct an Euclidean color difference formula:

$$\Delta E_{\mathrm{AN}} = \left[(0.23\,\Delta V_Y)^2 + (\Delta(V_X - V_Y))^2 + (0.4\ \Delta(V_Z - V_Y))^2\right]^{1/2} \tag{4.6}$$

This formula was thought to represent the scales of small perceived color differences as are represented in the interspacing of adjacent color specimens in the Munsell book of color in contrast to the threshold criterion used in the CIEUVW formula (4.3).

From the 1950s onwards, industry became very interested in the application of colorimetry and color difference evaluation to meet their needs to control color matching. Several papers dealt with problems of acceptability in the textile industry[11–14] or for paints.[15] In these papers several proposals for color difference evaluation were tested and the AN formula (4.6) was found to be among the better ones whereas the CIEUVW-formula (4.3) did not work well.

CONSTANT STIMULI EXPERIMENTS

For the coloring industries, the final release of a color-matched production depends on the customer's statement: "accept" or "not accept." For technical reasons the "accept" statement in many cases does not mean ideal color identity especially so in the textile industry. Small nuances of color may be tolerated because they may not be too conspicuous in the material in question. The colorist now judge on his production putting the production specimen beside the standard. This is a pair comparison task where both specimens form constant stimuli and the colorist judges whether to "accept" or "not accept." Near the tolerance limit, different observers will change their judgments between both the answers in a statistical way. McLaren[16] found that the relation between "not accept" answers and color difference metrics followed a Gaussian summation curve about the tolerance limit, which is defined by a 50% "accept/not accept" criterion.

In principle, the variable stimuli experiment of MacAdam and the constant stimuli experiments of pair comparison provide different observer tasks and must not give comparable results, especially when perceptibility and acceptability data of constant stimuli experiments are mixed. Kuehni[13] found some discrepancies when plotting ellipses of acceptability data in the chromaticity chart in comparison with MacAdam ellipses. In the blue color region, he found several MacAdam ellipses to be far too small. Wyszecki[17] as the chairman of a CIE Colorimetry Committee distrusted the acceptability experiments in industry to conform well to the perceived color differences because colorists could apply nonperceptual criteria such as technical difficulties and certain color preferences for their judgments. However, McLaren[18] found no statistically significant divergences between acceptability and perceptibility data. A detailed experiment comparing variable and constant stimuli methods for perceived color differences showed good conformity of both methods.[19] Nevertheless, as more data were published, the more diverse appeared the results of color difference experiments. Great variation appeared in different datasets either for variable stimuli experiments or for pair comparison methods. In general, the constant stimuli experiments were more common as these used practical materials similar to colorists' daily work. Whereas, the variable stimuli experiments required an optical apparatus, which in most cases had to be self-constructed in a laboratory. Nowadays, the computer controlled monitors may replace these.

CIE 1976 COLOR DIFFERENCE FORMULAS

CIE understood that the industry was unhappy with the CIEUVW formula (4.3) and started a new trial. A CIE technical committee discussed the industry concerns about improvement of color difference formulas. The form of the AN formula (4.6) received high attention. The only drawback was the need of a tedious fifth-order back transformation to calculate the V values from tristimulus values. An acceptable solution was to restrict the polynomial to the third order and calculate new variables by a cube root transformation. Moreover, to set a zero point of the new variables for the light source in question, tristimulus values of the object were divided by those of the light source (X_n, Y_n, Z_n), which defined relative values in contrast to the difference solution in the CIEUVW formula (4.2). After adapting a new lightness function to the Munsell value scale (see Figure 4.3), the new coordinate system, now termed as the CIE 1976 ($L^*a^*b^*$) color space (abbreviated CIELAB), was defined as follows[20]:

$$L^* = 116f(Y/Y_n) - 16 \tag{4.7}$$

$$a^* = 500[f(X/X_n) - f(Y/Y_n)] \tag{4.8}$$

$$b^* = 200[f(Y/Y_n) - f(Z/Z_n)] \tag{4.9}$$

where

$$f(X/X_n) = (X/X_n)^{1/3} \tag{4.10}$$

$$f(Y/Y_n) = (Y/Y_n)^{1/3} \tag{4.11}$$

$$f(Z/Z_n) = (Z/Z_n)^{1/3} \tag{4.12}$$

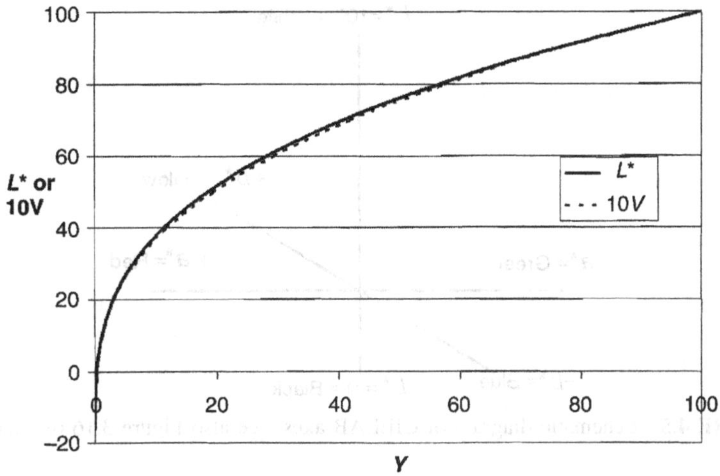

FIGURE 4.3 Intercomparison of L^* and Munsell value scale V.

Figure 4.3 indicates that the cube root formula of L^* is generally rather close to the Munsell value scale V (with $V = 10$ identical to $L^* = 100$). However, for values near zero a great disparity is shown in Figure 4.4: L^* drops to -16 and not to 0 for $Y = 0$.

Therefore, the definitions (4.10, 4.11, 4.12) apply only for values of

$$f(X/X_n), f(Y/Y_n), f(Z/Z_n) > (24/116)^3 \tag{4.13}$$

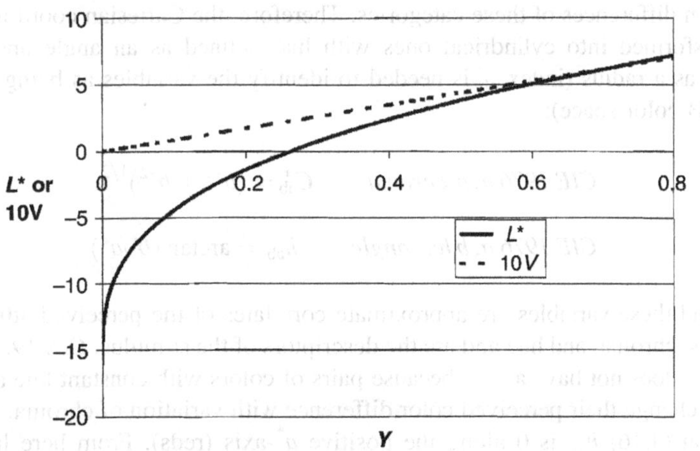

FIGURE 4.4 Deviation of L^* from Munsell value scale V at low Y values.

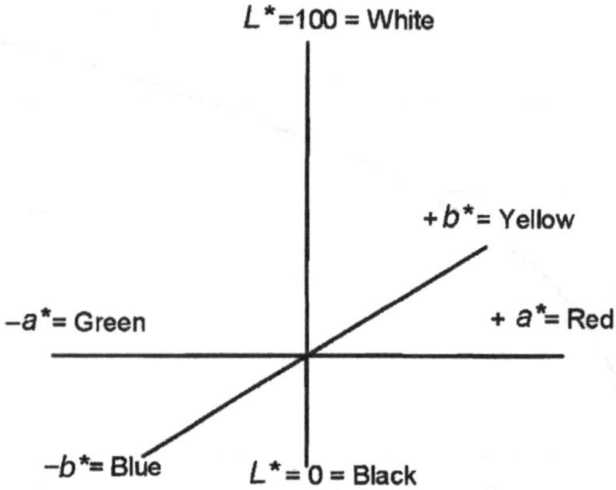

FIGURE 4.5 Schematic diagram of CIELAB axes (see also Figure 3.16 of Chapter 3).

CIE accepted Pauli's[21] proposal to use a tangent to the cube root function going through the zero point for colors below these values, which is written in the actual form[22]:

$$f(X/X_n) = (841/108)(X/X_n) + 16/116 \quad \text{if } (X/X_n) \text{ etc.} \leq (24/116)^3 \tag{4.14}$$

and similarly for $f(Y/Y_n)$ and $f(Z/Z_n)$.

This new color space forms a Cartesian, approximately uniform color space (see Figures 4.5 and 3.16) with the lightness coordinate $L^*(L^* = 0$ is black, $L^* = 100$ is white), a green–red-oriented coordinate $a^*(-a^*$ means greenness, $+a^*$ means redness), and a blue–yellow-oriented coordinate $b^*(-b^*$ means blueness, $+b^*$ means yellowness). a^* and b^* measure chroma in the sense of Munsell. Colorists are adapted to speak about lightness, chroma, and hue of a color and name the color shades in differences of these categories. Therefore, the Cartesian coordinates may be transformed into cylindrical ones with hue defined as an angle and chroma defined as a radius (index $_{ab}$ is needed to identify the variables as being from the CIELAB color space):

$$\textit{CIE 1976 a, b chroma} \qquad C^*_{ab} = (a^{*2} + b^{*2})^{1/2} \tag{4.15}$$

$$\textit{CIE 1976 a, b hue angle} \qquad h_{ab} = \arctan(b^*/a^*) \tag{4.16}$$

L^* and these variables are approximate correlates of the perceived attributes of lightness, chroma, and hue and are the descriptors of the stimulus. *CIE 1976 a,b hue angle* h_{ab} does not have a "*" because pairs of colors with constant hue angle difference change their perceived color difference with variation of chroma. From the definition (4.16) h_{ab} is 0 along the positive a^*-axis (reds). From here hue angle increases counter clockwise with 90° for $+b^*$-direction (yellows), 180°

for $-a^*$-direction (greens), 270° for $-b^*$-direction (blues). (For further details see Chapter 3, Section CIE 1976 uniform color spaces.)

If we denote a pair of color specimens with subscripts 0 and 1, which differ only by color and which were colorimetrically measured under identical conditions, the coordinate differences between them may be calculated as follows:

$$\textit{CIELAB lightness difference:} \quad \Delta L^* = L_1^* - L_0^* \tag{4.17}$$

$$\Delta a^* = a_1^* - a_0^* \tag{4.18}$$

$$\Delta b^* = b_1^* - b_0^* \tag{4.19}$$

$$\textit{CIELAB chroma difference:} \quad \Delta C_{ab}^* = C_{ab,1}^* - C_{ab,0}^* \tag{4.20}$$

$$\textit{CIELAB hue angle difference:} \quad \Delta h_{ab} = h_{ab,1} - h_{ab,0} \tag{4.21}$$

Δh_{ab} shall not lie outside the range of ±180°, which may happen if the line joining the two colors crosses the $+a^*$- axis. In this case the value of $\Delta\ h_{ab}$ shall be corrected by adding or subtracting 360°.

$$\textit{CIELAB hue difference:} \quad \Delta H_{ab}^* = 2(C_{ab,1}^* C_{ab,0}^*)^{1/2}\sin(\Delta h_{ab}/2) \tag{4.22}$$

ΔH^*_{ab} is the correlate of the perceptual magnitude of a hue difference and thus has a superscript "*". For alternative formulas to calculate ΔH^*_{ab}, see CIE 15:2004.[22] ΔC^*_{ab} and ΔH^*_{ab} become less useful if Δh_{ab} approaches 180°.

CIE 1976 a,b (CIELAB) color difference, ΔE^*_{ab} between the color pair, may now be calculated as an Euclidean distance between the two points in the color space representing them. Two equivalent forms of this color difference formula exist:

$$\Delta E_{ab}^* = [(\Delta L^*)^2 + (\Delta a^*)^2 + (\Delta b^*)^2]^{1/2} \tag{4.23}$$

$$\Delta E_{ab}^* = [(\Delta L^*)^2 + (\Delta C_{ab}^*)^2 + (\Delta H_{ab}^*)^2]^{1/2} \tag{4.24}$$

CIE was not completely satisfied with this formula as the application of a cube root transformation meant a nonlinear distortion of the chromaticity diagram, and additivity rules in the lighting industry could no longer be easily applied in the new coordinate system. To keep these capabilities as in the CIE 1960 UCS diagram, the definition in (4.1) was changed a little (elongation in the direction of v, which improved uniformity without disturbing additivity rules) and a new CIE 1976 ($L^*u^*v^*$) color space (abbreviated CIELUV) was created:

$$u' = u \quad v' = 3/2\,v \tag{4.25}$$

The *CIE 1976 lightness* L^* (4.7) was adopted and used to calculate chroma variables as in (4.2):

$$u^* = 13L^*(u' - u'_n), \qquad v^* = 13\,L^*(v' - v'_n) \tag{4.26}$$

Now *CIE 1976 u,v chroma* and *hue angle* may be calculated as in (4.15) and (4.16). In addition, a *CIE 1976 u,v saturation* is easily defined as chroma divided by lightness:

$$s_{u,v} = 13[(u' - u'_n)^2 + (v' - v'_n)^2]^{1/2} \tag{4.27}$$

Staring from the L^*, u^*, v^* coordinates, all the other definitions of the CIE 1976 CIELAB color space may be applied to calculate color differences in the CIE 1976 ($L^*u^*v^*$) color space.

CIE recommended both color spaces for further testing. Indeed, the CIELAB formula became widely adopted and tested and replaced numerous older color-difference formulas. This was a big step toward harmonization of color difference evaluation and also color description in the technical world.

TESTING AND IMPROVING CIELAB

The new CIELAB formula became quickly tested in the textile industry[23] and soon some inhomogeneities were described.[24,25] In the British textile industry, McDonald[26] collected a very large dataset from pass/fail experiments and found contours of equal tolerance limit were represented by ellipsoids of variable size in CIELAB space. This is in contrast to the expectation of spheres of equal size in a homogeneous color space. Keeping the CIELAB coordinate system as a start, changes were applied to the components of the color-difference formula (4.24) adapting weighting functions to the three component differences. The Color Measurement Committee (CMC) of the Society of Dyers and Colorists (SDC) proposed a new color difference formula named CMC(l:c)[27] which became an ISO standard for textile applications in 1995:

$$\Delta E_{\mathrm{CMC}} = \left[(\Delta L^*/l\,S_{\mathrm{L}})^2 + (\Delta C^*_{\mathrm{ab}}/c\,S_{\mathrm{C}})^2 + (\Delta H^*_{\mathrm{ab}}/S_{\mathrm{H}})^2\right]^{1/2} \tag{4.28}$$

$$S_{\mathrm{L}} = 0.040975\,L^*_1/(1 + 0.01765\,L^*_1), \tag{4.29}$$

unless $L^*_1 < 16$, when $S_{\mathrm{L}} = 0.511$

$$S_{\mathrm{C}} = 0.0638\,C^*_{\mathrm{ab},1}/(1 + 0.0131\,C^*_{\mathrm{ab},1}) + 0.638 \tag{4.30}$$

$$S_{\mathrm{H}} = S_{\mathrm{C}}(Tf + 1 - f) \tag{4.31}$$

where

$$f = \left\{ (C^*_{\text{ab},1})^4 / \left[(C^*_{\text{ab},1})^4 + 1900 \right] \right\}^{1/2} \quad (4.32)$$

$$T = 0.36 + |0.4 \cos(h_{\text{ab},1} + 35)| \quad (4.33)$$

unless $h_{\text{ab},1}$ is between 164° and 345° when $T = 0.56 + |0.2 \cos(h_{\text{ab},1} + 168)|$.

l and c in (4.28) are linear parametric factors to control relative sensitivities to lightness and chroma differences. For textiles $l{:}c$ is often chosen as 2:1. Subindex 1 (one) in the weighting functions refers to the values of the standard color. The lightness weighting function, S_L, depends on the lightness and reduces the effect of a lightness difference with increasing lightness beyond $L^*_1 = 16$. The chroma weighting function, S_C, reduces the effect of a chroma difference with increasing chroma, as is needed because in an a^*, b^*-diagram the size of ellipses of constant color difference increase significantly with the increasing chroma. The hue weighting function, S_H, is the most complex. Here variations with hue angle $h_{\text{ab},1}$ and chroma $C^*_{\text{ab},1}$ are used to cope with the general size-dependence on chroma and irregularities due to hue angle.

The mathematical form of the CMC formula indicates an important deviation from that of the CIELAB formula. In CIELAB, the color difference is the vector length between the two points in color space. This vector definition no longer holds in the CMC formula as weightings are applied to the vector components and, hence, the calculated color difference no longer is represented as a vector in a coordinate system. No tests were done to apply similar weightings to the CIELAB coordinates directly.

Independently, Robertson chairing the CIE Technical Committee on color difference evaluation proposed a new research project[28] with main topics as follows:

- intensive studies in the range of five selected color centers,
- evaluation of parametric effects on perception of color differences,
- formation of a reliable reference dataset on color differences,
- development and adoption of a new color difference formula.

The five colors consisted of a gray at medium lightness, a light yellow, medium light red and green, and a darker blue, these latter four colors at medium chroma. The idea behind the restriction to only five color centers was to receive directly comparable results from different laboratories because the variation of data in the past seemed to be due to different experimentation. There was no clear understanding of the effects of experimental parameters because of confounding with the ever-changing reference colors. Later Luo and Rigg[29] found that even after correcting datasets for certain parameters, unidentified factors remained, which made intercomparison difficult.

Indeed several laboratories adopted the new research project and started work.[29–34] New CIE Technical Committees were formed. One of them had to report on parametric effects in color difference evaluation with the outcome of reference conditions for experimentation as follows[35]:

- illumination — source simulating D65
- illuminance — 1000 lx
- observer — with normal color vision
- background field — uniform, neutral gray with $L^* = 50$
- viewing mode — object
- sample size — greater than 4° subtended visual angle
- sample separation — minimum, sample pairs with direct edge contact
- magnitude of ΔE — 0–5 CIELAB units
- sample structure — homogeneous without apparent pattern or nonuniformity.

If experiments deviate from these reference conditions, results may vary more or less. Known serious effects come from change of lightness of background field (e.g., high background contrast for dark sample pairs reduces sensitivity to color difference tremendously), separating sample pairs by a gap (reduction of sensitivity to color differences), changing illuminant source, for example, from standard illuminant D65 to standard illuminant A (nonlinear effects), and using textured in place of uniform surfaces (reduction of sensitivity to lightness differences). The effect of variation of such parameters should be quantified.

Another CIE Technical Committee tried to find an optimization of the CIELAB formula mainly based on new experiments under well-controlled reference conditions. The resulting recommendation[36] followed the general form of the CMC(l:c) formula (4.28):

$$\Delta E^*_{94} = \left[(\Delta L^*/k_L S_L)^2 + (\Delta C^*_{ab}/k_C S_C)^2 + (\Delta H_{ab}{}^*/k_H S_H)^2\right]^{1/2} \tag{4.34}$$

The weighting functions S_L, S_C, S_H are defined differently compared to the CMC(l:c) formula:

$$S_L = 1 \tag{4.35}$$

$$S_C = 1 + 0.045\, C^*_{ab} \tag{4.36}$$

$$S_H = 1 + 0.015\, C^*_{ab} \tag{4.37}$$

k_L, k_C, k_H are the parametric factors that describe the effect of change from reference conditions. For reference conditions, they all are set at 1.

These weighting functions were much more simpler than those in CMC(l:c). Could the CIE94 formula replace the CMC(l:c) formula effectively? Melgosa et al.[37] used published data on threshold color differences, normalized them for parametric effects, and optimized the CIE94 formula for the parameters in Equations (4.36) and (4.37). They found some variation of these parameters about their formula values and could not exclude possible further dependencies in the CIE94 formula of, for example, hue weighting on hue angle or lightness weighting on lightness. Rigg[38] compared different color difference formulas using a combined set of perceptibility and acceptability data and found the results of the CIE94 formula to be not far from the results of the CMC(l:c) formula; however, a lightness weighting function was needed. Others[39–43] said that CIE94 was not better than CMC(l:c) or that especially in the yellow region both the formulas did not work well.

COLLECTION OF NEW DATASETS

Since the formulation of CIE guidelines for coordinated research in color difference evaluation,[28] an ever-increasing number of new datasets were developed. These were collected by members of a new CIE Technical Committee, which had to investigate whether the CIE94 formula needed any extension.

These datasets consist of parts as follows:

- *BFD-P*: Perceptibility data on 2776 pairs of different materials (textile, paint) from various laboratories, mean $\Delta E^*_{ab} = 3.0$. Five CIE color centers with painted pairs near threshold were investigated under CIE reference conditions.[30,44] Red CIE color center with 51 textile specimens, pair comparison, mean $\Delta E^*_{ab} = 2.0$.[45] Five CIE color centers with textile pairs, pair comparison[31]; gray scale method applied to textile pairs under CIE reference conditions.[29]
- *RIT-DuPont*[46,34]: One hundred and fifty-six pairs of painted specimens arranged around 19 color centers, pairs contiguous, pair comparison with a gray reference pair with $\Delta E^*_{ab} = 1.02$, background middle gray, illuminant similar to standard illuminant D65.
- *Leeds*: Kim and Nobbs[47] investigated 307 pairs of painted specimens of which 104 formed a pair comparison experiment and 203 a gray scale experiment, mean $\Delta E^*_{ab} = 1.6$.
- *Witt*[48]: Five CIE color centers, 418 painted pairs scaled along extended axes of threshold ellipsoids, mean $\Delta E^*_{ab} = 1.9$, pair comparison with gray scale under CIE reference conditions.

DEVELOPMENT OF CIEDE2000

Luo and Rigg[29] previously found that most earlier color difference datasets suffered from nonidentical conditions, bad selection of difference specimens around a color center and some other unidentified factors. Hopefully, the new datasets under well-controlled conditions should help to find a new color difference formula with more confidence.

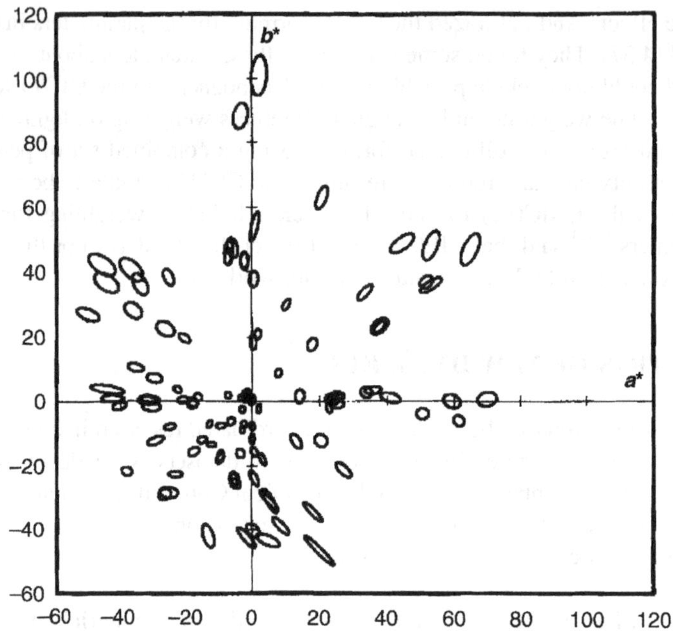

FIGURE 4.6 Experimental color discrimination ellipses plotted in a^*, b^*-diagram after Luo and Rigg,[29] source Ref. 49.

Figure 4.6 highlights the problem to be solved. Luo and Rigg[29,49] plotted experimental color discrimination ellipses in the a^*, b^*-diagram and showed serious eccentricities of ellipses in wide areas of color space appearing not to be better than the plot of MacAdam ellipses in the chromaticity diagram shown in Figure 4.1. CIELAB thus is a poor color space for calculating small color differences. Some other essential features can be easily told from Figure 4.6:

- The size of the ellipses is smallest near the origin, the gray colors, and increases with larger distance from the origin, that is, with increasing chroma.
- The orientation of the main ellipse axis is more or less directed to the origin with the exception of ellipses in the blue region, $-b^*$ with a^* around 0.

The nonideal form of CIELAB color space for small color differences has some consequences for an improved color difference formula. The simple Euclidean form (4.24) may only be kept if the difference components in the formula conform to the main axes of the three-dimensional ellipsoids. This appears to be true in the lightness direction and more or less in most cases of chroma and hue directions with the exception of the blues. If ellipses are tilted their mathematical formulation needs a so-called covariance term of the main axes.

The CIE Technical Committee 1–47 was founded in 1997 with the terms of reference:

"to investigate the hue and lightness-dependence of industrial color-difference evaluation using existing experimental data."

This meant testing the CIE94 formula (4.34) for the weighting functions S_L (4.35) and S_H (4.37). However, eventually the S_C weighting function also needed a correction after given changes of the two other weighting functions.

Color discrimination contours for achromatic colors are still ellipses with main axis near the b^*-axis and not the expected circles. Therefore, in the first place a new coordinate system was calculated that transformed the near-achromatic ellipses into circles by elongating the length of the a^*-coordinate:

$$a' = a^*(1 + G) \tag{4.38}$$

with $G = 0,5[1 - ({C^*_{ab,m}}^7/({C^*_{ab,m}}^7 + 25^7))^{0,5}]$, and $C^*_{ab,m} = (C^*_{ab,1} + C^*_{ab,0})/2$

$$b' = b^* \tag{4.39}$$

In addition to be consistent with a new coordinate set

$$L' = L^* \tag{4.40}$$

The effect of this transformation dies out with increasing chroma up to the medium chroma values near 30 (see Figure 4.7). The new coordinates are termed L', a', b'. The next step is to apply the CIELAB formalism for chroma and hue angle (4.15, 4.16) and for the difference terms (4.17, 4.20, 4.22).

In the CMC(l:c) formula, a lightness weighting function was introduced, however, in the CIE94 formula this was neglected. New data on 280 pairs of the near gray painted specimens exhibiting mainly lightness differences[50] indicated that indeed such a weighting function is needed (see Figure 4.8). The minimum in Figure 4.8 is at $L^* = 50$ identical with the background lightness. This is consistent with the so-called crispening effect, which says that color difference perception is

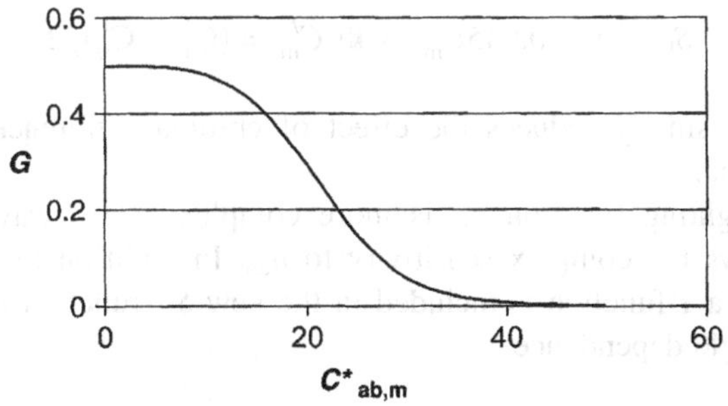

FIGURE 4.7 Dependence of function G on $C^*_{ab,m}$.

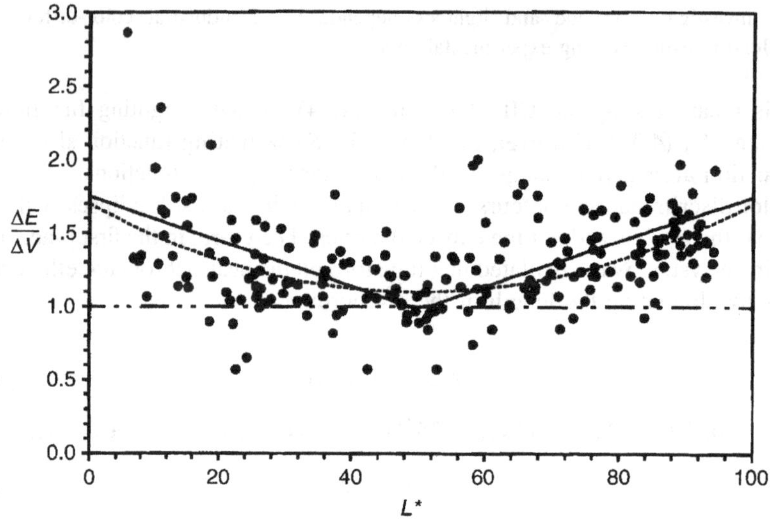

FIGURE 4.8 Dependence of relative color difference on lightness[49,50] with line drawings according to different lightness functions.

most sensitive when the background color is near the color of the object sample pair. The best curve fitting for Figure 4.8 is for a lightness weighting function S_L:

$$S_L = 1 + 0.015\,(L'_m - 50)^2/\left[20 + (L'_m - 50)^2\right]^{1/2} \tag{4.41}$$

with

$$L'_m = (L'_1 + L'_0)/2$$

The S_L weighting function (4.41) is plotted in Figure 4.8. This weighting function differs from the equivalent one in the CMC(l:c) formula (4.29) mainly for low lightness values.

The chroma weighting function S_C is equivalent to that in CIE94 (4.36):

$$S_C = 1 + 0.045C'_m, \text{ with } C'_m = (C'_1 + C'_0)/2 \tag{4.42}$$

This function simply reduces the effect of chroma differences linearly with increasing chroma.

The hue weighting function S_H is more complex. The "wavy" structure in Figure. 4.9 shows the complex sensitivity to h_{ab}. In addition to the S_H function of CIE94 (4.37) a T-function is included in the new S_H-function to cope with the complex hue angle dependence:

$$S_H = 1 + 0.015\,C'_m\,T \tag{4.43}$$

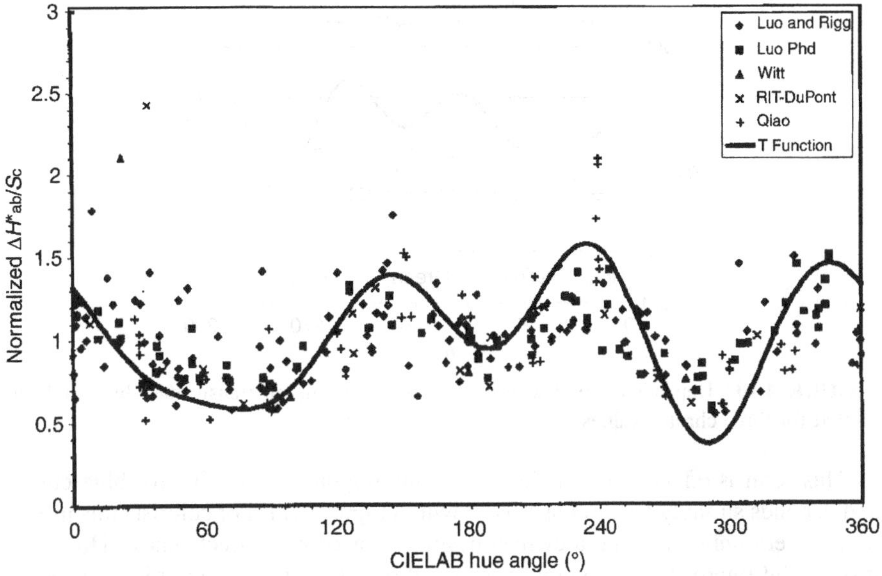

FIGURE 4.9 Normalized CIELAB hue-difference data against h_{ab}.[49]

with

$$T = 1 - 0.17\cos(h'_m - 30) + 0.24\cos(2h'_m) + 0.32\cos(3h'_m + 6) \\ - 0.20\cos(4h'_m - 63),$$

and

$$h'_m = (h'_1 + h'_0)/2$$

The increasing effect of the T-function with increasing chroma is shown in Figure 4.10.

The last step to cope with anomalies of the CIELAB color space is to look at those color regions where the main axes of tolerance ellipses do not point at the coordinate origin $(a^*,b^*) = (0,0)$. As discussed earlier, this case was restricted to the blue region. In that region a good color difference formula needs an interaction term of the type $\Delta'\ \Delta'$. A lengthy elaboration of different routes resulted in the definition of a so-called rotation term R_T[49] as a multiplicative function in the mixed term $R_T\ \Delta C'\Delta H'$, which is defined as follows:

$$R_T = -\sin(2\Delta\Theta)R_C, \tag{4.44}$$

With

$$\Delta\Theta = 30\exp\left[-((h'_m - 275)/25)^2\right] \tag{4.45}$$

$$R_C = 2(C'^7_m/(C'^7_m + 25^7))^{1/2} \tag{4.46}$$

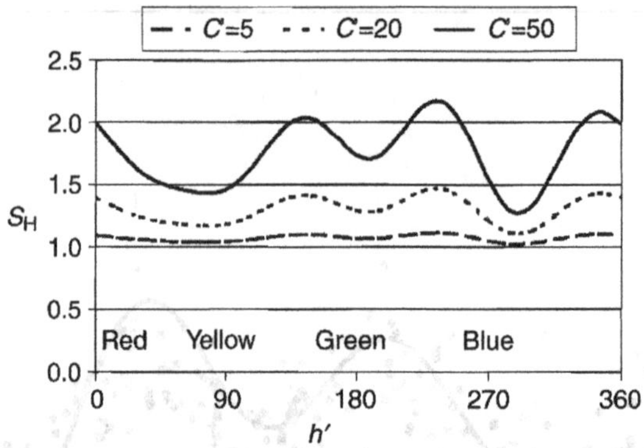

FIGURE 4.10 Dependence of hue weighting function S_H on transformed hue angle h' plotted for three chroma values C'.

This term is effective in a rather small hue region around 270° for blue colors and depends strongly on chroma as is shown in Figure 4.11. For colorists this mixed term is very unfamiliar and they may hesitate to accept its effectiveness. However, it is a valid improvement bringing perceptual data into the correct order. The color differences of blue color pairs are now much better estimated than with older color difference formulas, where such a term was missing.

Now, the complete color difference formula CIEDE2000 is written as follows:

$$\Delta E_{00} = [(\Delta L'/k_L S_L)^2 + (\Delta C'/k_C S_C)^2 + (\Delta H'/k_H S_H)^2 + R_T (\Delta C'/k_C S_C)(\Delta H'/k_H S_H)]^{1/2} \quad (4.47)$$

k_L, k_C, k_H are parametric factors, which may be chosen other than 1 if experimental conditions deviate from reference conditions. The formula (4.47) was developed from datasets nearly under reference conditions and is thus thought to be exactly valid only for these conditions. If other conditions are chosen, parametric factors should be estimated from previous experiments or should be newly elaborated.

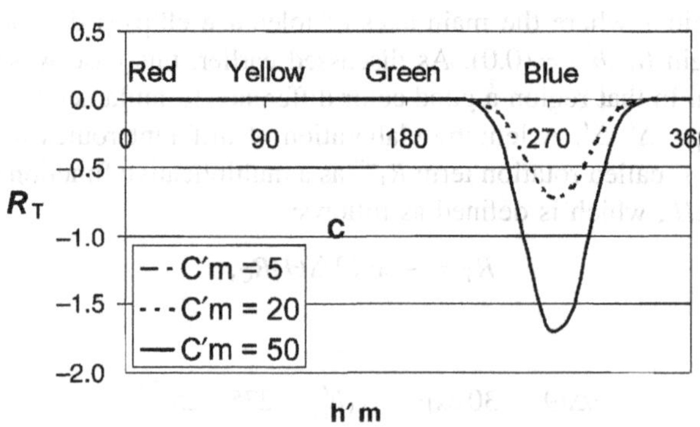

FIGURE 4.11 Dependence of rotation term R_T on transformed hue angle h'_m plotted for three chroma values C'_m.

An important result now is that CIEDE2000 outperforms the older CMC(l:c)- and CIE94-formulas which become historic and should no longer be applied. On the contrary, a further testing phase with other data is welcome to receive independent information on the validity of the formula.

FURTHER DEVELOPMENTS

The CIEDE2000 formula stands as the last in a long series of developments improving the CIELAB formula, which is an outcome of the older AN formula. The basis of this development was the value function, which was adapted to the lightness function in the Munsell color order system. This function did not obey the later-defined reference conditions as it resulted from variable background lightness: dark gray for dark colors, medium gray for medium light colors and light gray for light colors. Adams assumed that this function could be applied to the other tristimulus functions X and Z. With these issues, the poor quality of the CIELAB color space as an approximate AN-space for small color differences is not astonishing.

The outcome of a very complex CIEDE2000 formula as an improvement of the CIELAB color difference formula is thought to be the consequence of poor coordinate definition, which has nothing to do with true perceptual coordinates for small color differences.

An ideal color-difference formula should be based on physiological terms of color difference sensation. However, what do we know about such a physiological basis? Are tristimulus values the right start? A question to the physiologist Barry Lee[51] during the AIC 2005 conference in Granada, Spain about a physiological basis of color difference evaluation was answered: for thresholds possibly yes, for scales no. CIEDE2000 is a color difference formula for small color differences and, hence, the physiologist's answer offers little hope.

Is that an end of CIE's interest in further improvement of color difference evaluation? The answer is: no. CIEDE2000 is a formula that adapts a certain component basis of color difference evaluation and thus destroys the vector definition of the prime CIELAB-color difference formula. Another possible route is adaptation of the basic coordinates: transform the CIELAB coordinates immediately to a new color space and define a color difference as a vector distance in the transformed coordinate system. This idea was proposed with the new DIN99 formula.[52] This formula was later optimized with the dataset of the CIEDE2000 formula[53] and turned out to be nearly as effective as the CIEDE2000 formula and better than older color-difference formulas. This result opens discussions about other colorimetric solutions of a color difference formula based on coordinates which define a color space that may be used for color ordering and description. CIE formed two new Technical Committees: CIE TC 1–55 "Uniform Colour Space for Industrial Colour-Difference Evaluation" and CIE TC 1–63 "Validity of the Range of CIEDE2000." The work of both the TCs shall clarify whether further improvement of color-difference evaluation is possible.

REFERENCES

1. MacAdam DL (1942) Visual sensitivities to color differences in daylight. *J. Opt. Soc. Am.*, **32**, 247–274.
2. MacAdam DL (1943) Specification of small chromaticity differences. *J. Opt. Soc. Am.*, **33**, 18–26.
3. Simon FT, Goodwin WJ (1957) *Rapid Graphical Computation of Small Color Differences*, Union Carbide Corp., South Charleston, USA.
4. Brown WRJ, MacAdam DL (1949) Visual sensitivities to combined chromaticity and luminance differences. *J. Opt. Soc. Am.*, **39**, 808–834.
5. Brown WRJ, Howe WG, Jackson JE, Morris RH (1956) Multivariate normality of the color matching process. *J. Opt. Soc. Am.*, **46**, 46–49.
6. Brown WRJ (1957) Color discrimination of twelve observers. *J. Opt. Soc. Am.*, **47**, 137–143.
7. Judd DB (1935) A Maxwell triangle yielding uniform chromaticity scales. *J. Opt. Soc. Am.*, **25**, 24.
8. Wyszecki G (1963) Proposal for a new color-difference formula. *J. Opt. Soc. Am.*, **53**, 1318–1319.
9. Adams EQ (1942) X-Z-planes in the 1931 I.C.I. system of colorimetry. *J. Opt. Soc. Am.*, **32**, 168–173.
10. Nickerson D, Stultz KF (1944) Color tolerance specification. *J. Opt. Soc. Am.*, **34**, 550–570.
11. Davidson HR, Friede E (1953) The size of acceptable color differences. *J. Opt. Soc. Am.*, **43**, 581–589.
12. Thurner K, Walther V (1969) Untersuchungen zur Korrelation von Farbabstandsbewertungen auf visuellem Wege und über Farbdifferenzformeln, in: *Proceedings of the AIC Color 69*, Stockholm, pp. 671–687.
13. Kuehni RG (1971) Acceptability contours and small color difference formulas. *J. Col. App.*, **1**, 30–42.
14. Jaeckel SM (1973) Utility of color-difference formulas for match-acceptability decisions. *Appl. Opt.*, **12**, 1299–1316.
15. Robinson FD (1969) Acceptability of color matches. *J. Oil Chem. Assoc.*, **52**, 15–45.
16. McLaren K (1969) The precision of textile color matches in relation to color difference measurements, in: *Proceedings of the AIC Color 69*, Stockholm, pp. 688–706.
17. Wyszecki G (1974) CIE colorimetry committee—working program of color differences. *J. Opt. Soc. Am.*, **64**, 896–897.
18. McLaren K (1981) The development of improved color difference equations by optimisation against acceptability data, Golden Jubilee of Color in the CIE. *J. Soc. Dyers Col.*, **97**, 156–173.
19. Indow T, Robertson AR, von Grunau M, Fielder GH (1992) Discrimination ellipsoids of aperture and simulated surface colors by matching and paired comparison. *Color Res. Appl.*, **17**, 6–23.
20. CIE (1978) Supplement No. 2 to *CIE Publication* No. **15**-1978.
21. Pauli H (1976) Proposed extension of the CIE recommendation on "Uniform color spaces, color difference equations, and metric color terms". *J. Opt. Soc. Am.*, **66**, 866–867.

22. CIE Technical Report (2004) *Colorimetry*, 3rd ed., CIE Publication **15**:2004, CIE Central Bureau, Vienna.

23. McLaren K, Rigg B (1976) The SDC recommended color difference formula: change to CIELAB. *J. Soc. Dyers Col.*, **92**, 337–338.

24. Kuehni RG (1976) Color-tolerance data and the tentative CIE 1976 $L^*a^*b^*$ formula. *J. Opt. Soc. Am.*, **66**, 497–500.

25. McLaren K (1980) CIELAB hue-angle anomalies at low tristimulus ratios. *Color Res. Appl.*, **5**, 139–143.

26. McDonald R (1980) Industrial pass/fail color matching. Part I—Preparation of visual color matching data. *J. Soc. Dyers Col.*, **96**, 372–376; Part II—Methods of fitting tolerance ellipsoids. *J. Soc Dyers Col.*, **96**, 418–433; Part III—Development of a pass/fail formula for use with instrumental measurement of color difference. *J. Soc Dyers Col.*, **96**, 486–497.

27. Clarke FJJ, McDonald R, Rigg B (1984) Modification to the JPC79 color difference formula. *J. Soc. Dyers Col.*, **100**, 128–132 and 281–282.

28. Robertson AR (1978) CIE guidelines for coordinated research on color difference evaluation. *Color Res. Appl.*, **3**, 149–151.

29. Luo MR, Rigg B (1986) Chromaticity-discrimination ellipses for surface colors. *Color Res. Appl.*, **11**, 25–42.

30. Witt K, Döring G (1983) Parametric variations in a threshold color difference ellipsoid for green painted samples. *Color Res. Appl.*, **8**, 153–163.

31. Cheung M, Rigg B (1986) Color difference ellipsoids for five CIE color centers. *Color Res. Appl.*, **11**, 185–195.

32. Luo MR, Rigg B (1987) BFD(l:c) color difference formula, Part I—Development of the formula. *J. Soc. Dyers Col.*, **103**, 86–94; Part II—Performance of the formula. *J. Soc. Dyers Col.*, **103**, 126–132.

33. Witt K (1990) Parametric effects on surface color difference evaluation at threshold. *Color Res. Appl.*, **15**, 189–199.

34. Berns RS, Alman DH, Reniff L, Snyder GD, Balonon-Rosen MR (1991) Visual determination of suprathreshold color-difference tolerances using probit analysis. *Color Res. Appl.*, **16**, 297–316.

35. CIE Technical Report (1993) *Parametric effects in color difference evaluation*, CIE Publication **101**:1993 CIE Central Bureau, Vienna.

36. CIE Technical Report (1995) *Industrial color difference evaluation*, CIE Publication **116**:1995 CIE Central Bureau, Vienna.

37. Melgosa M, Hita E, Pérez MM, El Moraghi A (1995) Sensitivity differences in chroma, hue, and lightness from several classical threshold datasets. *Color Res. Appl.*, **20**, 220–225.

38. Rigg B (1995) Color difference formulae—recent developments. *J. Soc. Dyers Col.*, **111**, 267–271.

39. McDonald R, Smith KJ (1995) CIE94—a new color difference formula. *J. Soc. Dyers Col.*, **111**, 376–379.

40. Oglesby S (1995) The effectiveness of CIE94 compared with the CMC equation *J. Soc. Dyers Col.*, **111**, 380–381.

41. Berns RS (1995) To optimize or not to optimize (l:c) ratios. *AIC Interim Meeting Colorimetry*, Abstracts, 11.

42. Heggie D, Wardman RH, Luo MR (1996) A comparison of the color differences computed using the CIE94, CMC(l:c) and BFD (l:c) formulae. *J. Soc. Dyers Col.*, **112**, 264–269.

43. Berns RS, Qiao Y, Reniff LA (1997) Visual determination of hue suprathreshold tolerances. *AIC COLOR 97*, Kyoto, 223–226.

44. Witt K (1987) Three-dimensional threshold of color-difference perceptibility in painted samples: variability of observers in four CIE color regions. *Color Res. Appl.*, **12**, 128–134.

45. Strocka D, Brockes A, Paffhausen W (1983) Influence of experimental parameters on the evaluation of color-difference ellipsoids. *Color Res. Appl.*, **8**, 169–175.

46. Alman DH, Berns RS, Snyder GD, Larsen WA (1989) Performance testing of color-difference metrics using a color tolerance dataset. *Color Res. Appl.*, **14**, 139–151.

47. Kim DH, Nobbs JH (1997) New weighting functions for the weighted CIELAB color difference formula, in: *Proceedings of the AIC Color 97*, Kyoto, 1, pp. 446–449.

48. Witt K (1999) Geometric relations between scales of small color differences. *Color Res. Appl.*, **24**, 78–92.

49. Luo MR, Cui G, Rigg B (2001) The development of the CIE 2000 color-difference formula: CIEDE2000. *Color Res. Appl.*, **26**, 340–350.

50. Chou W, Lin H, Luo MR, Westland S, Rigg B, Nobbs J (2001) The performance of lightness difference formulae. *Col. Tech.*, **117**, 19–29.

51. Lee BB (2005) Pathways to color in the retina, in: Proceedings of 10th Congress of the International Color Association, Granada (Granada), pp. 163–168.

52. DIN, 2000, DIN 6176, Colorimetric determination of color differences of surface colors using the DIN99 formula.

53. Cui G, Luo MR, Rigg B, Roesler G, Witt K (2002) Uniform color spaces based on the DIN99 color-difference formula. *Color Res. Appl.*, **27**, 282–290.

5

SPECTRAL COLOR MEASUREMENT

YOSHI OHNO

National Institute of Standards and Technology, 100 Bureau Drive, Stop 8442, Gaithersburg, MD 20899-8442, USA

INTRODUCTION

Measurement of color is important for manufacturers and users of many products such as lamps for general lighting, light emitting diodes (LEDs), displays, traffic signals, signs, printing, paint, plastics, fabrics, and so on. The term *color* is used with different meanings in different applications. Lamp engineers use color as a property of light sources. Graphic arts engineers use color as a property of an object surface. To be exact, color is a perception, and color measurement is the measurement of color stimulus (see Chapter 3). In this chapter, the term color is used as the short form for color stimulus, covering both light source color and object surface color. In either case, color (stimulus) must be physically measured in order to ensure that the products meet the specification.

Physical measurement of color is based on the CIE colorimetry system as described in Chapters 3 and 4. Color of light stimulus is determined by the spectrum of light, which determines the tristimulus values of the light. The tristimulus values of light can be physically measured in two ways; one with a tristimulus colorimeter, and the other with a spectrometer with spectral computation using the color-matching functions. Tristimulus colorimeters are fast, handy, and inexpensive, and are suitable for production control and measuring color differences. However, spectral mismatch errors are inevitable with tristimulus colorimeters, and they are generally not suitable for high-accuracy absolute color measurement of various light sources of dissimilar spectral distributions or various different

Colorimetry: Understanding the CIE System, Edited by János Schanda

object colors. Spectrometers, on the contrary, theoretically do not have this problem, and generally provide a more accurate way of measuring various different colors though they are generally more expensive. Spectroradiometers also provide more information than tristimulus colorimeters, such as Color Rendering Index (CRI) of light sources. This chapter describes the color measurement using spectrometers for light sources and object surfaces. Tristimulus colorimetry is described and discussed in Chapter 6.

Measurements using spectrometers are subject to various sources of error such as wavelength scale shifts, stray light, bandwidth, scanning interval, detector nonlinearity, and imperfection of input optics. The uncertainties of color measurements depend not only on the type of instrument used but also on how it is set up and calibrated and how measurements are performed for test artifact. This chapter describes how to achieve accurate color measurements with spectrometers, what sources of error are critical, and how they can be corrected or minimized. A particular focus is on the effect of bandpass and scanning intervals and how they can be set up for acceptable uncertainty in measured color while allowing for efficient spectral measurement.

The discussion on spectrometers in this chapter is limited to those in the visible spectral region and for the purpose of color measurement. For general treatment of spectroradiometers and spectrophotometers, see Refs. 1–4.

GENERAL PRACTICE IN SPECTRAL COLOR MEASUREMENTS

Type of Instruments

Spectrometers are the instruments to measure spectral quantities of optical radiation or materials, such as spectral irradiance, spectral radiance, total spectral radiant flux, spectral reflectance, spectral reflectance factor (radiance factor), bidirectional reflectance distribution function (BRDF), and spectral transmittance. Instruments to measure spectra of light sources are called spectroradiometers, and instruments to measure spectral reflectance and transmittance of objects are called spectrophotometers (or spectroreflectometers for reflectance measurement). Spectrometers are also used to measure spectral responsivity of detectors, though it is not a subject of this chapter.

There are two types of spectrometers, the mechanically scanned type and the detector-array type. The mechanically scanned type of spectrometer selects the wavelength by mechanically rotating the dispersive element (such as a grating) and takes spectral readings sequentially. This is a traditional type of spectrometer and the measurements are slow. However, an important advantage is that two monochromators can be connected in series to reduce the stray light significantly. This type of spectrometer is called a double-monochromator type, and is considered the most accurate type of color-measuring instrument. Spectrometers using a single dispersive element (including detector-array types) are categorized as a single-monochromator type. Detector-array type spectrometers use a photodiode array or CCD detector array as a detector and take all the spectral readings simultaneously. This type of spectrometer, often called a spectrograph or multichannel spectrometer, is gaining popularity for its high-speed measurement. A disadvantage of detector-array type spectrometers is that they can be constructed only as a single-monochromator type, thus they are subject to much larger stray-light levels compared to the mechanically scanned double-monochromator type. Nonlinearity of the detector array is another notable source of error. These errors, however, can be corrected to some extent, and it is possible to establish high-accuracy color measurements using array-type spectrometers, for example, utilizing the stray-light correction techniques recently developed (see Section "Methods for Corrections of Error").

Use of Spectroradiometers for Light Source Color Measurement

Spectroradiometers measure spectral quantities of light sources, such as spectral irradiance, spectral radiance, spectral radiant intensity, and total spectral radiant flux. Depending on which quantity is to be measured, spectroradiometers are configured with appropriate input optics to introduce light into the entrance slit of a monochromator to measure the desired spectral quantity. Spectral irradiance and spectral radiance are measured in a given direction from the source, and such configurations of a spectroradiometer are referred to as irradiance mode and radiance mode, respectively. Total spectral radiant flux is measured as a spatial integral of radiation in all directions from the source, and such a configuration is referred to as the total flux mode. The color of a light source can be measured either in one direction (irradiance or radiance mode) or as an average of all directions (total flux mode). There are some differences in measured chromaticity between the two conditions because the color is not uniform in all directions for most of the light sources, especially the discharge lamps. The color of lamps for general lighting is often measured as the spatial average because it corresponds to the same geometry as the total luminous flux, which is the most important quantity for lamp products and both can be measured in an integrating sphere. Colors of LEDs are normally measured in one direction, but white LEDs for general illumination, often not spatially uniform in color, may be measured in total flux mode. Displays are normally measured in radiance mode.

Some spectroradiometers are sold with no input optics—for example, only a fiber bundle connected to the entrance slit. The end of such a fiber bundle has similar characteristics as the entrance slit of the spectroradiometer, and it requires appropriate input optics for accurate color measurement of sources. The light source under test should never directly illuminate the entrance slit or the fiber-optic input of a spectroradiometer because the spectral throughput of the spectrometer at the entrance slit (or at fiber-optic input) is nonuniform (spatially and angularly) and can be strongly polarized. Such direct illumination would cause serious errors when a test lamp is different from the standard lamp in physical size, shape, and polarization state. It is unlikely that the standard lamp and test lamp are always the same type with the same characteristics. It is essential to

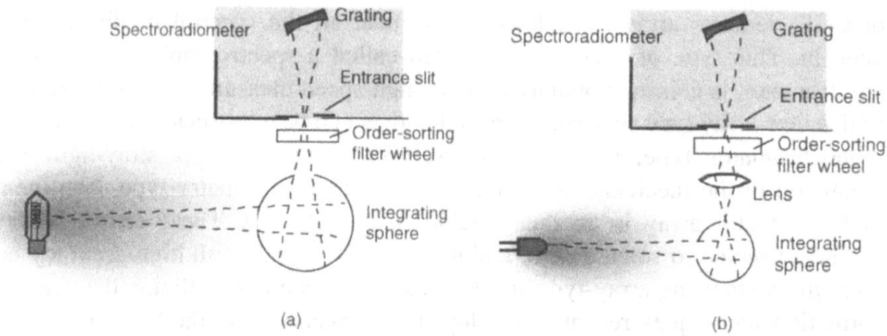

FIGURE 5.1 Typical input optics for an irradiance-mode spectroradiometer: (a) Direct coupling; (b) using a relay lens.

set up input optics such that the polarization sensitivity is removed and that the dispersive element is illuminated exactly the same way by the standard lamp and the test source. It is also required that the entrance slit should be overfilled and illuminated uniformly. Nonuniform illumination on the entrance slit will lead to a distorted bandpass function and thus to irregular shifts in the wavelength scale.

Irradiance Mode

Spectroradiometers configured in irradiance mode measure spectral irradiance or spectral radiant intensity of a light source, and are often used to measure lamps and LEDs. Figure 5.1 shows the typical input optics of a mechanically scanned spectroradiometer. The use of a small integrating sphere is the most effective way of removing polarization sensitivity and realizing spatially uniform responsivity. In Figure 5.1(a), the size of the exit port of the sphere and the distance to the entrance slit of the spectroradiometer determine the input solid angle of the monochromator, which should ideally match the solid angle formed by the grating and the entrance slit. If not matched, the grating should be underfilled to avoid stray light. If radiation hits the inner wall of the monochromator, it will create stray light. It is important that this solid angle is constant and the radiation within the solid angle is uniform no matter what type of source is measured. An integrating sphere works well to achieve this requirement. If the exit port of the integrating sphere is small and cannot be placed close to the entrance slit of the monochromator, the input solid angle would be very small resulting in low sensitivity of the spectroradiometer. In such a case, a relay lens (or concave mirrors) can be used to increase the input solid angle and thus increase the sensitivity, as shown in Figure 5.1(b). An order-sorting filter that is necessary to cut off the second- and higher-order diffraction from the grating (e.g., at wavelength setting of 700 nm and longer, a filter to cut off 350 nm and shorter wavelengths) is to be used.

A disadvantage in the use of an integrating sphere is that there is a significant loss of input light and the sensitivity of the spectroradiometer is reduced. When higher throughput is required, a reflecting diffuser such as a pressed or sintered polytetrafluoro-ethylene (PTFE) plaque or a transmitting diffuser such as opal glass

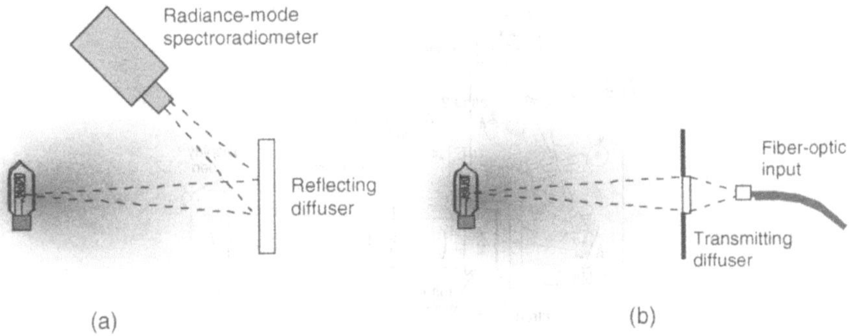

FIGURE 5.2 Other examples of input optics for irradiance mode: (a) Use of a reflecting diffuses; (b) use of a transmitting diffuses.

can be used though they are not as effective as integrating spheres in achieving spatially and angularly uniform responsivity and removing polarization sensitivity. Figure 5.2 shows such examples. Figure 5.2(a) is an example of a radiance-mode spectroradiometer converted into irradiance mode operation. Note that the reflecting diffuser should be aligned perpendicularly to the optical axis of the lamp; otherwise the radiance on the diffuser would be nonuniform depending on the distance to the source. Figure 5.2(b) is an example of a fiber-optic input of an array spectroradiometer. Note that it is generally more difficult to achieve good spatial uniformity of responsivity by using a transmitting diffuser. A diffuser exhibiting poor spatial uniformity would work only if it is illuminated uniformly by the source being measured.

Irradiance-mode spectroradiometers are calibrated against a spectral irradiance standard lamp. Such standard lamps are available from commercial calibration and testing laboratories as well as from national laboratories. For color measurement where only relative spectral distribution is measured, the standard lamp can be set at a distance where it produces similar irradiance levels as the test light source to be measured to minimize possible nonlinearity errors.

Radiance Mode

Spectroradiometers in radiance mode measure spectral radiance of a surface emitting or reflecting light, and is widely used to measure displays. Many commercial spectroradiometers for display measurements are the detector-array type. The input optic is formed using imaging optics. An example of such input optics used in the reference spectroradiometer for display measurements[5] at NIST is shown in Figure 5.3. For display measurement, it is essential to incorporate a depolarizer (such as a double-wedged crystalline quartz scrambler[6*]) to remove the polarization because liquid crystal displays (LCDs) are strongly polarized. Radiance-mode

*Specific firms and trade names are identified in this paper to specify the experimental procedure adequately. Such identification does not imply recommendation or endorsement by the National Institute of Standards and Technology or by the CIE.

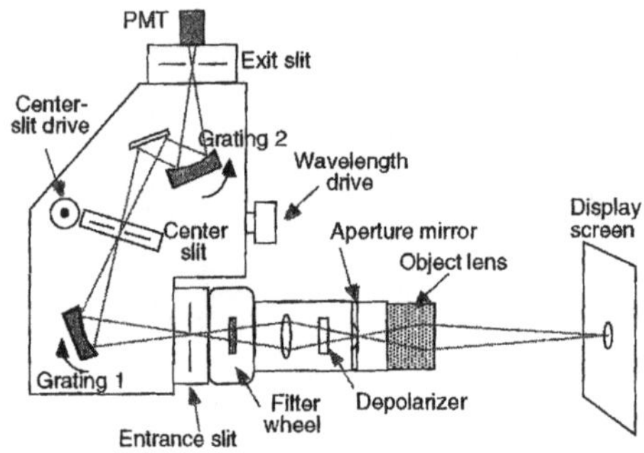

FIGURE 5.3 An example of a radiance-mode double-grating spectroradiometer.

spectroradiometers are often calibrated against standard integrating sphere sources though they are not stable over a long period of time. A more stable spectral radiance transfer standard is a tungsten ribbon filament lamp. Spectral radiance calibration and standards are available from calibration and testing laboratories as well as national laboratories (e.g., see Ref. 7).

Total Flux Mode

Spectroradiometers in total flux mode measure total spectral radiant flux (unit: W/nm) of a light source. The chromaticity of discharge lamps, which tend to have nonuniform color distribution, should be measured in this mode. Such measurement is typically made with a large integrating sphere using a spectroradiometer as the detector, as illustrated in Figure 5.4. The spectroradiometer together with the integrating sphere is calibrated against a total spectral radiant flux standard lamp. In other words, test lamps are measured spectrally in substitution with the standard lamp. Total spectral radiant flux standards are not widely available but are provided from some national laboratories.[8,9] For measurement of LEDs, smaller integrating spheres are used, and thus standard lamps of a smaller size and power level are required.

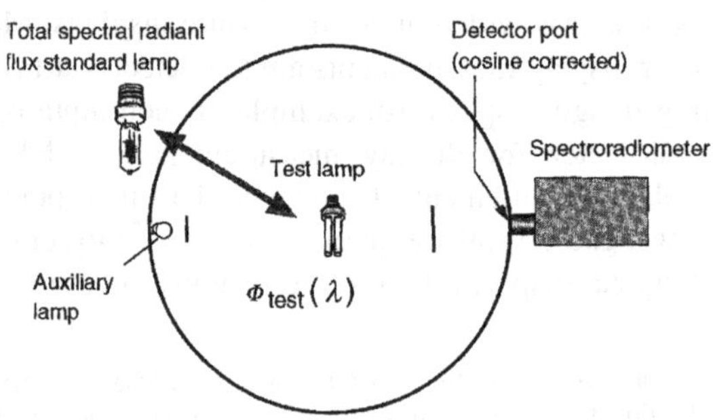

FIGURE 5.4 An integrating sphere system for total spectral radiant flux measurement.

Colorimetric Calculation

The tristimulus values are first calculated from the measured spectral distribution data and the CIE color-matching functions (those for 2° field-of-view are normally used for light source color). The calculation should be done with the spectral interval at which the measurement data is provided. If the wavelengths are not integer numbers as provided in the tables of the CIE publications,[10] the color-matching functions should be interpolated between 1 nm intervals to prepare tables that match the measured data. Interpolation of spectral data for light sources is not recommended because it can increase colorimetric errors if the spectral distribution has sharp features. Once the tristimulus values are obtained, the chromaticity coordinates are calculated. Both the (x, y) and (u', v') chromaticity coordinates are the current CIE recommendations and are widely used for light sources. When chromaticity differences of colored lights are evaluated, the (u', v') chromaticity coordinates should be used because the color differences are significantly nonuniform in the (x, y) diagram.

Use of Spectrophotometers for Object Color Measurements

Spectrophotometers measure spectral quantities of objects, such as spectral reflectance (specular, diffuse), spectral reflectance factor, spectral transmittance (regular, diffuse), spectral transmittance factor, and BRDF. One instrument may be configured to measure several of these quantities under different illumination/viewing conditions. In this section, spectrophotometers for object color measurement are briefly discussed. For general treatment of spectrophotometers in greater details, see Refs. 3,4.

The principal components of a spectrophotometer are the radiation source, the dispersing system, the sample compartment, the collecting optics and detector, and the data presentation system. The source is usually a tungsten halogen lamp or a discharge lamp such as xenon lamp. The dispersive element in the monochromator is typically a diffraction grating. The monochromator may be between the source and the sample (monochromatic illumination) or between the sample and the detector (polychromatic illumination). The first method has the advantage of reducing the radiant heat incident on the sample. The second method must be used, however, if the sample is fluorescent or if the detector is a photodiode array. The detector will usually be a photomultiplier, a silicon cell, or a photodiode array. The sample compartment and collecting optics vary considerably with the type of measurement to be made. For regular transmittance measurements, the sample compartment is usually a simple box-shaped enclosure between the source and the detector optics, with a lens or concave mirror as the collecting optics. Some instruments have regular reflectance accessories that can be mounted inside the sample compartment. If the sample is diffusing, then an integrating sphere or a 0°:45° radiance factor collector is used. Geometric tolerances for colorimetric measurements using integrating sphere or 0°:45° collectors have been defined in a CIE recommendation[11,12] and have also been described in Chapter 3.

The instrument may incorporate a double monochromator to reduce stray radiation. Instruments may also be of the single- or double-beam type. In double-beam instruments, the beam alternates between a path that includes an incidence on the sample and one that bypasses the sample. The instrument records a ratio of the signal from the two paths, and in this way the problem of source or detector drift is overcome. Some portable spectroreflectometers use an array-type spectrometer and illuminate the reference sample and test sample alternately with a xenon flash in a given geometry.

Except for reference instruments used by national laboratories and some commercial BRDF instruments that can do absolute measurements, typical commercial spectrophotometers (for reflectance factor measurement) measure the sample in substitution with calibrated reference white reflectance standards. Such reference standards and calibration services are available from the manufacturers and national laboratories. There are many sources of error, and the detailed descriptions of these depend on the type of instrument. The critical parameters for color measurement, such as bandwidth, scanning interval, wavelength scale offset, and stray light, are discussed in the later sections.

For object color measurement, samples often exhibit fluorescence, and specific methods should be used to measure colors accurately with the effect of fluorescence. Such details are beyond the scope of this chapter. Refer to other appropriate references[13–15] and the coming CIE technical report.[16]

Geometries for Reflectance Color Measurement

The spectral quantities of object surfaces vary depending on irradiation and viewing geometry. For object color measurement, several standardized geometries are defined.[11,12] Standard geometries fall into two large categories: bidirectional and directional-hemispherical. Some examples of the standard geometries are illustrated in Figure 5.5. The standard geometries for the bidirectional geometry are the (0°:45°) geometry (illumination at 0° and viewing at 45°) and its reversal (45°:0°). For the latter, the irradiation can be from one direction (45°x:0°) or by annular illumination (45°a:0). The annular irradiation is used for surfaces having directionally nonuniform texture. The standard geometries for the directional-hemispherical geometry are (d_i:8°) and (d_e:8°), their reversal, (8°:d_i) and (8°:d_e), and (d:0°). The subscript "i" and "e" represent specular included and specular excluded, respectively. The half-cone angle for the irradiation and for

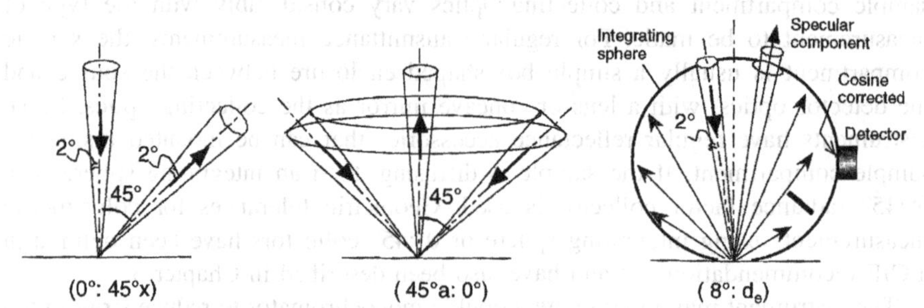

FIGURE 5.5 Some of the standard geometries for reflectance color measurement.

the viewing in these standardized geometries is recommended to be 2° in the new CIE technical report.[12] The half-cone angle for the specular exclusion is 4°. The tolerances for these angles are also given in the technical report (see also Chapter 3).

Color Calculation

The tristimulus values of a specimen are first calculated from the measured spectral reflectance factors, spectral power distribution of the illuminant, and the CIE color-matching functions (2° or 10° field-of-view appropriate for the application). The tristimulus values are computed with the spectral interval at which the measurement data is provided. If the wavelengths do not fall on exact 1 nm steps provided in the tables of CIE publications, the color-matching functions should be interpolated to match the measured data interval. Interpolation of measured spectral data may also work for spectral reflectance data. See also a recent CIE publication[17] on related issues. If the interval is 10 nm or 20 nm, the tables provided in ASTM E308[18] may be used to obtain tristimulus values with correction of bandpass and sampling interval. For details, see Section "Methods for corrections of error." Once tristimulus values are determined, the three-dimensional color coordinates (such as L^*, a^*, b^* in the CIELAB object color space[11]) are calculated. See Chapter 3 also for general recommendations on colorimetric calculation.

CRITICAL PARAMETERS OF SPECTROMETERS FOR COLOR MEASUREMENT

Many different types of spectroradiometers and spectrophotometers are used, having different specifications of bandwidth, scanning interval, wavelength range, wavelength accuracy, stray light, and other parameters. The accuracy of color measurement depends on how accurately tristimulus values are obtained, and is affected by all these parameters. This section discusses important aspects of these parameters for color measurement of light sources and object surface.

Sampling Interval and Bandpass of Instruments

Among these parameters, the spectral data interval (and thus the interval for spectral summation) has been an issue in CIE TC1-48, TC1-38, and some ISO committees. Although the CIE's current position is not to recommend intervals greater than 5 nm for colorimetry,[11] there have been debates to propose color calculation at 1 nm intervals, or to endorse ASTM E308[18] that is applied to spectral data at 10 nm and 20 nm intervals. Depending on the applications, the bandwidth and the scanning interval need to be matched for continuously scanned spectra, for example, for measurement of discharge lamps having emission lines, or to use bandpass error correction techniques given in Refs. 18,19. Thus, the data interval is often tied to an instrument's bandwidth and cannot be discussed separately.

The requirements are different for object color measurement, which deals with smoothly varying spectra, and for light-source color measurement, which deals with narrow-band spectral peaks. Therefore, for practical colorimetry, evaluation needs to be made not for the data interval alone but for combined effects of bandwidth and scanning interval. Different recommendations are needed for object color measurement and light-source color measurement.

In this section, analyses of colorimetric errors associated with an instrument's bandpass and scanning interval for measurement of light sources as well as object color are presented and discussed.

Sampling Interval for Object Color Measurement

The spectral reflectance factors of object surfaces are smoothly varying functions; thus the goal of measurement is to obtain the spectral reflectance curve as accurately as possible. For this purpose, it is ideal to use infinitely small bandwidth, and with intervals as small as possible. The question is how small the sampling interval should be. It is not important to match the bandwidth and the scanning interval for object color measurement unless the bandpass correction methods[18,19] are applied. Below, the effect of data interval is first discussed assuming infinitely small bandwidth, and then the effect of bandwidth is discussed.

Figure 5.6 shows the colorimetric errors caused by larger sampling intervals calculated for measurement of the 14 Munsell samples used in the CRI calculation.[20] These are simple calculations with the reflectance spectra of these color samples, originally prepared at 1 nm intervals, then abridged to 5 nm, 10 nm, and 20 nm, and the colors are calculated at these intervals using the abridged tables of D65 and Illuminant A. Therefore, this simulation is assuming negligible bandwidth, and evaluates purely the effect of sampling. The results are shown for errors in CIELAB unit, ΔE^*_{ab}. The samples, TCS9–TCS12, are strongly saturated colors, with TCS12 being strong blue. The same calculations were made for the BCRA tiles[21] and ColorChecker samples,[22] with the results similar and less than the maxima shown in these figures. The level of errors at 5 nm interval (<0.03 in ΔE^*_{ab}) is shown to be practically negligible, which demonstrates that 5 nm data interval is sufficient for object color measurement (though measurements at smaller intervals would have some benefits in reducing colorimetric errors for random noise in signal). If original data are given at 5 nm interval, interpolation to 1 nm interval is not necessary, and it would not reduce the measurement uncertainty. The results at a 10 nm interval (<0.15 in ΔE^*_{ab}) are still very small. The level of error at 20 nm interval ($\approx 1\Delta E^*_{ab}$) is not acceptable in many applications. It is also noted that there are large differences between D65 and Illuminant A results at 5 nm and 20 nm intervals. This may be due to the fact that the structured spectral distribution of D65 was linearly interpolated from original 10 nm interval data. The errors due to the abridgement at 10 nm and 20 nm intervals can be reduced by using Tables 5 of the ASTM E308.[18]

The analysis reported above assumed zero bandwidth of each spectral radiance factor value of the samples and analyzed only the effect of sampling interval. In this respect, this calculation can also be interpreted as a simulation of spectral

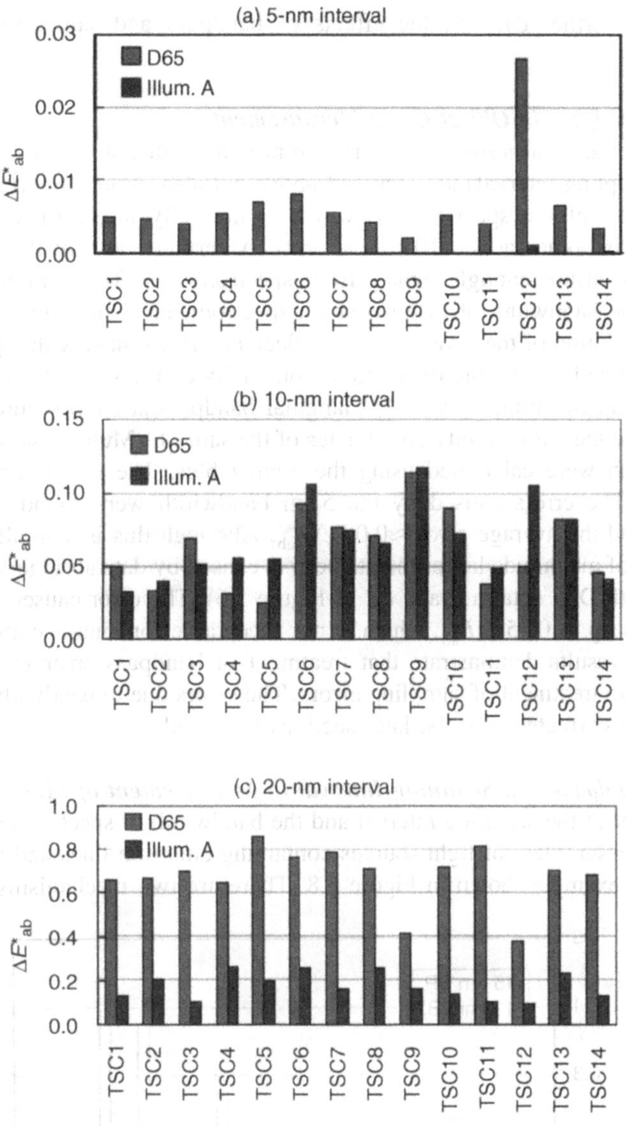

FIGURE 5.6 Errors in color difference ΔE^*_{ab} caused by the abridgement of data at intervals of 5 nm, 10 nm, and 20 nm, for the 14 color samples used in CIE 13.3.

measurements using a spectrometer with a negligible bandwidth (e.g., 1 nm) at the given intervals. Measurements at 5 nm intervals provide accurate results though the bandwidth and the scanning interval are not matched. This confirms that the matching of bandwidth and scanning interval is not necessary in object color measurement unless the bandpass error correction methods given in Refs. 18,19 are to be applied. Further data on the effects of bandpass and sampling interval is available.[23]

Effect of Bandpass in Object Color Measurement

When a spectral measurement is made, errors will occur not only due to the data interval (sampling interval) as discussed above but also, more importantly, due to the bandwidth of the spectrometer, which is normally larger than 1 nm, often ≈5 nm, and in some cases ≈10 nm or larger. A certain width of the bandpass is necessary to ensure enough signals from spectrometers. To examine the errors caused by the bandwidth of instruments, a spectrometer is simulated by carrying out the convolution of the given spectral reflectance factor data with a given bandpass function and calculating measured chromaticity coordinates. A 5 nm or 10 nm (full-width half-maximum: FWHM) triangular bandpass at 5 nm scanning interval was used, and the chromaticity coordinates of the same 14 Munsell samples illuminated by D65 were calculated using the 5 nm tables. The results are shown in Figure 5.7. The errors caused by the 5 nm bandwidth were found to be up to 0.1 ΔE^*_{ab}, and the average error ≈0.05 ΔE^*_{ab}. Although this is a small number, it is one order of magnitude larger than the errors caused by data abridgement at 5 nm (compare with D65 data in graph (a) of Figure 5.6). The error caused by a 10 nm bandwidth is up to 0.45 ΔE^*_{ab}, which is not acceptable for high-accuracy applications. These results demonstrate that treatment of bandpass error is much more important than treatment of sampling errors. The errors due to bandwidth of instruments can be corrected. See the later sections for details.

Effect of Bandpass and Scanning Interval in Measurement of Light Sources

The treatment of the scanning interval and the bandwidth of spectrometers is more critical in measurement of light sources containing emission lines and narrowband peaks, as an example shown in Figure 5.8. There are two mechanisms that cause

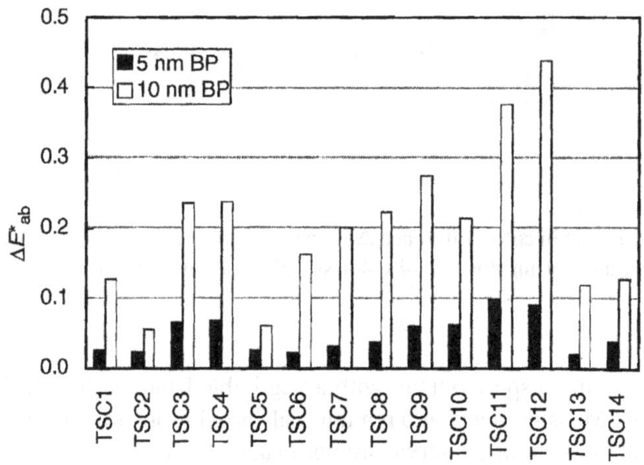

FIGURE 5.7 Errors in ΔE^*_{ab} due to a 5 nm and 10 nm bandwidth of a spectrometer measuring the 14 samples in CIE 13.3.

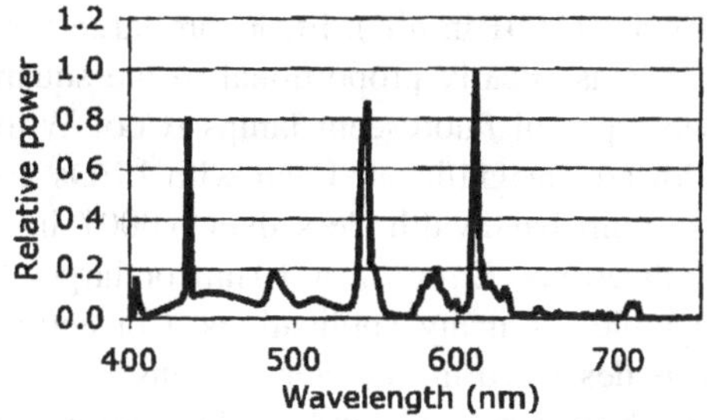

FIGURE 5.8 An example of spectral power distribution with sharp peaks (triphosphor fluorescent lamp).

error due to bandpass. One is broadening of measured spectra, which occurs regardless of scanning interval. The other is due to the mismatch between bandwidth and scanning interval, which causes error in the measurement of emission lines and narrowband peaks.

First, the effect of broadening of spectra on color measurement is analyzed. A simulation was performed for a spectroradiometer with a triangular bandpass of 5 nm and 10 nm (FWHM), measuring several different light sources at 5 nm and 10 nm intervals (under the condition that the bandwidth and scanning interval are perfectly matched). Figure 5.9 shows the results for errors in chromaticity (u', v'). $\Delta u'v'$ represents the distance in chromaticity coordinates (u', v') as

$$\Delta u'v' = \sqrt{(u'_{\mathrm{m}} - u'_0)^2 + (v'_{\mathrm{m}} - v'_0)^2} \tag{5.1}$$

where (u'_0, v'_0) is the original chromaticity and $(u'_{\mathrm{m}}, v'_{\mathrm{m}})$ is the measured chromaticity. The errors for a Planckian source (very smooth spectral distribution) are shown

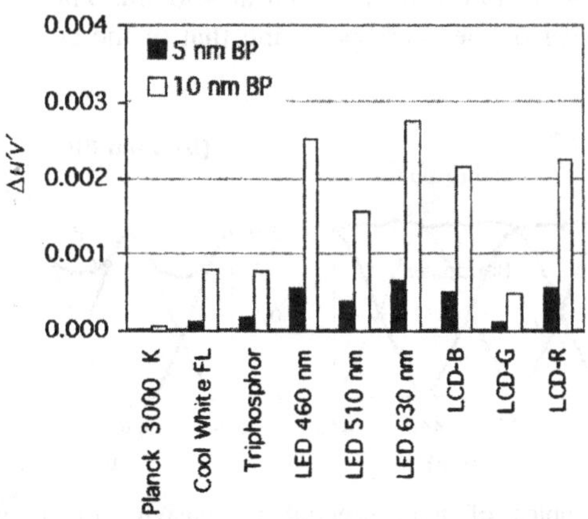

FIGURE 5.9 Errors in (u', v') due to a 5 nm and 10 nm triangular bandpass of a spectroradiometer at 5 nm scanning intervals.

to be negligible (less than 0.0001 in $u'v'$). Errors are large for sources having narrow-band peaks and increase nearly proportional to the square of the bandwidth. The errors for the two types of fluorescent lamps (Cool White and Triphosphor) with a 5 nm bandwidth are insignificant. Even with LEDs and LCD spectra, the level of errors with a 5 nm bandwidth (less than 0.0007 in $\Delta u'v'$) is acceptable for most practical applications. Errors at a 10 nm or larger bandwidth for these sources are not acceptable for many applications. Correction for such bandpass errors is possible and is described in the later sections.

For measurement of discharge lamps and displays having emission lines and/or very narrowband peaks, the whole spectrum needs to be scanned evenly without gaps. Emission lines falling anywhere between the peaks of the bandpass function at each measurement point should be properly weighted at neighboring scanning points to produce a flat response as a total of the signals. This can be ideally achieved by a triangular bandpass function that is matched with (an integer multiple of) the scanning interval. Real instruments, however, do not have a perfect triangular bandpass, nor is the bandwidth perfectly constant over the visible region, in which case a significant error in measured intensity of narrow-band peaks can occur. To examine such effects, the total spectral responsivity of a spectrometer—a sum of the normalized bandpass functions at all sampling points—can be checked. An example of such data of a real spectroradiometer (a double-grating spectroradiometer[24]) is shown in Figure 5.10. Graph (a) is a case of nearly perfect match (5 nm bandwidth and 5 nm interval), and graph (b) is a case of ≈20% mismatch (≈4 nm bandwidth with 5 nm interval). In case (a), the total spectral responsivity is kept to within ±2%, and in case (b), it deviates up to ±15% from the average value. The error in measured color of a source depends on where emission lines fall on the varying total spectral responsivity curve. A simple calculation demonstrates that a 10% error in the measured intensity of the 436 nm mercury line of a Cool White fluorescent lamp (4200 K) would lead to a chromaticity error of 0.005 in distance in (x, y) or 0.0024 in (u', v'). The effect of the 546 nm line is about one fourth of that at 436 nm. The effects depend on the ratio of spectral power of the sharp peak and that of the continuum.

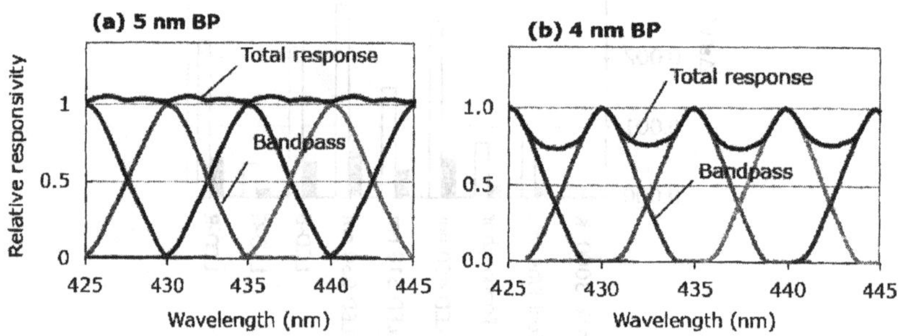

FIGURE 5.10 Examples of total spectral responsivity of a spectroradiometer. (a) Bandwidth and scanning interval are matched (both 5 nm). (b) Bandwidth (≈4 nm) is mismatched with the scanning interval (5 nm).

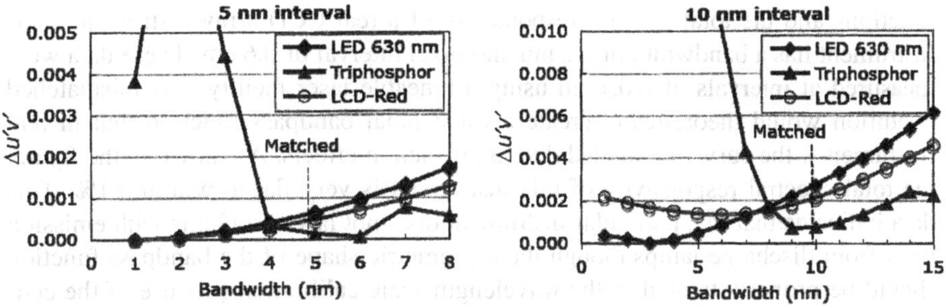

FIGURE 5.11 Colorimetric error of light sources measured with a varied bandwidth at 5 nm interval (left) and 10 nm interval (right).

To analyze further the errors when bandwidth is not matched with scanning interval, a simulation was performed for several light source spectra measured at fixed 5 nm and 10 nm intervals with a varied bandwidth. The results are shown in Figure 5.11. The effect of the bandwidth mismatch is dramatic for triphosphor fluorescent lamp and is not so relevant for the LED and LCD (narrowband phosphor emission). Simulations with other real LED spectra indicated that the bandwidth matching is not important for typical LEDs having spectral half-widths of 20 nm to 30 nm, and rather, use of a smaller bandwidth is important for LEDs in reducing bandpass errors.

The situation with array-type instruments is slightly different. Due to the large number of detector elements (e.g., 1024 pixels for 350 nm to 900 nm), recent array-type instruments typically have fairly small pixel intervals such as 1 nm or less with much larger bandwidth, typically ≈2.5 nm to 10 nm. Therefore, these instruments generally have a condition of extreme oversampling, and the requirement for matching of bandwidth and sampling interval is different. Due to the small pixel size, the bandpass shape of each pixel is typically trapezoidal or bell-shaped and not triangular. Also, there are small physical gaps between the pixels. However, these do not cause problems for color measurement. The extreme oversampling condition and small sampling intervals tend to negate the errors caused by the mismatch and the effect of the gaps. Figure 5.12 shows an example of the bandpass

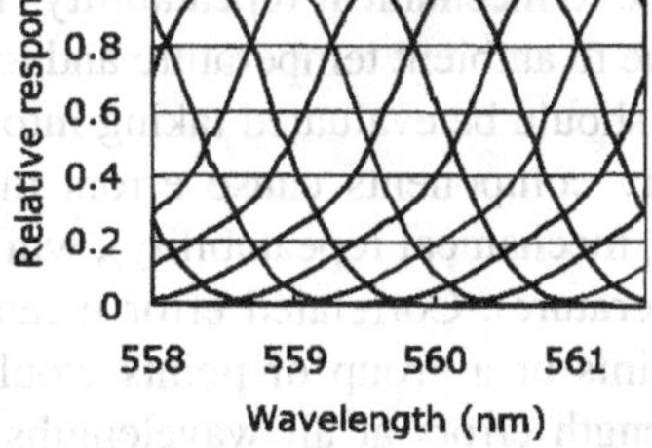

FIGURE 5.12 An example of an oversampling bandpass and total spectral responsivity of an array spectroradiometer.

functions and the total spectral responsivity of a real CCD array instrument. This instrument has a bandwidth of 1.2 nm and pixel interval of 0.6 nm. These data were measured at intervals of $\approx$0.1 nm using a tuneable laser facility.[25] A mismatched condition would theoretically create a trapezoidal bandpass function, but in real instruments, the curve is rounded due to diffraction effects. As shown in the figure, the total spectral responsivity of this instrument is very flat to within $\pm$1%. This data indicates that this particular instrument does not have problems with emission lines from discharge lamps though the asymmetric shape of the bandpass function should be properly treated in the wavelength scale calibration (the use of the centroid wavelength of the bandpass function generally gives good results, or the extended bandpass correction method described in the later sections may also be used to deal with the asymmetric bandpass).

For mechanically scanned spectrometers, larger sampling intervals are used to reduce measurement time. It is sometimes proposed that such data at larger spectral intervals be interpolated to smaller intervals to improve accuracy. This might work for object color measurement (but not necessary for intervals of 5 nm or less), but interpolation does not work for spectral distributions of light sources, particularly, discharge lamps with narrowband peaks. Experimental calculations have shown that if spectral data of a fluorescent lamp at 5 nm intervals is interpolated to 1 nm intervals, the errors often increase. Interpolation would also confuse the uncertainty information of original spectral data. The uncertainty in color quantities calculated from interpolated spectra will not be reduced (in spite of increased data points) because interpolated data points are strongly correlated.[26] It is recommended that color calculation of light sources from their spectral distribution be performed at the original spectral intervals.

Wavelength Scale Error

Another major source of error (uncertainty) in spectral color measurement is the wavelength error of spectrometers. The wavelength scale of a spectroradiometer is normally calibrated with line emissions from discharge lamps and gas lasers whose wavelengths are exactly known. For spectrophotometers, emission line sources and wavelength standard materials are used to check the wavelength scale. The wavelength correction as a function of wavelength is often given with a polynomial function. After the correction, however, there are residual errors, some random variations due to mechanical repeatability (for mechanically scanned types), and some effects due to ambient temperature and long-term drift. The uncertainty of wavelength scale should be evaluated taking into account all these effects.

Some of the uncertainty components cause errors that are uncorrelated with other scanning points (e.g., mechanical repeatability), while others cause correlated errors (e.g., ambient temperature). Correlated error means that the wavelength of the spectrometer at all points or a group of points would shift together in some relationship. If the wavelength errors at all wavelengths are fully correlated, the whole wavelength scale would simply shift in one direction or another. If the errors are uncorrelated, wavelength shift at each scanning point is random. The real cases are complex, and errors are combinations of different factors, and wavelength errors are partially correlated. As it is difficult to evaluate such partial correlations, simpler analyses are presented here, showing the results for a fully correlated case and a totally uncorrelated case. It is useful to evaluate both cases, as either one can be larger than the other, depending on the type of measurement artifacts.

The calculation for fully correlated case is simple. Prepare the spectral distribution data, either the spectral reflectance factors of a sample (multiplied by a reference illuminant) or a spectral distribution of a source, say, at 1 nm intervals. Calculate the original $L^*a^*b^*$ values, chromaticity coordinates, or any other color quantities. Then, shift the spectral data by 1 nm, and then repeat the color calculation. The differences in results show the colorimetric error for 1 nm wavelength shift. The color uncertainty of an instrument with a different wavelength uncertainty is obtained by prorating the results for the stated wavelength uncertainty of the instrument.

The calculation for the uncorrelated case is available[27,28] and discussed in the later section, "Uncertainty analysis" (see also Appendices 1 and 2). This case assumes that the wavelength errors occur randomly at each scanning point. This type of error would affect significantly the spectral distributions with narrow band peaks because a small shift in wavelength at a sharp slope of a spectral distribution curve would result in a large change in the measured spectral power distribution, and thus in color quantities. The uncertainty in this case depends on the scanning interval. For continuous spectral distributions, random uncertainties at smaller scanning intervals tend to cancel out the colorimetric errors. Thus, the color uncertainty for this case is roughly in proportion to the square root of the scanning interval.

Figure 5.13 shows the results of the calculations of both cases for spectral measurement of the 14 color samples in CIE 13.3 at 5 nm intervals, for a wavelength uncertainty of 0.2 nm, which is the level of high-quality commercial instruments. The uncorrelated results were calculated using the numerical method presented in Ref. 28. From these results, wavelength uncertainties of 0.2 nm do not cause significant errors in color differences (mostly less than 0.2 units in ΔE^*_{ab}) at 5 nm interval. For 10 nm interval, the results would be the same for fully correlated case, and roughly $\sqrt{2}$ times larger for the uncorrelated case.

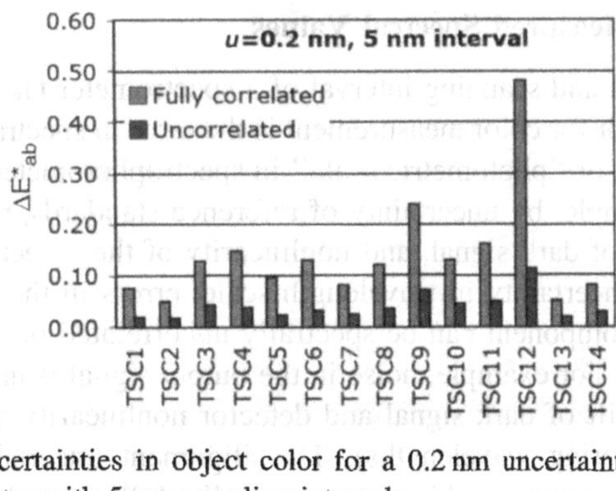

FIGURE 5.13 Uncertainties in object color for a 0.2 nm uncertainty of the wavelength scale of a spectrometer with 5 nm sampling intervals.

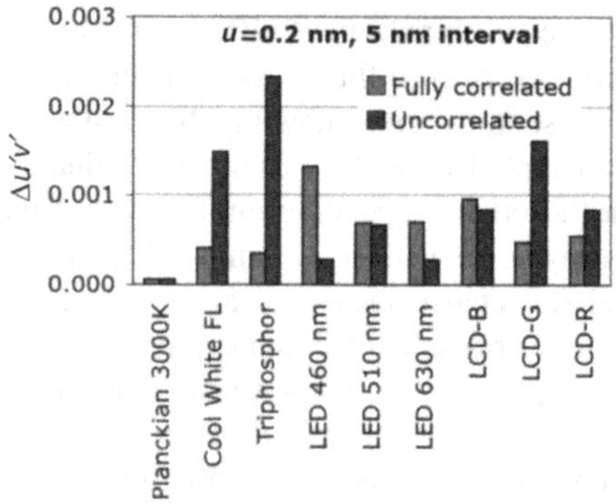

FIGURE 5.14 Uncertainties in light source color due to a 0.2 nm uncertainty of the wavelength scale of a spectrometer with 5 nm sampling intervals.

Figure 5.14 shows the results of the calculations for measurement of several light sources for a wavelength uncertainty of 0.2 nm (throughout the visible region) at a scanning interval of 5 nm, for the fully correlated case and uncorrelated case. Compared to Figure 5.13 for the results of object color where fully correlated uncertainty is always higher, the uncorrelated uncertainties are much higher for some of the sources that contain strong narrowband peaks (e.g., fluorescent lamps). The color uncertainties for a Planckian source, a smooth broadband spectrum, are least affected by wavelength uncertainty. The real cases may be somewhere between the two cases.

For the case of very small intervals as found in array-type instruments, the calculation for uncorrelated case would produce very small uncertainty values, which are likely to be under-estimated because the wavelength errors for neighboring pixels are probably strongly correlated. Determination of such correlations between neighboring pixels is a subject of future study. If such correlations are determined, calculation techniques are available as described in Appendix 2.

Uncertainties in Measured Spectral Values

With the bandwidth and scanning interval of a spectrometer taken care of, another major source of error for color measurement is the error in spectral values measured at each wavelength, or "photometric scale" in spectrophotometers. Such errors can be caused, for example, by uncertainty of reference standards, random noise from spectrometer, drift of dark signal, and nonlinearity of the detector.

Similar to the uncertainty in wavelength scale, errors in the measured spectral values from each component can be spectrally uncorrelated or fully correlated, or partially correlated. For example, noise in the output signal is uncorrelated at each wavelength, and drift of dark signal and detector nonlinearity are partially correlated (with neighboring wavelengths). The alignment uncertainty of a spectral irradiance standard lamp would cause spectrally fully correlated uncertainties,

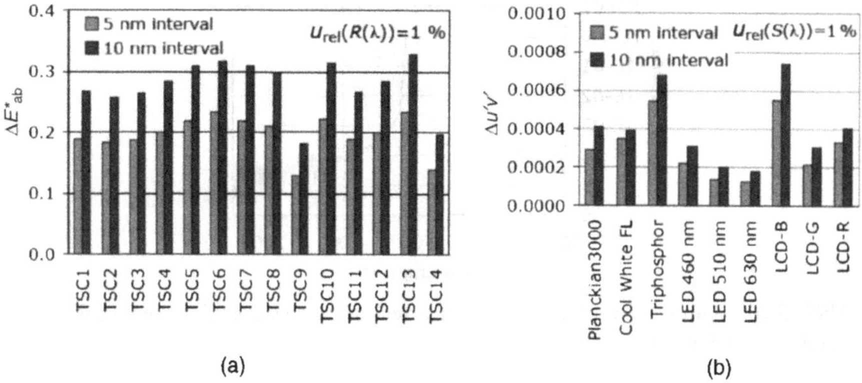

FIGURE 5.15 Uncertainty contributions in (a) $L^*a^*b^*$ for object color samples, and in (b) chromaticity (u', v') for light sources, caused by a 1% relative uncertainty (uncorrelated) in the spectral values at all wavelengths, at 5 nm and 10 nm scanning intervals.

which would lead to the same relative error in spectral values at all wavelengths, and does not contribute to error in color quantities. The uncertainties in reference standards (e.g., spectral irradiance standard lamp or white reference standard plaque) include many uncertainty components, and if possible, it is useful to separate correlated and uncorrelated components.

Spectrally fully correlated components that scale the relative data by a fixed amount at all wavelengths do not contribute to light source color, and thus, can be removed for source color measurement. However, for object color, correlated components in uncertainty in absolute radiance factor contribute to L^* and thus to ΔE^*_{ab}. For spectrally uncorrelated components, color uncertainties can be calculated using the methods described in the section "Uncertainty Analysis." For partially correlated components, if the covariance matrix can be obtained, calculation methods are available (see Appendix 2). In many cases, however, the correlation is complex and it is difficult to determine the covariance matrix. In such cases, color uncertainties may be evaluated by modeling based on the measurement equations.

Examples of color uncertainty calculation for uncorrelated components are presented in Figure 5.15. The figure shows the uncertainty contributions (standard uncertainty) in object color and light source color caused by a 1% relative standard uncertainty (uncorrelated) of spectral values at all wavelengths measured by a spectrometer at 5 nm and 10 nm scanning intervals. Similar to the wavelength uncertainty, the color uncertainties from spectral values depend on scanning interval and are roughly proportional to the square root of the interval.

Stray Light in the Monochromator

Due to imperfections of a monochromator, such as scatter from the grating, mirror, and reflections from detector array, among others, a spectrometer, at a given

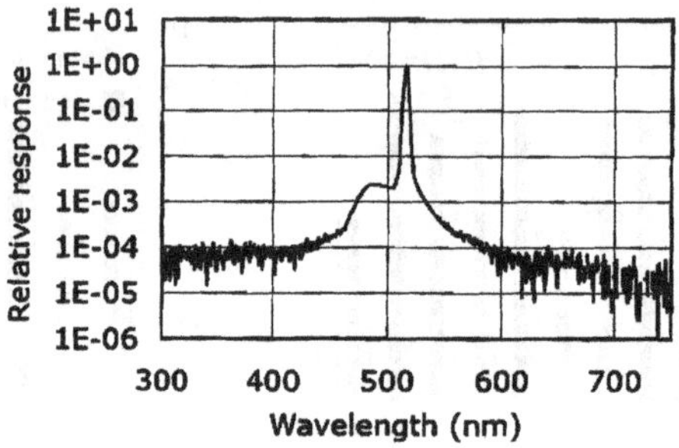

FIGURE 5.16 An example of a response of a real array spectroradiometer measuring a laser emission at 516 nm.

wavelength, responds to the radiation of the wavelengths other than from the bandpass. Such unwanted radiation within a monochromator is called (spectral) stray light. Stray light is prominent in single-grating instruments, in particular, array spectrometers. Stray light can be evaluated by measuring purely monochromatic radiation by the spectrometer if the instrument has a sufficient dynamic range. If the dynamic range is not sufficient, a "bracketing technique" can be used, where radiation lower than one count of A/D conversion can be measured with the high-level peak saturated (a care should be taken for possible "smearing" effects in the pixels neighboring a strong peak). An example of such data from a real array spectroradiometer measuring a laser emission at 516 nm is shown in Figure 5.16. Such a normalized relative response for monochromatic emission is called the line spread function (LSF). The response other than the sharp peak in the center of the figure, at the level of $\approx 10^{-4}$, is due to stray light. Although this number looks very small, stray light accumulates from broadband radiation and can cause significant errors. In this particular figure, a shoulder on the immediate left of the sharp peak is caused by interreflections of the detector-array surfaces.

The effect of stray light in color measurement becomes significant when measuring spectral distributions having high levels in some region and very low levels at other regions. Such an example is shown in Figure 5.17, comparing spectral

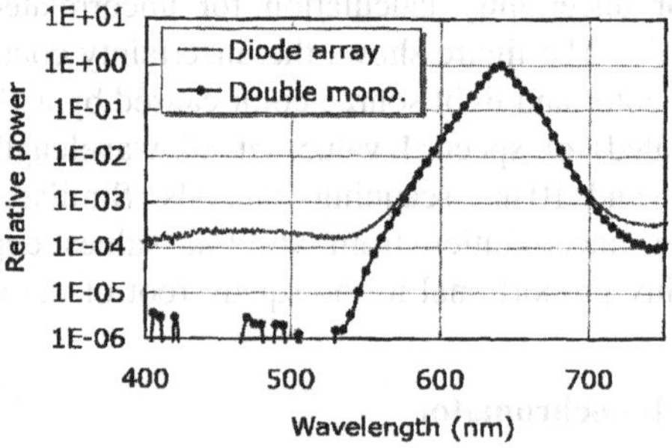

FIGURE 5.17 Spectral distribution of a red LED measured with an array spectroradiometer and a double monochromator.

distributions of a red LED measured with a diode-array instrument and with a double-grating spectroradiometer,[24] plotted in a log scale. The large differences between the two curves at shorter than 550 nm are due to stray light. In this case, the stray light causes a chromaticity error of $\Delta u'v' = 0.0026$. Stray-light error will be much less significant for broadband sources that have a considerable level of emission at all visible wavelengths.

In the case of a spectroradiometer, stray-light error occurs both when the instrument is calibrated against a standard lamp and when it measures a test light source. If the test light source is similar to the standard lamp, the stray-light error tends to be cancelled out. When the standard lamp is an incandescent lamp, the stray-light error is larger at shorter wavelength region where the signal is lower. The stray-light error can be significant when the spectral distributions of standard lamp and test lamp are dissimilar.

To understand how stray light affects the color measurement uncertainties, a spectroradiometer simulation was performed assuming a slit-scattering function (SSF)[2] and the instrument's relative detector responsivity as shown in Figure 5.18. The bandpass is a triangular 5 nm (FWHM) and the base stray-light level of 10^{-4} as shown in the figure. To reduce the stray light for color measurement, it is important that the total system does not have infrared (IR) response because an incandescent standard lamp has strong IR emission, which only causes stray light. Graph (b) is such an example of a system with the IR response cut off. A simulation was performed for the nine light sources and 14 color samples analyzed in the previous sections, assuming a calibration source of Planckian radiation at 2856 K (Illuminant A) and at 6000 K (though such a real source does not exist). For color samples, simulation was performed such that the spectral reflectance factors are measured with polychromatic illumination (2856 K and 6000 K) for the white reflectance standard and a color sample, and the ratios of reflected light measured with the spectroradiometer are obtained as reflectance factor. The color was then calculated with D65 data. The results are shown in Figure 5.19. The results show that errors are significant for saturated color sources (LEDs) and saturated color samples (TSC9–TSC12), while errors are much smaller for white light

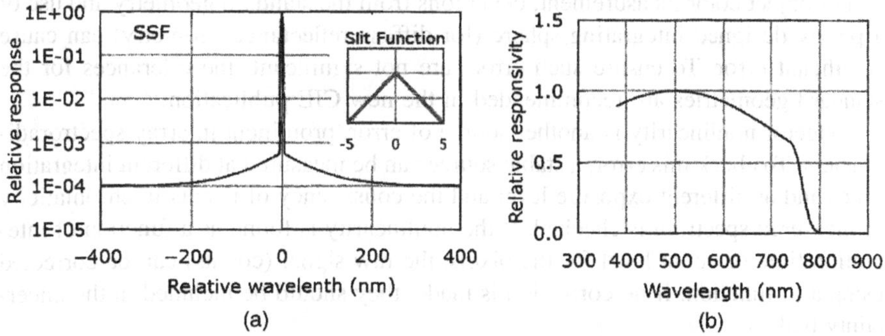

FIGURE 5.18 Conditions of the spectroradiometer simulation for stray light: (a) Slit scattering function and bandpass; (b) Relative spectral responsivity.

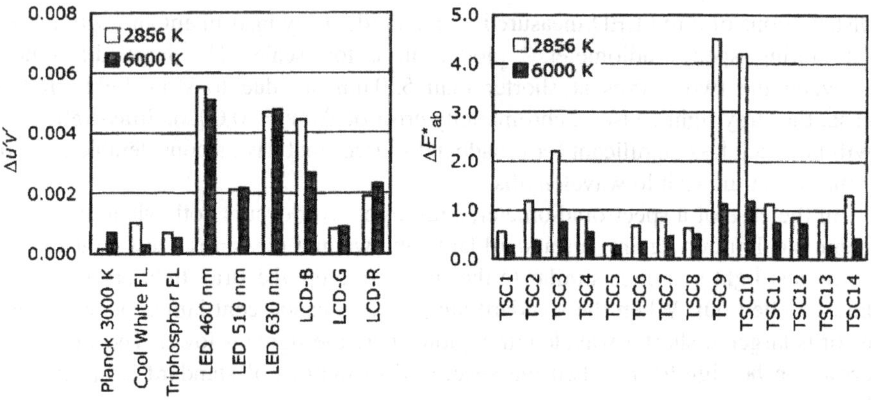

FIGURE 5.19 Errors in measured color caused by stray light of a modeled spectroradiometer having characteristics shown in Figure 5.18, with a calibration source of 2856 K and 6000 K Planckian radiation.

sources (the first three sources) and low saturation color samples. In the case of TCS9 (strong red) and TCS10 (strong yellow), the errors are large with the 2856 K calibration source because these samples have high reflectance in the red and IR region, and the strong red component of Illuminant A intensifies the stray light falling in the blue region. This error is much reduced with the 6000 K calibration source. A Planckian source at low color temperature tends to cause significant stray light at blue and UV region, and such error is introduced when the instrument is calibrated.

Other Sources of Error

There are several other sources of error. If polarization sensitivity is not completely removed, errors can occur, especially for light sources having strong polarization such as LCDs. To minimize polarization errors, recommended input optics as discussed in an earlier sections should be used.

For object color measurement, deviations from the standard geometry and use of a poorly designed integrating sphere (for diffuse reflectance geometry) can cause significant error. To ensure such errors are not significant, the tolerances for the standard geometries are recommended in the new CIE publication.[12]

Detector nonlinearity is another source of error, prominent in array spectroradiometers. To check this error, a stable source can be measured at different integration times and at different exposure level and the consistency of results in chromaticity or measured spectrum is checked. If the nonlinearity is found as a function of integration time or count level for the pixels, the raw signal (counts) can be corrected using a fit function. If no correction is made, they should be included in the uncertainty budget.

A grating has a higher-order wavelength diffraction, normally suppressed using order-sorting filters, but some poorly designed instruments have a leaked higher-order response. It may be useful to check nonexistence of the second-order diffraction (e.g., measure a 350 nm narrow-band emission and see if there is any signal measured at 700 nm).

Some commercial array spectroradiometers do not allow negative values of measured spectral distribution. In the spectral region where there is no source emission (e.g., blue region for a red LED), the noise signal should produce positive and negative values. If the negative noise is all truncated to zero, the remaining positive noise would cause similar effects as stray light and can cause significant error in chromaticity. This is critical for color sources that have no emission in some spectral region such as LEDs.

For object color, fluorescence can be another large source of error if an appropriate method is not used. For measurement of fluorescent samples, refer to other appropriate references.[14–16]

METHODS FOR CORRECTIONS OF ERROR

As discussed in the earlier sections, bandpass and stray light of a monochromator are among major sources of error for color measurement of light sources and object color. In this section, practical methods are described for correcting for these errors.

Correction of Bandpass Error

Bandpass of spectrometers can cause significant errors in color measurement if the bandwidth is larger than 5 nm. Even with a 5 nm bandwidth, the errors can be considerable for special samples and for applications that require low uncertainties. Some methods of correction for bandpass error are introduced below.

ASTM E308

This method can be applied only for object color measurement and only for the data interval of 10 nm or 20 nm, and it requires that the bandwidth of the instrument is triangular and its width is equal to the data interval. The weighting factor tables (Tables 6.1–6.36) of ASTM E308[18] are used to reduce bandpass errors effectively. The corrected tristimulus values are obtained by multiplying the spectral weighting factors (selected for the desired reference illuminant) by the reflectance factor data of the sample. It should be noted that a real instrument's bandpass is not exactly matched to the scanning interval, and there are some deviations. Figure 5.20 shows the results of the bandpass corrections using ASTM E308, for simulated measurements at 10 nm intervals and with the instrument's bandwidth of 10 (perfect match) and 9 nm (≈10% deviation). The correction is very effective for the matched condition but is fairly sensitive to the deviation from the matching condition.

Note that there are two sets of tables in ASTM E308. Tables 5 are for correction of sampling errors only, which are much less than bandpass errors as discussed in the previous section, and thus are not so useful. Also, correction by Tables 5 does not always work well and can increase error in some cases.

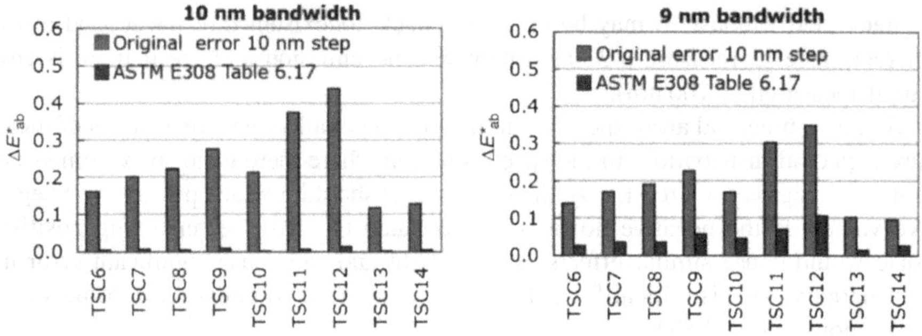

FIGURE 5.20 Results of bandpass correction using Table 6.17 of ASTM E308, for measurement (simulation) at 10 nm intervals and with the instrument's bandwidth of 10 (left) and 9 nm (right) bandwidth.

Stearns and Stearns' Method

This method can be used for light sources as well as for object color. This method, referred to as the S–S method, was developed as an analytical solution of the relationship between the true spectral values and the neighboring points of measured values, with the spectral distribution within the instrument bandpass modeled as quadratic with wavelength.[19] The correction is to give values at zero bandwidth. Tables 6 of ASTM E308 are based on this method. Similar to ASTM E308, the S–S method requires that the bandpass is a symmetric triangular function and its bandwidth $\Delta\lambda_{0.5}$ is matched with the scanning interval λ_{step}. This requirement can also be met with the oversampling conditions ($\Delta\lambda_{0.5} = n \cdot \lambda_{\text{step}}$; n: integer). The corrected spectral value S_i is simply calculated from the neighboring five points of the measured values M_i as

$$S_i = (M_{i-2} - 12M_{i-1} + 120M_i - 12M_{i+1} + M_{i+2})/98 \tag{5.2}$$

For example, when the bandwidth is 5 nm and the value at 450 nm is to be corrected, the corrected value is calculated from the measured values at 440 nm, 445 nm, 450 nm, 455 nm, and 460 nm with each weighting factor shown in the equation. In the case of oversampling conditions, for example, with scanning interval of 2.5 nm or 1 nm (and with 5 nm bandwidth), these points between the 5 nm intervals are skipped, and the calculation is applied for data points at 5 nm intervals. Equation (5.2) can be applied to any bandwidth that is matched with the scanning interval. This correction is very effective if the bandwidth and scanning interval are well matched. Even though the equation is derived for the spectral power distribution modeled as a quadratic, this method works well to reduce colorimetric errors of light sources (having narrowband peaks) as well as object color. Figure 5.21 shows an example of results for the S–S method correction applied to some light source measurements (simulation) at 5 nm intervals, with an instrument bandwidth of 5 nm (perfect match) and 4.5 nm (10% deviation). The results demonstrate that errors are mostly removed for the matched condition, but it is sensitive to the devia-

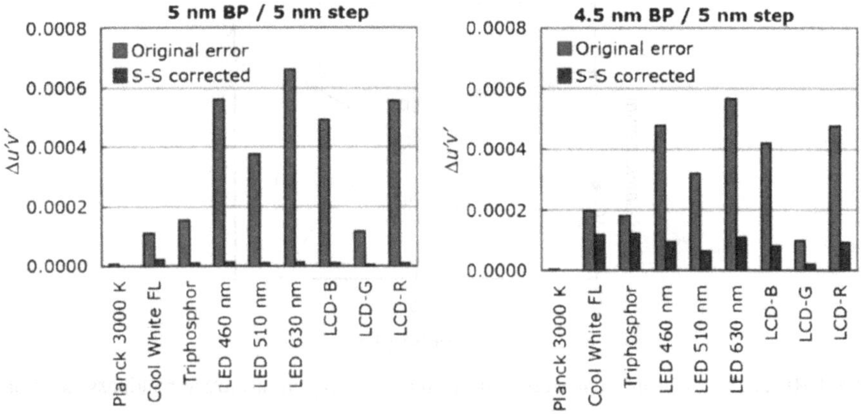

FIGURE 5.21 Results of bandpass correction for light sources using the S–S method, for measurement (simulation) at 5 nm intervals and with the instrument's bandwidth of 5 nm (left) and 4.5 nm (right) bandwidth.

tion of the bandwidth. Note that, although chromaticity is corrected, spectral power distribution of narrowband peaks are not corrected.

Even if a bandpass correction is applied, uncertainties due to residual errors should be evaluated as demonstrated in Figure 5.21 (right); because a real spectrometer's bandwidth is not perfectly matched to the scanning interval and is not perfectly triangularly shaped.

Extended Method for Bandpass Correction

Both the ASTM E308 and the S–S method require that the instrument bandpass is a triangular shape and its width is matched with the scanning interval. In real instruments, this requirement is not perfectly satisfied. It is often seen that the bandwidth of a spectroradiometer, designed to have a constant bandwidth, can vary as much as 20%. At 20% deviation, the reduction of the error with these methods will be about half or less. In some cases, the bandpass function might resemble a Gaussian function or a trapezoid or asymmetric, which would also add deviation from the requirement. In such conditions where the bandwidth and scanning interval are not well matched, ASTM E308 or the S–S method cannot be used.

An improved method is now available that can be applied to any arbitrary bandpass function and does not require the bandwidth and scanning interval of the spectrometer to be matched.[29,30]

Let the true spectral values of the source be S_{-1}, S_0, and S_1 at the neighboring wavelengths λ_{-1}, λ_0, and λ_1, as illustrated in Figure 5.22. The measured values are given as M_{-1}, M_0, and M_1. When the bandpass $s(\lambda, \lambda_0)$ encloses $\lambda_{-1}$to λ_1, the measured value M_0 is related to the true values S_{-1}, S_0, S_1 as

$$M_0 = a_{-1} S_{-1} + a_0 S_0 + a_1 S_1 \tag{5.3}$$

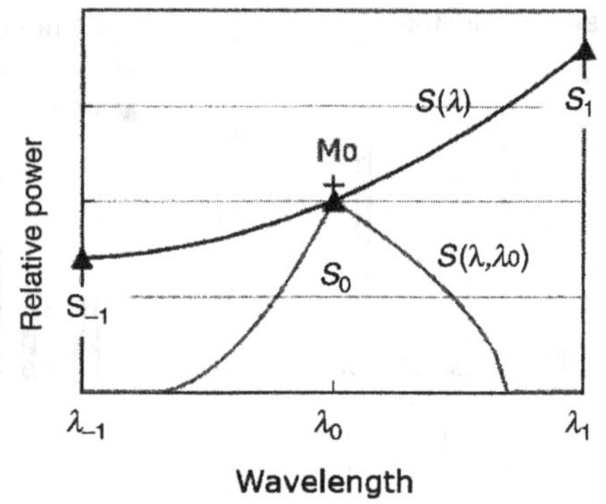

FIGURE 5.22 Measurement with a spectrometer having an arbitrary bandpass function.

and the spectral distribution of the source is modeled as a quadratic

$$S(\lambda) = a + b\lambda + c\lambda^2 \tag{5.4}$$

Then, the three coefficients are obtained by

$$a_{-1} = \frac{1}{2}\left(\frac{I_2}{\Delta\lambda^2} - \frac{I_1}{\Delta\lambda}\right), a_0 = \frac{1}{2}\left(I_0 - \frac{I_2}{\Delta\lambda^2}\right), a_1 = \frac{1}{2}\left(\frac{I_2}{\Delta\lambda^2} + \frac{I_1}{\Delta\lambda}\right) \tag{5.5}$$

where

$$I_0 = \int s(\lambda, \lambda_0)\ \mathrm{d}\lambda,\ I_1 = \int s(\lambda, \lambda_0)\lambda\ \mathrm{d}\lambda,\ I_2 = \int s(\lambda, \lambda_0)\ \lambda^2\ \mathrm{d}\lambda \tag{5.6}$$

Equation (5.6) is calculated numerically for any shape of a given bandpass. Once the three coefficients a_{-1}, a_0, a_1 are obtained, simultaneous equations are formed for the five neighboring points, M_{-2}, M_{-1}, M_0, M_1, M_2. With the approximation $S_{-3} = M_{-3}$ and $S_3 = M_3$, as was done in the derivation of the S–S method, the value of S_0 (corrected to zero bandwidth) is obtained by

$$S_0 = b_{-2} \cdot M_{-2} + b_{-1} \cdot M_{-1} + b_0 \cdot M_0 + b_1 \cdot M_1 + b_2 \cdot M_2 \tag{5.7}$$

with $b_{-2} = \dfrac{a_{-1}^2}{X}, b_{-1} = -\dfrac{a_{-1}}{X}, b_0 = \dfrac{a_0}{X}, b_1 = -\dfrac{a_1}{X}, b_2 = \dfrac{a_1^2}{X}$, and $X = a_0^2 - 2a_{-1}a_1$.

The calculation can be verified by checking that the values of the five coefficients b_{-2}, b_{-1}, b_0, b_{+1}, b_{+2} should be nearly equal to those for the S–S method (1/98, −12/98, 120/98, −12/98, 1/98) for a triangular bandpass with its width equal to the data interval.

As this is a numerical approach, the solution can be given for any shape of bandpass, which may be nonlinear, asymmetric, and/or not matched with the scanning interval, with the assumption that $S(\lambda)$ is a smooth function (quadratic within the bandpass). This method also provides accurate wavelength calibration for asymmetric bandpass (often found in array instruments).[30] As with the S–S method, this method works well for light sources as well as object color, and it can also work for the oversampling conditions. Note that this method does not correct sampling errors for emission lines of discharge lamps when the bandwidth and scanning interval are not matched.

Summary for Bandwidth and Scanning Interval Requirements

For object color measurement:

- Errors due to sampling intervals of less than 10 nm are insignificant. However, a smaller interval is advantageous in reducing color uncertainty due to measurement noise.
- Bandwidths of 5 nm or less are recommended. The narrower the bandwidth, the better. Bandwidths need not be matched to the scanning interval if no correction is applied. Undersampling will not be a problem.
- For bandwidths of 10 nm or larger, bandpass correction using one of the three methods discussed above is recommended.
- The bandwidth and scanning interval need to be matched if ASTM E308 or the S–S method is to be applied.

For light sources containing emission lines and narrowband peaks (discharge lamps):

- The bandwidth and scanning interval need to be matched. Matching of bandwidth within 10% is recommended. The matching condition is important even for very small bandwidths and intervals.
- Bandwidth of 5 nm or less is recommended. For bandwidth larger than 5 nm, bandpass errors should be corrected using one of the correction methods.
- Oversampling conditions relax the bandwidth-matching requirement. The matching condition will not be important for very small sampling intervals with several times larger bandwidth (as often found in array spectroradiometers).

For light sources that do not contain emission lines (such as LEDs):

- It is safe to use the recommended conditions for discharge lamps. But for typical LEDs having spectral half widths of 20 nm to 30 nm, mismatch conditions does not affect the results at 5 nm scanning intervals or less. This may not apply to LEDs having narrower spectral widths.

Correction of Stray Light

Some theories for stray-light correction are available in the literature, for example, Ref. 2. These methods require accurate characterization of the SSF of the spectrometer at all the scanning points, a daunting task. In addition, iterative solution for deconvolution often does not converge due to effect of measurement inaccuracies. A practical method has recently been developed that works well for array spectrometers.[31] This method is briefly described below. See the reference for further details.

First, a spectroradiometer under test measures monochromatic radiation at about 20 nm intervals (e.g., using tunable lasers) spanning the whole spectral range of the spectroradiometer. A function of measured relative signals is called the *LSF*. The LSF is then normalized to the total signals in the in-band region (e.g., all signals higher than 1% of the peak), and the signals in the in-band region are removed. The resulting function is called the *stray-light signal distribution function* (*SDF*). Figure 5.23 illustrates the LSF and SDF.

Then, the SDFs (measured at ≈20 nm intervals of laser emission lines) are interpolated for the measurement intervals of the instrument. With the number of pixels n of the instrument, this interpolation produces a $n \times n$ matrix, called the *SDF matrix*, $\boldsymbol{D}$. Each column of $\boldsymbol{D}$ is the SDF function at a given excitation wavelength. Each row of $\boldsymbol{D}$ forms the spectral stray-light response function for each array pixel. With $\boldsymbol{Y}_{\text{meas}}$ as a column vector of measured signals, the stray-light-corrected signals $\boldsymbol{Y}_{\text{corr}}$ (a column vector) is simply given by

$$\boldsymbol{Y}_{\text{corr}} = [\boldsymbol{I} + \boldsymbol{D}]^{-1}\boldsymbol{Y}_{\text{meas}} \tag{5.8}$$

where $\boldsymbol{I}$ is an identity matrix. The inverted matrix $\boldsymbol{C} = [\boldsymbol{I} + \boldsymbol{D}]^{-1}$ is called the *stray-light correction matrix*. Once this matrix is obtained, a stray-light correction is

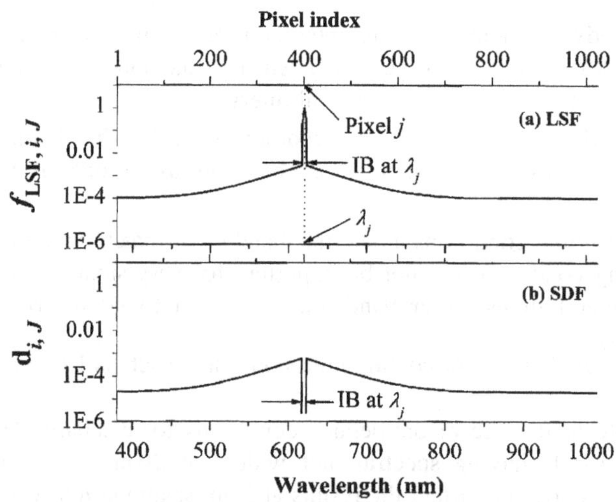

FIGURE 5.23 An illustration of the LSF and SDF.

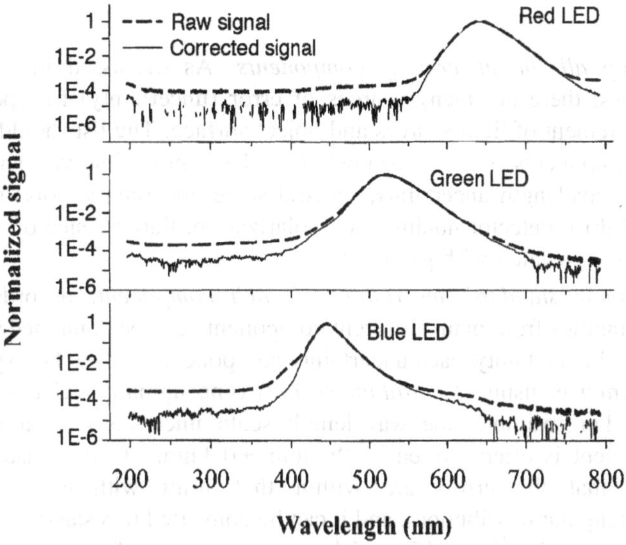

FIGURE 5.24 Plots of the relative raw signals and the stray-light-corrected signals of LEDs (red, green, blue) measured with an array spectroradiometer.

achieved by a simple multiplication of matrix C to the measured signals. The correction with this method is applied simply to the measured signals of a spectrometer; therefore, in case of a spectroradiometer, corrections need to be applied for the measurement of a calibration source (typically a tungsten halogen lamp) and the measurement of a test source. Figure 5.24 shows examples of the results of correction for measurement of LEDs. This case demonstrates that stray-light errors are reduced by one order of magnitude. Note that this method uses the spectrometer itself to obtain the correction matrix. Therefore, stray light from radiation outside the spectral range of the instrument is not corrected. To ensure that this method works, incoming radiation outside the spectral range of the instrument should be filtered out.

UNCERTAINTY ANALYSIS

When measurements are performed, the results should always be reported with a statement of uncertainty. Even if measurements are for internal use, it is always useful to know the uncertainty of the measurements. General guidelines for the uncertainty evaluation are available.[32] Also, the basic terms of uncertainty evaluation are introduced in Appendix 1 and the detailed treatment of uncertainty in color from spectral values is provided in Appendix 2. In this section, basic steps for uncertainty evaluation for color quantities are introduced, and a numerical method to estimate sensitivity coefficients required for uncertainty propagation is introduced.

Basic Steps

(1) *List up all the uncertainty components*: As discussed in the previous sections, there are many sources of error (uncertainty) for spectral color measurement of light sources and object surface. The list should include all the components, such as bandwidth and sampling interval, spectral stray light, wavelength uncertainty, spectral scale uncertainty, noise signal, dark signal drift, detector nonlinearity, polarization, fluorescence of sample, and factors associated with geometry.

(2) *Determine standard uncertainty of each component*: In order that the uncertainties from many different components can be combined to calculate the total uncertainty, each uncertainty component needs to be expressed in a uniform way, using *standard uncertainty* (one sigma in a Gaussian distribution). For example, the wavelength scale uncertainty of a commercial instrument is often written as "within ±0.3 nm." In this case, it may be taken that the errors are within that limit with equal probability (a rectangular distribution), and it can be converted to a standard uncertainty by $u(\lambda) = 0.3/\sqrt{3} = 0.17$ nm. $1/\sqrt{3}$ is a conversion factor from the limit of a rectangular distribution to a standard uncertainty. If the uncertainty of reference standards is given as the *expanded uncertainty* with a coverage factor $k = 2$ (as normally reported by national laboratories), standard uncertainty is obtained by dividing the expanded uncertainty by $k(= 2)$. Note that uncertainty values are all given as a deviation from the center value, so ± sign is not needed.

(3) *Determine the uncertainty contribution from each component*: *Uncertainty contribution* is the uncertainty in the final measured quantity (e.g., in color quantity) contributed from each component of uncertainty, and is given as a standard uncertainty. The color uncertainty from such factors as bandpass and scanning interval, as well as spectral stray light, can be estimated by simulations as presented in the previous sections or from such data. The color uncertainty due to uncertainty in wavelength scale and spectral values (reference standard and transfer measurement) can be propagated from sensitivity coefficients estimated by an analytical approach[27] (derive partial derivatives analytically) or by a numerical approach[28] (briefly introduced in the subsection below). Monte Carlo simulations may also be used for a rigorous analysis.[33]

(4) *Combine the uncertainty contributions*: Assuming that all the components of uncertainty are not correlated, the total uncertainty in a color quantity is obtained from each uncertainty contribution by

$$u_c(X) = \sqrt{\sum_{i=1}^{N} u_i(X)^2} \tag{5.9}$$

where $u_c(X)$ is the combined standard uncertainty in color quantity X, and $u_i(X)$ is the uncertainty contribution in color quantity X from each uncertainty component i. When there are correlations between the components, a more complex form of the equation is required. For such rigorous analysis, refer to Appendix 2.

(5) *Report the uncertainty*: To report the uncertainty of measurements, expanded uncertainty, U, is used:

$$U = k \cdot u_c(X) \tag{5.10}$$

where k is the coverage factor, and $k = 2$ (corresponding to a 95% confidence interval) is normally used. Report U together with the value of k used.

Note that the color uncertainties depend very much on the type of light sources or the color of samples. The uncertainty values should be stated with the type of artifacts. Commercial spectroradiometers often specify uncertainty in chromaticity (e.g., 0.002 in x, y) for Illuminant A. Spectroradiometers (for color measurements) are normally calibrated with a tungsten halogen standard lamp, so when a CIE Source A (tungsten lamp spectrum at similar color temperature) is measured, most of the sources of error would cancel out. Therefore, such specification for Illuminant A represents only the reproducibility and stability of the instrument and does not indicate the uncertainty of measurement of various different light sources.

Numerical Method for Sensitivity Coefficient

A color quantity q (any color quantity such as chromaticity coordinates or CRI R_a) is given as a function of the spectral quantity $S(\lambda)$ by

$$q = f\{S(\lambda)\} = f\{S(\lambda_1), S(\lambda_2), \ldots\ldots, S(\lambda_n)\} \tag{5.11}$$

The partial derivative of q with respect to $S(\lambda_i)$ is numerically obtained at each wavelength by

$$\frac{\partial q}{\partial S(\lambda_i)} = \frac{1}{\Delta S}[f\{S(\lambda_1), S(\lambda_2), \ldots\ldots, S(\lambda_i) + \Delta S, \ldots\ldots, S(\lambda_n)\} - f\{S(\lambda)\}] \tag{5.12}$$

where ΔS is chosen to be small enough relative to the average value of $S(\lambda)$, for example, 10^{-4} to 10^{-6} of the maximum value of the spectra. This partial derivative is called the *sensitivity coefficient* (of q with respect to $S(\lambda_i)$). This can be calculated easily by using the computer program itself to calculate the color quantity from the spectral data. For example, for the chromaticity (x, y) uncertainty, just add a small value ΔS (like 0.001) to the relative spectral distribution value at one wavelength, and look at the change in chromaticity x and y. The sensitivity coefficient (at that wavelength) is the change in x or y divided by ΔS.

For random effects in the spectral measurements, uncorrelated at different wavelengths, the combined standard uncertainty $u(q)$ of the color quantity is given from the standard uncertainties $u\{S(\lambda_i)\}$ of the spectral values by

$$u(q) = \sqrt{\sum_{i=1}^{n} \left\{ \frac{\partial q}{\partial S(\lambda_i)} \right\}^2 u^2\{S(\lambda_i)\}} \tag{5.13}$$

For random wavelength uncertainties, Equation (5.12) is replaced by

$$\frac{\partial q}{\partial \lambda_i} = \frac{1}{\Delta\lambda} [f\{S(\lambda_1), S(\lambda_2), \ldots\ldots, S(\lambda_i + \Delta\lambda), \ldots\ldots, S(\lambda_n)\} - f\{S(\lambda)\}] \tag{5.14}$$

and Equation (5.13) is replaced by

$$u(q) = \sqrt{\sum_{i=1}^{n} \left\{ \frac{\partial q}{\partial \lambda_i} \right\}^2 u^2(\lambda_i)} \tag{5.15}$$

As this is a numerical approach, it can be easily applied to any color quantities with any complex calculation, such as CIELAB coordinates, correlated color temperature, distribution temperature, and CRI, where the analytical solution of the partial derivatives is difficult. Further details of this numerical method are available in Ref. 28. The same sensitivity coefficients can be used to estimate uncertainties for effects completely correlated between wavelengths, as described in Appendix 2.

ACKNOWLEDGMENT

The author thanks Dr. Jim Gardner, Dr. János Schanda, and Dr. Maria Nadal for reviewing this chapter and providing many useful comments and discussions.

REFERENCES

1. CIE (1984) *Spectroradiometric Measurement of Light Sources*, CIE Publ. **63**-1984.
2. Kostkowski HJ (1997) *Reliable Spectroradiometry*, Spectroradiometry Consulting, La Plata, MD, USA.
3. CIE, *The Determination of Errors of Measurement in Spectrophotometry*, A Technical Report from TC2-28 (to be published).
4. *Applied Spectroscopy* (1997) (Eds., J. Workman, Jr. and A. W. Springsteen), Academic Press, Chestnut Hill, MA and San Diego, CA, USA.
5. Brown SW, Ohno Y (1999) *NIST Colorimetric Calibration Facility for Displays*, Society for Information Displays 99 DIGEST, pp. 794–797.

6. For example, see http://www.klccgo.com/cqmscramb.htm.
7. Walker JH, Saunders RD, Hattenburg AT (1987) *NBS Measurement Services: Spectral Radiance Calibrations*, NBS Spec. Publ. 250-1.
8. Goodman TM, Moore JR, Pearce NC', Murry DK (1991) The establishment of a new national scale of spectral total flux, in: *Proceedings of the, 22nd Session of CIE*, Melbourne 1991, Vol. 1, Part 1, 50–53.
9. Ohno Y, Zong Y (2004) NIST facility for total spectral radiant flux calibration, in: *Proceedings of the Symposium of Metrology 2004*, CENAM, Mexico, Oct. 2004 (published in CD).
10. CIE (2004) *Colorimetry—Part 1: CIE standard colorimetric observers*, CIE S014-1/E:2006.
11. CIE (2004) *Colorimetry*, 3rd ed. CIE Publication, **15**: 2004.
12. CIE (2006) *Geometric Tolerances for Colour Measurements*, CIE Publication 176:2006.
13. CIE (1988) *Intercomparison on measurement of total spectral radiance factor of luminescent specimens*, CIE Publication **76**-1988.
14. ASTM (1998) *Standard Practice for Colour Measurement of Fluorescent Specimens*, ASTM **E991**-98.
15. ASTM (2001) *Standard Practice for Obtaining Bispectral Photometric Data for Evaluation of Fluorescent Colour*, ASTM **E2153-01**.
16. CIE, *Calibration Methods and Photoluminescent Standards for Total Radiance Factor Measurements*, A Technical Report from TC2-25 (to be published).
17. CIE (2005) *Recommended practice for tabulating spectral data for use in color computations*, CIE Publication **167**:2005.
18. ASTM (1999) *Standard Practice for Computing the Colours of Objects by Using the CIE System*. ASTM **E308** – 99 (1999).
19. Stearns EI, Stearns RE (1988) An example of a method for correcting radiance data for bandpass error. *Color Res. Appl.*, **13**-4, 257–259.
20. CIE (1995) *Method of measuring and specifying colour rendering properties of light sources*, CIE Publication **13.3**-1995.
21. Clarke FJJ, Malkin F (1981) Development of a New Series of Ceramic Colour standards. *J. Soc. Dyes Colour.*, **97**, 503–504.
22. ColorChecker, Product No. 50105, GretagMacbeth LLC, 617 Little Britain Road, New Windsor, NY 12553[#].
23. Ohno Y (2003) Obtaining spectral data for colorimetry, in: *Proceedings of the* 25th *Session of the CIE, San Diego*, Vol. 1, D2, pp. 44–47.
24. Ohno Y, Kránicz B (2001) Spectroradiometer characterization for colorimetry of LEDs, in: *Proceedings of the* 2nd *CIE Expert Symposium on LED Measurement*, May 11–12, 2001, Gaithersburg, Maryland, USA, pp. 56–60.
25. Brown SW, Eppeldauer GP, Rice JP, Zhang J, Lykke KR (2004) Spectral irradiance and radiance responsivity calibrations using uniform sources (SIRCUS) facility at NIST, *Proc. SPIE*, **5542**, 363–374.
26. Gardner JL (2003) Uncertainties in interpolated spectral data, *J. Res. Natl. Inst. Stand. Technol.*, **108**, 69–78.
27. Gardner JL (2000) Uncertainty estimation in colour measurement, *Colour Res. Appl.*, **25**, 349–355.

28. Ohno Y (2001) A numerical method for colour uncertainty, in: *Proceedings of the CIE Expert Symposium 2001 on Uncertainty Evaluation*, Jan. 2001, Vienna, Austria, pp. 8–11.
29. Ohno Y (2005) A flexible bandpass correction method for spectrometers, in: *Proceedings of the AIC'05* Granada, Spain, May 9–13, 2005, pp. 697–700.
30. Gardner JL, Bandwidth correction for LED chromaticity, *Colour Res. Appl.* (to be published).
31. Zong Y, Brown S W, Johnson BC, Lykke K R, Ohno Y (2006) Simple stray light correction method for array spectroradiometers, *Appl. Opt.* **45**/6, 1111–1119.
32. ISO, *Guide to the Expression of Uncertainty in Measurement* (1993).
33. JCGM, Evaluation of measurement data — Supplement to the "Guide to the Expression of Uncertainty in Measurement" — Propagation of distributions using a Monte Carlo method (to be published).

6

TRISTIMULUS COLOR MEASUREMENT OF SELF-LUMINOUS SOURCES

János Schanda,[a] George Eppeldauer,[b] and Georg Sauter[c]

[a]*University of Pannonia, Egyetem u. 10., H-8200 Veszprém, Hungary*

[b]*National Institute of Standards and Technology (NIST), 100 Bureau Drive, Mail-Stop 8441, Gaithersburg, MD 20899, USA*

[c]*Physikalisch-Technische Bundesanstalt, AG 4.12 Photometrie, Postfach 3345, D-38023, Braunschweig, Germany*

INTRODUCTION

In Chapter 3, we discussed that there are two methods to determine the tristimulus values of a color stimulus: by measuring the spectral power distribution (SPD) of the stimulus and then multiplying, wavelength by wavelength, this SPD with the color-matching functions (CMFs), or by building a tristimulus colorimeter whose spectral responsivity mimics the CMFs of the standard observer. The spectral method was discussed in Chapter 5. Tristimulus colorimeters were extensively used in color-difference meters, as it provided a much faster measuring cycle, compared with the tedious measurement using a (nonautomated) spectroradiometer and doing subsequent calculations. With the introduction of computer-controlled spectrometric instrumentation, where the computer performs the calculations, the advantage of speed diminished, especially when spectrometers with array detectors* became available, where the complete spectrum could be measured in one step. Nevertheless,

*The CIE International Lighting Vocabulary distinguishes among different types of detectors. In this chapter we deal only with photoelectric detectors, and for the sake of brevity will often not spell out the adjective photo- or photoelectric.

Colorimetry: Understanding the CIE System, Edited by János Schanda

tristimulus colorimeters still have their useful place in colorimetric practice, mainly in measuring self-luminous objects (light sources) and especially if this measurement has to be made as an area-resolved measurement (image-taking colorimetry).†

Due to the highly automated character of the modern colorimeters, the user often does not get enough information on what the (systematic) errors of the instrument are and with what amount of uncertainty he has to calculate; thus, it is important to determine (or request from the manufacturer) the characteristics of the instrument. The present chapter intends to help the user of tristimulus colorimeters to acquire the necessary background information (further details on the quantities that are used to determine the performance of a tristimulus colorimeter can be found in the CIE publication dealing with the characterization of such instruments[1]).

Normal practice for the calibration of light-measuring tristimulus colorimeters is the use of a calibrated light sources. A further subchapter will deal with the detector-based calibration technique, an item still under development.

BASIC STRUCTURE OF A TRISTIMULUS COLORIMETER

Modern tristimulus colorimeters for self-luminous sources are usually built in such a form that they provide the possibility to measure a photometric quantity (usually either luminance or illuminance) and the tristimulus values or chromaticity coordinates at the same time. In the case of illuminance-measuring systems, one frequently finds the detectors filtered for the three CMFs side by side under one

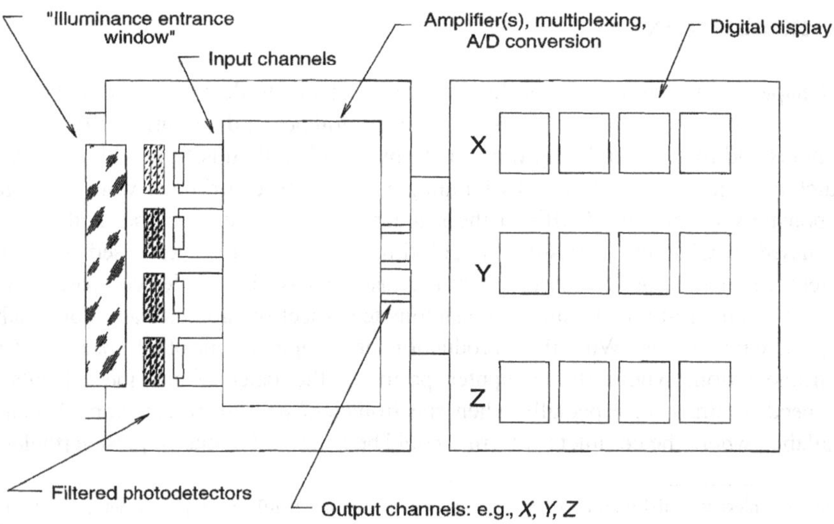

FIGURE 6.1 Schematic diagram of an illuminance-measuring tristimulus colorimeter for light-source measurement.

†For terms and definitions used in conjunction with tristimulus colorimeters see the Glossary at the end of this chapter.

diffuser, as shown schematically in Figure 6.1. In luminance-measuring systems, especially in image-capturing luminance-measuring ones, often only one detector is used and the filters to shape the spectral responsivity of the detector to the $\bar{x}(\lambda)$, $\bar{y}(\lambda)$, and $\bar{z}(\lambda)$ functions are brought consecutively into the light path. Frequently—especially in the case of the illuminance-type instruments—the input optics, filter package, and detector is in an individual compartment (sometimes also some preamplifiers), called a colorimeter head. For luminance-type instruments, it is a common practice to place the electronics into the colorimeter head itself because the input optics and some means to view the object, the luminance of which is to be measured, are bulky enough and the observer has to be in the vicinity of the instrument to aim the colorimeter onto the object to be measured.

INPUT OPTICS OF A COLORIMETER FOR SELF-LUMINOUS OBJECTS

As already mentioned, we distinguish between the three main input optics of a colorimeter for self-luminous objects: illuminance-, luminance- and image-taking colorimeter.

Illuminance-Meter-Type Input Optics

The illuminance-meter-type input optics of a tristimulus colorimeter head is similar to that of an illuminance meter; the main difference is that if the channels for simulating the CMFs are placed parallel to each other then the input optics has to divide the radiation in such a form among the channels that a noneven illumination of the head should not cause any measurement error. The spatial inhomogeneity error index (see Refs. 1, 4) gives an estimate on how well the radiation is distributed among the channels.

For an illuminance meter, it is critical to know the input plane of the measuring head, from which the distance to a source has to be measured. This has to be shown in the case of the colorimeter head to be of any value for absolute illuminance measurement.

For the practical realization of the input optics, one often finds that the arrangement is a transmitting diffuser surrounded by a ring. By proper selection of the diffuser material and the size of the ring, a good cosine response (needed for illuminance measurement) can be achieved. For the goodness of the cosine correction, the CIE introduced the directional response error index[4]:

$$f_2 = \int_{\varepsilon=0}^{85^\circ} \left| \frac{Y(\varepsilon)}{Y(0^\circ) \cdot \cos\varepsilon} - 1 \right| \cdot \sin 2\varepsilon \, d\varepsilon \tag{6.1}$$

where ε is the angle of incidence measured to the normal of the measuring plane or to the optical axis, and Y is the signal output (e.g., the Y tristimulus value).

The definition supposes the rotational symmetry about the axis of normal incidence. Integration is performed only up to 85°, and not 90°, because the determination of $Y(\varepsilon)$ between 85° and 90° becomes increasingly difficult and uncertain. Usually, the error index is presented in the form of percentages.

Luminance-Meter-Type Input Optics

The luminance-meter-type tristimulus colorimeter measures the luminance of the object envisaged as the Y tristimulus value. To view the same part of the scene as that which hits the detector, the instrument contains an optical system in which the measured scene can be seen.

Figure 6.2 shows the schematic structure of a luminance-measuring-type colorimeter input optics. The section of the scene to be measured (the measured object) is focused by the input optics onto a field stop. Behind the field stop is the color-correcting filter set (for the three or four channels of color measurement) and the (photo-)detector. Characteristic quantities of the input optics are the acceptance area—the area of the lens which contributes to the response of the instrument—and the measuring field angle, the angular space within which the directional dependence of the responsivity of the instrument is at least 10% of the maximum responsivity.

To test the directional response of the luminance-measuring colorimeter, the following error index can be used:

$$f_2(\varepsilon_{1/100}) = \left(1 - \frac{\bar{\varepsilon}_{1/10}}{\bar{\varepsilon}_{1/100}}\right) \tag{6.2}$$

where $\bar{\varepsilon}_{1/10}$ is the average 10% measuring angle and $\bar{\varepsilon}_{1/100}$ is the average 1% measuring angle.

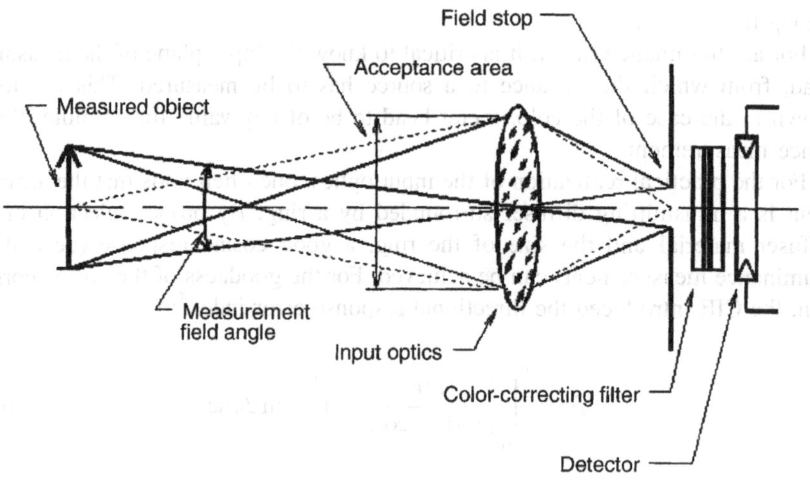

FIGURE 6.2 Schematic arrangement of the input optics of a luminance-measuring-type instrument.

The extent of the luminous area, within which the ε angles are determined, has to be smaller than 5% of the measurement field angle (for further details see Ref. 4).

Image-Taking Colorimeters

A group of colorimeters that is becoming more and more popular is that of the image-taking colorimeters. The input optics of these instruments is very similar to that of the luminance-meter-type colorimeters, the main difference being that they measure the luminance not in one small field angle, but image the scene onto a charge coupled device (CCD) or other image-recording two-dimensional detector array, so that the luminance and the colorimetric coordinates in each pixel of the detector can be determined. Information from a number of pixels has to be processed; therefore, these instruments are always coupled to a computer. As the scene is visible on the computer screen, no second-viewing optical system is needed.

Computer control provides observation of the scene on the computer screen with a subsequent image capture and processing operation. Although tuneable filters exist, which permit the capture of the image in narrow spectral bands, the processing of very large amounts of the captured data and the still-not-perfect spectral selectivity of such tuneable filters (considerable stray radiation) make the tristimulus approach attractive. As mentioned in the introduction, partly based on the higher cost of the parallel image capture and processing, the image-taking colorimeters usually have one CCD detector matrix and change the color-matching filters in front of this detector.

The manufacture of filters for tristimulus colorimeters is not a simple task, but to produce good filter sets for image-taking colorimeters is even more demanding: The filters have to have equal optical thickness to avoid refocusing for every channel and have to be extremely free of any optical inhomogeneity, as this would distort the image. A further problem that limits the maximum achievable accuracy of such filter–detector combinations is that the length of the light path through the filter to a pixel in the middle of the matrix is different from the length of the path through the filter to a pixel in the corners of the CCD array. This limits the maximum spectral correction that can be achieved.[2]

The flexibility of an image-taking colorimeter can be increased if different imaging optics can be attached to the colorimeter. As the spectral transmission of different optics is usually different, either sets of color-correcting filters optimized for each input optic have to be supplied, or the set of optics to be used with a given colorimeter has to be corrected individually to the colorimeter.

SPECTRAL MATCHING OF THE COLORIMETER

The four detector channels shown in Figure 6.1 realize the two peaks of the $\bar{x}(\lambda)$ function and the $\bar{y}(\lambda)$ and $\bar{z}(\lambda)$ functions. In some older instruments, one finds only three filter–detector combinations: Instead of a separate filter to realize the

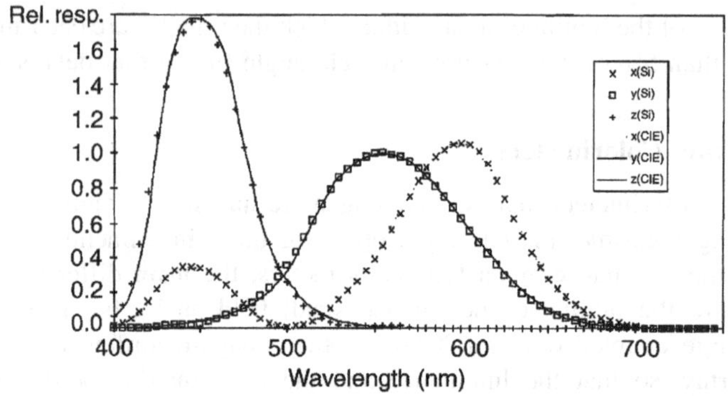

FIGURE 6.3 Spectral responsivity functions of a tristimulus colorimeter with Si-cell and full-filter correction ($f_1'(x) = 2.6$; $f_1'(y) = 2.2$; $f_1'(z) = 4.3$; courtesy of InPhoRa Corp.[3]

short-wavelength sensitivity peak of the $\bar{x}(\lambda)$ function, a filter to realize the $\bar{z}(\lambda)$ function is used.

Just as for a photometric instrument, an important characteristic of the tristimulus colorimeter is how well the filtered responsivity of the instrument resembles the $\bar{x}(\lambda)$, $\bar{y}(\lambda)$, and $\bar{z}(\lambda)$ CMFs (or the $\bar{x}_{10}(\lambda)$, $\bar{y}_{10}(\lambda)$, and $\bar{z}_{10}(\lambda)$ CMFs). Two techniques are used to achieve the proper spectral responsivity: the so called full-filter and the partial-filter technique. In the case of full filtering, different colored filter glasses are cut and ground to the proper thickness and then are polished and glued together to change the spectral responsivity of a detector, nowadays usually a Si photovoltaic cell,[†] to match the spectral sensitivity to the CMFs. Figure 6.3 shows the CIE CMFs and typical tristimulus colorimeter spectral responsivities for the four channels of a colorimeter. To achieve even better spectral correction in some instruments, chips of filter glasses are glued side by side, called partial filtering, see Figure 6.4.

There is no internationally accepted method for characterizing the goodness of fit of the spectral match of a colorimetric detector head. At present, many manufacturers use the method suggested by the CIE for radiometers and photometers.[4] This method is based on the spectral mismatch error index f_1' that has to be provided for each channel of the colorimeter:

$$f_{1,i}' = \frac{\int_0^\infty \left| s_{\mathrm{rel},i}^*(\lambda) - \bar{t}_i(\lambda) \right| \cdot \mathrm{d}\lambda}{\int_0^\infty \bar{t}_i(\lambda) \cdot \mathrm{d}\lambda} \cdot 100\% \tag{6.3}$$

where $\bar{t}_i(\lambda)$ is one of the CMFs ($\bar{t}_1(\lambda) = \bar{x}_s(\lambda)$, $\bar{t}_2(\lambda) = \bar{x}_l(\lambda)$ (index "s" describes the short-, "l" the longwave peak of the $\bar{x}(\lambda)$ function), $\bar{t}_3(\lambda) = \bar{y}(\lambda)$, and

[†]Also called Si-cell or Si-photodiode.

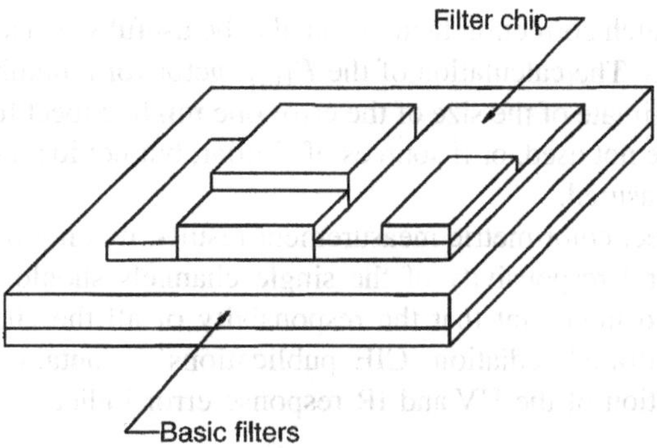

FIGURE 6.4 Schematic structure of a partially filtered filter set.

$\bar{t}_4(\lambda) = \bar{z}(\lambda)$); i has similar meaning in the following equations. $s^*_{\text{rel},i}(\lambda)$ is the normalized relative spectral responsivity:

$$s^*_{\text{rel},i}(\lambda) = \frac{\int_0^\infty S_\text{m}(\lambda) \cdot \bar{t}_i(\lambda) \cdot \text{d}\lambda}{\int_0^\infty S_\text{m}(\lambda) \cdot s_{\text{rel},i}(\lambda) \cdot \text{d}\lambda} \cdot s_\text{rel}(\lambda) \quad (6.4)$$

$S_\text{m}(\lambda)$ is one of the standard illuminants, in photometry usually Standard Illuminant A, in colorimetry often Standard Illuminant D65, and some manufacturers use the equienergy spectrum (illuminant E). $s_{\text{rel},i}(\lambda)$ is the spectral responsivity of the detector–filter combination. $f'_{1,3}$ values of the $\bar{y}(\lambda)$-function approximation can be better than 1.5%, for the $\bar{x}(\lambda)$ and $\bar{z}(\lambda)$ functions, the approximations are usually poorer, depending to a large extent on the permissible total absorption.

The $f'_{1,i}$ value neither provides the measurement error in a quantitative way nor states the measurement uncertainty contribution due to the mismatch; it is only an index giving information on the magnitude of the mismatch between the responsivity of a filtered detector and the CMFs. These indices are useful for specification and selection purposes and can also be used by manufacturers for controlling and improving the performance of their instruments. The error for a given light source can only be determined if both the relative spectral responsivities and the relative SPD of the source are known.

If the relative SPDs of the test, $S_\text{T}(\lambda)$, and calibration, $S_\text{C}(\lambda)$, sources, and the relative spectral responsivities of the output channels ($s_{\text{rel},i}(\lambda)$), are known, the output of the channels can be corrected by multiplying the output of each of the channels by the appropriate spectral mismatch correction factor $F_{\text{TC},i}$:

$$F_{\text{TC},i} = \frac{\int S_\text{T}(\lambda) \cdot \bar{t}_i(\lambda) \cdot \text{d}\lambda}{\int S_\text{C}(\lambda) \cdot \bar{t}_i(\lambda) \cdot \text{d}\lambda} \Bigg/ \frac{\int S_\text{T}(\lambda) \cdot s_{\text{rel},i}(\lambda) \cdot \text{d}\lambda}{\int S_\text{C}(\lambda) \cdot s_{\text{rel},i}(\lambda) \cdot \text{d}\lambda} \quad (6.5)$$

Spectral mismatch correction factors can also be useful when assessing measurement uncertainties. The calculation of the $F_{TC,i}$ factor for a number of test sources can provide an estimate of the size of the error one might expect for those sources if the corrections are not used, or if sources of similar, but not identical, spectral characteristics are measured.

To obtain correct colorimetric measurement results, it is not only necessary that the relative spectral responsivity of the single channels should mimic the CMFs well, but it is also important that the responsivity of all the channels is zero for ultraviolet and infrared radiation. CIE publications[1,4] contain recommendations for the determination of the UV and IR response error indices.

ELECTRONICS

Most illuminance- and luminance-measuring tristimulus colorimeters use silicon photovoltaic cells as detectors. Usually a current-to-voltage converter is applied to the detector, the detector being connected between the inverting and noninverting inputs of an operational amplifier. The voltage drop between these two inputs is small enough—in the range of a few microvolts or less—for linear operation of the detector (short-circuit mode). The amplified input offset voltage of the operational amplifier together with the stray-light-produced current create nondesired signal components in the output offset voltage. A stable output offset voltage can be compensated by a "dark measurement."

The subsequent electronics is straightforward analog and digital electronics, with analog-to-digital conversion and digital manipulation of the captured quantities. Si-photodiodes with proper design of the current-to-voltage converters of the colorimeter channels can produce linear operation over many orders of magnitude.[6]

With image-taking colorimeters, an interesting approach to obtain meaningful colorimetric results both in shadow areas and at highlights is to take pictures with different integration times; CCD arrays can be corrected to be linear in a decade signal range[21]. Smart digital electronics can adjust several pictures taken with different integration times and artificially produce high dynamic range pictures, even if the resolution is only 1 in 1000 in each subimage.

Care should be taken with electronic calculations. The computer can calculate and display many more digits than that justified by the errors and the uncertainty of the measurement.

The electronic design issues of standards quality tristimulus colorimeters to satisfy low-uncertainty spectral responsivity calibrations (as discussed in the next section) have been published earlier.[7]

CALIBRATION

Calibration With a Standard Source

Tristimulus colorimeters for self-luminous sources are designed to measure light sources; thus, the standard technique for their calibration is to use a standard lamp (or luminance standard). The output signal of the instrument will be

$$T'_{C,i} = k_i \int_{\lambda=380\,\text{nm}}^{780\,\text{nm}} S_C(\lambda) \cdot s_i(\lambda) \cdot d\lambda \tag{6.6}$$

instead of the $T_{C,i}$ tristimulus value, k_i is a proportionality factor, $S_C(\lambda)$ is the relative SPD of the calibration source, and $s_i(\lambda)$ is the relative spectral responsivity of channel i of the colorimeter; $i = 1$ for the short wavelength peak of the x channel, $i = 2$ for the long wavelength peak of the x channel, $i = 3$ for the y, and $i = 4$ for the z channel.

In the course of calibration, the k_i factors have to be determined. If calibration is performed with a CIE source A, the k_i factors have to be set so that

$$T'_{A,1} : T'_{A,2} : T'_{A,3} : T'_{A,4} = 5.380:104.470:100:35.585 \tag{6.7}$$

because the tristimulus values of CIE standard illuminant A are $X = 109.85$, $Y = 100$, $Z = 35.58$, and the minimum between the two peaks of the $\bar{x}(\lambda)$ function lies at 505 nm, and divides the X tristimulus value of CIE standard illuminant A into the two parts: $X_s = 5.380$ and $X_l = 104.470$. For an illuminance/luminance value differing from the relative value 100, both sides of Equation (6.7) have to be multiplied with the illuminance/luminance value divided by 100.

Similar calculations can be performed for standards having a different relative SPD.

If the four input values (X'_1, X'_2, Y', and Z') are available simultaneously, as in the example of Figure 6.1, it is possible to perform calibrations with more sources and calculate optimized k_i factors.[8–10] These can be then implemented in the form of matrixing the input quantities to get the output signals corresponding better to the tristimulus values.

If $V_{i,k}$ are the input signals ($i = 1$ to 4, referring to the input channels as before) for n illuminants ($k = 1, \ldots, n$) used to calibrate the colorimeter, then the X'_k, Y'_k, Z'_k output signals can be determined using the following matrix equation:

$$\begin{vmatrix} X'_k \\ Y'_k \\ Z'_k \end{vmatrix} = \begin{vmatrix} a_{x1} & a_{x2} & a_{x3} & a_{x4} \\ a_{y1} & a_{y2} & a_{y3} & a_{y4} \\ a_{z1} & a_{z2} & a_{z3} & a_{z4} \end{vmatrix} \cdot \begin{vmatrix} V_{1,k} \\ V_{2,k} \\ V_{3,k} \\ V_{4,k} \end{vmatrix} \tag{6.8}$$

The $a_{i,k}$ matrix elements can be optimized in such a form that the ΔE^* (ab or uv) difference between the X'_k, Y'_k, Z'_k values and the tristimulus values of a number of sources gets minimized. For four sources, and the additive mixture of the lights from these four sources, the equation can be solved exactly. Thus, for example, calibrating the luminance channel of a tristimulus colorimeter with a CIE source A and using the three primary colors of a display as additive calibrating sources—in principle—one can correct the readings of the instrument, even if the spectral responsivity curves of the four detector channels differ considerably from the CMFs. (For other sources the measurement errors will naturally not diminish; they may be even larger than using k_i factors from a standard calibration.)

Calibration Based on Standard Detectors

Introduction

Improvements in detector technology over the past decade have opened a new era in radiometric and photometric calibrations.[11] Lower measurement uncertainties can be achieved with modern detector standards than with traditionally used source standards. The detector-based approach has been extended to colorimetry as well.[12,13] Based on the spectral responsivity of the tristimulus colorimeter channels, a colorimetric scale can be realized and maintained. With low responsivity uncertainty determination of the colorimeter channels, x, y chromaticity coordinate measurement uncertainties of 0.0002 can be achieved when Planckian radiators are measured. These uncertainties are smaller than the measurement uncertainties obtained with current primary lamp standards. Using the detector-based method for special test sources, such as nonwhite LCD or CRT displays, chromaticity coordinate measurement uncertainties of smaller than 0.001 can be performed.

The Spectral Responsivity Based Calibration Method

The $k_{X,i}$ channel calibration factors of a tristimulus colorimeter can be determined from the ratio of the CIE tristimulus values to the measured $I_{X,i}$ output currents of the four channels[12]:

$$k_{X1} = \frac{X_1}{I_{X1}} = \frac{K_m \int_\lambda S(\lambda) \cdot \bar{x}_1(\lambda) \cdot d\lambda}{\int_\lambda S(\lambda) \cdot s_{X1}(\lambda) \cdot d\lambda} \qquad k_{X2} = \frac{X_2}{I_{X2}} = \frac{K_m \int_\lambda S(\lambda) \cdot \bar{x}_2(\lambda) \cdot d\lambda}{\int_\lambda S(\lambda) \cdot s_{X2}(\lambda) \cdot d\lambda}$$

$$k_Y = \frac{Y}{I_Y} = \frac{K_m \int_\lambda S(\lambda) \cdot \bar{y}(\lambda) \cdot d\lambda}{\int_\lambda S(\lambda) \cdot s_Y(\lambda) \cdot d\lambda} \qquad k_Z = \frac{Z}{I_Z} = \frac{K_m \int_\lambda S(\lambda) \cdot \bar{z}(\lambda) \cdot d\lambda}{\int_\lambda S(\lambda) \cdot s_Z(\lambda) \cdot d\lambda} \tag{6.9}$$

where $S(\lambda)$ is the spectral distribution of the reference source chosen to calculate the calibration factors, $\bar{x}_1(\lambda)$, $\bar{x}_2(\lambda)$, $\bar{y}(\lambda)$, and $\bar{z}(\lambda)$ are the CIE CMFs, K_m is the maximum spectral luminous efficacy, 683 lm/W (needed only in the calculation if $S(\lambda)$ is inserted in absolute units and the Y tristimulus value should correspond to a photometric quantity), $s_{X1}(\lambda)$, $s_{X2}(\lambda)$, $s_Y(\lambda)$, and $s_Z(\lambda)$ are the measured spectral responsivities of the realized channels, and λ is the wavelength. Both the responsivity measurements and the integrals are to be made between 360 nm and 1000 nm (for the responsivity range of the silicon detectors applied in the colorimeter).

The tristimulus values of test light sources are determined using the above calibration factors:

$$X' = X'_1 + X'_2, \text{ where } X'_1 = k_{X1} I'_{X1} \text{ and } X'_2 = k_{X2} I'_{X2}, Y' = k_Y I'_Y, \text{ and } Z' = k_Z I'_Z \tag{6.10}$$

where I'_{X1}, I'_{X2}, I'_Y, and I'_Z' are the output signals of the channels for a given test source.

The calibration procedure can be applied to various measurement geometries (e.g., illuminance, luminance, luminous flux, or luminous intensity) depending on the units in which the spectral responsivities are expressed. The lowest responsivity uncertainty can be achieved if the channels are calibrated against a silicon trap-detector with tuneable narrow band sources like a monochromator or tuneable lasers. Trap detector standards operate in either radiant power or irradiance measurement mode.

Calibration and Measurement Considerations

Realization of the channel responsivities of a tristimulus colorimeter with a small spectral mismatch to the CIE CMFs and the low-uncertainty spectral responsivity measurements of the channels make it possible to realize a color temperature scale with low uncertainty, especially for tungsten lamps. For example, the four channels of the reference tristimulus colorimeter of NIST[14] could be calibrated for spectral irradiance responsivity with a relative expanded uncertainty of 0.15% $(k = 2)$. This SIRCUS-reported uncertainty applies to the single-responsivity points at each laser wavelength where measurements were taken. In the spectrally integrated responsivity measurements of the four colorimeter channels, the responsivity values at neighboring wavelengths are affected in (nearly) the same way. Therefore, in addition to the SIRCUS-reported independent (uncorrelated) components, systematic (correlated) components across the wavelength are present. When accurate color (and responsivity) measurements are needed, it is usually the systematic components that determine the measurement uncertainty. Research is being continued to describe the propagation of these uncertainty components in the most correct way.[25] The main goal is to keep the responsivity measurement uncertainty at the lowest level because the combined relative uncertainty of the spectrally integrated responsivity (the reciprocal of the channel calibration factor) of the four channels will determine the minimum uncertainty in the x and y chromaticity coordinates.

As the internal quantum efficiency (output electrons per input photons) of the trap detector in the NIST colorimeter can be modelled,[15] its spectral responsivity can be interpolated between 406 nm and 920 nm with a relative expanded uncertainty of 0.06% $(k = 2)$. When this model is applied, instead of the full spectral responsivity calibrations, only a few trap-detector responsivity (tie) points are needed. The spectral transmittance of each color-correcting filter combination can be multiplied with the interpolated spectral responsivity of the common trap detector. (Interpolation can produce responsivity function values with correlated uncertainties at least between the neighboring values.) The filter combinations are temperature stabilized; thus their spectral transmittances remain stable between calibration and use. The electronic system provides a signal dynamic range for the instrument of 12 decades.[16]

With the above calculated (one set of) calibration factors, the output signals from the individual colorimeter channels were calculated using the integrals in the denominators of Equation (6.9) for the distributions of test Planckian sources with temperatures varying from 2000 K to 3200 K. The tristimulus values and the chromaticity coordinates for the different blackbody temperatures were calculated according to Equation (6.10). Thereafter, chromaticity coordinate differences between the true (calculated from the 2856 K source distribution using the CIE CMFs) and the calculated values for the different temperature test blackbodies were determined. The results in Figure 6.5 show that the errors in the chromaticity coordinates using the NIST reference colorimeter are less than 0.001 within a temperature range of about 650 K (at 2200 K). These errors dominate over the (above discussed) responsivity-determined chromaticity uncertainty of 0.0004 ($k = 2$). The results show, that the detector-based calibration method, when applied to a tristimulus colorimeter with spectrally well-matched channels to the CIE functions, is highly invariant to source distribution changes.

When sources other than CIE source A are measured, spectral mismatch errors can introduce chromaticity errors even with well-realized (matched) channel spectral responsivities. In this case, the uncertainty of color measurement increases. As an example, the chromaticity errors of the NIST reference tristimulus colorimeter were calculated for different test white-source distributions, such as the LCD-white,

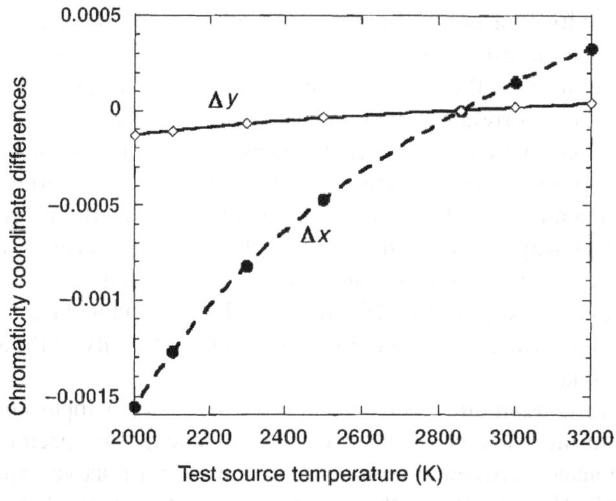

FIGURE 6.5 Chromaticity errors of a reference tristimulus colorimeter[14] for different distribution temperature Planckian test sources. The Δx and Δy chromaticity errors are between the calculated (with Equation (6.9)) and true values (using the CIE functions and a 2856 K reference source).

CRT-white, HMI, cool-white, tri-phosphor, and daylight fluorescent lamps. Using the CIE standard illuminant A as $S(\lambda)$ in Equation (6.9), the Δx and Δy errors were smaller for all of the above test sources than ± 0.0015. When the same calculations were repeated to nonwhite test sources, such as LCD and CRT blue, green, and red, the chromaticity errors increased to a maximum of 0.0058 (Δy of CRT-blue).

Even if the channel calibration factors are highly insensitive to source distribution changes, an approximate spectral distribution of test sources with structured spectral distribution (called here as a special test source) is needed for the calculation of the calibration factors to avoid large spectral mismatch and to keep the chromaticity errors small. In this case, an inexpensive and fast spectrograph with large measurement uncertainty associated to the measured spectrum values (such as an array spectrometer) can be used to obtain a rough spectral distribution of the special test source to be measured. Thereafter, the tristimulus colorimeter calibrated on a detector base (using the array spectrometer measured special test source distribution) can measure the x, y chromaticity coordinates with errors of less than 0.001. The simple and fast detector-based method is very competitive to the traditionally applied scanning spectroradiometer based methods. In the expensive, complicated, and slow scanning spectroradiometer based methods, lamp standards are used for calibration. To avoid using any kind of spectrometers, variable reference source (spectral distribution) models have been worked out to further simplify the detector-based calibration method. These models can adjust the reference spectral source distribution (used to calculate the channel calibration factors) with iterative calculations until it becomes nearly equal to the test source distribution. The model works extremely well for tungsten test lamps. With this approach, spectral mismatch errors can be removed from the tristimulus color measurement of test sources (see e.g., Ref. 17).

Transfer of Calibration

Test (field) colorimeters can be calibrated against a reference colorimeter under illumination by a reference (transfer) source. The transfer source can be a CIE source A for certain (mostly tungsten lamp) applications, but when the test colorimeter is used to measure various sources other than tungsten, the spectral mismatch error can be significant. To avoid such errors, the transfer calibration needs to be done with the same type of source that is to be measured by the test colorimeter to perform strict substitution. The transfer calibration needs to be performed with a known approximate spectral distribution $E_a(\lambda)$ of the source to be measured. The reference colorimeter is calibrated with $E_a(\lambda)$ replaced in Equation (6.9), and determines the tristimulus values X_r, Y_r, Z_r of the source. The test colorimeter is calibrated against the X_r, Y_r, Z_r values. Such transfer calibrations for various sources can be done effectively by using a spectrally tuneable source (STS).[18] The STS is set to simulate a given test source spectrum $E_m(\lambda)$, and the test colorimeter is calibrated against the reference colorimeter under that illumination. The STS can produce as many spectra as needed, and transfer calibration can be done for each type of source with very small spectral mismatch errors, without the knowledge of the relative spectral responsivity of the test colorimeters.

UNCERTAINTY ESTIMATION OF A TRISTIMULUS COLORIMETER MEASUREMENT

Principle of the Tristimulus Calibration for a Self-Luminous Object Measuring Tristimulus Instrument

As discussed in Chapter 3, the tristimulus values X, Y, Z are calculated as convolution of the CMFs $\bar{x}(\lambda), \bar{y}(\lambda), \bar{z}(\lambda)$ with the relative SPD $S(\lambda)$. In this section, we will discuss the uncertainty evaluation when calibrating a tristimulus colorimeter, as an example of a multiple input, multiple output problem discussed in Appendix 1. The symbols X_S, Y_S, Z_S will be used for the tristimulus values and the x_S, y_S, z_S symbols for the chromaticity coordinates of a source with an SPD of $S(\lambda)$. Thus the tristimulus values and chromaticity coordinates are calculated as

$$
\begin{aligned}
X_S &= k\int S(\lambda)\cdot\bar{x}(\lambda)\cdot \mathrm{d}\lambda; \qquad & x_S &= X_S/(X_S+Y_S+Z_S)\\
Y_S &= k\int S(\lambda)\cdot\bar{y}(\lambda)\cdot \mathrm{d}\lambda; \qquad & y_S &= Y_S/(X_S+Y_S+Z_S)\\
Z_S &= k\int S(\lambda)\cdot\bar{z}(\lambda)\cdot \mathrm{d}\lambda; \qquad & z_S &= 1-x_S-y_S
\end{aligned}
\tag{6.11}
$$

The $\bar{x}(\lambda)$-CMF shows a relative minimum at wavelength $\lambda_{\mathrm{sl}} \cong 505\,\mathrm{nm}$ between two maxima, which divides the function in two parts $\bar{x}(\lambda) = \bar{x}_{\mathrm{s}}(\lambda \leq \lambda_{\mathrm{sl}}) + \bar{x}_{\mathrm{l}}(\lambda > \lambda_{\mathrm{sl}})$ referred as "short" and "long." Accordingly, the X tristimulus value can be composed from two parts: $X_S = X_{S\mathrm{s}} + X_{S\mathrm{l}}$. The related chromaticity coordinate $x_S = x_{S\mathrm{s}} + x_{S\mathrm{l}}$ is also a sum of two parts. The shape of the $\bar{x}_{\mathrm{s}}(\lambda)$ function is quite similar to that of the $\bar{z}(\lambda)$ function; thus frequently in tristimulus colorimeters, the $\bar{z}(\lambda)$ channel is used instead of the $\bar{x}_{\mathrm{s}}(\lambda)$ channel, with a ratio $\alpha_S = X_{S\mathrm{s}}/Z_S$ of the tristimulus values.

For a general and traditional calibration of a tristimulus colorimeter, the SPD of a CIE standard illuminant A is used as spectral distribution, which is approximated by a relative Planck-function $P(\lambda, T_{\mathrm{A}})$ with temperature $T_{\mathrm{A}} = 2856\,\mathrm{K}$. The chromaticity coordinates $x_{\mathrm{A}}, y_{\mathrm{A}}, z_{\mathrm{A}}$ are evaluated from a *relative* spectral distribution, and the factor k in Equation (6.11) can be adjusted to give the tristimulus value Y a value of the illuminance E_{A}. That means, the vector $\boldsymbol{D} = \{x_{\mathrm{A}}, y_{\mathrm{A}}, z_{\mathrm{A}}\}$ has to be multiplied with the ratio $E_{\mathrm{A}}/Y_{\mathrm{A}}$. The chromaticity coordinates $x_{\mathrm{A}}, y_{\mathrm{A}}, z_{\mathrm{A}}$ and the ratio α_{A} have fixed values for CIE standard illuminant A as given in the next section.

$$
\begin{aligned}
X_{\mathrm{A}} &= k\int P(\lambda,T_{\mathrm{A}})\cdot\bar{x}(\lambda)\cdot\mathrm{d}\lambda; & x_{\mathrm{A}} &= X_{\mathrm{A}}/(X_{\mathrm{A}}+Y_{\mathrm{A}}+Z_{\mathrm{A}}); & x_{\mathrm{As}} &= X_{\mathrm{As}}/(X_{\mathrm{A}}+Y_{\mathrm{A}}+Z_{\mathrm{A}})\\
Y_{\mathrm{A}} &= k\int P(\lambda,T_{\mathrm{A}})\cdot\bar{y}(\lambda)\cdot\mathrm{d}\lambda; & y_{\mathrm{A}} &= Y_{\mathrm{A}}/(X_{\mathrm{A}}+Y_{\mathrm{A}}+Z_{\mathrm{A}}); & x_{\mathrm{Al}} &= x_{\mathrm{A}}-x_{\mathrm{As}}\\
Z_{\mathrm{A}} &= k\int P(\lambda,T_{\mathrm{A}})\cdot\bar{z}(\lambda)\cdot\mathrm{d}\lambda; & z_{\mathrm{A}} &= 1-x_{\mathrm{A}}-y_{\mathrm{A}}; & \alpha_{\mathrm{A}} &= X_{\mathrm{A}}/Z_{\mathrm{A}}
\end{aligned}
\tag{6.12}
$$

In principle, a tristimulus head is build from three channels with relative spectral responsivity functions $s_x(\lambda), s_y(\lambda), s_z(\lambda)$ matched to the CMFs $\bar{x}(\lambda), \bar{y}(\lambda), \bar{z}(\lambda)$. In some heads four channels have to be calibrated because the $\bar{x}(\lambda)$-function is build from two separate channels with responsivities $s_{xs}(\lambda)$ and $s_{xl}(\lambda)$ matched to the two parts $\bar{x}_s(\lambda \leq \lambda_{sl})$; $\bar{x}_l(\lambda > \lambda_{sl})$ separated for the short- and long-wavelength ranges, as discussed above. A third type of tristimulus heads has the long-wavelength range with responsivity $s_{xl}(\lambda)$ matched to $\bar{x}_l(\lambda)$ and a missing short-wavelength part. The latter is approximated by a fraction α of the signal from the z channel.

A CIE source A illuminating the acceptance area of a typical tristimulus head produces an illuminance E_A and creates photocurrents, which are converted to voltages V_x, V_y, V_z and measured by repeated observations. The standard deviations of the means are taken as variances $\{u^2(V_x), u^2(V_y), u^2(V_z)\}$ and for stable sources no statistical correlations $u(V_x, V_y) = u(V_x, V_z) = u(V_y, V_z) = 0$ will be detected. The output signals multiplied with calibration factors $\boldsymbol{C} = \{c_x, c_y, c_z\}$ change linearly with the illuminance. It should be noted that calibration factors are used instead of the inverse responsivities s_x, s_y, s_z to get a linear model $\boldsymbol{F}(\boldsymbol{x}, \boldsymbol{y}) = \boldsymbol{G} - \boldsymbol{D} = \boldsymbol{0}$. The linear system of equations $\boldsymbol{G}$ is solved by adjusting the values of calibration factors $\boldsymbol{C} = \{c_x, c_y, c_z\}$ to fit the values of the chromaticity coordinates in the vector $\boldsymbol{D}$. The variances and covariances of the input quantities are summarized in the covariance matrix $\boldsymbol{u_x}$.

The matrices for the three different types of tristimulus heads are given below, starting with the first type, the typical three-channel tristimulus head

$$\boldsymbol{G}_{\mathrm{I}} = \begin{bmatrix} V_x \cdot c_x \\ V_y \cdot c_y \\ V_z \cdot c_z \end{bmatrix}; \quad \boldsymbol{D}_{\mathrm{I}} = \begin{bmatrix} x_A \\ y_A \\ 1 - x_A - y_A \end{bmatrix}; \quad \boldsymbol{u}_{x\mathrm{I}} = \begin{bmatrix} u^2(V_x) & 0 & 0 \\ 0 & u^2(V_y) & 0 \\ 0 & 0 & u^2(V_z) \end{bmatrix} \tag{6.13}$$

followed by the four-channel tristimulus head

$$\boldsymbol{G}_{\mathrm{II}} = \begin{bmatrix} V_{xs} \cdot c_{xs} \\ V_{xl} \cdot c_{xl} \\ V_y \cdot c_y \\ V_z \cdot c_z \end{bmatrix}; \quad \boldsymbol{D}_{\mathrm{II}} = \begin{bmatrix} x_{As} \\ x_{Al} \\ y_A \\ 1 - x_{As} - x_{Al} - y_A \end{bmatrix};$$

$$\boldsymbol{u}_{x\mathrm{II}} = \begin{bmatrix} u^2(V_{xs}) & 0 & 0 & 0 \\ 0 & u^2(V_{xl}) & 0 & 0 \\ 0 & 0 & u^2(V_y) & 0 \\ 0 & 0 & 0 & u^2(V_z) \end{bmatrix} \tag{6.14}$$

and finally the three-channel type with a responsivity limited to $\bar{x}_1(\lambda > \lambda_{sl})$

$$G = \begin{bmatrix} V_x \cdot c_x + \alpha \cdot V_z \cdot c_z \\ V_y \cdot c_y \\ V_z \cdot c_z \end{bmatrix}; \quad D = \begin{bmatrix} x_A \\ y_A \\ 1 - x_A - y_A \end{bmatrix};$$

$$u_x = \begin{bmatrix} u^2(V_x) & 0 & 0 \\ 0 & u^2(V_y) & 0 \\ 0 & 0 & u^2(V_z) \end{bmatrix} \tag{6.15}$$

The calibration of this type is explained as an example in more details. The model $\boldsymbol{F}$ and the matrix of the derivatives with respect to the output quantities $\boldsymbol{F_y} = \partial \boldsymbol{F}/\partial \boldsymbol{C}$ and the inverse matrix (exponent "−1") are determined.

$$F(x,y) = G - D = \begin{bmatrix} V_x \cdot c_{x0} + \alpha \cdot V_z \cdot c_{z0} - x_A \\ V_y \cdot c_{y0} - y_A \\ V_z \cdot c_{z0} - (1 - x_A - y_A) \end{bmatrix}$$

$$F_y = \begin{bmatrix} V_x & 0 & \alpha \cdot V_x \\ 0 & V_y & 0 \\ 0 & 0 & V_z \end{bmatrix}; \quad F_y^{-1} = \begin{bmatrix} 1/V_x & 0 & -\alpha/V_x \\ 0 & 1/V_y & 0 \\ 0 & 0 & 1/V_z \end{bmatrix} \tag{6.16}$$

The vector $\boldsymbol{C}$ of calibration factors is found from the product of the inverse matrix $\boldsymbol{F_y^{-1}}$ and the nonhomogeneous part $\boldsymbol{D}$ multiplied with the factor E_A/y_A.

$$C = F_y^{-1} \bullet D \cdot \frac{E_A}{y_A}; \quad \begin{bmatrix} c_x \\ c_y \\ c_z \end{bmatrix} = \begin{bmatrix} (x_A - \alpha \cdot z_A)/V_x \\ y_A/V_y \\ (1 - x_A - y_A)/V_z \end{bmatrix} \cdot \frac{E_A}{y_A} \tag{6.17}$$

The covariance matrix $\boldsymbol{u_y} = \boldsymbol{Q} \bullet \boldsymbol{u_x} \bullet \boldsymbol{Q}^T$ associated to the calibration factors $\boldsymbol{C}$ is calculated from the variance matrix $\boldsymbol{u_x}$ of the input data in Equation (6.13) and the sensitivity matrix $\boldsymbol{Q} = \boldsymbol{F_y^{-1}} \bullet \boldsymbol{F_x}$ given below, where the factor on the right is the partial derivative $\boldsymbol{F_x} = \partial \boldsymbol{F}/\partial V$ with respect to the input quantities V_x, V_y, V_z of the model $\boldsymbol{F}$.

$$F_x = \begin{bmatrix} c_x & 0 & \alpha \cdot c_z \\ 0 & c_y & 0 \\ 0 & 0 & c_z \end{bmatrix}; Q = F_y^{-1} \bullet F_x = \begin{bmatrix} -c_x/V_x & 0 & 0 \\ 0 & -c_y/V_y & 0 \\ 0 & 0 & -c_z/V_z \end{bmatrix} \tag{6.18}$$

The variance–covariance matrix $\boldsymbol{u}_y$ holds the squared uncertainties in the main diagonal, and the uncertainties of the output quantities are noncorrelated.

$$\boldsymbol{u}_y = \boldsymbol{Q} \bullet \boldsymbol{u}_x \bullet \boldsymbol{Q}^{\mathrm{T}} = \begin{bmatrix} (u(V_x)(x_{\mathrm{A}} - \alpha \cdot z_{\mathrm{A}})V_x^{-2})^2 & 0 & 0 \\ 0 & (u(V_y)y_{\mathrm{A}}V_y^{-2})^2 & 0 \\ 0 & 0 & (u(V_z)z_{\mathrm{A}}V_z^{-2})^2 \end{bmatrix} \cdot \left(\frac{E_{\mathrm{A}}}{y_{\mathrm{A}}}\right)^2 \tag{6.19}$$

The uncertainties $\boldsymbol{u}(\boldsymbol{C})$ associated to the calibration factors $\boldsymbol{C}$ with values solved in Equation (6.15) are determined as the square root of the variances in the variance–covariance matrix $\boldsymbol{u}_y$:

$$\boldsymbol{C} = \begin{bmatrix} \dfrac{x_{\mathrm{A}} - \alpha \cdot z_{\mathrm{A}}}{V_x} \\ \dfrac{y_{\mathrm{A}}}{V_y} \\ \dfrac{1 - x_{\mathrm{A}} - y_{\mathrm{A}}}{V_z} \end{bmatrix} \cdot \frac{E_{\mathrm{A}}}{y_{\mathrm{A}}}; \qquad \boldsymbol{u}(\boldsymbol{C}) = \begin{bmatrix} u_{\mathrm{rel}}(V_x)\dfrac{(x_{\mathrm{A}} - \alpha \cdot z_{\mathrm{A}})}{V_x} \\ u_{\mathrm{rel}}(V_y)\dfrac{y_{\mathrm{A}}}{V_x} \\ u_{\mathrm{rel}}(V_z)\dfrac{1 - x_{\mathrm{A}} - y_{\mathrm{A}}}{V_x} \end{bmatrix} \cdot \frac{E_{\mathrm{A}}}{y_{\mathrm{A}}} \tag{6.20}$$

Numerical Example for a Tristimulus Calibration

An incandescent lamp is the reference for the calibration of a tristimulus colorimeter. The source and the colorimeter are mechanically aligned to the optical axis, electrically supplied and in thermal equilibrium after burning-in. The radiation incident on the acceptance area of the colorimeter head has a relative spectral distribution characterized by a distribution temperature $T_{\mathrm{A}} = 2856\,\mathrm{K}$ denoted as CIE standard illuminant A and the value of illuminance E_{A} is also stated in the certificate of the standard lamp.

The three channels of the colorimeter with Si photodiodes are matched by glass filters to the CMFs $\bar{x}_1(\lambda), \bar{y}(\lambda), \bar{z}(\lambda)$. The mean values V_{x1}, V_y, V_z of the three photocurrents plus the related standard deviations taken as standard uncertainties $u(V_{x1}), u(V_y), u(V_z)$ are calculated from 30 readings measured simultaneously. No significant statistical correlation was detected.

Table 6.1 shows input data of the example. The first column gives the reference to the channels x_1, y, z. The mean values and related uncertainties of photocurrents are listed in the columns marked V and $u(V)$. The chromaticity coordinates calculated

TABLE 6.1 Measurement results of the example

	V (nA)	$u(V)$ (nA)	$(x_{\mathrm{A}}, y_{\mathrm{A}})$, $\alpha = 0.1512$
x_1	6.543	0.0065	0.4475
y	5.432	0.0043	0.4074
z	4.321	0.0087	0.1451
E_{A}			50.0 lx

for the distribution temperature of the radiation are given in the column headed (x_A, y_A). The illuminance E_A is stated additionally.

The values of the table have to be inserted into Equation (6.18). With the value of $\alpha = 0.1512$, this gives the values of the output quantities, which are the calibration factors in the vector $\boldsymbol{C}$.

$$\boldsymbol{C} = \begin{bmatrix} c_{xl} \\ c_y \\ c_z \end{bmatrix} = \begin{bmatrix} (x_A - \alpha \cdot z_A)/V_{xl} \\ y_A/V_y \\ z_A/V_z \end{bmatrix} \cdot \frac{E_A}{y_A} = \begin{bmatrix} 7.982 \\ 9.205 \\ 4.121 \end{bmatrix} \tag{6.21}$$

From the input values and Equation (6.18), the values in the uncertainty matrix of the output quantities are determined. It should be noted, that all nondiagonal elements are zero that means the calibration coefficients and the associated uncertainties are not correlated.

$$\boldsymbol{u}_y = \begin{bmatrix} u^2(c_x) & 0 & 0 \\ 0 & u^2(c_y) & 0 \\ 0 & 0 & u^2(c_z) \end{bmatrix} = \begin{bmatrix} 0.63\ 10^{-4} & 0 & 0 \\ 0 & 0.53\ 10^{-4} & 0 \\ 0 & 0 & 0.69\ 10^{-4} \end{bmatrix} \tag{6.22}$$

The expanded uncertainties are determined for a coverage factor of 95.45% probability as the square root from the values in the diagonal of Equation (6.20) multiplied with $k = 2$ as $\boldsymbol{U} = k \cdot \boldsymbol{u}(\boldsymbol{C})$, and the result is reported as:

The calibration factors of the colorimeter with tristimulus head were determined at a level of approximately 50 lx with light having a spectral distribution similar to CIE standard illuminant A:

$$\boldsymbol{C} = \begin{bmatrix} 7.982 \pm 0.016 \\ 9.205 \pm 0.015 \\ 4.121 \pm 0.017 \end{bmatrix} \cdot \frac{\text{lx}}{\text{nA}} \tag{6.23}$$

The expanded uncertainties associated to the calibration factors are stated as the standard uncertainties multiplied by the coverage factor $k = 2$, which corresponds to a coverage probability of approximately 95%. The standard uncertainty has been determined in accordance with the "Guide to the Expression of Uncertainty in Measurement"[19] as described in the Appendix 1 of the book.

Calibration for Selected Spectral Distributions

The general calibration of a tristimulus head is the basis for a wide use of the head without any specialization. If only a specific spectral distribution has to be measured, then also a specific set of calibration factors can be determined using the same procedure as before with a vector **D** containing chromaticity coordinates specific for the spectral distribution of the test objects, see Equation (6.11), different of those of CIE standard illuminant A.

Often neither a general calibration nor a unique spectral distribution has to be measured, but a certain selection of spectral distributions characterized by the index "*i*." Then the values of the calibration factors should be determined as a "best fit" for this selection. In the following extended model, three more equations are entered for each new spectral distribution in the system of equations $\boldsymbol{G}$, in the vector $\boldsymbol{D}$ of chromaticity coordinates, and three more lines and rows in the variance matrix u_x of the input quantities.

$$\boldsymbol{G} = \begin{bmatrix} V_{x,i} \cdot c_x + \alpha \cdot V_{z,i} \cdot c_z \\ V_{y,i} \cdot c_y \\ V_{z,i} \cdot c_z \\ \ldots\ldots \\ V_{x,n} \cdot c_x + \alpha \cdot V_{z,n} \cdot c_z \\ V_{y,n} \cdot c_y \\ V_{z,n} \cdot c_z \end{bmatrix}; \quad \boldsymbol{D} = \begin{bmatrix} x_{A,i} \\ y_{A,i} \\ 1 - x_{A,i} - y_{A,i} \\ \ldots\ldots \\ x_{A,n} \\ y_{A,n} \\ 1 - x_{A,n} - y_{A,n} \end{bmatrix}; \tag{6.24}$$

$$u_x = \begin{bmatrix} u^2(V_{x,i}) & 0 & 0 & 0 & 0 & 0 & 0 \\ 0 & u^2(V_{y,i}) & 0 & 0 & 0 & 0 & 0 \\ 0 & 0 & u^2(V_{z,i}) & 0 & 0 & 0 & 0 \\ 0 & 0 & 0 & \ldots\ldots & 0 & 0 & 0 \\ 0 & 0 & 0 & 0 & u^2(V_{x,n}) & 0 & 0 \\ 0 & 0 & 0 & 0 & 0 & u^2(V_{y,n}) & 0 \\ 0 & 0 & 0 & 0 & 0 & 0 & u^2(V_{z,n}) \end{bmatrix}$$

The linear model $\boldsymbol{F}(\boldsymbol{x},\boldsymbol{y}) = \boldsymbol{G} - \boldsymbol{D} = \boldsymbol{0}$ for the least mean square fit for the determination of the calibration factors $\boldsymbol{C} = \{c_x, c_y, c_z\}$ is the same as mentioned earlier and a set of start values $\boldsymbol{C}_0 = \{c_{x0}, c_{y0}, c_{z0}\}$ is modified by iteration to a better solution, which is taken as the start value for the next iteration. The index "0" at the matrix product indicates that the valid start values have to be used for calculation in each step.

$$\boldsymbol{C}_1 = \boldsymbol{C}_0 - \lambda \cdot \Delta\boldsymbol{C}; \quad \Delta\boldsymbol{C} = (\boldsymbol{F}_y^{-1} \bullet \boldsymbol{F})_0; \quad \sqrt{\Delta\boldsymbol{C} \bullet \Delta\boldsymbol{C}} < \varepsilon \tag{6.25}$$

The value $0 < \lambda \leq 1$ controls the effective width of each step during the iteration, to reduce the possibility of divergence. The iteration stops when the correction $\sqrt{\Delta\boldsymbol{C} \bullet \Delta\boldsymbol{C}} < \varepsilon$ is small enough. The variance matrix of the calibration factors is found as before.

$$\boldsymbol{U}_y = \boldsymbol{Q} \bullet \boldsymbol{U}_x \bullet \boldsymbol{Q}^{\mathrm{T}}; \quad \boldsymbol{Q} = -\boldsymbol{F}_y^{-1} \bullet \boldsymbol{F}_x \tag{6.26}$$

Note: The solution was found with the assumption that all values of the input quantities are determined with (nearly) the same associated uncertainties. If a weighted least mean square approximation is needed, then the steps in Equation (A1.11) of Appendix 1 should be used.

GLOSSARY

Basic Terms

The definitions presented are consistent with the CIE Vocabulary[20] (entries in the CIE-ILV are marked by { } brackets) as far as they are enumerated there.

Acceptance area (of a colorimeter head)
: The area of the colorimeter head over which the incident radiation is evaluated, and which is the location for distance measurements.

Colorimeter head
: Part of a physical colorimeter containing the detector and means for spectral and spatial corrections of the detector responsivity. Typically, the colorimeter head will contain several separate detectors and associated filters, where, for example, each detector and filter combination might be designed to match one of the colorimetric functions.

Detector (of optical radiation)
: Device in which the incident optical radiation produces a measurable physical effect.

Note: Several other terms are used like Si-photodiode, silicon photodiode, photovoltaic cell, photo cell, photodetector, and so on.

Image-taking colorimetry
: Colorimetry where each tristimulus value of all pixels of an image are taken simultaneously.

Limiting aperture
: Aperture limiting the incoming radiation. In the case of an illuminance-measuring colorimeter, the plane of the aperture from which the inverse squared law is valid.

Optical axis of the colorimeter head
: In the case of illuminance-measuring input optics: the axis perpendicular to the middle point of the surface of the limiting aperture of the colorimeter head;
In the case of luminance-measuring input optics: the axis through the middle of the input field.

Responsivity; (of a detector) [s] {845-05-54}
: quotient of the detector output signal Y by the detector input quantity X

$$s = Y/X$$

Note: If the detector signal is Y_0 in the absence of input and is Y_t when there is a detector input quantity X, the responsivity is $s = (Y_t - Y_0)/X$.

Spectral responsivity {845-05-56}

(Absolute) spectral responsivity (of a detector) $[s(\lambda)]$: quotient of the detector output $dY(\lambda)$ by the monochromatic detector input $dX_e(\lambda) = X_{e,\lambda}(\lambda)d\lambda$ in the wavelength interval $d\lambda$ as a function of the wavelength λ.

$$s(\lambda) = dY(\lambda)/dX_e(\lambda)$$

Relative spectral responsivity (of a detector) $[s_r(\lambda)]$ {845-05-57}

Ratio of the spectral responsivity $s(\lambda)$ of the detector at wavelength λ to a given reference value s_m.

$$s_r(\lambda) = s(\lambda)/s_m$$

Note: The given reference value s_m can be an average value, a maximum value, or an arbitrarily chosen value of $s(\lambda)$. Sometimes, it may be helpful to choose

$$s_m = \frac{\int_{\lambda_1}^{\lambda_n} s(\lambda)d\lambda}{\lambda_n - \lambda_1} = \frac{\sum_{i=1}^{n} s(\lambda_i)\Delta\lambda}{\lambda_n - \lambda_1}$$

where λ_1 and λ_n are the wavelength limits, for which the normalization is performed.

Tristimulus colorimeter

instrument for measuring the tristimulus values of a color stimulus using broad-band input channels.

Specific Terms

The following terms are used for the characterization of the performance of tristimulus colorimeters and influence the uncertainty of measurements. As the different factors might influence the measurement in the single channels of the colorimeter to a different extent, these quantities have to be determined for each output channel of the tristimulus colorimeter independently. This is shown either using an x, y, z index for the given quantity or, in shortened form by the use of an index, i, where $i = 1$ refers to the x channel, $i = 2$ to the y channel and $i = 3$ to the z channel[§].

Calibration standard uncertainties $u_{c,x}$, $u_{c,y}$, $u_{c,z}$,: $u_{c,i}$,

Quantities describing the uncertainty of the calibration factors of the colorimeter.

Spectral mismatch error indices $f'_{1,x}$, $f'_{1,y}$, $f'_{1,z}$: $f'_{1,i}$

indices that characterize the magnitude of the spectral mismatch of the individual colorimetric channels compared to the respective CMFs.

[§]For more details see Ref.1.

UV response error indices $f_{UV,x}$, $f_{UV,y}$, $f_{UV,z}$: $f_{UV,i}$
Indices that characterize the magnitude of the UV responsivity of the individual colorimetric channels.

IR response error indices $f_{IR,x}$, $f_{IR,y}$, $f_{IR,z}$: $f_{IR,i}$
Indices that characterize the magnitude of the IR responsivity of the individual colorimetric channels.

Directional response error indices $f_{2,x}$, $f_{2,y}$, $f_{2,z}$: $f_{2,i}$
Indices that characterize the deviation of the directional response of the colorimeter head from its prescribed form (e.g., in the case of a tristimulus colorimeter intended for use in illuminance geometry, this index characterizes the deviation of the directional response from the ideal cosine law behavior).

Linearity error indices $f_{3,x}$, $f_{3,y}$, $f_{3,z}$: $f_{3,i}$
Indices that describe the change in the colorimeter responsivity from its nominal response as the input illuminance or luminance is varied.

Display unit error indices $f_{4,x}$, $f_{4,y}$, $f_{4,z}$: $f_{4,i}$
Indices that describe the deviation from linearity of the analog or digital display units of the colorimeter.

Fatigue indices $f_{5,x}$, $f_{5,y}$, $f_{5,z}$: $f_{5,i}$
Indices that describe the stability of the colorimeter responsivity under constant irradiation over long periods of time.

Temperature-dependent error indices $f_{6,x}$, $f_{6,y}$, $f_{6,z}$: $f_{6,i}$
Indices that describe the influence of changes in the ambient temperature on the colorimeter responsivity.

Modulated radiation error indices $f_{7,x}$, $f_{7,y}$, $f_{7,z}$: f_7
Indices that describe the errors produced in the colorimeter responsivity by modulated radiation of various frequencies.

Polarization error indices $f_{8,x}$, $f_{8,y}$, $f_{8,z}$: $f_{8,i}$
Indices that describe the errors produced in the colorimeter responsivity by polarized radiation.

Spatial inhomogeneity error indices $f_{9,x}$, $f_{9,y}$, $f_{9,z}$: $f_{9,i}$
Indices that describe the errors produced in the colorimeter responsivity by nonuniform distribution of irradiation.

Range change error indices $f_{11,x}$, $f_{11,y}$, $f_{11,z}$: $f_{11,i}$
Indices describing the errors produced by the nonexact adjustment of the range setting of the display units or amplifiers.

REFERENCES

1. CIE (2007) *Methods for characterizing tristimulus colorimeters for measuring the color of light, CIE* **179**: 2007.

2. Fryc I (2003) Distribution of the spectral correction error of the CCD, in *Proceedings of the CIE Symposium on Temporal and Spataial Aspects of Light and Colour Perception and Measurement*, **CIE x025**:2003 39–43, CIE Central Bureau, Vienna.
3. Technical leaflet: Colorimeter head, InPhoRa California, USA.
4. CIE Technical Report (1987) *Methods of characterizing illuminance meters and luminance meters*, CIE **69** - 1987, CIE Central Bureau, Vienna.
5. Eppeldauer GP, Sauter G, Gardner JL (2006) Uncertainties of spectral responsivity measurements, in: *Proceedings of the 2nd CIE expert symposium on Measurement Uncertainty*, 133–138. CIE Central Bureau, Vienna.
6. Thompson A, Chen H-M (1994) Beamcon III, a linearity measurement instrument for optical detectors. *J. Res. NIST*, **99**, 751–755.
7. Eppeldauer GP (2000) Noise-optimized silicon radiometers. *J. Res. NIST*, **105**, 209–219.
8. Schanda J, Lux G (1973) On the electronic correction of errors in a tristimulus colorimeter, in: *Proceedings of the AIC Colour 73*, York, Hilger, London, pp. 466–469.
9. Ohno Y, Hardis JE (1997) Four-color matrix method for correction of tristimulus colorimeters, in: SID Fifth Color Imaging Conference, pp. 301–305.
10. Ohno Y, Brown SW (1998) Four-color matrix method for correction of tristimulus colorimeters—Part 2. SID Sixth Color Imaging Conference, pp. 65–68.
11. Cromer CL, Eppeldauer G, Hardis JE, Larason TC, Ohno Y, Parr AC (1996) The NIST detector-based luminous intensity scale, *J. Res. NIST*, **101**(2), 109–132.
12. Eppeldauer GP (1998) Spectral response based calibration method of tristimulus colorimeters. *J. Res. NIST* , **103**(6), 615–619.
13. Eppeldauer GP, Brown SW, Miller CC, Lykke KR (2003) Improved accuracy photometric and tristimulus-color scales based on spectral irradiance responsivity, in: *Proceedings of the 25th Session of the CIE*, **Vol. 1**, D2-30–D2-33.
14. Eppeldauer GP, Racz M (2004) Design and characterization of a photometer-colorimeter standard. *Appl. Opt.*, **43**, 2621–2631.
15. Gentile TR, Houston JM, Cromer CL (1996) Realization of a scale of absolute spectral response using the NIST high-accuracy cryogenic radiometer. *App. Opt.*, **35**, 4392–4403.
16. Eppeldauer GP. Lynch DC (2000) Opto-mechanical and electronic design of a tunnel-trap Si-radiometer, *J. Res. NIST.*, **105**(6), 813–828.
17. Eppeldauer GP, Brown SW, Lykke KR, Ohno Y (2005) Realization and application of a detector-based tristimulus color scale at the National Institute of Standards and Technology, USA, in *Proceedings of the AIC Colour 05* Part I (Ed., J. L. Nieves and J. H-Andres), pp. 693–696.
18. Fryc I, Brown SW, Eppeldauer GP, Ohno Y (2004) A spectrally tunable solid-state source for radiometric, photometric, and colorimetric applications. *Proc. SPIE*, **5530**, 150–159.
19. International Organisation for Standardisation, *Guide to the Expression of Uncertainty in Measurement*, Geneva, 1993.
20. CIE (1987) International Lighting Vocabulary. CIE Publication 17.4–1987.
21. Brown SW, Larason TC, Habauzit C, Eppeldauer GP, Ohno Y, Lykke KR (2001) Absolute radiometric calibration of digital imaging systems, in: *Proc. SPIE*, **4306**, 13–21.

7

COLOR MANAGEMENT

JÁN MOROVIČ and JOHAN LAMMENS
Hewlett-Packard, S.L. (LFP) Avda Graells 501, 08174 Sant Cugat del Valles, Barcelona, Spain

INTRODUCTION

"The eye is the window of the human body through which it feels its way and enjoys the beauty of the world," (Leonardo Da Vinci). And as that beauty is presented in full color to the sense of vision, particular attention needs to be paid whenever color is brought about, especially as color can today be created by a plethora of means. Because these mean encode, control, and address color in different ways, color information needs to be managed.

The relatively recent emergence of new color imaging technologies, for example, digital cameras, printers, displays, and projectors, their widespread availability, and the use of color information from one technology in another have made dealing with how color is communicated a ubiquitous need. Take, for example, a family who use a digital camera to take pictures during their holidays, then view these pictures on their television and home PC's display, print out some of the pictures on their desktop printer, share the pictures with their friends and relatives via a Web site, and get a commercial print service provider to produce a poster from one of their holiday snaps.

On closer inspection it can be seen that the above example involves at least five distinct types of color imaging technologies (each capable of addressing a different range of color appearances), two or three operating systems, more than three ways of encoding color information, and at least six instances

Colorimetry: Understanding the CIE System, Edited by János Schanda

of interfacing color information between technologies with their different capabilities and encodings.

The processes that translate and communicate color information at such interfaces can be referred to as color management. In other words, color management is, for example, used to take color information captured by the digital camera and translate it to make it suitable for display on a television or a computer display. Color management is also used to take color information viewed on a display and translate it for a desktop or commercial large format printer or translate it to make it suitable for sharing via the Internet.

In summary, color management can be defined as "the process of providing a chosen relationship between colors generated using different imaging devices by translating color information from a source device to color information for a destination device."

Color Reproduction Objectives

A key question therefore is what the relationships between source and destination colors can and should be. While there is a natural inclination to say that "the colors should match," it will be shown next that this is rarely possible and even when it is possible it might not be desirable.

In his analysis of reproduction objectives, Hunt[1] presents a hierarchy of degrees of matching between a pair's of colors (or color images) and shows clearly why in most cases even a weak degree of matching is not possible.* He starts with discussing differences in the *spectral* properties of source and destination colors† and goes on to point out the effect of differences in light source intensity and chromaticity between the color pair's viewing conditions, making a match in colorimetry impossible in most cases and undesirable in the rest. A match even in, for example, relative CIE XYZ‡ tristimulus values[2] between a source color from a display with a white point of 9300 K and a printed destination color viewed under a D50 simulator would not preserve the source's appearance but result in a printed color that looked bluish when compared with the displayed source.

The objectives that Hunt puts forward as practically meaningful are those of "corresponding" and "preferred" color reproduction. Corresponding color reproduction is equivalent to the CIE technical committee TC 8-03's "subjective accuracy."[19] In subjectively accurate color reproduction, the destination is "as close to the [source] as possible, this similarity is determined psychophysically and [there are] no image enhancing aims." In other words, this objective is about obtaining

*Please note that unless explicitly stated otherwise, color reproduction will be used here to refer both to the reproduction of individual colors and of color images.

†For example, a CRT's red has the "spiky" spectrum of emissions from a rare-earth phosphor whereas printers typically obtain red by combining their magenta and yellow inks, which are very much smoother spectrally.

‡In this chapter - to follow the general practice of the colour management community - the XYZ tristimulus desgnation will not be printed in Italic.

such a destination color that looks as similar to the source color as possible given the differences in the range of color appearances (i.e., color gamuts) of the source and the possible color range of the destination under their respective viewing conditions. (See Chapter 11 "CIE color appearance models and associated color spaces" and CIE publication 159:2004[3] for further details.) Note that this implies first that a source image that looks unpleasant (e.g., has some defects) will result in a destination image that also looks unpleasant and second that the destination device's color capabilities might not be used to their full potential. An example of the second implication is the reproduction of a source image printed on uncoated paper by a destination image on coated, glossy paper that has a larger color gamut. In this case the destination image would not make use of the entire available color gamut if the "subjectively accurate" reproduction objectives were followed.

Because subjective accuracy is not what is needed in many color reproduction scenarios, Hunt and the CIE also specify the "preferred" color reproduction objective, which in Hunt's words intends to "give a more pleasing result to the viewer" than the source did. For example, when printing holiday snaps there is less concern about the accuracy of representing captured images than about having "nice" looking photos and the preferred color reproduction objective is appropriate. Achieving this can include changes to the source that make memory colors more like their ideal prototypes[4] (e.g., by changing the sky in an image's reproduction to look "bluer" than it was in the source), changes to the tonal and chromatic distribution of an image[5], and other modifications of the destination image that adapt it to the destination gamut's properties.

Viewing a Pair of Colors

Before addressing the question of how to manage color information when interfacing it between two devices, it is useful to look at a pair of colors generated using different imaging devices and consider the range of factors that contribute to their perceived relationship (e.g., whether they match or how they differ).

The first thing to notice in Figure 7.1 is that there are many factors that are involved in determining the outcome of comparing the color appearances of two stimuli and that if any of these factors change, the outcome of the comparison can change too.

More specifically it can be seen from Figure 7.1 that there are several factors besides the pair of color stimuli that affect color appearance:

1. The presence of other stimuli generated using the two devices that are simply viewed alongside them (e.g., Are the white points corresponding to the two devices visible? Is there a neutral more luminous than the two white points present when the color stimuli are viewed? etc.—affects adapted white and hence perceived appearance attributes);

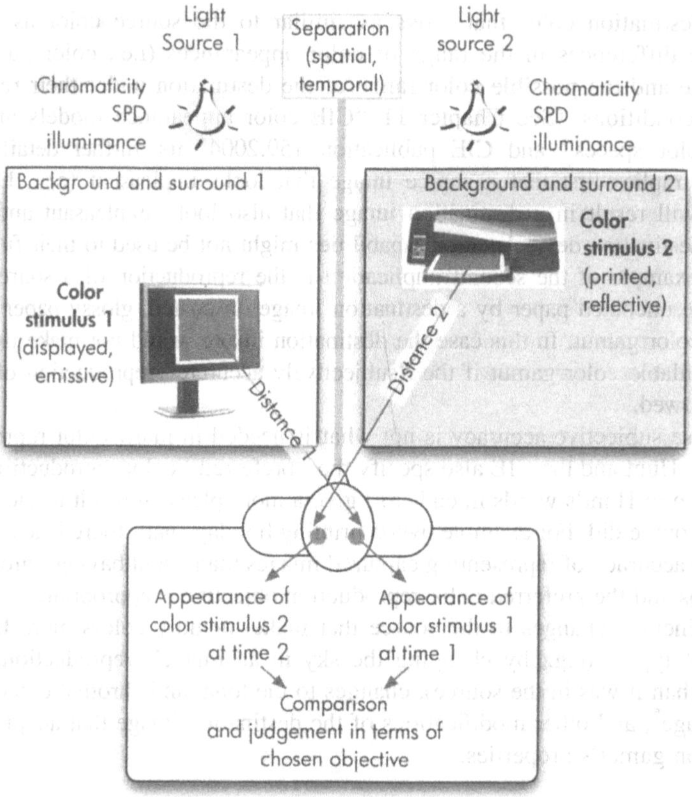

FIGURE 7.1 Viewing a pair of colors generated using different imaging devices. See color insert.

2. The background and surround against which the stimuli are viewed (affecting adaptation, involving simultaneous contrast‡ and crispening§ effects);
3. The spectral power distributions, chromaticities, and illuminances of the light sources under which the stimuli are viewed (strongly affecting adaptation, setting limits to possible colors perceptible under them, determining the Hunt effect**);

‡*Simultaneous contrast* refers to the phenomenon whereby the color surrounding another affects it in the opposite direction of their difference (e.g., a darker surrounding color will make the central color appear lighter). See Chapters 11 and 12 for further detail.

§*Crispening* refers to the phenomenon whereby a pair of similar colors looks more different from each other against a background similar to them than against a dissimilar background (e.g., the difference between a slightly lighter and a slightly darker than mid-gray color will appear greater when they are seen against a mid-gray background than when they are seen against a white background). See Chapters 11 and 12 for further detail.

**The Hunt effect describes the phenomenon of surface colors appearing to be more chromatic as illumination level increases. See Chapters 11 and 12 for further detail.

4. The distances from which they are seen (affecting perceived lightness and chroma[6]).
5. The nature of separation between the two stimuli's viewing environments (i.e., stimuli can be seen simultaneously under mixed viewing conditions where the state of adaptation can be complex to determine, or at different times in which case the judgment about the relationship of stimulus appearances also involves memory[††]);
6. The specific physiology of the individual observer (determining the particular color matching functions, affecting perceptibility thresholds);
7. The experience the observer has for making the requested observations, comparisons, and judgments (affecting perceptibility thresholds, judgment tolerances, and repeatability).

The following are the key points to take away from this analysis: First, the relationship between the ways in which two color stimuli appear to an observer is not a property solely of those two stimuli but rather of a complex constellation of numerous states. Second, it is not color stimuli that are compared and judged, but the mental representations of their appearances, which clearly points to the inherently subjective nature of the color comparison task.

These observations, however, lead to the question of how color can be managed given that color management only deals with interfacing and translating color information between devices and therefore has potential to influence only a small part of a color reproduction setup—the color stimuli. To address this question, at least the following four aspects need to be considered: First, many of the factors have limited impact on the final judgment if their states change little (e.g., a viewing distance of 60 cm versus 70 cm will lead to similar experiences). Second, if the comparison of color stimuli is not simultaneous, the limits of color memory set wide thresholds on judgments and diminish the impact that factor changes and differences have. Third, for color critical tasks stimuli need to be compared under standardized or at least controlled viewing conditions (e.g., ISO 3664:2000 specifies how to view prints and transparencies in the graphic arts context). Fourth, the complexity of color reproduction setups introduces fundamental limits to what can be obtained.

In other words, color can be managed closely only if other factors of a color reproduction setup, beyond the imaging devices, are also controlled. However, color management can also lead to acceptable results even if such control is not possible either when comparisons are not done simultaneously or when the state of a color reproduction setup is close to what the color management process assumes.

Conceptual Stages of Color Reproduction

Given the above discussion of color reproduction objectives and the factors involved in the task of viewing and comparing colors, let us now turn to the conceptual stages

[††]A time difference of even just 15 s introduces a ΔE^*_{ab} color differences of around 5 units between the color seen and the color remembered, and this difference slowly increases with time.[7]

that are involved in translating color data from a source device into color data for a destination device.

Before going into the details of the color reproduction process, it is worth making a distinction between a *color imaging device* and a *color reproduction medium.* The difference between the two can readily be seen when considering a computer display device and a printing device. While in the former case the device is what is viewed to see the generated color stimulus (i.e., we look at the stimulus generated by changing the properties of parts of the display), in the latter case it is a separate object that is altered with the help of the device and constitutes the color stimulus that is viewed (i.e. the printed pattern on a substrate). Here the printer is an imaging device, the print is a color reproduction medium, and the display is both a device and a medium. An imaging device is involved in generating a color reproduction medium, which is what is viewed in the end (e.g., the display, the print (but not the printer), the projection (but not the projector)).

Device Color Spaces

Coming back to the process of translating color information between a pair of devices (as it is these that can be controlled directly), the starting point is an encoding of source colors in terms of a device color space of the source device (Figure 7.2).

The term *device color space* refers to a space that does not in itself have a colorimetric interpretation, but is simply used to address whatever colors can be generated using an imaging device. For example, the same red, green, and blue (RGB) values can be sent to a range of devices. Depending on which device they are sent

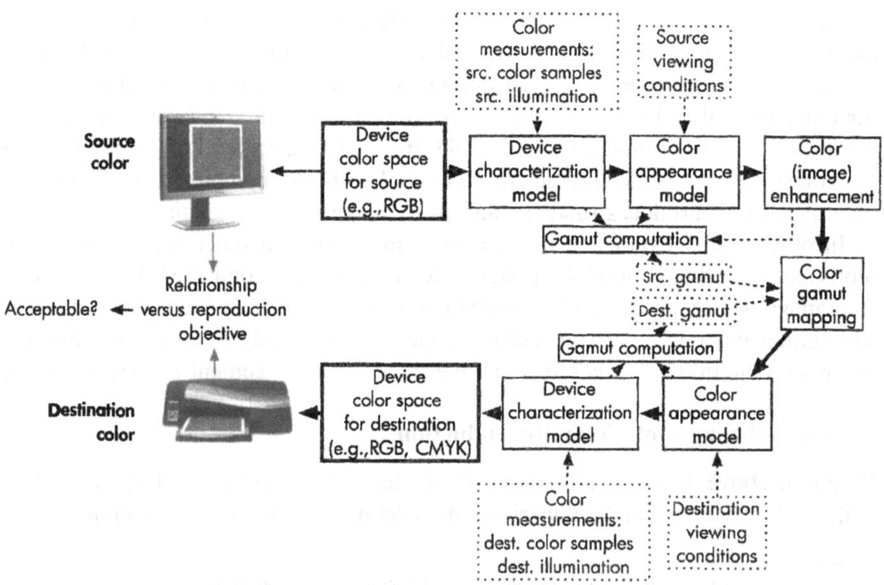

FIGURE 7.2 Conceptual stages of color reproduction.

to, they will result in different color stimuli. Hence, such device RGB values do not in themselves represent specific color stimuli, but only allow for the addressing of the color stimuli that can be generated on the device they are sent to.

A key advantage of device color spaces is that they address all of the range of stimuli that a device can generate and nothing but that range, and this makes them very well suited for addressing device color capabilities. Of increasing relevance is also that device color spaces provide a layer of abstraction between the inputs to a device and the colorants that the device uses. For example, device RGB (dRGB) can be used to address all of the color stimuli that can be generated using a display with red, green and blue phosphors; a projector that in addition to RGB channels also has a white channel, and a printer that uses cyan, magenta, yellow, and black inks or also other inks like dilutions of some of CMYK. Each of these devices then deals internally with the assignment of colorant combinations to each device RGB triplet so that the device RGB space addresses all of its color range. Analogously, CMYK can also be used as a device color space.

Device Characterization and Calibration

Given a set of device color space coordinates (e.g., dRGB $= [10, 20, 30]$) from the source device, the first step of the color reproduction process is to determine what stimulus the device will generate when receiving it as an input.

This translation of device color space inputs to stimulus outputs is the role of *device characterization*, and there exist a wide variety of models for this stage of the process that will be introduced in following sections of this chapter. In all cases, characterization models have their parameters determined from pairs of (a) device color space inputs and (b) color measurements of stimuli generated by the device when receiving them. Given these parameters a characterization model can then perform two predictions:

- for a device color space input, predict the color stimulus that would be measured if it were sent to the device (e.g., given $dRGB = [10, 20, 30]$ predict CIE XYZ tristimulus values that would be measured were it displayed on an LCD);
- for a stimulus that is to be generated using a device, predict what device color space inputs to send to it (e.g., given CIE $XYZ = [30, 20, 10]$ predict what dCMYK values to send to a printer to match it).

In addition to the level of accuracy achievable using a given characterization model, a strong condition to its predictive powers is the state in which the device is to which it applies. If the device has not changed since measurements were made from which model parameters were determined, then all is well. However, if the device has changed—either because of changes to its components (e.g., aging), to its environment (e.g., change of temperature, relative humidity), or to its settings (e.g., having a display's brightness setting increased)—then the relationships that the model attempts to represent may no longer hold and while its predictions might have worked in the original state, they now cease to hold.

To address the potential mismatch between the state in which a device has been characterized and the state in which it is at a later date, the process of *device calibration* is used. Simply put, the role of device calibration is to take a device in whatever state it is and restore it to a predefined state. Here the predefined state can be a standard one (e.g., sRGB[8] for displays) or the state of the given device at an earlier time. For a more detailed discussion of device calibration and characterization, see Bala.[9]

Color Appearance Model

The next stage is to start with the prediction of the source stimulus and predict from it its color appearance under the source's viewing conditions, as it is this appearance that is more closely involved in color communication than the stimulus itself.

Although there are several available models for predicting color appearance, currently the most advanced model, able to predict the appearance of color stimuli and suitable for use in color reproduction, is CIECAM02 (see Chapter 11). In addition to CIECAM02, work on extending it to take into account some spatial phenomena and make it perform better for complex images (e.g., photographs, etc.) has also been done[10] (see also Chapter 12) and is being further promoted within the CIE's TC8-08 on *Spatial Appearance Models*.

Finally, it is also important to be aware of the work done on dealing with the viewing of color reproduction media where the state of adaptation is a mixture of component adaptation states. A scenario where this would be the case is the simultaneous viewing of a display and a print, where there is a significant difference between the white points of the two media (e.g., D93 for the display and D50 for the print). The color appearance of the displayed colors will be different to what they would be if the display was viewed separately as a viewer would not be fully adapted to the display's white point and the actual, mixed state of adaptation—affected by the simultaneous viewing of a print—needs to be estimated. To find more information about this issue, see the work of CIE TC 8-04,[11] and for further detail on color appearance models consult Fairchild.[12]

Color and Image Enhancement

Given the appearance of the source color under its viewing conditions, it is in some cases desirable to alter it, before considering its reproduction using a destination device. Note that by definition this stage is not used if the reproduction objective is subjective accuracy as there it is indeed the appearance of the source color that is the aim.

However, when preferred reproduction is wanted there can be reasons for changing color appearance before its reproduction in the destination. One such change that can lead to more preferred reproductions is a change in hue whereby source hues are moved based on the primaries and secondaries[‡‡] of the source and

[‡‡]A "primary" of an imaging medium is a color obtained by fully applying one of its colorants (e.g., a print's 100% yellow, a display's 100% red), and a secondary is a color obtained by fully applying two of a medium's colorants (e.g., printing 100% of both yellow and magenta, displaying 100% of both red and green). The importance of these colors is that they play key roles in determining the shape of a medium's gamut.

destination media so as to make them more similar.[13] For example, if a source display medium's yellow secondary is greener than the yellow primary of a printed destination medium then keeping hue unchanged can result in a reproduction of a bright, pure source yellow as a darker, less chromatic greenish yellow. If, instead, the source hues are changed so as to move the source yellow towards the destination yellow then the reproduction can preserve more of the brightness and purity of the source color at the cost of some hue change.

Finally, note that enhancements to the source content can also be performed in color spaces other than those of color appearance models. However, in any case their final output can be expressed in color appearance terms and serve as the input to the next stage of the process.

Color Gamut Mapping

Given the color appearance that is desired in the destination—it can be either the appearance of the source or a modified, enhanced version of it—the next step is to ensure that it can be matched. To do this, it is first necessary to know at least the destination gamut, and a means of determining it is required.

A number of techniques can be used to determine the color gamut of a color reproduction medium, including alpha shapes,[14] mountain range,[15] and segment maxima.[16,17] In all these cases the starting point are color appearances corresponding to a sampling of device color inputs to the device that is involved in the color reproduction medium whose gamut is to be determined. For example, when dRGB is used, a uniform sampling of all combinations of 10 steps per dimension (resulting in 10^3 samples) can be used. The gamut computation technique then uses this set of samples to generate a surface that delimits them from the rest of color space.

Next, a transformation needs to be applied to all source colors that results in each of them ending up inside the destination gamut and this transformation is called *color gamut mapping*.

Where the destination gamut is smaller than the source gamut, gamut reduction needs to be applied, which can either be a kind of clipping or compression. Clipping here refers to a gamut mapping where the source colors that are already inside the destination gamut are left unchanged and each of the source colors outside the destination gamut is mapped onto its surface (e.g., to the closest point on it). Compression on the contrary can change all colors—even ones that are already in gamut—so as to distribute gamut differences across a wider part of color space and allow for preserving more of the color differences that were present in the source than is possible with clipping. Finally, where the destination gamut is larger than the source, gamut expansion can be applied to make use of some of the additional color space.

Note that all three kinds of gamut mapping may be applied to a single source destination gamut pair as the destination may be smaller in some parts of color space (e.g., around red), but larger in other parts (e.g., around cyan). For a more detailed look at gamut mapping see Morovič.[18] For the evaluation of gamut mapping algorithms see CIE.[19]

Completing the Process

Starting from source device color values, the process has taken us to a color appearance that is desired in the destination medium, and it is next necessary to determine what stimulus, under the destination's viewing conditions, has that appearance. A color appearance model is used in the inverse direction[§§] and results in the stimulus that is to be produced in the destination medium. The characterization model of the device that generates it is used again in the inverse direction to predict the appropriate device color inputs, which are then sent to the destination device. The pair of colors—, that is, the source, which is the starting point of the reproduction process, and the destination, which is its end—is finally viewed and the relationship is judged with respect to the chosen reproduction objective.

The ICC Color Management Framework

While the previous section outlined the conceptual stages of a color reproduction process, we will now consider how such a process can be implemented in practice. This in turn leads directly to the *International Color Consortium* (ICC), whose color management framework is currently the *de facto* standard, at least as far as the reproduction of still images is concerned.

The ICC was established in 1993 by eight imaging companies, "for the purpose of creating, promoting and encouraging the standardization and evolution of an open, vendor-neutral, cross-platform color management architecture and components."[20] To this end, the solution proposed by the ICC is one where the color reproduction process is divided into two transformations: First, a forward one that takes device color data and transforms it into a colorimetric description for specific viewing conditions (called the profile connection space—PCS). Second, an inverse one that takes such a colorimetric description and transforms it back into device color space data.

Color interchange between devices is then achieved by being able to perform both parts of the transformation for each of the color reproduction media among which color is to be managed (Figure 7.3). The parameters based on which the forward and inverse transformations are performed for a given color reproduction medium[***] are stored in a data file referred to as the "ICC device profile" and its detailed specification as well as the specification of the entire architecture can be found in ICC.[21] An overview of what the key parameters are for different types of imaging devices will be discussed in following sections of this chapter.

The PCS, through which all color communication takes place, is defined by the ICC as "the reference color space in which colors are encoded in order to provide

[§§]That is, the forward direction is to predict appearance from information about a stimulus and its viewing conditions and the inverse is to predict a stimulus given a desired color appearance and viewing conditions under which it is desired.

[***]Note that a specific profile is needed for each color reproduction medium rather than just for each imaging device. For example, a printer printing on plain paper will need a different profile to that same printer printing on glossy paper.

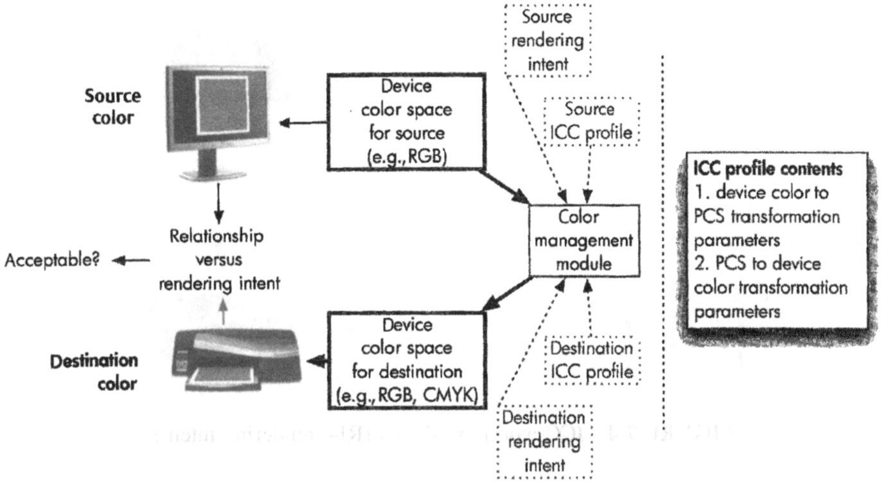

FIGURE 7.3 Overview of ICC color management architecture.

an interface for connecting source and destination transforms" (see p. 8 of ICC[21]). The color spaces that can be used for the PCS are CIE XYZ and CIELAB for a reference viewing environment, defined for the graphic arts by ISO 3664[22] viewing condition P2 standard (D50 light source; 500 lx illuminance; 20% surround reflectance).

As color reproduction objectives are essential to color reproduction, the ICC too specifies alternatives for them and refers to them as *rendering intents*. The four rendering intents defined in the current version of the specification are

First, the *media-relative colorimetric* intent rescales in-gamut tristimulus values to map a medium's white point to the PCS white point and is useful for reproductions between media to which observers are fully adapted. A popular use of this rendering intent is also in conjunction with black point compensation (BPC)[23] where the source luminance range is linearly scaled to the destination luminance range before gamut clipping is performed.

Second, the *ICC—absolute colorimetric* intent leaves tristimulus values of in-gamut colors unchanged and is useful for reproducing, for example, spot colors and for proofing.

Third, the *perceptual* intent is useful for a preferred or pleasing reproduction of images, particularly pictorial or photographic-type images—especially where source and destination media are substantially different. To allow for more control in providing preferred color reproduction, the ICC specifies a reference medium for this rendering intent. This medium is an ideal reflection print with a specific dynamic range and its purpose is to allow for improved results when performing gamut mapping. There are also proposals within the ICC to define a gamut for this reference medium to provide further control over the rerendering and gamut mapping process that has to occur in two stages via the PCS.[24]

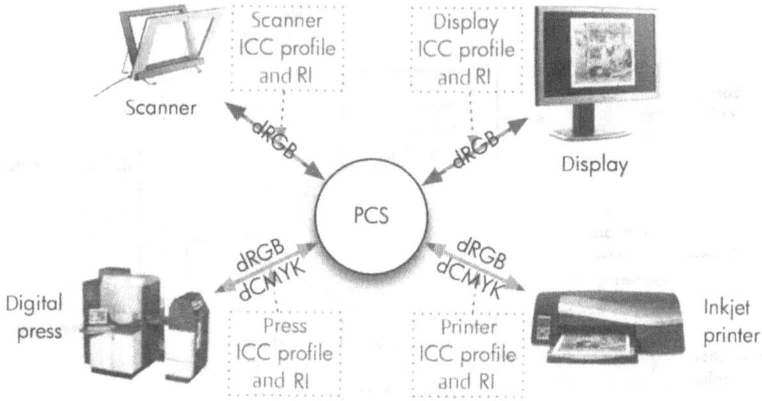

FIGURE 7.4 ICC workflow sketch (RI—rendering intent).

Fourth, the *saturation* intent is also vendor specific, involves compromises such as trading off preservation of hue in order to preserve the vividness of pure colors and is useful for images containing objects like charts or diagrams.

In summary, color transformations in the ICC framework are performed between devices on the basis of device profiles via the PCS and rendering intent choices, and when color data is to be communicated it is necessary to provide them alongside the color data itself (Figure 7.4). For further detail see the ICC Web site (http://www.color.org), which also includes useful white papers on a number of color management topics.

sRGB Color Management

An important, complementary approach to the ICC color management architecture, where the color imaging behavior of each device is characterized with reference to colorimetry, is to base all color communication on a single device-related, but colorimetrically defined, color encoding.

Specifically, color communication workflows can also provide good results by taking two decisions: first, RGB will be used to communicate color information between devices and second, RGB will be given a unique colorimetric interpretation, for example, sRGB.[8] Each device then does the best it can to either encode its native color information in sRGB so that the result is pleasing (e.g., scanners, digital cameras) or to provide pleasing color output given sRGB input (e.g., printers, displays, projectors).

The key properties of this approach are that only RGB content gets passed between devices and that each device internally does the best it can to relate the colors it generates or captures to sRGB. A clear advantage of such a setup is that it is very simple and transparent to other elements in color reproduction workflows, such as operating systems and software applications (Figure 7.5). The flip side though is that only a single reproduction objective can be followed by each device,

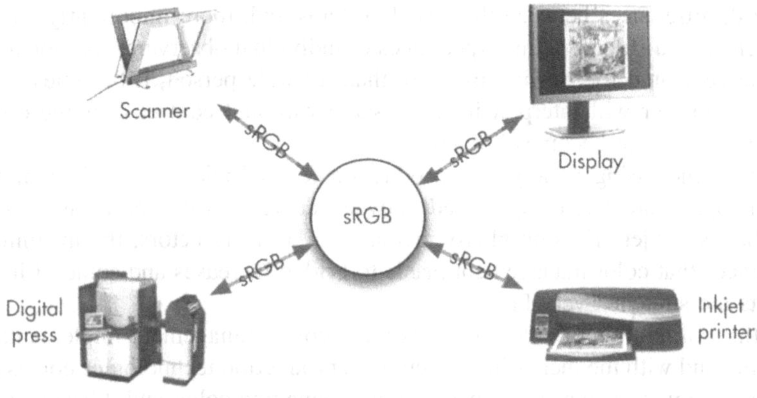

FIGURE 7.5 sRGB workflow sketch.

which is significantly more challenging than if a specific reproduction objective were communicated alongside color data.

Nonetheless, the approach works very well where pleasant color reproduction is the aim and where users do not need to or want to customize the reproduction process. Finally, it is also worth bearing in mind that sRGB workflows can also interface with ICC ones via an sRGB ICC profile and that other colorimetrically defined encodings can be used instead of sRGB to set up analogos workflows (e.g., Adobe RGB[25]).

Challenges of Color Management

Even though the above introduction to color management has already presented a high degree of complexity, there are a number of additional challenges that make control over the relationships of source and destination colors even more challenging.

First, the relationship between device color space data and corresponding stimuli for a given device is not constant. Instead, as components of devices vary temporally and also spatially across the device's imaging area and as components of devices are replaced from time to time, so does the way their device-specific color data relates to stimulus colorimetry. Hence characterization models and profiles are valid only for the state in which the device was when it was characterized or profiled. For example, a profile generated for a device today may not describe it well in two weeks' time.

Second, the output of a given device may be viewed under multiple viewing conditions. For example, a print may be viewed in a graphic designer's studio under daylight, in a press room under a D50 simulator and in an end user's home under tungsten light. However, that print is made with a single specific viewing condition setup in mind and its appearance in other conditions is not considered.

Third, differences between the visual systems and, more importantly, the color preferences and color judging experiences of individual observers: As soon as color is to be generated for viewing by more than a single person, it becomes unlikely that each viewer will interpret it in the same way and control over the effect of the generated color becomes limited.

Fourth, increasing variety of reproduction technologies and their gamuts. As color image content is being reused between devices as dissimilar as newsprint, cell phones, inkjet prints on glossy media, and laser projectors, the magnitude of differences that color management needs to bridge increases and some of its components are stretched very far.

Fifth, proliferation of color encodings and color management implementations. Hand in hand with the increasing variety of reproduction technologies comes also a greater variety in the means of encoding and managing color, and it becomes a significant challenge to ensure that the way a color reproduction setup is arranged is consistent and communication is effective. For example, as color management itself is possible at different points in a workflow (e.g., in a software application, an operating system, an imaging device), it is easy for it to happen more than once and therefore for transformations to be applied to other data than is appropriate. sRGB can be wrongly assumed to be device RGB, and device CMYK can be thought to be SWOP CMYK, and so on.

DOES COLOR NEED TO BE MANAGED?

Merriam–Webster's 11th Collegiate Dictionary lists a number of potentially relevant definitions of "manage": to handle or direct with a degree of skill; to make and keep compliant; to treat with care; to work upon or try to alter for a purpose.[26] All of these would seem to imply active involvement on the part of the manager, and with a certain degree of difficulty. Indeed, looking at Figures 7.2 and 7.3 one gets the impression that this is difficult business, whereas the setup of Figure 7.5 looks a lot simpler. One often hears and reads that color management is hopelessly complicated and something for eggheads, whereas a simple system like the one based on sRGB has proven to work well and does not need color management. As will be clear from the introduction, we do consider both ICC-based and sRGB-based systems to be examples of color management, albeit different in their apparent complexity and degree of user involvement needed. We will illustrate this below. One might also think that color management is a monster created by the move from analog to digital systems on one hand and by the simultaneous move from closed and single vendor to open and multivendor systems on the other. We will argue against this view of the genesis of color management.

Despite their apparent differences, the ICC and sRGB systems as illustrated in Figures 7.3 and 7.5 actually have two important things in common: first, the definition of a common interchange "color language," and second, the translation of that common language into or out of each individual device's private "color dialect."[†††] Some of the differences reside in the presence or absence of

- explicit color "dictionaries" (profiles) or descriptions of a device's "private color language";
- an explicit "translator" component (color management module) to convert one device's color language into another's; and
- the use of explicit parameters (rendering intent) to modify that translation for some particular purpose.

As explained in the introduction, the common color language in the ICC architecture is either CIE XYZ or CIELAB, which in the case of the *perceptual rendering intent* is output referred[27] to an *ideal reflection print* as seen under well-defined viewing conditions. The common color language for the sRGB architecture is also output referred, but in this case to an *ideal CRT monitor* as seen under well-defined viewing conditions. The color space of that ideal monitor is itself described in terms of the CIE XYZ coordinates of its primaries, among other things. So even though ICC can be thought of as more printcentric (or traditional Graphic Arts centric) and sRGB can be thought of as more monitorcentric (or perhaps computer centric), they both use an explicit output-referred exchange language that in both cases relies on fundamental CIE color spaces for its definition.

In both architectures, participating system components (devices or software applications) need to be able to translate their own color languages into or out of the common one (the PCS). This requires device calibration and characterization, as explained in the introduction. In the ICC architecture, the result of characterization is stored in an explicit and standard format, which can be exchanged freely within ICC compliant systems. A distinction is made among input, display, and output devices, with corresponding differences in the type and complexity of characterization formats (profiles) used. Input profiles are unidirectional (device to PCS only) and can use simple or more complex device models (see below). Display profiles typically use rather simple device models that can be inverted and can hence be considered bidirectional (device to PCS as well as PCS to device). Output profiles tend to use more complex device models and are required to be bidirectional, among other things to enable proofing (the simulation of one device on another, different one). In the sRGB architecture, device characterizations (or the corresponding color transformations) are typically fixed and built directly into the devices themselves, which makes them invisible from a user point of view—yet they do exist. In some cases, particularly CRT monitors, no explicit device characterization or color transformations are needed because the device's color space and behavior correspond closely to those of the ideal monitor that the common color language is referred to. This has obvious efficiency benefits for devices like

[†††]Wittgenstein might not take lightly to the suggestion that devices speak a private language or dialect, but we will ignore this for the sake of the discussion.

monitors with their high data throughput needs. The built-in characterizations or color transformations of the sRGB architecture are typically unidirectional only, which follows from the fact that they are built-in, hidden, and not exchangeable.

The "translator component" is known as CMM or Color Management Module[‡‡‡] in the ICC framework. Its task is to connect two (or more) color profiles via the defined connection space (converting between CIE XYZ and CIELAB as needed), accept rendering intent parameters specifying which of the defined ICC rendering intents to use, and construct a "color world" transform that can subsequently be used to transform color coordinates from input to output color spaces. Actual transformation of color coordinates typically involves interpolation using different size and precision lookup tables (LUTs), and possibly different interpolation algorithms (tetrahedral and tri-linear being the most popular ones). The CMM is a "pluggable" component in the ICC architecture, that is, in principle it can be replaced at will by any other ICC-compliant CMM. In typical sRGB implementations (there is no standard architecture defined for sRGB systems), the "translator component" is a fixed part of a device or software program and invisible to both the system and the user. Yet it performs much the same task of transforming between input and output color spaces and interpolating values using LUTs and interpolation algorithms.

As mentioned, ICC systems allow (and in fact require) explicit parameters and components to perform color management operations. The profiles that describe device color behavior, the rendering intent parameters needed to perform conversions, and the CMM itself are all explicitly defined using standard formats and interfaces, and meant to be replaceable and exchangeable. sRGB systems, on the contrary, are more "closed" systems in the sense that only the interchange color space is explicitly defined, standardized, and accessible, but little else. Nevertheless they perform very similar tasks.

ANALOG COLOR MANAGEMENT

How did things work in the analog age, and did color management exist, or was it needed at all? Let us consider only a few representative examples: traditional silver halide photography and traditional graphic arts production of printed publications. Traditional silver halide based photography involves (very schematically) the following steps:

- An image is captured by exposure to light of a film substrate coated with a photosensitive emulsion. Light interacts with the photosensitive chemicals in the emulsion, altering their properties in function of the amount of light received (exposure), and the spectral characteristics of the light as filtered through coupled color filters. Both negative (print) and positive (transparency) films exist and have been widely used.
- The film is "developed" using chemical means, which "fixes" the changes brought about by exposure to light and makes the film insensitive to further

[‡‡‡]Sometimes also glossed as "Color Management Method."

changes, that is, makes it no longer be light sensitive. This results in a negative or positive daylight viewable representation of the image on the developed film.

- The developed film is used to create an image on paper by shining a light through it and projecting the resulting image onto photosensitive paper. The process is very similar to the one that created the negative or positive: the paper contains photosensitive chemicals coupled with color filters that form a latent image after exposure, which is then developed and fixed through chemical means, resulting in a daylight viewable image on paper (and the paper no longer being photosensitive). Positive and negative photo papers exist, analogous to positive and negative films.

Is there any color management in all this? There is indeed, although not in the form of digital color profiles and CMMs. Starting with the capture process, one has to choose an appropriate film type for the lighting conditions (daylight or indoors), which mostly refers to the spectral composition of the illuminant. Each type (and often brand) of film has to be developed with a specific process. The paper type used must match the film characteristics. Photosensitive chemicals and coupled color filters must be carefully designed and manufactured to exacting standards. Device calibration is ever present, in the form of exposure measurement and control for the camera and enlarger; temperature, timing, and concentration control for development agents, and so on. The main reasons that the process seems simple and automatic (at least to consumers) are that it has been under development for about two centuries now and a large degree of standardization has taken place. Digital color management is still in diapers by comparison, and standardization is still in its early stages. Nevertheless, an all-sRGB digital photo capture, editing, and printing system as is common for consumer systems today comes quite close to the ease of use and thoughtlessness of the famous Kodak "Brownie" systems. In Brownie days, all the complexity and science was hidden in film R&D and manufacturing plants on one end and in photo processing laboratories on the other end—but there can be no doubt that it was there. If sRGB systems can be compared to Brownie consumer systems, could ICC systems perhaps be compared to the much more sophisticated and difficult to use professional analog photography systems? Perhaps so, but there is no reason to assume that ICC systems could not be made as easy to use as sRGB systems, if that were the objective. The move from analog to digital photography has opened up many new possibilities, many of which still have to be worked out and "brought under control" (standardized).

Now let us have a brief look at traditional (analog) graphic arts print production systems, using the watercolor reproduction scenario described in the next section. Completely photographic systems would follow much the same flow as described above, so let us concentrate on electronic but analog graphic arts systems. The basic steps involved would be (very schematically):

- The original artwork is scanned in a high end (drum) scanner, which directly produces analog CMYK separation signals (voltages), as calculated by an embedded analog "color computer."

- The separation signals are used one at a time to expose pieces of photo sensitive separation film, one for each ink color to be used in printing. Both positive and negative separation films have been used, analogous to traditional photography.
- The separation films are developed and fixed, and used to expose photosensitive printing plates, much like photographic negatives or positives are used to expose photosensitive paper.
- After development and fixing, the printing plates are mounted on a printing press, inked, and used to sequentially print each ink color onto the final substrate (paper or other).

Is there any color management going on here? There certainly is. The color computer that calculates separations is converting images from RGB to CMYK color spaces, which is one of the things that CMMs do in the digital domain. It typically does not use explicit device profiles for this, although some models have used programmable LUTs, not unlike the LUTs used in ICC profiles today. Everything related to film and plate exposure and development uses the same kind of chemistry-based color management that we hinted at above. Process control (calibration) is essential to the proper functioning of a printing press, as is precise control of the composition of printing inks, papers, and other consumables used. The process may seem less complex and more robust than modern day heterogeneous digital systems, but the main reason for this is that they were essentially closed systems, relying on a large degree of standardization. Again the move to digital systems has opened up many more possibilities, many of which still have to be worked out and "brought under control" (standardized). But we see no reason why digital color management ought to be more complex or less robust than its analog predecessors, given the necessary time and effort.

WATERCOLOR REPRODUCTION SCENARIO

To make the following detailed discussion of color management more concrete, we will introduce an example scenario, which will serve as a backdrop for the remainder of this chapter. This scenario (Figure 7.6) revolves around the production of promotional material for an exhibition of watercolors and involves the following stages: scanning of a watercolor original; viewing and editing of the scanned image; page layout of a poster and a leaflet for the exhibition and their proofing and production.

Please note that the focus in the following sections will be on the color management aspects of the scenario rather than on a comprehensive description of the technologies and processes that it would involve. Hence when addressing the scanning of the watercolor original, the discussion will revolve around how to relate scanned data to the original artwork rather than how to best scan it (i.e., scanner requirements, settings, original handling, etc.).

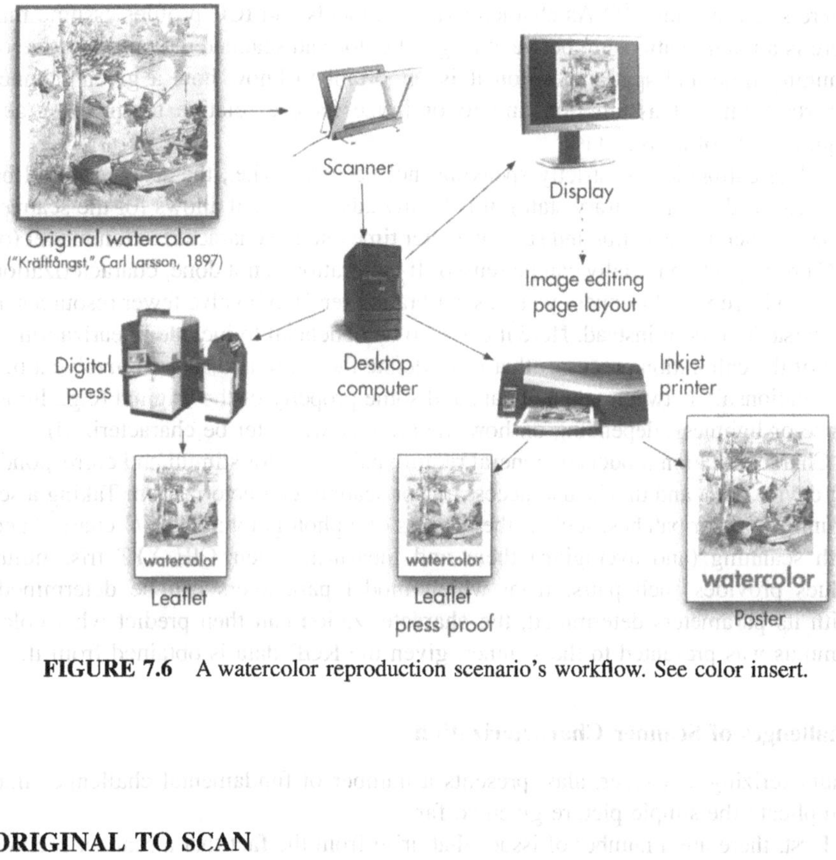

FIGURE 7.6 A watercolor reproduction scenario's workflow. See color insert.

ORIGINAL TO SCAN

The first task in our watercolor scenario is to obtain color data that encodes the appearance of the original watercolor artwork and to color-manage this process. A scanner, which illuminates a transmissive or reflective original and then samples and digitizes the light modulated by it both spatially and spectrally, is therefore used.

To relate data obtained from a scanner to the original's color properties, it is necessary to characterize or profile the scanner. This requires that it be in a state in which its output is stable, and it is also advisable to calibrate it. The outcome of having a stable and characterized scanner is that an original's color properties will be predictable from scanned data, which is needed when the original is to be reproduced using an output-imaging device.

In general it is advisable to first warm up a flat-bed scanner before use as the built-in light source's output varies most in the initial minutes of operation and its output subsequently stabilizes. When accuracy is important, it is also worth determining the uniformity of scanner response across its scanning area, as some scanners tend to have nonuniformity, parallel to the scanning sensor, in the region where scanning starts.[§§§] As characterization models and ICC profiles assume that there is a fixed relationship between original color and scanned data irrespective of scanning time and spatial position it is important to know how a given scanner departs from that assumption and to optimize its use relative to the accuracy requirements of a given task.

While calibration is, strictly speaking, not necessary (i.e., the scanner could be characterized in an arbitrary state), it is highly advisable as it allows for the scanner to be returned to the calibrated state at a later time and the characterization model (or ICC profile) set up initially can be reused. If calibration is not done, characterization needs to be repeated frequently and as calibration tends to involve fewer resources, it is advisable to do it instead. Here it can also be beneficial to include linearization as part of the calibration process—that is, to define the calibrated state to involve a linear relationship between scanned data and some property of the original (e.g., luminance or lightness, depending on how the scanner will later be characterized).

Characterization models in general require pairs of color stimuli and corresponding device data, and this is also necessary for scanner characterization. Taking a set of uniform color patches, such as the patches of a photographic IT8.7/1 chart,[28] and both scanning (and averaging) them and measuring their CIE *XYZ* tristimulus values provides such pairs, from which model parameters can be determined. With its parameters determined, the characterization can then predict what color stimulus was presented to the scanner, given the RGB data is obtained from it.

Challenges of Scanner Characterization

Characterizing a scanner, alas, presents a number of fundamental challenges that complicate the simple picture given so far.

First, there are a number of issues that arise from the fact that a scanner records only three values for each spatial location in the source, while the color-related properties of the source (i.e., its spectral reflectance or transmittance) are not three-dimensional. This in turn means that a whole range of source spectral properties will result in the same scanned RGB values (Figure 7.7). In other words, there is a many-to-one relationship between source spectra and scanned RGB values and the question of what the source was like, given particular scanned values has many answers.

This many-to-one nature of the scanning process has three key implications: (a) That comprehensive characterization models need to predict a source color set given a single RGB response—such a set is referred to as a *metamer set.*[29] (b) That if the spectral response of the scanner is different from that of the human visual system[****] (as represented, e.g., by the $\bar{x}(\lambda), \bar{y}(\lambda), \bar{z}(\lambda)$ color matching functions (see Chapter 3) then metameric problems arise from: the scanner–human

[§§§]A cause of this can be the fact that some scanners only have their light source on when scanning and they therefore switch it on shortly before each scan. The first part of the scanned image can therefore exhibit warm-up changes.

[****]Or, more precisely, if the scanner responsivities are not a linear combination of the human visual system's sensitivities.

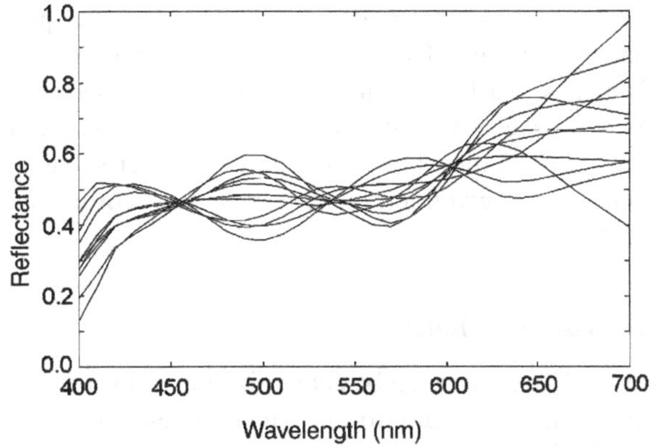

FIGURE 7.7 Sample spectral reflectances resulting in identical RGB responses on a given scanner under D50 (PM Morovič, personal communication, March 5, 2004).

observer difference directly†††† and from differences between the light source used in the scanner and the light source under which the source is viewed.

Where there is a difference in scanner and human spectral responses and where the characterization model is not of the many-to-one kind (e.g., ICC profiles), there is also the problem of different characterization parameters being needed for sources of different kinds (i.e., different types of photographic paper, digital output, original artwork materials, etc., all need different parameters). In other words, a characterization model that is accurate for some photographic material is likely to be inaccurate for a watercolor original.

Second, characterizing scanners also presents a problem specific to input imaging devices that follows from the difficulty of fully sampling the inputs to the device. In general it is important to sample all possible inputs to a device and then determine characterization model parameters from these samples and from corresponding outputs of the device.[30] While this is not difficult with most output imaging devices, where the inputs are digital values that can be generated easily, with input imaging devices, the range of possible inputs is the range of all possible color stimuli. Even restricting ourselves to reflective and transmissive materials, it becomes unfeasible to have a set of samples that cover the gamut of all possible color properties (i.e., the object color solid—OCS[31]). As a consequence a characterization model will only be valid and have known properties over the color gamut of the samples that it is based on and its predictions for stimuli outside that gamut will be unreliable. This is particularly a problem where there is a need to have a characterization model that works for a wide range of source kinds and less of an issue where samples are possible across the full gamut of a particular kind of source (e.g., when only photographs made on specific photographic paper are scanned and a sampling of that paper's gamut is available).

††††As a result of such observer metamerism, the scanner may record different RGBs for parts of the source that look the same to a human observer and the same RGBs for parts that look different.

Third, a number of scanner properties (e.g., tone response, quantization, noise) result in the scanner having an effective gamut[32] that is more limited than the object color solid. This means that for certain subvolumes of the OCS the scanner is not able to differentiate between distinct colors and records them as being the same. The most direct example of this is the effect of the scanner's dynamic range, which results in variation below a given luminance not being recorded and being mapped onto a single value instead.

Scanner Characterization Models

Scanner characterization models can be categorized in terms of two criteria: model-based versus empirical[9] and many-to-one versus one-to-one. Here model-based characterization, which tends to be many-to-one, attempts two parallel stages of the scanning process in its computational structure, whereas empirical approaches treat the process as a black box and are concerned primarily with prediction accuracy.

Model-based approaches use information about the spectral responsivities (or sensitivities) of a scanner, the scanning light source, and the scanner's tone reproduction[‡‡‡‡] to relate RGBs to scanned spectra. The forward direction (i.e., from surface reflectance to colorimetry) is in this case significantly simpler than the inverse as it involves a reduction in dimensionality from n to 3, where n is typically at least 31, and there are several detailed descriptions of how to perform the mapping (see pp. 315–316 in Bala[9]).

The inverse direction of a many-to-one model is, however, more complex as it needs to take a single RGB value and predict from it the set of all possible spectra that could have resulted in it under the given sensor and light source conditions. One such model is *Metamer Constrained Color Correction,*[33] which predicts a convex metamer set for a given set of scanned values and also provides methods for choosing a single representative of the set where necessary. This is important in an imaging context as, even though the RGB[§§§§] to spectral reflectance (and therefore also XYZ) relationship is a one-to-many one, in the end a given scanned RGB needs to be represented by a single XYZ, which is then further transformed into a color reproduction system. Furthermore, having the entire set of possible XYZs that correspond to a given RGB and knowing the XYZs obtainable on another device (e.g., a display) also allows for the choice to be made in a way that optimizes cross-device reproduction. In the absence of such information a choice can be made on statistical grounds (i.e., as some XYZs occur more frequently than others in nature or a particular original medium).

[‡‡‡‡] What is meant by tone reproduction is the relationship between device color space values (i.e., RGB here) and corresponding color or density attributes. For example, the relationship of scanned RGB values of a reflective grayscale's steps and the grayscale's lightness values.

[§§§§] In this section the X, Y, Z and R, G, B tristimulus values will be written in a short form as XYZ and RGB.

Empirical models, on the contrary, simply require a list of source XYZs and scanned RGBs based on which their parameters are computed. The simplest model, which is an application of Yule's Masking Equations[34] to scanners, first involves linearization of the scanned data by applying 1D transformations to the scanned RGBs so as to make them linear in terms of XYZ. Next, a 3×3 matrix is computed (e.g., using a linear least squares approach) and applied to the linearized RGBs to predict XYZs from them. This model works well only if the scanner sensor responsivities are close to being linear combinations of the $\bar{x}(\lambda), \bar{y}(\lambda), \bar{z}(\lambda)$ color matching functions (as the model assumes). As this assumption often does not hold, an extension of the simple model is to use higher order matrices that allow for nonlinear relationships between RGB and XYZ.[35] Alternatively neural networks can also be used to empirically model scanners[36,37] as can direct interpolation from the data obtained from a characterization chart.[38,39] Finally, note that many of the empirical techniques used for scanner characterization can (by their general nature) be used for modeling other types of imaging devices as well.

Scanner ICC Profiles

The ICC defines two types of profiles for color input devices: *Three-component matrix-based*, which allow for the storing of the simplest model's parameters introduced in the previous section (i.e., three 1D lookup tables for linearization and a 3×3 matrix), and *N-component LUT-based* input profiles (see p. 22 of ICC[21]). The latter type's most important component is an ICC *AToB0Tag* data structure, which stores parameters for the following sequence of transformations: input data $\rightarrow$ 3×3 matrix $\rightarrow$ 1D input LUTs $\rightarrow$ multidimensional LUT $\rightarrow$ 1D output LUT $\rightarrow$ output data. In the case of scanner profiles the input data is scanned RGB and the output data is XYZ or LAB for the PCS, the multidimensional LUT is three-dimensional and the rendering intent is perceptual. Analogously *AToB1Tags* contain parameters for the colorimetric and *AToB2Tags* for the saturation rendering intents.

More specifically, the LUT contains PCS values for an even sampling of the input RGB space and a key challenge here is to populate all entries of the LUT as it is often the case that the RGB–XYZ pairs obtained by scanning a characterization target do not cover the entire RGB cube and some form of extrapolation is needed. Furthermore the CIE XYZs or LABs stored for each LUT entry are not simply measurements of source colors (or values interpolated from them) but values that represent a chosen rendering intent.

On a related subject, it is worth noting some issues when using ICC input profiles with images from digital cameras. Most cameras (or camera raw processing applications) render color images to some standard output-referred color encoding, like sRGB[8] or Adobe RGB[25], so the profile attached should be the appropriate color space profile. It may also be possible to obtain "raw" camera RGB, which has not undergone any color space transformations. In this case, it will be possible to use a new kind of ICC input profile in the future that will interpret the camera RGB values. Such profiles, however, will be image, not camera-specific. The colorimetric intents will depend on the scene illumination, and the perceptual intents will depend also on the desired color rendering of the scene. Although it is mechanically possible to use the colorimetric intent of a raw camera RGB profile to convert to a standard scene-referred color image encoding, such as RIMM RGB[40] or scRGB,[41] doing so today may result in interoperability issues and that is something that the ICC is currently addressing. Note that if the colorimetric intent of such a camera RGB profile is used to convert to an output-referred color encoding, this effectively indicates that the result is the desired colorimetry on the encoding reference medium. However, it is also possible to consider the initial result of such a transformation as the starting point for manually applied color rendering.

The key point to take away from this section is that ICC profiles are simply parameter containers and that the challenge is primarily in the computation of these parameters, which can in turn be done by performing various stages of the color reproduction process (e.g., characterization, color appearance modeling, color enhancement, gamut mapping). The LUT for the perceptual rendering intent will therefore contain values that are a reinterpretation (or rerendering) of the PCS colorimetry that corresponds to given scanned RGBs and that give a perceptually more pleasing reproduction on other media (e.g., changes to contrast, saturation, etc.). On the contrary, the colorimetric rendering intent's LUT will represent the colorimetry of a scanned original and will only deal with presenting the scanned data under PCS conditions as opposed to the viewing conditions in the scanner.

Scanned Watercolor

In terms of our scenario, the outcome of its first stage is an RGB image obtained by scanning the original watercolor, and to allow for the controlled reproduction of the scanned data, also an ICC profile of the scanner.

The key challenge here is to generate a profile that gives accurate predictions of the scanned original as it would ideally have to be based on watercolor color patches painted using the same paints and paper as the original that is to be scanned (and also aged in the same way as the original has aged). However, as this is likely not to be possible, an attempt would have to be made to at least use a characterization target that is as similar to the original watercolor as is practical. The scanned data together with the scanner's profile provide a colorimetrically based description of the original that will be the input to following stages of color manipulation and reproduction.

SCAN TO DISPLAY

The next stage of the watercolor scenario is to view the scanned image on a display to then be able to edit it and incorporate it into the page layout of a poster and a flyer. As the scanning stage of the scenario results in RGB data representing the original watercolor and as the associated scanner profile also provides a colorimetric representation of the scanned data (either with a perceptual or colorimetric rendering intent), the next color management task is to provide an appropriate rerendering of the scanned image on a display.

What will therefore be needed again is a characterization of the display that relates displayed colorimetry to the digital input that brought it about as this will be the basis for determining what display inputs to use to represent given scanned pixels.

Just as was the case with the scanner, a display too needs to be warmed up, as its output is different soon after being switched on from what it is like an hour later. Unlike with scanners though, the issue of sampling inputs to a display is trivial as there are no obstacles to driving it with digital values that sample the full range of possible digital inputs to it.

Challenges of Display Characterization

There are two key challenges to successfully characterizing a display. First, displays suffer from nonuniformity, either spatially across the displayed area (on a CRT the luminance of output from the center of the display can be up to 50% greater than from the edges for the same digital input) or as a function of viewing angle (color output of LCDs can vary dramatically with viewing angle changes).

Second, as displays are typically not viewed in complete darkness, it is also necessary to take into account the impact of ambient illumination on the appearance of display output and this is a much more serious challenge than the first. The difficulty of dealing with the impact of ambient conditions is twofold: First, there is a challenge to knowing how to take into account both the white point of the display itself and the white point of the ambient illumination to arrive at an effective white point that controls the color appearance of display stimuli. Here there has been a significant progress made by CIE TC8 - 04 that recently published a report on how to deal with such mixed-adaptation conditions[11]. Second, however, there is a practical challenge of knowing what the ambient illumination is like and also of a color reproduction setup staying up to date with ambient illumination changes (e.g., in a studio where daylight is used, that daylight is likely to change), and this is much more difficult to address. In practice displays are typically profiled in complete darkness and if these displays are viewed under other conditions then the mismatch between profiling and actual conditions contributes to shortcomings in cross-media color reproduction.

Display Characterization Models and Their Implementation in Profiles

As displays are in colorimetric terms simple devices whose channels' output is additive (i.e., the separate CIE XYZ values of the red channel's output and the green channel's output can simply be added to get the CIE XYZs of their combined use), their characterization models too are simple and can be very accurate. To predict the colorimetry of a display's output given the digital input to it, display characterization models perform two stages of transformation. First, the digital values of each channel are transformed to be linear in terms of luminance, and this can be done using a range of models that are specific to different display technologies. For example, the GOG model[42,43] is a popular solution for CRT displays, the sigmoidal model is suitable for LCDs,[44] while the PLCC model[45] works for any display technology as it simply used 1D LUTs for the task. The second step in all cases is to transform the luminance-linear RGBs to CIE XYZ, and this can be achieved using a 3×3 matrix obtained either directly from the RGB primaries of the system or optimized across the whole gamut. All of the above models can also be inverted to predict RGB display inputs from desired CIE XYZs. As more than three primary systems are emerging there is also a growing literature proposing solutions for their characterization (e.g., Murakami et al.[46]).

Specifically for the characterization of CRTs there is also a CIE recommendation[47] that provides full details of how to apply Berns' GOG model and also specifies how to take into account flare and other factors affecting display output.

Turning to the implementation of display characterization models in ICC profiles we find that they can be of the same two types as scanner profiles. Here the *Three-component matrix-based* type is most common and it allows for the storage 1D LUTs for the first part of the transformation and a 3×3 matrix for the second. These display profiles, by definition, use the CIE XYZ encoding of the PCS. However, using this simpler display profile type does not provide control over the gamut mapping to be performed from PCS colorimetry (simple 1D clipping in XYZ gets applied) and neither does it allow the implementation of perceptual or saturation rendering intents. *N-component LUT-based* display profiles can be used though to provide such control and also to differentiate between the ICC rendering intents.

Transforming Scanned Data to Data for Display

Given colorimetric data of the scanned image that is available either using a scanner characterization model or scanner ICC profile and a characterization model of the display on which it is to be rendered, it is necessary to perform the following transformations, as has been introduced in general previously: First, the color appearance of the original needs to be computed by taking original viewing conditions (e.g., the viewing conditions in the gallery where the original watercolor is displayed) into account using a color appearance model. Then, color and image enhancement may be applied, though in the case of a poster for the exhibition of artwork this is less desirable as the original artwork is of interest as it is. Next, it needs to be ensured that the original's colors are all inside the display's gamut and if they are not then gamut mapping needs to be applied. Finally, the viewing conditions of the display need to be taken into account to determine what stimuli to render on it and the display characterization model is used to compute digital inputs.

While these steps can be performed individually as described above, in practice they are more likely to be encoded in the scanner and display profiles, whereby the first provides PCS colorimetry of the scan and the second takes that colorimetry and computes digital inputs for the display from it.

The end result of the scenario stages described so far is that if scanner and display characterization models used take actual viewing conditions into account and if, for example, a simple gamut clipping algorithm is used, then the displayed

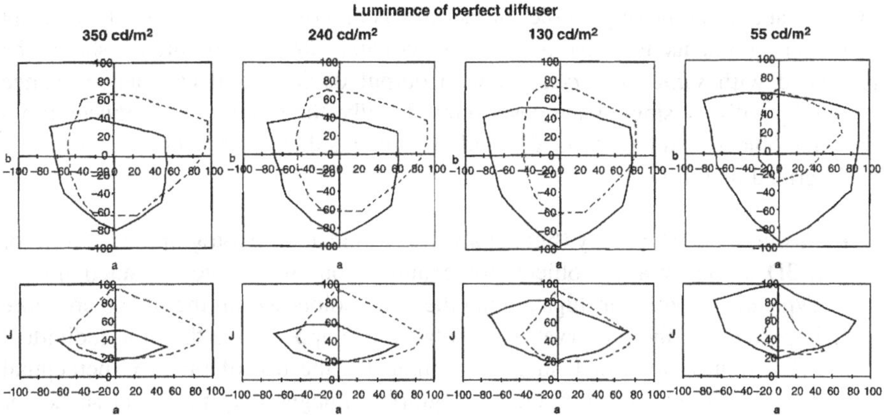

FIGURE 7.8 Display (solid line) and reflection print (dashed line) gamuts when viewed simultaneously under varying levels of illumination (CIECAM97s, D50).[30]

version of the watercolor will look like the original watercolor (especially as it is unlikely in the case of this media combination that there would be parts of the original that are significantly outside the display gamut). When actual viewing conditions are not taken into account, the relationship between what the profiles represent and what an observer would see can be very weak. To illustrate this point, Figure 7.8 shows the measured gamuts of a display and of a reflective original, such as a print or a watercolor, under a range of ambient illumination conditions. Here the profiles may implicitly assume one of the possible states, whereas the actual viewing conditions can be quite different.

EDITING AND PAGE LAYOUT

Let us assume that we have successfully scanned our watercolor original and displayed it on a monitor with an acceptable level of appearance matching between display and original. Our next task is to use the scanned image to create electronic versions of leaflets to be printed on a digital press, and posters to be printed on a large format inkjet printer (Figure 7.6). Naturally we want both to display the best possible appearance match to the original, and hence also to each other.

Photo editing programs can be used to adjust scanned images for color or other properties (scratch and dirt removal, etc.), but we will assume that no such edits are needed for our scanned image. Page layout programs are used to assemble complete page descriptions from their constituting elements, such as raster images (photos), text, lines, and other vector content (typically defined with solid area fill color). One of the challenges of page composition with respect to color is the different origin and intent of some of those elements, and the combination of them in a single coherent and good looking whole. In fact this is one of the reasons for the concept of rendering intents to exist in the ICC context. Rendering intents should be associated with *source objects*, not with output devices, and can hence change repeatedly within a single page description. Briefly, the relationship between types of source objects and the "typical" rendering intents they are associated with could be described as follows:

- *Raster images* typically (but not always) originate as photographic captures of a 3D scene or a 2D object, for example, our watercolor scanned image. Typically the relationships among the pixel colors within the image are more important for the preservation of total image appearance than the individual color values are, which in general indicates the use of the ICC perceptual rendering intent. In addition, photographic images may represent scenes or objects with colors that are outside the gamuts of typical hardcopy devices, which would again indicate the ICC perceptual rendering intent. Even though the latter may not be the case for our watercolor image, the former would still indicate the use of the perceptual intent.
- *Renderings* are raster images of a special kind, not originating from the capture of a scene or original, but rather having been generated (rendered) by computer models of objects or scenes. They are increasingly popular for architectural work, for instance, to give potential clients an idea of what buildings would look like in a rather realistic environment. Depending on the sophistication of the computer models used, the results can be almost indistinguishable from captured images of real scenes or objects, or can be quite recognizably "artificial" in shape, texture, lighting, or color properties. Although much depends on the objects or scenes being modeled and the intended use of the rendered images, in general it is safe to say that renderings typically go with the perceptual rendering intent also.
- *Text* is very different from images in that it is typically defined using primary or secondary (non-halftoned) colors, both for readability and for effect. Especially because of readability and because of the limitations of many printing technologies, that definition typically goes beyond color in the strict sense. The well known "black on white" way of rendering text does not only refer to a certain color of low lightness and chroma but also to the actual inks used to print it: 100% black ink, 0% of all other inks. Although one could typically achieve the same CIE XYZ values with different combinations of printing inks, the result is likely not the same in terms of the sharpness of the character outlines (due to issues such as color to color registration). Neither is it likely to be the same in terms of cost because black ink tends to be cheaper than color ink. So what rendering intent does this correspond to? Actually *none*, at least in terms of ICC rendering intents. A rendering intent serves to modify a color conversion down the line, but if an object is being defined using direct device color definitions, there is no color conversion or rendering intent involved.

- *Line art*, so called because it is typically made up of straight or curved line segments with the enclosed areas filled with "flat" or "solid" color, is also known as "vector content" in the technical jargon. In general it behaves more like text than like raster images in terms of color. For (output) device-independent color specifications the colorimetric or saturation intent is in principle more appropriate than the perceptual intent, but as is the case with text, direct device color definitions are quite frequent. Named color (e.g., Pantone®) specifications are also frequently used and represent a challenge of their own. In principle named colors represent certain fixed combinations of color coordinates in *any* color space, identified by name. In practice, they often refer to physical samples, often produced with specific ink sets on specific substrates. The translation of named colors into device color coordinates for display or printing is a complex topic that we will leave aside for now.
- *Logos* could be considered a special case of line art or vector content, often combined with text, sometimes even with images. Because they tend to represent corporate or other identities, requirements for accurate color reproduction are typically stringent. The use of named colors is common. Most often they would be associated with the colorimetric rendering intent.

A successful page layout application must be able to distinguish between the various object types, and let the user associate different rendering intents with them either on an individual object basis, or on an object type basis, or both. These rendering intents must be preserved in page descriptions generated by the layout application, together with the objects and their corresponding color definitions. Specific page description languages (PDLs) have been developed over the years for this purpose, for example, PCL, PostScript, or PDF.

We have seen that some objects may be defined using source color spaces (e.g., scanner RGB) and rendering intents, whereas others may be defined directly in device color space coordinates (e.g., text as 100% black ink, 0% all other inks). Both types of definitions must be recorded appropriately in the resulting page descriptions and must be honored by the proofing systems, printing systems, or raster image processors (RIPs) that interpret them down the line.

Two broadly different approaches can be followed with respect to color management as controlled by layout applications. The first one, which we might term *greedy color evaluation,*[*****48] attempts to convert all object color specifications to output device color coordinates right away. The result is a PDL file that uses only a single color space, namely the output device's, which makes it easy to interpret but by definition device dependent. The gamut mapping and possibly rerendering transforms that take place during the conversion to device color coordinates is in general

[*****]By analogy to *greedy algorithms* from computer science: a type of algorithm that makes the locally optimum choice at each stage with the hope of finding the global optimum, which results in making as many decisions as possible as quickly as possible.

not invertible, so in a sense the document is now *committed* to a specific output device and that commitment cannot easily be undone. The second approach, which we could term *lazy color evaluation*, attempts to maintain source object color specifications as long as possible, without converting to output device color coordinates. The result is a PDL file which may use many different color spaces, which makes it harder to interpret but (output) device-independent. Such a file can be *repurposed* more easily, if necessary.

Let us assume the second strategy for our watercolor reproduction example: Both the flyer and the poster compositions will maintain the original object color specifications to the extent possible.

PROOFING

Digital proofing is generally understood as "preparing a sample of printed output on a computer printer before the job is printed on a commercial press" (*Computer Desktop Encyclopedia*[49]). We might extend this definition to include the production of a sample of output of any kind to simulate a sample of output of any other kind. Some commonly distinguished subtypes are

- *Soft proofing*: a computer display is used to simulate or preview a piece of printed output. Only color appearance can typically be soft proofed, not other properties such as substrate texture or spatial halftone attributes such as moiré.
- *Hard proofing*: a hardcopy of some kind is used to simulate or preview a hardcopy of a different kind. For instance, a digital inkjet printer can be used to simulate the output of an analog offset press. All hardcopy proofing systems attempt to simulate color appearance, and some also attempt to simulate other properties such as substrate texture or spatial halftone attributes.
- *Press proofing*: rather than a simulation this is a sample of actual printing on the intended substrate, produced on the actual printing device. In fact this is not a simulation case but rather a (very) short run length sample of the real thing.

We will limit our discussion here to the simulation of color appearance, mainly in hard proofing contexts. In general the issues involved with proofing will be easier to resolve, and the simulation more effective, to the extent that the proofing system and the target (final) output system resemble each other. Important dimensions of resemblance include substrates (if any), colorants, marking technology, halftoning characteristics (if any), and last but not the least viewing conditions. An extreme example of this is a press proof, where all these characteristics are identical between proofing and target system, even including the viewing conditions. In that case there is obviously no need for any kind of explicit color management (excepting the usual calibration and process control procedures). It is not too much of an exaggeration to say *that the best color management is no color management*, if conditions allow. The more dissimilar the proofing system is to the target system, the more explicit color management will need to be involved. An extreme example of this is a soft proof, where all the characteristics mentioned are different between proofing and target system, including the viewing conditions. We will now consider the case of (partly) dissimilar proofing and target systems in somewhat greater detail.

Proof Printer Calibration

Device calibration is essential for repeatable results both from the same printer (*intra*device consistency) and from different printers (*inter*device consistency), and the production of digital proof prints is no exception (see Introduction). A typical calibration process consists of the following steps:

- A calibration target is printed;
- The printed target is allowed to dry and/or stabilize completely;
- The printed target is measured with a densitometer, colorimeter, or spectrophotometer;
- Printer calibration parameters are calculated from the measurements;
- Calibration parameters are applied to the printer.

Depending on the calibration algorithm used, the calibration target may consist of density ramps of single colorants or combinations of different colorants. The most traditional calibration algorithms attempt to calibrate both the maximum amount and the tone response of each individual colorant, using optical density-based measurements. Because this obviously does not take interactions among colorants into account, more sophisticated algorithms have also been developed that do take interactions into account, typically using colorimetric or even spectral measurements.[50] Which type of algorithm is deemed more suitable for a given proof printer depends in part on the characteristics of its marking engine, in part on characteristics of the hardware and software controlling the printer, and in part also on personal preferences and experience.

Depending on the characteristics of colorants and substrates used, it may be necessary to let printed calibration targets dry or stabilize for a certain amount of time, to make sure that no further density and/or color changes will occur after measurement. The aim is to calibrate the final appearance of proof prints, not their appearance immediately after printing. Naturally the same criteria should be applied to actual proof prints, in addition to calibration targets.

The calculation and application of calibration parameters naturally depends on the calibration algorithms used. For the traditional maximum density and dot gain (tone response) calibration method, it is normally sufficient to calculate one-dimensional *transfer functions* to be applied to continuous tone color separated versions of input images (one channel or separation per printing colorant). A transfer function simply maps an input value in a given domain (say [0,1] for simplicity) to an output value in a given range (say [0,1] also) in a deterministic fashion, and without taking any other inputs into account. It should be noted that in order to avoid extrapolation beyond measured data, this type of calibration can only be used to modify the tone response of individual colorants and to *reduce* the maximum amount of colorant, but not to *increase* it beyond the range represented by the calibration target. The latter should therefore represent a larger range than that needed for actual proof prints. To calibrate interactions among colorants a more sophisticated mechanism is needed, for instance multidimensional LUTs.

Proof Printer Characterization

Device characterization or profiling is essential for colorimetrically accurate results by establishing a correspondence between device-dependent and device-independent color representations, and the production of digital proof prints is no exception. A typical characterization process consists of the following steps:

- A characterization target is printed;
- The printed target is allowed to dry and/or stabilize completely;
- The printed target is measured with a colorimeter or spectrophotometer;
- A printer profile is calculated from the measurements;
- The printer profile is applied in the generation of proof prints.

For most if not all proof printers, characterization targets consist of patches combining different amounts of several colorants, possibly in addition to patches containing single colorants only. The amount and exact kind of the patches used depends on the characterization algorithms used, and in particular on their underlying printer models. Some of the most common printer models are

- The *Neugebauer* model and its many variants and derivatives; see for instance Mahy.[51] The essence of the Neugebauer model is the summation of spectral reflectances of unprinted substrate and solid (full area coverage) single and overprinted colorants (also known as *device states*) to predict the spectral reflectance of non-solid combinations of any number of colorants, weighted by their expected area coverage.
- *Masking equations*—see for instance Berns.[52] Masking equations (whether linear or nonlinear) attempt to establish a closed form correspondence between device color values (e.g., normalized area coverage of individual colorants) and resulting colorimetric or spectral characterizations.
- LUTs with interpolation; see for instance Kasson and Plouffe.[53] LUTs are collections of precomputed (or measured) function values (one- or multi-dimensional) that together with interpolation (linear or nonlinear) can be used to estimate the function value for any input within a predetermined domain.

As used in printer modelling they are typically three or four dimensional, mapping for instance, CMY or CMYK input values to measured CIE XYZ or CIELAB values, often using multilinear or tetrahedral interpolation.[54]

In all cases, a *forward* printer model is first established, which maps device color values to colorimetry. The forward model then gets inverted to produce an *inverse* printer model, which relates colorimetry to device color values. In proofing contexts both types of model are used: a forward model is first applied to the device color values of the target device (e.g., an offset press) which results in desired colorimetry values, which then get mapped through an inverse model of the proofing device (e.g. an inkjet printer) to produce the device color values that when printed will result in the desired colorimetry. The colorimetric proofing process can obviously only work as expected if the proofing device gamut completely encompasses the target device gamut, which we will assume for the sake of our discussion. For more detail on printer characterization models see for instance Bala.[9]

Rendering Intents for Proofing

As we have seen, proofing is an example of strict colorimetric reproduction (and as such, probably one of the few such examples for practical applications). We have also seen that ICC profiles contain both forward and inverse printer models, usually in the form of LUTs, in different rendering intent versions. Which rendering intent(s) should we use for proofing then? Given the colorimetric reproduction goal we have two obvious choices: (media) relative colorimetric intent or absolute (diffuse white relative) intent. The same intent should always be used for both forward and inverse tables.

The most accurate colorimetric reproduction can in principle be obtained with *absolute colorimetric* rendering intents, so should not we just use those? In principle yes, but in practice *your mileage may vary.* Absolute colorimetric reproduction implies that even the color of unprinted media on the target system will be simulated on the proofing system, resulting in target unprinted media areas being reproduced as printed media areas in the proof print. There are a number of potential issues with this approach:

- The proofing media used must have a lighter (higher) white point than the target media, to accommodate the simulation of the latter. This condition is implicit in the general condition that proofing systems should have a gamut completely encompassing the target system gamut, but it is not always easy to achieve. This is especially true if we want the proofing system to use a media type that has a very similar appearance to the target media, which in general is a very good idea.
- Furthermore, the proofing media should preferably have a white point resembling the target media closely in terms of hue and chroma, to avoid unnatural looking and difficult to achieve and maintain target media white point simulations of a slightly different color cast.

- To allow proper viewer adaptation to the simulated white point, any nonprinted margins showing the original proofing media white point should be removed from proof prints (or from view), otherwise mixed adaptation and very likely unnatural looking simulated target media color will result.

All of these issues can be avoided or resolved by using the *relative colorimetric* rendering intent instead, which renders target media unprinted areas as proofing media unprinted areas. One would be well advised in this case to look for a proofing media with very similar appearance (in terms of brightness, white point, glossiness, etc.) as the target media, but this is a good advice for proofing in general. Any remaining small differences between target and proofing media, normally speaking, will be "absorbed" by viewer white point adaptation, unless original and proofs are being compared side by side. The latter may be habitual in the evaluation of proofing system capabilities, but not in actual proofing system use—for if one had the original available, there would be little point to producing a simulation of it.

It is interesting to compare the digital proofing case, which is what we have been discussing, to the analog proofing systems of yore. One of the best known examples is no doubt the *3M Matchprint* system, where proofs are "made by exposing the CMYK negatives onto four acetate films which are developed and laminated together"[55] and on top of a sheet of the actual printing stock that will be used on the target system. The resulting appearance (and colorimetry) will, of course, be influenced by the properties of the printing stock underneath, and unprinted areas on the target system will show as blank (un-imaged) areas on the proof. Both of these properties are comparable to what is obtained on digital proofing systems with the ICC relative colorimetric intent, and more so than with the ICC absolute rendering intent.

Evaluation of Proof Prints

In general, proof prints should be evaluated under conditions as similar as possible to the ones that will be used to view or evaluate the final target system prints.[22] It is worthwhile to note that when ICC profiles are being used for proofing, this in principle implies a single choice of illuminant (D50), illumination level (500 lx), measurement geometry (45/0), standard observer functions (CIE 1931), and so on. If there are significant differences between viewing conditions used for evaluation and measurement conditions used for constructing profiles, a color match between proof and final print cannot be guaranteed.

To express the differences between a proof and the target printing system that it is meant to proof, colorimetric data can be obtained for a sampling of the target's gamut and its proofing. The CIEDE2000(1:1) color difference equation[56] can then be used to express the differences between corresponding color pairs and the difference distribution's statistics can be reported. If the distribution is normal (Gaussian) then parametric statistics are appropriate—mean and standard deviation—and if not, then the median and a high percentile (e.g., 95th) can be reported.

In all cases the maximum error needs to be given. However, doing this only expressed how well the proof represents the target's colors and not how different the colors of a target and proofed images' will be. To approximate that, it is necessary to take into account factors other than just individual colors and it is more appropriate[57] to use a high percentile of the difference distribution computed using CIEDE2000(1:2) with optional spatial prefiltering (http://www.color.org/tc8-02/; MR Luo, personal communication, March 10, 2006).

Of course, there are many non-color-related issues involved with the evaluation of proofs, for instance, halftoning properties (moiré effects), media surface finish and appearance, the presence of fluorescence (due to the use of "optical brightener" media additives), media weight, and so on. We have limited our discussion here to the color-related ones.

Challenges and Opportunities

We have already hinted at a number of challenges and opportunities involving color proofing. Here we will briefly outline some more, without going into much detail due to space constraints.

- *Print metamerism.* Metamerism (see Chapter 3) is commonly defined as "the quality of some colors that causes them to appear differently under different light sources. For example, two color samples might appear the same in natural light, but not in artificial light."[58] Besides the light source, the colorants (inks) and substrates being used obviously also affect metamerism. Unless the substrates and colorants of the proofing system are exactly the same as the ones of the target system, they are likely to behave differently to changes in the viewing environment away from the standard viewing conditions. This may cause a proof print to match the target print well only under very constrained viewing conditions that can be difficult to maintain in practice. Even though they may produce matching colorimetric measurements, they can still look different.
- Related to metamerism is also the *color inconstancy* of a print, which refers to the changes of its appearance when viewed under different light sources. This may cause a print to be an acceptable visual proof under one set of viewing conditions, when viewed on its own (e.g., when sent to a client), and for it to be unacceptable under others. The degree of color inconstancy can be expressed using the color inconstancy index (CII).[59]
- *Fading and/or color stability.* Although proof prints are typically intended for fairly short term use, it is important for them to quickly reach a stable state after printing and to not be too susceptible to light fade or air fade under normal conditions of use.
- *Proofing actual versus standard devices.* In principle one can generate a color profile of a target printing device in whatever state it happens to be at a given moment, and use that profile to produce a color proof. But if the target system

is not stable over time or it cannot easily be returned to the condition it was in when the profile was made, the proof will serve little purpose. This observation points to the importance of device calibration for both proofing and target systems, as outlined above. In the commercial printing world this has also lead to a movement to target not actual physical printing presses but a small number of idealized devices that each represent a certain category of real devices and/or applications. The proof is relative to such an ideal device, and the task of actual printing press calibration then becomes to make it resemble the chosen ideal device as closely as possible. In a sense it is then the press that is matching the proof, not the other way around. While that may seem like getting things backward, in practice it has many advantages and is proving to be a workable approach.

The last point can be rephrased succinctly as "color management is not a substitute for process control." Process control remains essential for printing in general and includes such essentials as device calibration. It is not because we *can* describe arbitrary states of printing systems in a color profile that we *should.*

POSTER AND LEAFLET PRODUCTION

We have now almost come to the end of our watercolor reproduction scenario. We have captured an original, examined and/or edited it on a computer and electronic display, and have produced color proofs of leaflets to be printed on a digital press and posters to be printed on a large format inkjet printer. Now all that is left for us to do is print the leaflets and posters, distribute them all over town, and organize the opening reception for our art exhibition.

During our discussion of proofing above we have finessed a subtle but important point. Once the device color values of a target system are known, we can use them together with a color profile of that system to derive the corresponding colorimetry, which we can then use to generate a color proof. But how do we determine the desired device color values for the different systems involved? Assuming that we have calibrated each device correctly and that we have accurate color profiles available for each, we can map the desired colorimetry through each profile to obtain the corresponding device color values. So what is the desired colorimetry? If we have colorimetrically accurate capture data available, we can simply choose that. This would basically bring us back to the colorimetric proofing scenario, except that now we are proofing original colorimetry (*artist colorimetry*, as it were) rather than some target printing system colorimetry. As we have discussed, proofing implies using a colorimetric rendering intent, whether relative or absolute. But as we also discussed, colorimetric proofing really only makes sense under very constrained circumstances, including viewing environments and substrate types, which we most likely cannot maintain for our leaflet and poster production scenario.

If it is not colorimetric proofing that we are after, then what is it? Because it would lead us too far to discuss the issues involved in detail, conjunctive to say that we are striving for reproductions that are adapted to or optimized for each printing system and substrate type involved, while still remaining faithful to the *look and feel* of the original. In terms of ICC color management this indicates the use of perceptual rendering intents rather than colorimetric ones. In terms of print evaluation this suggests that each print should be taken on its own terms (and in its own viewing environment), while still keeping an eye on the original. A tall order for which unfortunately there are no well-defined recipes that will guarantee us the results we are after. In the next and final section of this chapter we will discuss some future opportunities that may help us achieve our goal.

FUTURE OPPORTUNITIES

No design is perfect, and although the current CIE-based color management systems can reach excellent levels of performance and sophistication, a number of opportunities for future improvement can be identified. We will just mention a few of these, sketching the expected benefit and indicating how they are related to past or ongoing work within the CIE.

Self-Calibrating and Self-Profiling Devices

Device calibration ("to adjust or bring into balance"[60]) is fundamental to reliable color management, as we have indicated above. Calibration brings a device into a known state, without which device characterization data as represented in a device profile is meaningless. Nearly all color caputre, display, and reproduction devices in use today have considerable analog subsystems, which tend to require periodic calibration to compensate for drift. To the extent that calibration procedures can be made automatic and unobtrusive, the reliability and quality of color management systems will increase, as will their user friendliness. Most modern desktop scanners have automatic calibration capabilities, for instance, using internal white reference strips. Displays can be calibrated, but typically not automatically and unobtrusively. Some high-end displays come with dedicated hardware (sensors) and software for calibration, but the onus is on the user to periodically take out the sensor, mount it on the display, and trigger the calibration procedure. To make progress, calibration sensors should be incorporated into the display itself (using wave guides or other appropriate means to measure part of the emitted light), and calibration procedures should be triggered and run their course without the need for user intervention (for instance, in the disguise of a "screen saver" that is run during computer idle time). Recently a number of inkjet and other printers have appeared with built-in calibration sensors (typically densitometers or derivatives), with the corresponding automatic calibration routines (as so-called "firmware"). Even though such devices can automatically trigger calibration routines during idle time or when the necessary conditions are met, the calibration procedures themselves can hardly be called unobtrusive. They typically involve the printing of a test pattern on the available media, its measurement using the built-in sensor, and the calculation and application of a set of calibration parameters. Nevertheless, this is a major progress compared to using off-line measurement and calibration procedures.

Once a device is calibrated, it is in a known state. Exactly what that state looks like needs to be described by collecting device characterization data and constructing a device model from it, typically in the form of a device profile. For capture devices, this involves recording a profiling target with known device-independent color specifications (using one of the CIE color spaces), and relating the latter to the device color values resulting from the capture. Conversely, for display and reproduction devices this involves (re)producing a profiling target with known device color values, measuring the corresponding device output in device-independent (CIE-based) color coordinates, and relating the two. Both types of profiling procedures could in principle be made automatic. For scanners one could imagine a color characterization target being built into the device itself, similar to the white references being used today. This, however, would not take into account the effect of different types of "original" media being scanned, which is necessary for best performance. For displays, the same type of "screen saver" approach could be used as described above for calibration, but the built-in measurement device may have to be upgraded to a colorimeter or spectrophotometer that can be traced to CIE standards (which is not necessarily required for calibration only). Likewise, printers would need a built-in colorimeter or spectrophotometer traceable to CIE standards, but otherwise the process would be comparable to the ones in use today for automatic calibration.

Workflow Automation

In technology, *workflow* typically means "the automatic routing of documents to the users responsible for working on them."[61] With respect to color management, we could perhaps rephrase this as *the routing of color content to the system components responsible for transforming it.* We can think of devices such as digital cameras, scanners, displays, and printers, and software for capturing, editing, and preparing for (re)production as some of these system components. If such transformations of color content are to be automatic, something desirable in all but the most high-end of application contexts, then they must be represented in an unambiguous way, and certain *division of labor* agreements must be in place among the system players.

To illustrate the first requirement, it is not enough to know that a certain file contains RGB data for it to be displayed correctly. At the very least it must be made clear what kind of RGB is involved, for instance by relating it to a CIE-based device characterization profile of the device that produced it. Likewise, it is not sufficient to know that a document contains CMYK data for it to be printed correctly. Reference could be made to another CIE-based device characterization profile of the device that the document was *intended* to be reproduced on. Such links between color encodings and device characterization profiles can be made via the embedding of the latter in digital file formats, or via other mechanisms such as metadata tags referring to a set of standardized profiles. Unless all data in a given system shares the same image state and expected viewing conditions,[27]

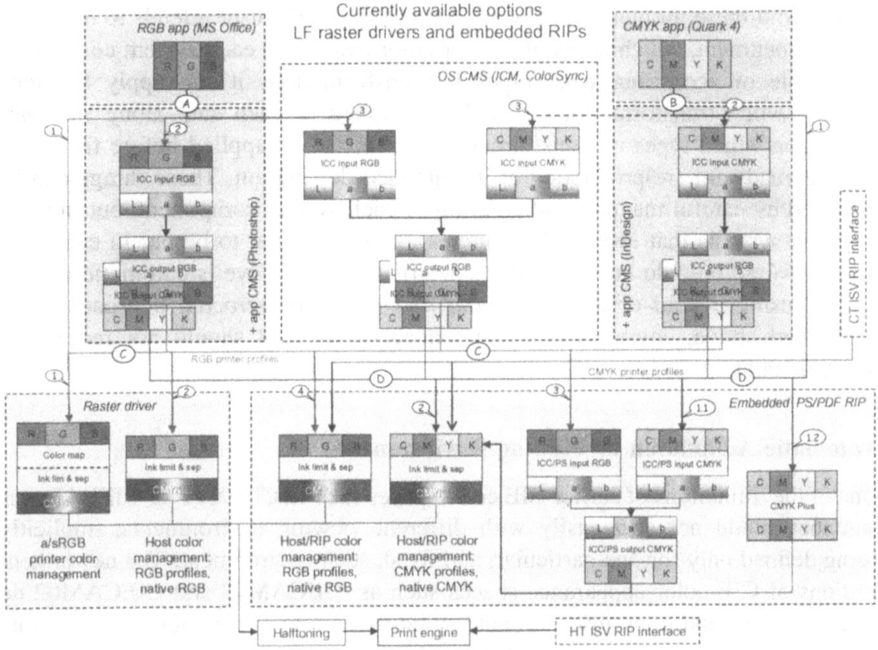

FIGURE 7.9 Proliferation and redundancy of color management functionality in heterogeneous color managed systems. The large shaded boxes represent RGB-based applications, operating system functionality, CMYK-based applications, raster printer drivers, and printer embedded PS/PDF RIPs respectively, in English reading order. Smaller boxes inside the former represent color profile and color data types, and lines among them the different ways in which they can in principle be connected. Although in principle there is only one color transformation from input to output color space involved, a system as depicted allows 27 different paths to be constructed. Almost half of those result in wrong output, and about 75% of them are redundant, resulting in the same output as some other one. See color insert.

such properties must also be made explicit, and system components must be prepared to deal with them.

Unambiguous representation of exchanged color data is necessary, but not sufficient for everything to work as expected. One of the most common causes of problems in color-managed systems is the proliferation and redundancy of color management functionality among different devices and software components of heterogeneous systems.[†††††] This often leads to unintended *double color management* (or worse) and poor results, which in turn need complex configurations of each component involved to try and prevent it. Figure 7.9 provides a schematic illustration of the issue.

[†††††]In the sense of components (hardware or software) from different manufacturers being involved.

To avoid these problems, one needs to know what the user intends to do with a certain document, which types of data or color encodings each system component is capable of accepting, what types of transformations it can apply to them, exactly which transformations have been applied at each step along the way, and hence which types of transformations remain to be applied before the document is ready for (re)production at its intended destination. These things can be achieved by careful manual configuration of each system component, but in more complex systems that eventually requires a color scientist to be put in each product box (and they do not come *that* cheap). The alternative is automatic configuration using a kind of *universal plug and play*[62] or *zeroconf* approach[63]. As mentioned above, color needs to be managed, but that should not require an MBA or PhD.

Automatic Adaptation to Viewing Environment

One of the limitations of earlier CIE color spaces such as CIE XYZ or CIELAB was that they could not deal easily with different viewing environments, implicitly being defined only for one particular, and fixed, such environment. The newer generations of CIE color appearance spaces such as CIECAM97 and CIECAM02 do take viewing conditions into account explicitly and in a parameterized fashion. What this means in practice is that if one knows the intended (or actual) viewing environment for a particular document, one can process its color data such that the result will appear "correct" in that environment. But where does one get the required information about the viewing environment? In some cases, for instance the sRGB systems discussed above, one can just assume that it is fixed and hope for the best. In other cases, for instance, in the current ICC-based systems, the implicit viewing environment parameters are normative rather than informative, and hence the results produced by such systems are only "valid" if actual viewing conditions match the prescribed ones. Unfortunately such constraints cannot be successfully imposed in many practical scenarios, and usually it is not much of a consolation to tell an unhappy user that "your viewing environment is wrong, hence your complaint is not admissible."

If viewing condition parameters could be established (measured) automatically, perhaps the necessary adaptations could also happen automatically and transparently to the user? In the case of digital cameras, the viewing environment is not independent of the scene being captured, in fact in a sense it *is* the scene being captured. Most modern digital cameras do record a number of types of metadata together with the image, such as time (and sometimes place) of capture, aperture and shutter speeds, and so on. Perhaps this would need to be extended to more scene-related parameters such as brightness level, estimated illuminant color temperature or spectral characteristics, and so on. For scanners the situation is rather different because they self-illuminate an original (whether reflective or transmissive) with a fixed light source using a fixed illumination geometry. A description (perhaps added to a scanner profile) of the characteristics of the light source and illumination geometry used might help in some cases, but the utility of this seems limited. Perhaps an "active scanner" concept would be more to the point: rather than using a single fixed "viewing environment," such a device would accept instructions to "view" an original being scanned under the conditions that the image requester would consider appropriate. One could think of brightness level, illumination geometry, and perhaps even color temperature or spectral composition of the light source used for scanning.[‡‡‡‡‡]

Electronic display or "softcopy" devices are unique when compared to input and hardcopy devices, in the sense that the images produced by them cannot be "transported" to different viewing environments than the one they were produced in.[§§§§§] The images produced are ephemeral and inextricably linked to the device itself (*the medium is the message*, quite literally perhaps). As such there is no need to worry about recording viewing environment parameters for later use, only for instantaneous use. As an example of this approach, a well-known color management hardware and software manufacturer has recently introduced a consumer level monitor calibration system that has an interesting and unusual feature. The sensor can be put into a small stand and left on the desk next to the monitor (i.e., in the same viewing environment as the monitor), and under software control will periodically measure the ambient light level and adjust the monitor's brightness to compensate for any changes. Although the idea is not new (some TV sets already used a "magic eye" for much the same purpose in the 1960s, and some clock radios include a similar circuit for dimming the display brightness at night), it is a nice example of automatic measurement of and adaptation to viewing environment parameters in the digital color management domain. Brightness control is probably the easiest thing to do, but perhaps more sophisticated types of measurement and adaptation will follow.

Hardcopy devices (printers) share with scanners and digital cameras that the images they produce can and typically are "transported" to other viewing environments than the one they have been produced in. Hence the measurement of and adaptation to production viewing environments would only serve a temporary purpose, and might actually make things worse for the "end user" of the images produced. Mirroring the suggested capability of scanners to "view things a certain way," one could imagine printers being capable of "producing images a certain way," to make them suitable for a particular viewing environment. Some of this is possible today, for instance via the use of different printer profiles intended for different viewing environments, but this type of control is rather coarse, discrete, and clumsy in use. Other options might exist, for instance, changing the way color separations are being calculated inside (or for) the printer to optimize them for certain intended viewing environments. An example might be the reduction of illuminant metamerism, or the increase of color constancy, for a certain set of

[‡‡‡‡‡]Many spectrophotometers use a primitive version of this already, in the form of a UV cut filter that can be placed into or removed from the illumination path.

[§§§§§]Ignoring corner cases like electronic displays being photographed and printed on paper, for the sake of the argument.

illuminant types. The most principled solution for the latter is more likely to be truly spectral image reproduction (see below), but lacking that there may be other possibilities.

Spatial Processing

At present the operations needed for managing color are typically performed on a color-by-color basis, which means that all instances of an original color are reproduced as the same reproduction color. Although this is certainly a reasonable starting point, benefits can be had from taking a color's spatial neighborhood in the original into account when determining its reproduction. Examples of such spatial approaches are spatial gamut mapping algorithms[64,65] as well as efforts to model color appearance of the parts of spatial complex stimuli, such as photographic images (see the iCAM model.[66] The CIE's TC8-08 on *Spatial Appearance Models* is also active in this area.[67]

Smart CMMs

CMMs are responsible for transforming color data from an input color space to an output color space, using input and output color profiles and additional parameters such as rendering intent. In the current ICC architecture, the role of the CMM is essentially limited to that of an interpolation engine, whereas the color profiles contain all the "value add" such as gamut mapping, rerendering transforms, and perhaps even viewing condition compensation. All of these things are "built in" at profile creation time and cannot be changed afterward. This type of arrangement has been described as a "smart profile, dumb engine" architecture. The opposite of this would then be a "dumb profile, smart engine" architecture, which might have certain advantages over the former:

- Device profiles would contain little more than colorimetric (or spectral) measurements, which would make them very easy to produce and liberate them of undocumented and vendor specific (incompatible and nonportable) "secret sauce."
- Viewing conditions could be specified explicitly and independently of device profiles, and taken into account by the smart engine when calculating color transforms.
- Gamut mapping and/or rerendering transforms could be specified explicitly and independently of device profiles, and applied with knowledge of the specific pair of devices or even images (and their respective gamuts) for which a color transform is being calculated.

Such a system would seem more flexible than the current one, with the potential for improved results especially considering heterogeneous systems and/or often or quickly changing device pairings. Also things like automatic adaptation to viewing environments, as discussed above, could in principle be accommodated more easily.

So what are we waiting for? As always, there are also a number of potential drawbacks or difficulties with such a radically different architecture:

- The "secret sauce" that would be eliminated from device profiles is actually there for a good reason. Calculating gamut mappings and rerendering transforms is usually not an easy task, and perhaps one that cannot be accomplished with sufficient quality in real time (or at least, an acceptably short time for a smart CMM scenario).
- It is not immediately evident how to specify viewing conditions in sufficient detail, or how to use such specifications in on-the-fly construction of color transformations between arbitrary pairs of devices, or color spaces.
- Gamut mapping is in fact a big part of the aforementioned "secret sauce," hence the same considerations apply to it. While it is certainly conceptually interesting to be able to separate gamut mapping algorithms from the data (and color spaces) they operate on, it is not evident how such a separation could be implemented in practice, and with sufficient quality and speed.

Although the concept of smart CMMs has been around for quite a while, until recently little progress had been made toward its realization in practical architectures and implementations. The recent announcement of Microsoft's *Windows Color System* (WCS)[68] has changed that. Scheduled for introduction in 2006, together with Microsoft's next generation operating system dubbed *Vista*, WCS effectively aims to implement a smart CMM color management system. The issues mentioned above are addressed in the following way:

- *Device profiles* are indeed simple XML-formatted "containers" of colorimetric device measurements, specified in CIE XYZ and related color spaces. *Device models* are responsible for turning those measurements into complete forward (and inverse) models of the devices at hand. Some types of device models are provided as a standard part of WCS, others can be provided as *plug-ins* by device manufacturers or other third parties.
- Viewing conditions are separate from device models and are specified using CIECAM02.
- Gamut mapping is separate from device models and viewing conditions. Some gamut mapping algorithms are provided as a standard part of WCS, and these are derived from the work of CIE TC 8-08 on gamut mapping. Other algorithms can be provided as plug-ins by device manufacturers or other third parties, much as is the case for device profiles.

It is not our intention here to evaluate the quality or potential of the WCS system, but merely to flag its existence as a potentially important step in the evolution of practical color management systems. Another worthwhile observation is that although WCS represents a fairly radical departure from existing systems, it is nevertheless clearly and explicitly based on the work of the CIE in its various shapes and forms.

It could be expected that the announcement of WCS might rekindle interest in smart CMM architectures within the ICC, where much discussion and even some prototyping has taken place on this subject in the past.

Multispectral Imaging (CIE TC8-07)

As more and more imaging devices are capable of more than trichromatic color reproduction, the possibility arises for reproducing an original's spectral properties (as opposed to only the tristimulus) (see Chapter 1 of Hunt[1]). For self-luminous originals this would give a match in the spectral power distribution of original and reproduction and as a consequence the two would look identical to observers with normal color vision, to those who have deficient color vision and also to those animals whose visual systems are sensitive to electromagnetic radiation in the range where the spectral match holds. For reflective and transmissive originals a spectral match would be in terms of spectral reflectance or transmittance and, in addition to the properties of spectral matches of self–luminous originals, it would result in the original and reproduction looking the same under any illumination (i.e., their appearance would change in the same way for each change of light source).

The key challenges in digital multispectral color reproduction include questions about the encoding of multispectral data (e.g., Uchiyama et al.[69]), the spectral characterization of imaging devices (e.g., Chen et al.[70]) and the potential benefits to gamut mapping from working in a spectral domain (e.g., Derhak and Rosen[71]). The CIE's TC 8-07 on *Multispectral Imaging* (http://www.colour.org/tc8-07/) too is active in this field.

CONCLUSION

We hope to have provided the reader with a reasonable overview of color management systems past, present, and future, and their importance to an ever-increasing number of applications. If anything is clear from our discussion, it is that the work of the CIE has been an essential part of most if not all of these developments. We sincerely hope that this will continue to be the case for the foreseeable future.

ACKNOWLEDGMENTS

Eugenio Martinez–Uriegas (Hewlett-Packard Company) for the idea of representing color comparisons as comparisons between observer-internal entities as used in Figure 7.1; Craig Revie (Fuji Film Electronic Imaging Ltd.), Eric Walowit (Color Savvy Systems Inc.), and Jack Holm (Hewlett-Packard Company) for advice on future developments within the ICC.

REFERENCES

1. Hunt RWG (1995) *The Reproduction of Colour*, 5th ed., Fountain Press, England.
2. CIE Technical Report (2004) *Colorimetry*, 3rd ed. Publication 15:2004, CIE Central Bureau, Vienna.
3. CIE Technical Report (2004) *A colour appearance model for color management systems*, CIECAM02 Publication CIE 159:2004 CIE Central Bureau, Vienna.
4. Yendrikhovskij SN (1998) *Color reproduction and the naturalness constraint*, PhD thesis, Technische Universiteit Eindhoven, The Netherlands, ISBN 90-386-0719-9.
5. de Ridder H (1995) Naturalness and image quality: influence of chroma variation at various lightness levels, in: *IS&T/SID 3rd Color Imaging Conference*, Scottsdale, AZ, pp. 140–143.
6. Xiao K, Li C, Luo MR, Taylor C (2004) Color appearance for dissimilar size, in: *IS&T 2nd European Conference on Color in Graphics, Imaging and Vision (CGIV)*, Aachen, Germany, pp. 65–69.
7. de Fez MD, Luque JM, Capilla P, Pérez-Carpinell J, DÀez MA (1998) Color memory matching analysed using different representation spaces, *J. Optics*, **29**, 287–297.
8. IEC (1999) IEC 61966-2-1 *Multimedia systems and equipment—Colour measurement and management—Part 2–1: Colour management—Default RGB colour space—sRGB*, IEC, Switzerland.
9. Bala R (2003) Device characterization, in: *Digital Color Imaging Handbook* (Ed., G. Sharma), CRC Press, pp. 269–384.
10. Fairchild MD, Johnson GM (2002) Meet iCAM: a next-generation color appearance model, in: *IS&T/SID 10th Color Imaging Conference*, Scottsdale, AZ, pp. 33–38.
11. CIE Technical Report (2004) *Chromatic adaptation under mixed illumination condition when comparing softcopy and hardcopy images*, Publication CIE 162:2004 CIE Central Bureau, Vienna.
12. Fairchild MD (2004) *Color Appearance Models*, 2nd ed., John Wiley & Sons.
13. Johnson A (1992) *Colour Appearance Research for Interactive System Management and Application—CARISMA, Work Package 2—Device Characterization*, Report WP2-19 Colour Gamut Compression (For a description of the CARISMA algorithm see also (Morovič, 1999, pp. 271–274)).
14. Cholewo TJ, Love S (1999) Gamut boundary determination using alpha–shapes, in: *IS&T/SID 7th Color Imaging Conference*, Scottsdale, AZ, pp. 200–204.
15. Braun GJ, Fairchild MD (1997) Techniques for gamut surface definition and visualization, in: *IS&T/SID 5th Color Imaging Conference*, Scottsdale, AZ, pp. 147–152.
16. Morovič J, Luo MR (1999) Developing algorithms for universal color gamut mapping, in: *Color Engineering: Vision and Technology* (Eds., M.W. MacDonald and M.R. Luo), John Wiley & Sons, Chichester, England, pp. 253–282.
17. Morovič J, Luo MR (2000) Calculating medium and image gamut boundaries for gamut mapping. *Color Res. Appl.*, **25**:394–401.
18. Morovič J (2003) Gamut mapping, in: *Digital Color Imaging Handbook* (Ed. G. Sharma), CRC Press, pp. 639–685.
19. CIE (2004) *Guidelines for the evaluation of gamut mapping algorithms*, Publication 156:2004, CIE Central Bureau, Vienna.

20. Stokes M (1997) Industry adoption of color management systems, in: *Proceedings of AIC Color 97 Kyoto*, pp. 126–131.
21. ICC (2004) Specification ICC.1:2004–10 (Profile version 4.2.0.0) *Image technology color management—Architecture, profile format, and data structure*, http://www.color.org/ICC.
22. ISO (2000) ISO 3664:2000 *Viewing conditions—Prints, transparencies and substrates for graphic arts technology and photography*, ISO.
23. Borg L (2004) *Black Point Compensation From Adobe Systems*, Adobe Systems, http://www.color.org/Adobe1bpc.pdf [accessed 15 November 2005].
24. Koh KW, Tastl I, Nielsen M, Berfanger DM, Zeng H, Holm J (2003) Issues encountered in creating a Version 4 ICC sRGB profile, in: *IS&T/SID 11th Color Imaging Conference*, Scottsdale, AZ, pp. 232–237.
25. Adobe Systems Inc. (2005) *Adobe® RGB (1998) Color Image Encoding*, Version 2005–05, Adobe Systems Inc.
26. Merriam–Webster Inc. (2003) *Merriam–Webster's 11th Collegiate Dictionary*, Merriam–Webster Inc.
27. ISO (2004) ISO 22028–1:2004 *Photography and graphic technology—Extended colour encodings for digital image storage, manipulation and interchange—Part 1: Architecture and requirements*, ISO.
28. ISO (1997) ISO 12641:1997 *Graphic technology—Prepress digital data exchange—Colour targets for input scanner calibration*, ISO.
29. Morovič PM (2002) *Metamer sets*, PhD thesis, School of Information Systems, University of East Anglia.
30. Morovič J, Sun PL, Morovič PM (2001) The gamuts of input and output color reproduction media, *Proc. SPIE*, **4300**, 114–125.
31. Schroedinger E (1920) Theorie der Pigmente von groesster Leuchtkraft, *Ann. Phys.* (Paris), **62**, 603–622.
32. Morovič J, Morovič PM (2003) Determining colour gamuts of digital cameras and scanners. *Colour Res. Appl.*, **28**(1), 59–68.
33. Finlayson GD, Morovič PM (1999) Metamer constrained colour correction, in: *IS&T/SID 7th Colour Imaging Conference*, Scottsdale, AZ, pp. 26–31.
34. Yule JAC (1938) The theory of subtractive colour photography. I. The conditions for perfect colour rendering, *J. Opt. Soc. Am.*, **28**, 419–430.
35. Kang HR (1992) Color scanner calibration, *J. Imaging Sci. Tech.*, **36**(2), 162–170.
36. Kang HR, Anderson PG (1992) Neural network applications to the color scanner and printer calibrations. *J. Electronic Imaging*, **1**(2), 125–135.
37. Gatt A, Morovic J, Noriega L (2003) Colorimetric characterization of negative film for digital cinema post–production, in: *IS&T/SID 11th Color Imaging Conference*, Scottsdale, AZ, pp. 341–345.
38. Kasson JM, Nin SI, Plouffe W, Hafner JL (1995) Performing color space conversions with three-dimensional linear interpolation. *J. Electronic Imaging*, **4**(3), 226–250.
39. Shepard D (1968) A two-dimensional interpolation function for irregularly spaced data, in: *Proc. ACM*, ACM Press, pp. 517–524.
40. Spaulding, KE, Woolfe GJ, Giorgianni EJ (2000) Reference input/output medium metric RGB color encodings (RIMM/ROMM RGB), in: *IS&T PICS 2000 Conference Proceedings*, Portland, OR, pp. 155–163.

41. IEC (2003) IEC 61966–2–2 *Multimedia systems and equipment—Colour measurement and management—Part 2–2: Colour management–Extended RGB colour space—scRGB*, IEC, Switzerland.
42. Berns RS, Motta RJ, Gorzynski ME (1993) CRT colorimetry. Part I: Theory and practice. *Color Res. Appl.*, **18**(5), 299–314.
43. Berns RS, Gorzynski ME, Motta RJ (1993b) CRT Colorimetry. Part II: Metrology. *Color Res. Appl.*, **18**(5), 315–325.
44. Kwak Y, MacDonald LW (2000) Characterization of a desktop LCD projector. *Displays*, **21**(5), 179–194.
45. Post DL, Calhoun CS (1989) An evaluation of methods for producing desired colors on CRT monitors. *Color Res. Appl.*, **14**, 172–186.
46. Murakami Y, Hatano N, Takiue J, Yamaguchi M, Ohyama N (2004) Evaluation of smooth tonal change reproduction on multiprimary display: comparison of color conversion algorithms. *Proc. SPIE*, **5289**, 275–283.
47. CIE Technical Report (1996) *The relationship between digital and colorimetric data for computer–controlled CRT displays*, Publication 122:1996. CIE Central Bureau, Vienna.
48. "Greedy algorithm" *Wikipedia*. Wikipedia, 2005. *Answers.com* 21 Dec. 2005. http://today.answers.com/topic/greedy–algorithm.
49. "Digital proofing" Computer Desktop Encyclopedia. Computer Language Company Inc., 2005. Answers.com 06 Mar. 2006. http://today.answers.com/topic/digital–proofing.
50. Bala R, Sharma G, Monga V, Van-de-Capelle JP (2005) Two-dimensional transforms for device color correction and calibration, *IEEE Trans. Image Proc.*, **14**(8), 1172–1186.
51. Mahy M (1997) Calculation of color gamuts based on the Neugebauer model. *Color Res. Appl.*, **22**(6), 365–374.
52. Berns RS (1997) A generic approach to color modeling, *Color Res. Appl.*, **22**(5), 318–325.
53. Kasson JM, Plouffe W (1992) An analysis of selected computer interchange color spaces. *ACM Trans. Graphics*, **11**(4), 373–405.
54. Hardeberg JY, Schmitt F (1997), Color printer characterization using a computational geometry approach, in: *IS&T/SID 5th Color Imaging Conference*, Scottsdale, AZ.
55. "Matchprint" Computer Desktop Encyclopedia. Computer Language Company Inc., 2005. *Answers.com* 09 Mar. 2006. http://today.answers.com/topic/matchprint.
56. CIE Technical Report (2001) *Improvement to industrial color difference evaluation*. Publication 142–2001 CIE Central Bureau, Vienna.
57. Uroz J, Morovic J, Luo MR (2002) Perceptibility thresholds of colour differences in large printed images, in: (Eds., L.W. MacDonald and M.R. Luo) *Colour Image Science: Exploiting Digital Media*, John Wiley & Sons, pp. 49–73.
58. "Metamerism" Computer Desktop Encyclopedia. Computer Language Company Inc., 2005. Answers.com 09 Mar. 2006. http://today.answers.com/topic/metamerism.
59. Luo MR, Li CJ, Hunt RWG, Rigg B, Smith KJ (2003) The CMC 2002 colour inconstancy index: CMCCON02, *Coloration Technol.*, **119**, 280–285.
60. "Calibrate" *Computer Desktop Encyclopedia.* Computer Language Company Inc., 2005. *Answers.com* 17 Dec. 2005. http://today.answers.com/topic/calibrate.
61. "Workflow." *Computer Desktop Encyclopedia.* Computer Language Company Inc., 2005. *Answers.com* 17 Dec. 2005. http://today.answers.com/topic/workflow.

62. "Universal plug–and–play" *Computer Desktop Encyclopedia.* Computer Language Company Inc., 2005. *Answers.com* 17 Dec. 2005. http://today.answers.com/topic/universal–plug–and–play.

63. "Zeroconf." *Wikipedia.* Wikipedia, 2005. *Answers.com* 17 Dec. 2005. http://today.answers.com/topic/zeroconf.

64. Bala R, DeQueiroz R, Eschbach R, Wu W (2001) Gamut mapping to preserve spatial luminance variations. *J. Imaging Sci. Technol.*, **45**(5), 436–443.

65. Morovič J, Wang Y (2003) A multi-resolution, full-color spatial gamut mapping algorithm, in: *IS&T/SID 11th Color Imaging Conference*, pp. 282–287.

66. Fairchild MD, Johnson GM (2004) The iCAM framework for image appearance, image difference, and image quality. *J. Electronic Imaging*, **13**, 126–138.

67. Johnson GM (2005) Cares and concerns of CIE TC8–08: Spatial appearance modeling & HDR imaging, in: *SPIE/IS&T Electronic Imaging Conference*, San Jose, pp. 148–156.

68. Microsoft (2005) *Windows Color System: The Next Generation Color Management System*, Microsoft Corporation.

69. Uchiyama T, Yamaguchi M, Haneishi H, Ohyama N, Nambu S (2004) A method for the unified representation of multispectral images with different number of bands. *J. Imaging Sci. Technol.*, **48**(2), 120–124.

70. Chen Y, Berns RS, Taplin LA, Imai FH (2003) A multi–ink color–separation algorithm maximizing color constancy, in: *IS&T/SID 11th Color Imaging Conference*, Scottsdale, AZ, pp. 277–281.

71. Derhak MW, Rosen MR (2004) Spectral colorimetry using LabPQR—an interim connection Space, in: *IS&T/SID 12th Color Imaging Conference*, Scottsdale, AZ, pp. 246–250.

8

COLOR RENDERING OF LIGHT SOURCES

JÁNOS SCHANDA

University of Pannonia, Egyetem u. 10., H-8200. Veszprém, Hungary

INTRODUCTION

The evaluation of color rendering is another example (besides that of color management, the subject discussed in Chapter 7), where CIE provided not only the general guidance but also the recommendations. CIE prepared a recommendation to evaluate the color-rendering characteristics of light sources. The word color rendering is also used in other areas of light and lighting, but we will restrict the subject in this chapter to light source characterization.

The problem of light source color rendering became important when the light source industry was able to prepare sources with different spectral power distributions (SPDs) but equal correlated color temperature (CCT) (and even chromaticity). Such "metameric light sources" (see Chapter 3) will provide different tristimulus values for a reflecting test sample if illuminated with one source or the other. The light source industry needed guidance on how to tailor the spectra of new sources, and the applied illuminating engineer wanted to compare the sources that rendered the colors in the environment differently.[1]

The first experimental methods to characterize the color rendering of light sources were based on the so-called spectral band methods: The SPD of the lamp was divided into eight, later six, bands, and the power in each band was compared to the required values. The first CIE color-rendering evaluation method was based on this principle.[2] But the purpose of a color-rendering index is to find a correlate of the visual impression the observer has when viewing the

Colorimetry: Understanding the CIE System, Edited by János Schanda

illuminated scene. Thus, researches looked for alternative methods. One of these was based on the color-difference calculation of test samples illuminated with a test and a reference source.[3,4]

The CIE Technical Committee responsible for the subject submitted in 1963 a recommendation to evaluate color rendering of light sources by the test sample method.[5,6] CIE published an updated, revised edition of this publication in 1974,[7] and republished it recently with minor editorial changes.[8] Therefore, basically the method prescribed in 1974 is still in force, despite the fact that the method has been criticized in a number of publications, and CIE itself has established several technical committees to update the color-rendering method (and a TC is also active at the time of writing this report). We will come back to these questions after reviewing the current official method.

THE OFFICIAL CIE TEST SAMPLE METHOD OF COLOR RENDERING EVALUATION

The CIE defined the color rendering in the International Lighting Vocabulary[9] as

> "Effect of an illuminant on the color appearance of objects by conscious or subconscious comparison with their color appearance under a reference illuminant."

The first and the most difficult problem of this definition is that it requires "a reference illuminant," but leaves open the selection of the reference illuminant. The CIE Technical Committee that was responsible to develop the test method had long discussions on this question, because the selection of the reference illuminant has profound influence on the calculation result.[10] Finally, it was decided to use illuminants of equal CCT*: a blackbody radiation below 5000 K, and a phase of daylight above this CCT. This meant on one side that there are an infinite number of reference illuminants, and that, for example, an incandescent lamp with a CCT of 2900 K will have the same good color rendering as natural daylight. Ever since making this decision, the question has been debated, but no acceptable solution was found. The argument in favor of the present definition is that, on one hand, we are accustomed to accept the color rendering of incandescent light as being quite good and would like to have, for example, our warm white fluorescent lamps to mimic the color rendering as experienced under tungsten light, and on the other hand, even if we would like to have only one reference light source—and a natural choice would then be CIE standard illuminant D65—the chromatic adaptation formulas that are available (or at least that were available in 1974) are not good enough to bridge this chromaticity difference reliably.[11]

---

*See Chapter 3 for the definition of correlated color temperature.

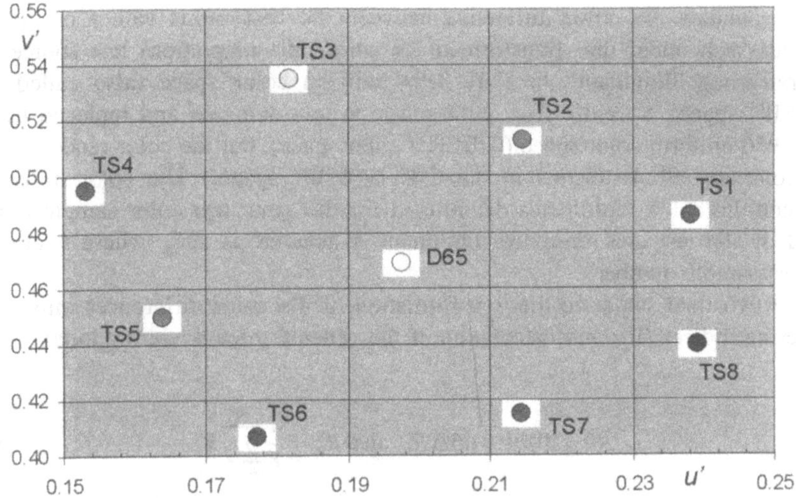

FIGURE 8.1 The CIE 13.2 test samples for D65 illumination and the CIE 2° observer in the u', v' diagram. See color insert.

Thus, one has to select the reference illuminant of equal CCT from the pool of the blackbody radiators and phases of daylight (the CIE publication gives tolerances as to how accurately the same CCT has to be selected, but with the present day computer techniques this can be done with a much smaller difference in CCT). As the chromaticity of the test light source might deviate from the chromaticity of the reference illuminant (along the line of constant CCT, the so-called isotemperature line) one should allow for chromatic adaptation by means of a Von Kries type of transformation[12,13] with reference stimuli given by Judd,[14] transforming the chromaticity of the test source to that of the reference illuminant.[†]

The next step of the calculation is to determine the tristimulus values of a number of test color samples for both the transformed test source and the reference illuminant. Fourteen test samples were chosen from the Munsell Book of Colors, where the first eight are of moderate chroma, and based on these a general color-rendering index is calculated. The other six color samples have high chroma values and serve to provide additional testing of the color-rendering properties of the test source. Figure 8.1 shows the first eight test color samples in the u', v' diagram for CIE standard illuminant D65 and CIE 1931 standard colorimetric observer. Four of the other six test samples represent strong red, yellow, green, and blue colors, and the last two were intended to mimic human (Caucasian) complexion and leaf green, as these colors are often encountered in real life.

[†]See Chapter 11 for more details.

To calculate the color difference between the tristimulus values of the test samples seen under the (transformed for chromatic adaptation) test source and the reference illuminant, the CIE 1964 uniform color space (also called CIE $U^*V^*W^*$ space) is used. (This color space is now outdated and replaced by the CIE 1976 uniform color space (CIELUV color space), but the color-rendering calculations are sill performed in the CIE $U^*V^*W^*$ space.) The color difference between the color coordinates determined for the same test color sample illuminated by the test and reference illuminant is denoted as ΔE_i, where i refers to the test sample number.

The two final steps are the transformation of the color differences into color-rendering indices (R_i) and calculation of the general color-rendering index (R_a):

$$R_i = 100 - 4.6\Delta E_i \quad \text{and } R_a = \frac{1}{8}\sum_{i=1}^{8} R_i \tag{8.1}$$

where the constant 4.6 has been selected to give a value of $R_a = 50$ for a traditional warm white fluorescent lamp.

Figure 8.2 shows the flowchart for the determination of the color-rendering indices.

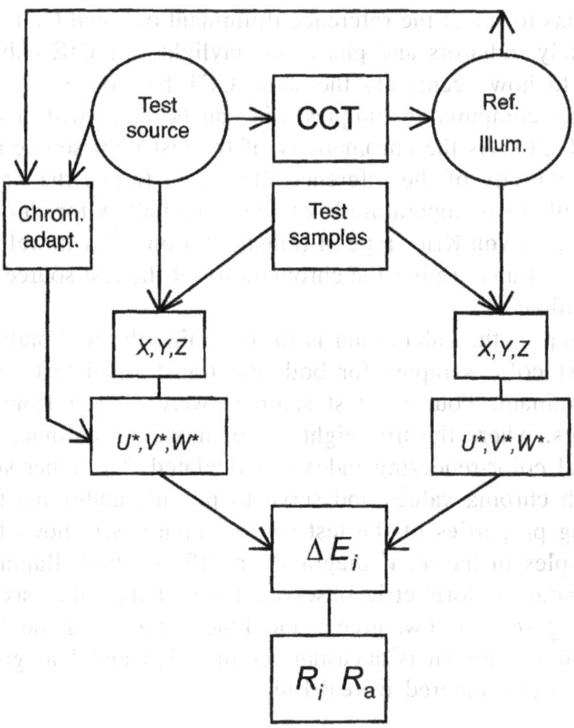

FIGURE 8.2 Flowchart for determining the color-rendering indices.

RECENT INVESTIGATIONS TO UPDATE THE COLOR-RENDERING INDEX CALCULATION

As discussed in the previous section, the CIE Test Sample Method[7] is based on colorimetric techniques, where in recent years CIE found better colorimetric methods. Thus, for example, the CIELUV and CIELAB uniform color spaces[15] are more equidistant than the CIE $U^*V^*W^*$ space, there are better chromatic adaptation transformations than the Von Kries transformation, even if CIE was still unable to recommend only one formula,[16] and the test samples—as shown in Figure 8.1—do not span the chromaticity diagram evenly.

During the past 30 years, a large number of papers were published that partly criticized the CIE Test Sample Method, and showed some evidence where the method breaks down and how a new method could be developed, but they were not conclusive enough to be able to come up with a better method (see, e.g.[17–23]).

On the contrary, lamp developers were happy to have a mathematical description of color rendering and used the technique to optimize the SPD of new lamp families for luminous efficacy and color rendering, a technique that also had its critics, as visual observations did not always support the calculated choices.[24–29]

Investigations were carried out to test the usefulness of the CIE Test Samples[30–33] and the chromatic adaptation formulas,[34,35] testing the method for practical sources as well.[20,36] CIE tackled the question several times, technical committees were established, and after 5–10 years closed down, as they could not find a solution that every party would have agreed upon. The last such committee, CIE TC 1–33, was established in 1991 and closed down in 1999. It was unable to recommend a new color-rendering index formula, but published its closing remarks,[37] and in this publication formulated some ideas that could provide guidance for future research.

First, the experts agreed that the test samples used in the CIE Test Method are not optimum for evaluating current lamp types, suggested to use eight samples from the Macbeth Color Checker Chart,[38] defined average complexion spectra for Caucasian and Oriental skin, and used these as further test sample spectra. Figure 8.3 shows the a^*, b^* coordinates of the test samples for D65 illuminant and 2° observer. The numbers on the graph show the test sample number and their CIE lightness values. The two reddish-yellow samples of small chroma (in the middle of the graph) are the two complexion colors.

As the definition of color rendering states "effect of an illuminant on the color appearance of objects..." the committee hoped to be able to use a color appearance model (see Chapter 11), but in the mid-1990 no such generally accepted model was available. Therefore, calculations were made using the CIELAB color space, and as this color space was developed to be uniform under D65 illumination, the idea was to transform both from the chromaticity of the test source and of the reference illuminant to D65 and perform the color-difference calculation in CIELAB under D65 illumination. The recommended flowchart for calculating color-rendering indices is seen in Figure 8.4. Here, Test refers to the test source, Reference to the reference illuminant, the n test samples are the eight Color Checker Chart samples plus the two skin tone samples.

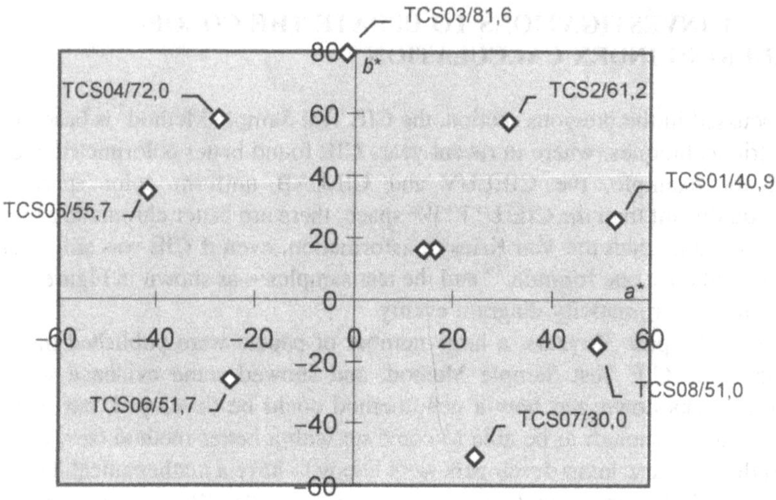

FIGURE 8.3 $L^*a^*b^*$ values for standard illuminant D65 and 2° observer, of the 10 test samples agreed by CIE TC 1–33 to be used in color-rendering calculations.

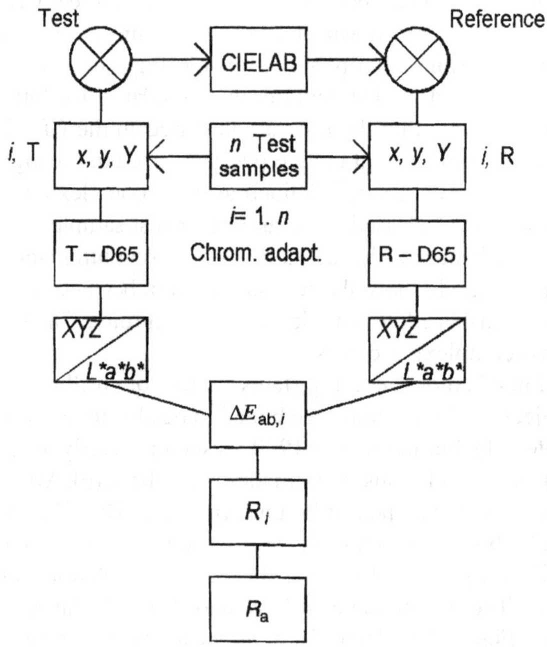

FIGURE 8.4 Flowchart of the recommended color-rendering index calculation method of CIE TC 1–33.

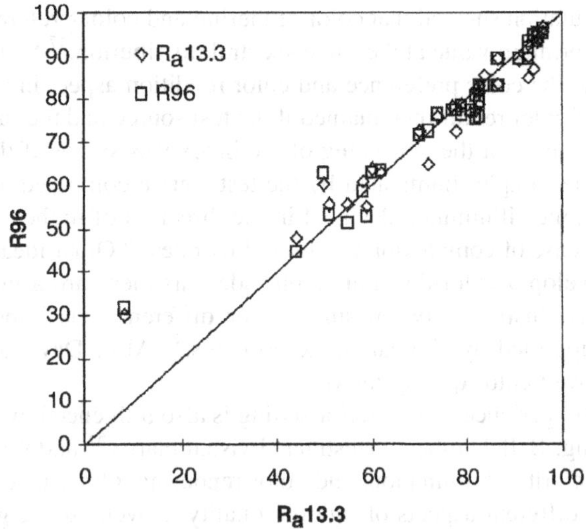

FIGURE 8.5 Correlation chart depicting the *R*96 values versus the R_a 13.3 ones.

A comparison between the traditional method (R_a 13.3)[8] and the new proposed one (R96) showed reasonable agreement, as seen in Figure 8.5. For some of the tested light sources, the rank order changed. Experts were unable to agree whether this was important or not, and whether the new method was so much better that a change was worth, and the final decision was that further visual experiments are needed to be able to decide on a new formula.

Such experiments had been performed already before the new proposal (see, e.g.[39]), and further experiments were started[40,41] partly to get an alternative description of color rendering.[42] These experiments have shown that there are certainly better ways to describe the color-rendering properties of light sources, especially of white LED sources, where the white color is produced by mixing the light of some red, green, and blue LEDs. The decision of the CIE Division 1, responsible for light, color, and vision was for the time being not to change the current color-rendering index calculation method, but develop a new descriptor, such as "color appearance rendering" or "color quality index." A Technical Committee was established in 2006 to investigate this question.

SUPPLEMENTARY METHODS TO DESCRIBE COLOR QUALITY OF LIGHT SOURCES

The wish to supplement color rendering with further quality descriptors is not new. Judd coined the term flattery index already in 1967.[43] The flattery index was intended to describe whether a light source renders colors in a more pleasant (flattery) way then an other. Jerome discussed the differences between flattery and rendition in detail.[44] Later the word preference was used instead of flattery.[45]

Thornton's calculation showed that color-rendering and color-preference indices do not have their optimum value at the same spectral distribution.[47] Some experiments tried to combine the color preference and color rendition aspect in such a way that the maximum of color rendition remained if the test source had the same SPD as the reference illuminant, but the worsening of the index was slower if the color difference between the sample illuminated by the test source compared to the illumination by the reference illuminant deviated in the direction of higher chroma, or, for example, in the case of complexion toward redder hues.[46] Other ideas went into the direction to develop a color-discrimination index, as there are a number of tasks where the discrimination between small color differences is important.[47,48] All these can be supported by simulation experiments.[49] Also, Davis and Ohno published on improved color quality metrics.[50]

The comfort experience in an interior setting is also influenced by the color quality of the lighting. Bellchambers investigated visual clarity[51] and found correlation between visual clarity, illumination, and color rendering. Other investigations tried to correlate the different aspects of lighting quality as well (see, e.g.[52]).

An interesting new approach is based on the hue shift of many colors that shows which hues are highly distorted compared to a reference and which are rendered correctly.[53,54] Our recent studies go in a similar direction by starting from the supposition that if a designer has carefully chosen the colors of an environment to be pleasant under one light source, that is, the observer gets a harmonious impression of the environment, then another light source will be accepted if after chromatic adaptation the colors of the environment stay harmonious.[55]

SUMMARY

Color rendering of light sources has interested color scientists and lamp manufacturers as nonincandescent lamps became available because based on the color-rendering index, the lamp engineer was able to optimize the lamp spectra to get the optimum luminous efficacy and color-rendering index at the same time. It was obvious from the very beginning that color rendering is a color appearance phenomenon, thus it will be difficult to find a color stimulus descriptor that will correlate well with the perception.

After a first trial to simply investigate the amount of power emitted in the different parts of the spectrum, the attention of the experts turned toward the investigation of color distortions that one light source will produce on surface colors compared to the color of the same samples under a reference illuminant. As no really good corollaries of color perception were available, first a chromatic difference and later a color-difference metric was used to characterize the color distortion that the single test samples suffered when the reference illuminant was changed to the test source. At present we already have reasonable color-appearance models that could cope with this problem.

Recent investigations have shown, however, that light color quality is more than the simple color differences determined on a number of test samples. The entire environment has to be considered, perhaps in a similar form as image color appearance models try to cope with the problem of the mutual influence of adjacent color samples (induction and assimilation effects).[56]

The above ideas seem to lead into the direction that the color quality of artificial interior lighting has to be considered as a total environmental question: We need different illumination and CCT of the light source in a dining room, in an office, or in a laboratory. For the different spaces, not only the lighting levels and the color rendering has to be specified, but the color quality of the light also has to fulfill different criteria for one environment and for another. It is a long way that lighting engineering has to go, before it will be able to define for each application area its special lamp quality index but one has the impression that one general color-rendering index will not be enough in all situations. Naturally, this will not only need the research activity by vision and color experts, but also a major rethinking of the concept of color rendering (and color preference—discrimination, etc.) as well as transferring the new concepts into practice, where at the present moment for example, several indoor lighting guides specify minimum color-rendering index values for different applications.

REFERENCES

1. Bouma PJ (1937) Colour reproduction in the use of different sources of 'white' light. *Philips Tech. Rev.*, **2**, 1–7.
2. CIE (1948) *Compte Rendu 11th. Session, Paris*, p. 5.
3. Nickerson D (1960) Light sources and colour rendering. *J. Opt. Soc. Am.* **50** 57–69.
4. Ouweltjes JL (1960) The specification of colour rendering properties of fluorescent lamps. *Die Farbe*, **9**, 207–246.
5. CIE (1963) *Compte Rendu 15th Session, Vienna*, Vol. A, pp. 115–159.
6. CIE Technical Report (1965) *Method of measuring and specifying colour rendering properties of light sources*, 1st ed., Publication CIE **13** (E-1.3.2).
7. CIE (1974) Publication CIE **13.2** (TC-3.2).
8. CIE (1995) *Method of measuring and specifying colour rendering properties of light sources*, Publication CIE **13.3** (TC-3.2). CIE Central Bureau, Vienna.
9. CIE *International Lighting Vocabulary*. Publication CIE **17.4** 1987.
10. Ouweltjes JL (1969) Chromatic adaptation and colour rendering of light sources, in: *Compte Rendu 1st AIC Colour Congress "Color 69" Stockholm*, pp. 831–838.
11. Jerome CW (1974) Absolute color rendering. *J. IES*. 1974 Oct. 25–28.
12. Von Kries J (1904) Die Gesichtsempfindungen, in: Nagel's *Handbuch d. Physiol. d. Menschen* **3. Band** Teil VI. Positive und negative Nachbilder. Lokale und Farbumstimmungen des Sehorgans, pp. 205–220.
13. Von Kries J (1902) Chromatic adaptation. From Selected papers on colorimetry—fundamentals. *SPIE Milestones Series* B 47–52 (Original published in Festschrift der Albrecht-Ludwigs-Universität, pp. 145–158 (1902), reprinted from Sources of colour science, pp. 109–119. MIT Press (1970)).
14. Judd DB (1945) Standard response functions for protanopic and deuteranopic vision. *J. Opt. Soc. Amer.*, **35**, 199–220.

15. CIE (2004) *Colorimetry*, 3rd ed., Publication **15:2004**, CIE Central Bureau, Vienna.
16. CIE (2004) *A review of chromatic adaptation transforms*, Publication **160**:2004, CIE Central Bureau, Vienna.
17. Schanda J (1981) Possibilities of colour rendering evaluation based on a single reference source. CIE, Spanish Committee, in: *IV Lux Europa 1981*, Granada, May 5–7, p. 9.
18. Seim T (1985) In search of an improved method for assessing the colour rendering properties of light sources. *Lighting Res. Technol.*, **17**, 12–22.
19. Pointer MR (1986) Measuring colour rendering—A new approach. *Lighting Res. Technol.*, **18**, 175–184.
20. Xu H (1997) Assessing the effectiveness of colour rendering. *Lighting Res. Technol.*, **29**, 89.
21. Van Trigt C (1999) Color rendering, a reassessment. *COLOR Res. Appl.*, **24**, 197–206.
22. Schanda J (2002) The concept of colour rendering revisited. in: *CGIV '2002 First European Conference on Color in Graphics Imaging and Vision*, Univ. Poitiers, France 2002. 04. 2–5.
23. Schanda J, Sándor N (2003) Colour rendering, past–present–future, in: *International Lighting and Colour Conference, Cape Town*, 2–5 Nov. 2003.
24. Thornton WA (1971) Luminous and colour-rendering capability of white light. *J. OSA*, **61**, 1155–1163.
25. Koedam M, Opstelten JJ, Radielovic D (1972) The application of simulated spectral power distributions in lamp development. *J. IES*, 1972 July 285–289.
26. Opstelten JJ, Radielovic D. Wanmaker WL (1973) The choice and evaluation of phosphors for application to lamps with improved color rendition. *J. Electrochem. Soc.*, **120**, 1400–1408.
27. Walter W (1977) Optimum lamp spectra. *IES Technical Conference* Aug. 1977 New York.
28. Verstegen MPJ, Radielovic D, Vrenkin LE (1974), A new generation of "deluxe" fluorescent lamps combining an efficacy of 80 lumens/w or more with a color rendering index of approximately 85. *J. Electrochem. Soc.*, **121**(12), 1627–1631.
29. Valberg A, Seim T, Sällström P (1979) *Colour rendering and the three-band fluorescent lamp*. Annex 6 to Circular 2/80, Meeting of the CIE 1979, Kyoto, August, 26 p.
30. Opstelten JJ (1980) The dependence of the general colour rendering index on the set of test colours, the standard observer and the colour-difference formula. *Lighting Res. Technol.*, **12**, 186–194.
31. Schanda J (1985) The influence of the test samples used for calculating colour rendering indices. *AIC Congress COLOUR* Monte Carlo, 6 p.
32. Rea MS, Robertson AR, Petrusic WM (1990) Colour rendering of skin under fluorescent lamp illumination. *COLOR Res. Appl.*, **15**, 80–92.
33. Hisdal B (1993) Colour samples and colour rendering of light sources. *Lighting Res. Technol.*, **25**, 13–17.
34. Fuchida T, Mori L (1982) Comparison of correcting methods for chromatic adaptation used for color-rendering specification. *COLOR Res. Appl.*, **7**, 294–301.
35. Schanda J (1982) Chromatic adaptation and colour rendering. *CIE J.*, **1/2**, 30–37.
36. Guo X, Houser KW (2004) A review of colour rendering indices and their application to commercial light sources, *Lighting Res. Technol.*, **36**, 183–197.

37. Commission Internationale de l'Eclairage (1999) *Colour rendering*, TC 1-33 closing remarks. *Publ. CIE*, **135**(2).
38. McCamy CS, Marcus H, Davidson JG (1976) A color-rendition chart. *J. Photographic Eng.*, **2**, 95–99.
39. Kambe N, Mori L (1971) Visual judgements of color rendering properties of light sources. *Proceedings of CIE*, Barcelona **21A**, pp. 135–140.
40. Sándor N, Bodrogi P, Csuti P, Kránicz B, Schanda J (2003) Direct visual assessment of colour rendering. *Proceedings of the CIE Session San Diego* 2003, CIE Publication CIE **152:**2003, pp. D1-42-45.
41. Sándor N, Csuti P, Bodrogi P, Schanda J. (2004) Visual observation of colour rendering. *Proceedings of the CIE Symposium '04, LED light sources*. CIE **x026:**2004 pp. 16–19.
42. Yaguchi H, Takahashi Y, Shioiri S (2001) A proposal of color rendering index based on categorical color names, in: *International Lighting Congress*, Istanbul 2001.
43. Judd DB (1967) A flattery index for artificial illuminants. *Illum. Eng.* **62**, 593–598.
44. Jerome CW (1972) Flattery versus rendition. *J. IES*, 208–211.
45. Thornton WA (1974) A validation of the color-preference index. *J. IES*, 48–52.
46. Schanda J (1985) A combined colour preference-colour rendering index. *Lighting Res. Technol.*, **17**, 31–34.
47. Thornton WA (1972) Color-discrimination index. *J. OSA*, **62**, 191–194.
48. Schanda J, Czibula G (1980) New description of color discrimination properties of light sources. *Acta Chrom.*, **3**(5) 209–211.
49. Davis W, Gardner JL, Ohno Y (2005) NIST facility for color rendering simulation, in: *Proceedings of the AIC Colour 05* Part 1. 519–522.
50. Davis W, Ohno Y (2005) Toward an improved color rendering metric, in: *Proceedings of the Fifth International Conference on Solid State Lighting*, SPIE Vol. **5941**, 59411G.
51. Bellchambers HE (1972) Illumination, colour rendering and visual clarity. *Lighting Res. Technol.*, **4**, 104–106.
52. Schanda J (1978) Colour rendering and the impression of comfort with artificial illumination. *Information Couleur*, **3**(2), 23–28.
53. Kemenade JTC, Burgt PJM (1988) Light sources and colour rendering: additional information to the Ra-index. *National Lighting Conference* Cambridge.
54. Kemenade van JTC, Burgt van der PJM (1995) Towards a user oriented description of color rendition of light sources. *Proceedings 23rd Session of the CIE*, CIE **119/**1, pp. 43–46.
55. Schanda J, Madár G, Sándor N, Szabó F. (2006) Colour rendering – colour acceptability, in: *Sixth International Lighting Research Symposium* On Light and Color, Florida Feb. 2006.
56. Fairchild MD, Johnson GM, Kuang J, Yamaguchi H (2004) *Image appearance modelling and high-dynamic-range image rendring*. MSCL research report.

9

COLOR-MATCHING FUNCTIONS: PHYSIOLOGICAL BASIS

FRANÇOISE VIÉNOT[a] and PIETER WALRAVEN[b]
[a]*Museum National d'Histoire Naturelle, CRCC, 36 rue Geoffroy Saint-Hilaire, 75005 Paris, France*
[b]*TNO, Kampweg 5, 3769 DE Soesterberg, The Netherlands*

THE LINK BETWEEN COLORIMETRY AND PHYSIOLOGY

Over more than 70 years, the CIE has provided users with data and methods that accomplish color specification and which are widely used in the industry. Colorimetry is based on visual experiments. As a visual stimulus, light is specified by its tristimulus values.

When colorimetry was established, it was only possible to formulate hypotheses about visual processes. These last decades, considerable progress has been made in the understanding of vision. Color vision begins with the absorption of photons by the photopigments contained in the cones of the retina. Three families of cones, short- (S), middle- (M), and long-wavelength (L) sensitive, preferentially absorb photons in the short-, middle- and long-wavelength range of the visible spectrum. Cone signals are processed through the retinal neuron network, being added or subtracted to form specific information channels: a luminance channel and two chromatic channels.[1] Therefore, as a visual stimulus, light can be specified in terms of three numbers that are related to the bioelectrical signals elicited in the retina.

Colorimetry: Understanding the CIE System, Edited by János Schanda

The Definition of Cone Fundamentals

Young's theory leads to the supposition that a certain particular choice of colorimetric primaries, called "fundamental sensations" by Donders already in 1880, has a real basis in physiology.

We call "cone fundamentals" the spectral response functions of the long-, middle-, and short-wavelength sensitive cone receptor mechanisms, measured in the corneal plane. In other words, we call "cone fundamentals" the spectral sensitivity of the cones embedded in the eye. Any inert pigment included in the eye acts as a filter that transmits a fraction $\tau(\lambda)$ of the light at wavelength λ that reduces the sensitivity of the eye compared to the sensitivity of the cones.

A color match is obtained only when two stimuli produce equal quantum catches in the three kinds of cones. Any linear transformation of the color-matching functions describe the color-matching properties of the eye. Therefore, the cone fundamentals are obtained from one of these transformations.

Historical Background

For many years, the cone responses had not been accessible by objective experimental methods.

As early as 1886, König and Dieterici[2] derived a remarkably good estimate of the cone fundamentals from color matches of normals and abnormals. They produced the eponym "König" hypothesis that dichromats lack one of the three cone families, which form the basis of modern cone fundamental proposals. Later, Judd[3] made a proposal using, with some modification, the experimental values by Pitt for dichromats. Recent estimates have been constructed from CIE 1931, CIE 1964, or the Judd–Vos 2° color-matching functions.[4–10] Every attempt has contributed to an improved prediction of the cone fundamentals.

Decision by CIE

In 1991, following a suggestion already expressed by Boynton in 1979, the CIE decided to form a technical committee to "Establish a fundamental chromaticity diagram of which the coordinates correspond to physiologically significant axes." The scope of this chapter is to present the work of that technical committee. We intend to demonstrate the connection between the color specification and the underlying physiology. For this, we will review the experimental databases that can be exploited to construct a physiologically significant colorimetry. Then, we will develop the concepts and the derivation of cone fundamentals. Finally, we will present the advantages of referring to physiologically significant colorimetry and will show perspectives.

AVAILABLE EXPERIMENTAL DATA

State of the Art in Physiology

In Vitro Measurements

Precise data of the spectral sensitivity of the cones have been obtained using two different *in vitro* techniques.

In *microspectrophotometry*, a retinal patch is positioned on a microscope slide. Two narrowly defined beams of light pass transversely through the outer segment of a single receptor and through the surrounding liquid, respectively, and are compared. From the relative spectral transmission curve, one can derive the spectral absorbance (or transmission optical density) of the receptor. Although slightly noisy, the technique has been used for many animal species. In 1983, *Dartnall* et al.[11] provided numerical tables for the average spectral absorbance of each type of cone, and showed histograms representative of their numerosity in the eyes of seven persons. Besides one family of *rods*, they identified three families of cones with adjusted maximum sensitivity at 419.0 ± 3.6 nm, 530.8 ± 3.5 nm, and 558.4 ± 5.2 nm.

Recording photocurrents, as small as a few picoampares, elicited in individual cones has allowed the measurement of the spectral response to the illumination of a few *L-cones* and *M-cones* of humans.[12] The technique is very precise. Measurements can be obtained over a six-decade range, yielding fine information about the width of the spectral response.

In 1999, an elegant method using adaptive optics allowed visualization of individual cones at an eccentricity of 1° in the nasal retina and revealing the three families of cones in the living eye.[13,14]

The Principle of Univariance

The *principle of univariance*[15] states that the response of a photoreceptor depends on the number of photons absorbed, whatever be the wavelength of the absorbed photons. The wavelength merely affects the probability that an incident photon be absorbed. In other words, because every captured photon produces the isomerization of one molecule of *11-cis retinal* to *all-trans retinal* and because every *isomerization* converts to the same sequence of events in a cone, the cone acts as a counter of isomerizations produced by absorbed photons. Accordingly, the wavelength that determines the energy of the photon controls the suitability of that energy to produce isomerization.

Note that the intensity of light can be expressed either in terms of energy or in terms of photons. When light produces a chemical change in tissue or in a material, it is instructive to express the intensity of the light in photon units. This is the case when one wants to count the number of photoisomerized *rhodopsin* molecules. When one is interested in the propagation of energy in a medium or in the production or absorption of energy, it is more useful to describe the intensity of the light in terms of energy at different wavelengths. This is the tradition in colorimetry.

Dartnall Nomogram: Dilute Pigment: Effective Transmission Optical Density

All rhodopsin molecules contain the same molecule of *11-cis retinal.* The absorbance of photopigments usually varies smoothly as a function of wavelength.

After Dartnall,[16,17] there is empirical evidence that the low-density spectral absorbance of mammal cone photopigments can be represented by a unique template or "nomogram," when expressed in quantum units and plotted on a logarithmic scale versus some function of wavelength—either the frequency ν, the logarithm of the wavelength,[9] the fourth root of wavelength,[11,18] the logarithm of the wave number,[12] or the normalized frequency $\nu/\nu_{\max}$.[19] The nomogram rule greatly facilitates the modeling because the low-density spectral absorbance of an unknown photopigment can simply be calculated from the wavelength of the peak sensitivity.

So far, the shapes of the spectral absorbance curves would differ only in their lateral positions.

Note a few definitions. The spectral absorptance of the cone is the ratio of the absorbed energy to the incident energy. The absorbance of the cone is the logarithm to base 10 of the inverse of the transmission factor of the cone. Precisely, the low-density absorbance of the photopigment is the absorbance of an infinitely dilute quantity of photopigment. Although, its absolute value tends to be zero, its relative spectral value may be different from zero.

Available Psychophysical Measurements

Psychophysics provides cone fundamental spectral sensitivities that are much more accurate than data provided by physiology.

Spectral Sensitivity Functions of Dichromats and the König Hypothesis

The spectral sensitivities of the three cone types overlap extensively throughout the spectrum. Consequently, the measurement of the spectral sensitivity of a single cone type in the normal trichromatic observer requires special procedures to isolate its response from the responses of the other two unwanted cone types. Isolation is greatly facilitated with dichromatic vision.

The first theoretical basis for deriving cone fundamentals was by König and Dieterici,[2] who postulated that dichromatic vision is a reduced form of trichromatic vision where one cone response is missing and the two others are left unchanged in the spectral sensitivity. Cone isolation can be simplified in these cases. Protanopes are missing the L-cone function, and deuteranopes are missing the M-cone function. The spectral sensitivity of cones can be directly measured with dichromats when one of the functioning cone mechanisms is not active, for one reason or another. For instance, S-cones are insensitive to rapid flicker, and thus in this condition, protanope and deuteranope vision reduces to M-cone or L-cone vision, respectively. Then, the spectral luminous efficiency function corresponds to the sensitivity of the only cone fundamental that is present.

The group of Stockman and Sharpe[9,20] measured the spectral sensitivity of the isolated cones of dichromats, after sequencing and identifying their photopigment genes to ensure that these observers possessed only two pigment types. Stockman et al.[20] measured S-cone thresholds in blue-cone monochromats that possessed only rods and S-cones and derived the S-cone spectral sensitivity. Precise sensitivity functions are obtained over a six-decade range.

Spectral Sensitivity Functions of Isolated Cone Mechanisms

The "two-color threshold" technique proposed by Stiles (all papers by Stiles have been collected in a single volume published in 1978[21]) consists of measuring the threshold of a target at a wavelength to which one cone type is preferentially sensitive against a background at another wavelength to which the other cone types are preferably sensitive. In practice, complete isolation of one cone type is difficult to obtain, but the method can be improved by further suppressing unwanted cone types. S-cones that are insensitive to high temporal frequencies would not respond to flickering targets. Stockman et al.[22] found that the M-cone spectral sensitivity could be measured within half a second following the exchange from a blue to a deep-red, while the L-cone spectral sensitivity could be measured following the reverse exchange. Precise sensitivity function estimates are given over a three-decade range.

Therefore, the proposed cone fundamentals should reflect:

- visual responses similar to those of dichromatic vision in the cases where normal vision naturally reduces to dichromacy as in foveal tritanopia;
- the spectral response of normal trichromatic vision if a correct and total cone isolation has been achieved by some specially designed experimental protocol.

Short Description of Colorimetric Databases

Despite the great progress that has been achieved in physiology and psychophysics, the color-matching functions remain the most reliable data for colorimetric purposes. Several authors[23–26] have given an historical account of the experimental data that were collected during the twentieth century and which form the experimental basis for standard or supplementary colorimetry. Their analyses allow us to understand the foundations and the development of colorimetry. Milestones papers in colorimetry are available in a volume edited by MacAdam in 1993.[27]

Up to the early twentieth century, several experimental investigations had founded the concepts of trichromacy.

Historic Summary

Helmholtz clarified in 1860 the difference in nature between the production of color with pigments that selectively absorb some part of the spectrum, allowing only the remaining radiations to exit the surface (subtractive mixture), and the production of color by the superposition of lights that add their effect on the eye (additive mixture).[28] He identified complementary wavelengths that yield a given white when the lights are mixed in suitable amounts. He understood that the spectral sensitivity of the eye photopigments should overlap to account for trichromacy.

With his color box, Maxwell was able in the same year to match daylight with a mixture of three spectral lights.[29] By replacing an appropriate amount of one of the three spectral lights, he could obtain a series of what we now call Maxwell color matches. He derived the absolute amounts of three fixed primaries required to match any monochromatic light and plotted their relative amounts in a color triangle. By adjusting the weights of the red, green, and blue primaries, sunlight could lie at the center of the triangle. For two observers, he showed that the spectrum locus is concave and lies outside the triangle.

Concurrently, Grassmann formulated the laws of color mixture,[30] and Ives (1915) dealt in detail with the transformation of one trichromatic system to another.[31]

An important insight into the sensitivity of the visual photopigments was gained from investigating the dichromatic form of color blindness. With the assumption that color blind people retain only two of the normal receptors and lack the third one, which allow them to accept a color match with only two primaries, König and Dieterici[2] measured the spectral sensitivities of the cones, data that are incredibly close to modern determinations.

The activity at the beginning of the twentieth century was mainly devoted to photometry. In 1924, the CIE adopted a standard relative visibility function. Although it has been recognised that additivity of photometric comparisons holds only under special experimental conditions, the function incorporates at least six different sets of data. It aggregates four branches, each of them being dominated by a different experimental method. It has now been proved that some of these methods do not yield additive results. Furthermore, a wide range of luminous levels were mixed together. It appears now that the CIE 1924 $V(\lambda)$ function seriously underestimates the sensitivity at short wavelengths.

The color-matching data of Guild and the color-matching data of Wright have been used to define the CIE 1931 standard colorimetric observer. They measured the proportions of three primaries that were required to match spectral lights throughout the visible spectrum. This method is called the maximum saturation method. Guild published mixtures of nonspectral primary lights and normalized the units for each observer based on a match to the NPL white.[32] Seven observers participated. Wright independently collected data from 10 observers.[33] He used spectral primaries and, for each observer, adjusted the units based on two reference wavelengths. Such standardization eliminates part of the variability due to the individual preceptoral absorption. The two independent sets of data were appropriately transformed, compared, and recognized to be in very close agreement.[32]

Guild and Wright obtained spectral chromaticity coordinates. Their color-matching experiments did not include measurements of the spectral luminous efficiency curve of their individual observers but, and I quote Wright, "in the climate of opinion at that time, it was regarded as essential that the standard observer for photometry and the standard observer for colorimetry should be one and the same person."[24] This explains why the spectral chromaticity coordinates derived from the original color-mixture data from Guild and Wright were combined with the spectral luminous efficiency function of the CIE standard photometric observer $V(\lambda)$, in order to reconstruct the color-matching functions of the CIE 1931 standard colorimetric observer. It was assumed that the $V(\lambda)$ luminous efficiency function is a linear combination of the three color-matching functions. In a further step, a linear transformation yielded to an all positive set of functions $\bar{x}(\lambda)$, $\bar{y}(\lambda)$, $\bar{z}(\lambda)$. The $V(\lambda)$ curve was incorporated into the luminance values of the three primaries. Unfortunately, the defects of the CIE 1924 $V(\lambda)$ were propagated into the color-matching functions.

In 1951, Judd, having reviewed more recent measurements of visibility functions, proposed a revision of $V(\lambda)$.[34] Values at wavelengths shorter than 460 nm were considerably increased. Then Judd used the same basic colorimetric data as the CIE 1931 colorimetric standard observer and proposed a revised set of color-matching functions $\bar{x}'(\lambda)$, $\bar{y}'(\lambda)$, $\bar{z}'(\lambda)$. In 1978, Vos completed Judd's work, refining the projective transformation between the CIE chromaticity diagram (x,y) and Judd's chromaticity diagram (x',y') and including Brindley's color reversal data.[35] Corresponding color-matching functions were published, sometimes referred as the CIE-*Judd–Vos functions*, $\bar{x}_M(\lambda)$, $\bar{y}_M(\lambda)$, $\bar{z}_M(\lambda)$.

Stiles and *Burch* undertook a completely new colorimetric study. They matched three spectral primaries to other monochromatic lights. They measured the radiant power $\{P_\lambda d\lambda\}$ of the monochromatic test stimulus and of each primary stimulus. They directly obtained from the observations the color-matching functions for equal energy spectral lights. No appeal to heterochromatic brightness measurements or to any luminous efficiency function was required. Ten observers served in a matching experiment on a 2° field. The spectral transmission of the ocular media of individual observers was also measured. The individual data of the 10 observers were recovered by Trezona.[36] Values of the color-matching functions are given in logarithmic units so that the low values are accurately reported. This pilot experiment on a 2° field was originally designed to validate the matching method. Stiles and Burch reported that his results agreed with Wright's very closely.[37] These 2° data were considered as the 2° pilot data only, but they are of great interest to the derivation of cone fundamentals.

Their other matching data were obtained on 10° fields and were used to define the CIE 1964 supplementary standard colorimetric observer. Stiles and Burch used the same method as for the 2° field, measuring directly the 10° color-matching functions of 49 observers.[38] These were instructed to ignore the central 1° or 2° of vision. The individual functions of 20 observers have been published in the second edition of Wyszecki and Stiles.[25] The superiority of the data is examined in a further paragraph of this chapter. In Russia, *Speranskaya* measured directly the 10° color-matching functions of 18 observers on a 10° field from which the central 2° were occluded.[39] The luminance of the field was very much lower than in the experiment of Stiles and Burch and was at a level where the rod mechanism systematically affected the results. A few unexplained discrepancies appear between the two studies. The color-matching functions obtained in the two studies were weighted and averaged to derive the CIE 1964 supplementary standard colorimetric observer, now known as the CIE 1964 standard colorimetric observer.

Although the color-matching functions of the CIE 1931 standard colorimetric observer, of the Judd revised colorimetric observer, and of the Stiles and Burch pilot study all fall within normal physiological variability, they fail to be perfect.

Superiority of Stiles and Burch Colorimetric Data Basis In 1959, Stiles and Burch produced the measurement of 10° color-matching functions, which constitutes the most comprehensive set of color-matching data, from a large number of subjects (49 subjects from 392.2 nm to 714.3 nm, and in nine subjects from 392.2 nm to 824.2 nm).[38] Because the data are based on the records of many individuals, they are likely to be the representative of the average population. The available data, $\bar{r}_{10}(\lambda)$, $\bar{g}_{10}(\lambda)$, and $\bar{b}_{10}(\lambda)$, refer to the experimental red, green, and blue *primaries*.

- The Stiles and Burch 10° colorimetric data were measured directly. Absolute radiometry was carried out. Because they were not reconstructed from chromaticity data and luminous efficiency data, they were not contaminated by photometric defaults.
- The Stiles and Burch 10° colorimetric data have been safely derived. The calibration at NPL has been carefully completed. Stiles and Burch corrected the low-illuminance color matches for rod intrusion.
- Together with the independently collected data from Speranskaya, that conform to the values from Stiles and Burch, they form the basis of the CIE 1964 supplementary standard colorimetric observer that is an internationally accepted standard for many industries.

Recognizing the high standard of the Stiles and Burch color-matching data, the CIE technical committee TC 1–36 has approved the choice of Stockman and Sharpe[9] to base the recommended cone fundamentals on the large-field 10° color-matching functions $\bar{r}_{10}(\lambda)$, $\bar{g}_{10}(\lambda)$, and $\bar{b}_{10}(\lambda)$ of Stiles and Burch.[38]

Extending Colorimetric Data From 10° Field to Any Field Size From 10° to 1° The possibility to derive the 2° cone fundamentals from the CIE 2° color-matching functions as modified by Judd[5,7] or from Stiles and Burch 2° color-matching functions[8] was suggested by several authors. Unfortunately, the Stiles and Burch 2° functions have never been recognized by the CIE, and the 1931 colorimetric observer and its further corrections suffer from drawbacks.

The possibility is to derive the 2° cone fundamentals from the CIE 1964 10° color-matching functions[8] or from Stiles and Burch 10° function[9] has been achieved successfully. The procedure will be explained in the following sections. Stockman and Sharpe[9] incorporated the new spectral sensitivity data and TC 1–36 has approved this choice. Finally, TC 1–36 has agreed on a method for deriving color-matching functions and proposing cone fundamentals as a function of field size from 10° to 1°. The 10° color-matching measurements of Stiles and Burch[38] constitute the starting point of the derivation.

Consequently, neither the CIE 1931 nor the CIE 1964 color-matching functions will be used to construct cone fundamentals. Nevertheless the newly derived color-matching functions, compared to those of the standard observers, should only moderately modify the customs and habits of the CIE colorimetry. The higher quality is promising for future colorimetric development.

THE CONE FUNDAMENTALS

Following the proposal by Stockman, Sharpe, and colleagues, the CIE has adopted the *10° cone fundamentals* derived from the 10° color-matching functions of Stiles and Burch[38] and has reconstructed functions as the *2° cone fundamentals*.[40] As a modified derivation scheme applies to S-cone fundamentals at longer wavelengths, the CIE recommends cone fundamentals in tabular form. For tutorial purpose, we present in this section the details of the derivation.

Linear Transformation That Yields the 10° Cone Fundamentals

From all possible sets of the three primaries obtained from a linear transformation of color-matching functions $\bar{r}_{10}(\lambda)$, $\bar{g}_{10}(\lambda)$, and $\bar{b}_{10}(\lambda)$, there is only one set that specifies the spectral response of the actual cone fundamentals $\bar{l}_{10}(\lambda)$, $\bar{m}_{10}(\lambda)$, and $\bar{s}_{10}(\lambda)$ (Figure 9.1).

$$\begin{pmatrix} \bar{l}_{10}(\lambda) \\ \bar{m}_{10}(\lambda) \\ \bar{s}_{10}(\lambda) \end{pmatrix} = \mathbf{A} \begin{pmatrix} \bar{r}_{10}(\lambda) \\ \bar{g}_{10}(\lambda) \\ \bar{b}_{10}(\lambda) \end{pmatrix} \tag{9.1}$$

with

$$\mathbf{A} = \begin{pmatrix} 0.192325269 & 0.749548882 & 0.0675726702 \\ 0.0192290085 & 0.940908496 & 0.113830196 \\ 0 & 0.0105107859 & 0.991427669 \end{pmatrix} \tag{9.2}$$

Note that several numerical values can be found in the literature depending on how the normalization of the curves was made.

With such a linear transformation, the S-cone fundamental is insensitive to the red primary. Because the linear transformation presents some weakness for deriving the S-cone fundamental beyond 505 nm, a special procedure was used by Stockman et al.,[20] which is to be explained in a further paragraph.

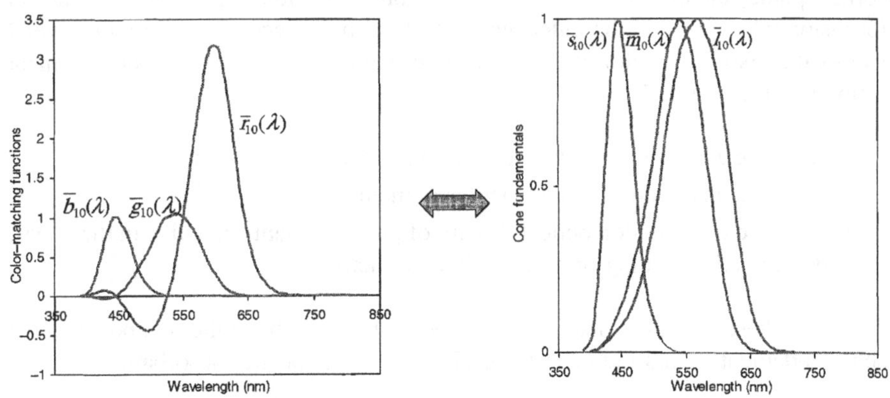

FIGURE 9.1 A linear transformation links color-matching functions and cone fundamentals, except at the end of the longwave branch of the S-cone fundamental.

Stockman and Sharpe optimized adjustments in macular, lens, and photopigment densities in order to reconstruct the 2° M- and L-cone fundamentals that best fit the Stiles and Burch based[37] 2° M- and L-cone fundamentals and dichromat cone spectral sensitivities.

Validation of Cone Fundamentals

Although measured at the corneal level, *cone fundamentals* are determined by the absorption of photons at the cone level.

The choice of the cone fundamentals comes from their ability to represent the actual responses of cones and their similarity to the directly obtained cone spectral sensitivities from psychophysical experiments.

- the spectral sensitivity curves overlap, they peak around 440 nm, 540 nm, and 565 nm, and yield spectral sensitivities of individual cones that are close to those recently measured by objective methods on individual cones, the difference being explained by pre-retinal absorption,
- they yield pigment action spectra that conform to a common spectral template when plotted on an appropriate scale (usually normalized frequency), the long-wave branch descends rapidly as the probability of quantum catch by the pigment decreases,
- they are consistent with 10° normal color matches and 2° normal and tritanopic color matches.

Calculation Scheme From Dilute Photopigment Spectral Absorbance to Color-Matching Functions, and Reverse

As the fundamental spectral sensitivities are measured outside the eye, in the corneal plane, and as the actual absorption of photons initiating the electrical signal for vision takes place in the outer segment of the photoreceptors, there are several factors that explain the deviation of the cone fundamentals from the photopigment action spectra (Figure 9.2):

- selective absorption by the lens and other preretinal media,
- selective absorption by the macular pigment,
- effective transmission optical density of photopigments included in the cones, due to self-screening or other, unknown factors.

Taking these factors into account, it is possible to establish the link between the cone fundamentals and the underlying photopigment spectral absorbance.

Lens and Other Preretinal Media

The lens transmission optical density increases at very short wavelengths. Stockman et al.[20] proposed slight modifications to the van Norren and Vos *lens pigment*

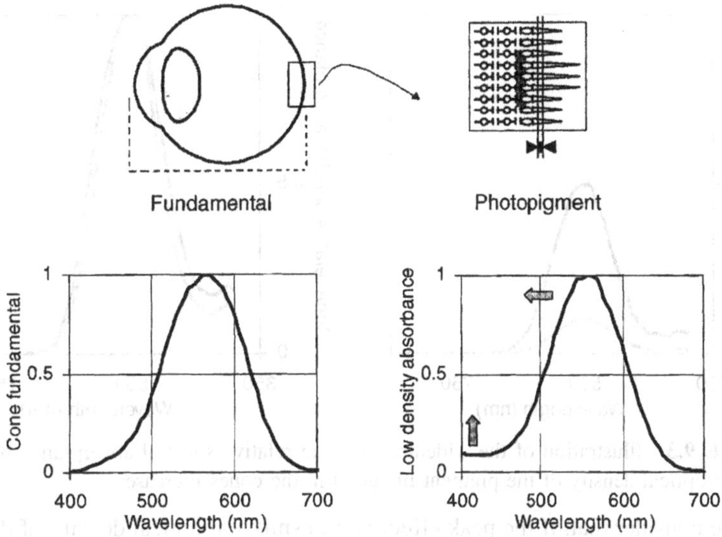

FIGURE 9.2 Illustration of cone fundamental and photopigment absorptance.

transmission optical density spectrum.[41] Although certainly dominated by lens pigment, this transmission optical density spectrum is likely to reflect filtering by the lens and some absorbing media of unknown origin.

Given this template for lens and other preretinal media, the transmission optical density at 400 nm associated with the 10° or 2° cone fundamentals is fixed at 1.7649.

Macular Pigment

The macular pigment transmission optical density is maximum at about 460 nm. The spectral template given by Wyszecki and Stiles needed slight correction. The macular pigment spectrum adopted by Stockman et al.[20] is the spectrum of lutein and zeaxanthin mixed in the same ratio as found in the foveal region. Besides, the macular pigmentation varies considerably among observers. Representative values of the macular pigment transmission optical density at 460 nm are 0.095 for 10° field and 0.350 for 2° field.

Calculation Scheme From Dilute Photopigment Spectral Absorbance to Cone Spectral Absorbance, and Reverse

In general, $A(\lambda)$ being the spectral absorbance that characterizes a unitary solution of the pigment (one unit length, one unit concentration), the Beer–Lambert law states that for a solution of concentration c and optical path length l, the optical density of the pigment included in cones is (Figure 9.3)

$$D(\lambda) = c \cdot l \cdot A(\lambda) \tag{9.3}$$

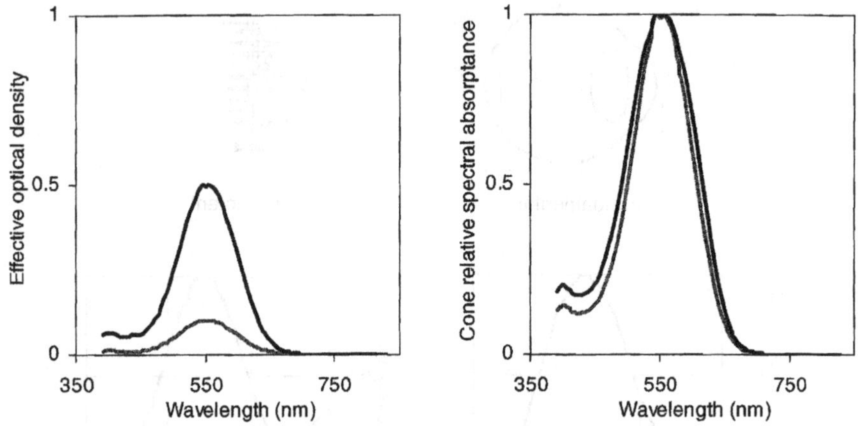

FIGURE 9.3 Illustration of the widening of cone relative spectral absorptance when the effective optical density of the pigment included in the cones increases.

If we consider that, if the peak effective transmission optical density of the cone is comparable to $c \cdot l$, then the cone absorptance $\alpha(\lambda)$ as a function of wavelength and is described by

$$\alpha(\lambda) = 1 - 10^{-D(\lambda)} = 1 - 10^{-c \cdot l \cdot A(\lambda)} \tag{9.4}$$

The reverse procedure to obtain the action spectra of the dilute photopigment was proposed by Lamb.[19]

Normalizing the spectral absorbance of the photopigment in dilute concentration and the absorptance of the cones to a value of unity at the top of the curve gives

$$\frac{D(\lambda)}{D(\lambda_{max})} = \frac{A(\lambda)}{A(\lambda_{max})} \tag{9.5}$$

and

$$\frac{\alpha(\lambda)}{\alpha(\lambda_{max})} = \frac{1 - 10^{-D(\lambda)}}{1 - 10^{-D(\lambda_{max})}} \tag{9.6}$$

Replacing the term $D(\lambda)$ from Equation (9.5), it gives

$$\frac{\alpha(\lambda)}{\alpha(\lambda_{max})}(1 - 10^{-D(\lambda_{max})}) = 1 - 10^{-D(\lambda_{max})\frac{A(\lambda)}{A(\lambda_{max})}} \tag{9.7}$$

Isolating the power term $A(\lambda)/A(\lambda_{max})$ for variable wavelength on one side and then taking the logarithmic value on each side, one obtains

$$-D(\lambda_{max})\frac{A(\lambda)}{A(\lambda_{max})} = \lg\left(1 - \frac{\alpha(\lambda)}{\alpha(\lambda_{max})}(1 - 10^{-D(\lambda_{max})})\right) \tag{9.8}$$

$$\frac{A(\lambda)}{A(\lambda_{\max})} = \lg(1 - \frac{\alpha(\lambda)}{\alpha(\lambda_{\max})}(1 - 10^{-D(\lambda_{\max})}))/(-D(\lambda_{\max})) \tag{9.9}$$

For the derivation of the low-density spectral absorbance of the photopigments in terms of quanta, the calculation scheme just described can be applied with the following peak effective optical densities

$$D_{\text{L-cone}}(\lambda_{\max}) = D_{\text{M}}(\lambda_{\max}) = 0.50; \quad D_{\text{S}}(\lambda_{\max}) = 0.4, \text{for } 2^\circ \text{ fields}$$
$$D_{\text{L}}\text{-cone}(\lambda_{\max}) = D_{\text{M}}(\lambda_{\max}) = 0.38; \quad D_{\text{S}}(\lambda_{\max}) = 0.3, \text{for } 10^\circ \text{ fields}$$

S-cone Fundamental From 510 to 615 nm (2° field and 10° field)

Above 505 nm, the linear transformation given in Equation (9.1) yields incorrectly the estimated S-cone sensitivity, compared to the direct psychophysical measurements and to the result to be expected from the absorption of rhodopsin at longer wavelength.

To produce the 2° S-cone fundamental from 510 nm to 615 nm, Stockman et al.[20] used the 10° color-matching functions of Stiles and Burch adjusted to 2° assuming the same photopigment-effective transmission optical density and macular pigment transmission optical density as for shorter wavelengths, and their experimental threshold data measured in normal and blue-monochromat observers. They produced a sensible expansion to the 2° S-cone fundamental by fitting a Gaussian function to all the data expressed in terms of quantum units versus a frequency scale. They additionally smoothed some irregularities of the curves. Finally, the S-cone fundamental is set to zero at wavelengths longer than 620 nm.

In a further step, the 2° S-cone fundamental was readjusted back to 10°, using exactly the same parameters as above for shorter wavelengths.

The S-cone fundamental is so small at longer wavelengths that corrections, although useful for fundamental consistency, probably have little consequence in colorimetric applications. For longer wavelengths, CIE recommends tabulated $\bar{s}_2(\lambda)$ and $\bar{s}_{10}(\lambda)$ values (Figure 9.4).

Extension to Any Field Size

Once it is admitted that reconstructing 2° cone fundamentals from 10° cone fundamentals is a valuable procedure, the exercise can be extended to derive fundamentals as a function of field size.

Given the change of macular pigment transmission optical density and of photopigment effective transmission optical density with eccentricity, it is possible to link the cone fundamental and the photopigment low-density relative spectral absorbance and to extend the derivation of the cone fundamentals to any field size from 1° to 10°.

The distribution of transmission optical density $D_{\tau,\text{macula}}$ of the macular pigment over the retina, as assessed by color matching on circular fields, has been described

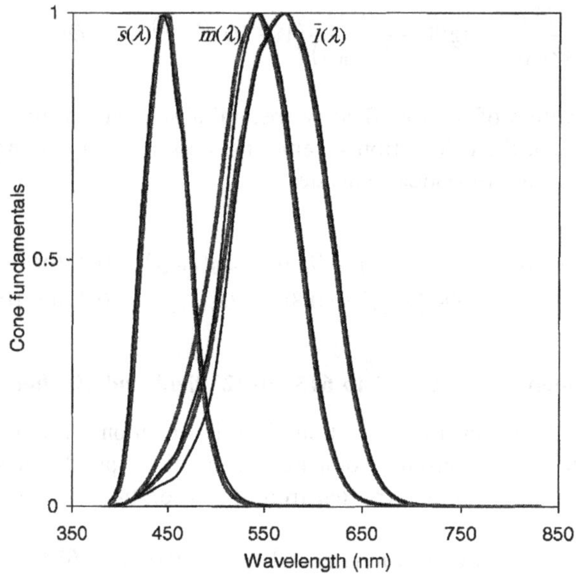

FIGURE 9.4 Normalized L-, M-, and S-cone fundamentals for 2° (—) and 10° (—) fields.

by an exponential function of the field diameter f.[42]

$$D_{\tau,\mathrm{macula}} = a \cdot \mathrm{e}^{(-f/b)} \tag{9.10}$$

where $a = 0.485$ and $b = 6.132$.

The length of the cones increases in the fovea. The *Beer–Lambert law* states that as the thickness of a solution increases, its transmission optical density proportionally increases. Applied to the cones, this results in a broadening of the relative spectrum as the length of the photoreceptors increases. This is a source of variation in the color-matching functions. The exponential decay of the transmission optical density of the photoreceptors with field diameter has been formulated by *Pokorny and Smith*[43] as

$$D_{\mathrm{cone}} = a + b \cdot \mathrm{e}^{(-f/c)} \tag{9.11}$$

with $a = 0.38$ and $b = 0.54$ for L- and M-cones, $a = 0.30$ and $b = 0.45$ for S-cones, and $c = 1.333$.

Note that in general, when considering the relationship between color-matching functions and cone spectral sensitivity, *energy units* are used, and when considering raw spectral sensitivity data of photopigment spectra, *quantum units* are used.

The Aging Observer

Most of the observers in the Stiles and Burch experiments were young and the color-matching functions that were derived are considered to be those of a 32-year-old observer.

The *lens transmission optical density* $D_{\tau,\text{ocul}}(\lambda)$ increases dramatically with age.[44] It has been characterized as having two components with one varying with age. Therefore, it is possible to take into consideration age dependence. The transmission optical density of the lens of an average observer may be estimated by a formula with one or another set of a, b, and c values depending whether the age A is between 20 and 60 or over 60 years.

$$D_{\tau,\text{ocul}}(\lambda) = D_{\tau,\text{ocul2}}(\lambda) + D_{\tau,\text{ocul1}}(\lambda)[a + b(A - c)] \tag{9.12}$$

with $a = 1$; $b = 0.02$; $c = 32$ if $20 \le A \le 60$
with $a = 1.56$; $b = 0.0667$; $c = 60$ if $A > 60$

The Calculation of Tristimulus Values

Once the essential step of deriving cone fundamentals is made, colors can be specified in a three-dimensional *LMS color space*. Given a stimulus **Q** with its spectral distribution of light $P_\lambda(\lambda)$, its tristimulus values L_Q, M_Q, and S_Q are obtained as

$$\begin{aligned} L_Q &= k_L \int P_\lambda(\lambda) \cdot \bar{l}(\lambda) d\lambda \\ M_Q &= k_L \int P_\lambda(\lambda) \cdot \bar{m}(\lambda) d\lambda \\ S_Q &= k_L \int P_\lambda(\lambda) \cdot \bar{s}(\lambda) d\lambda \end{aligned} \tag{9.13}$$

All real colors are included within a volume limited by the optimal colors that are the purest achievable object colors at a given luminance factor (Figure 9.5).

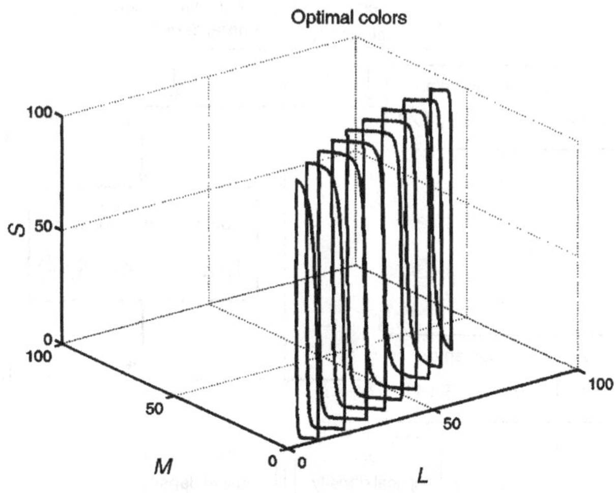

FIGURE 9.5 Isoluminant contours of optimal colors in LMS color space. *LMS* tristimulus values are scaled according to the luminance as described in the discussion.

CIE Recommendations From CIE and Final Tables

CIE proposes a continuous fundamental observer with a continuous design from 10° to 1°. The 10° color-matching measurements of Stiles and Burch[38] provide the basis for this continuous fundamental observer. Consequently, neither the CIE 1931 nor the CIE 1964 color-matching functions will be used. Tables are provided for 2° and 10° fields, also downloadable from www.cvrl.org Web site. Fundamentals can be derived for any field size from 1° to 10°, taking into account the change of the macular pigment transmission optical density and of the photopigment transmission optical density.

The procedure for deriving fundamental curves as a function of field size is as follows (Figure 9.6):

1. Start with the 10° cone fundamentals at the corneal level
2. Transform 10° cone fundamentals into the spectral absorptance of the cones
   - correcting for ocular media spectral absorption
   - correcting for macular pigment spectral absorption on 10° field
3. Transform the 10° spectral absorptance of the cones into the spectral absorbance curves of the dilute photopigments
   - correcting for the effect of transmission optical density on the spectral absorbance of the photopigment for 10° field
4. Carry out the reverse computation to find the cone fundamental curves at the corneal level for another field diameter, introducing appropriate corrections of macular pigment transmission optical density and of photopigment effective transmission optical density for this field size, and of ocular media optical density.

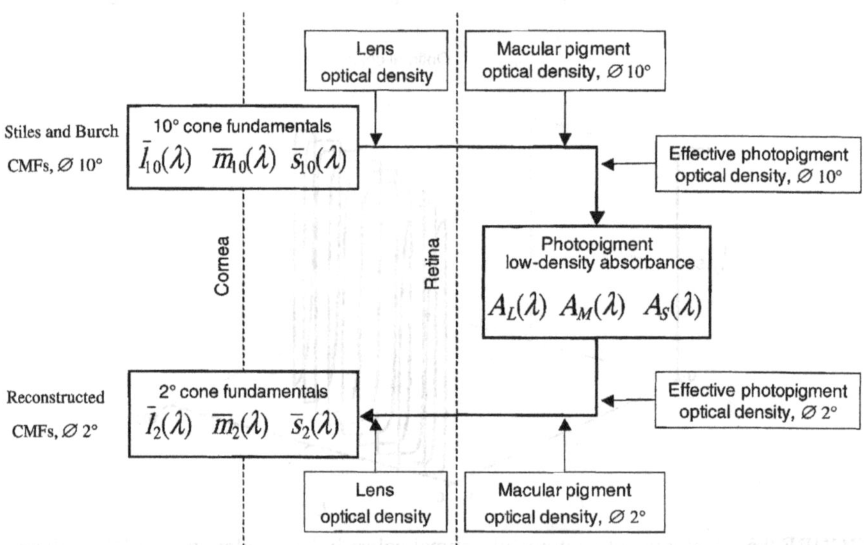

FIGURE 9.6 Procedure for deriving cone fundamentals as a function of field size.

DISCUSSION AND PERSPECTIVES

An Isoluminant Fundamental Chromaticity Diagram

The main decision in proposing a chromaticity diagram is the choice of units.

Units and Luminous Efficiency Function

Much psychophysical and physiological evidence supports the hypothesis that the shortwave sensitive cones make no contribution to luminance. Thus the L and M tristimulus values should be scaled to add up to the luminance.

Further, it has been verified that the spectral luminous efficiency function can be represented by a weighted sum of $\bar{l}(\lambda)$ and $\bar{m}(\lambda)$. Thus, a new spectral luminous efficiency function related to fundamentals V_F could be introduced where $w_{L/M}$ is the relative weight of the L-cone versus the M-cone contribution, and χ_q is the normalizing factor to give a value of unity at the maximum, when all functions are expressed in terms of quanta.

$$V_{Fq}(\lambda) = (w_{q,L/M}\bar{l}_q(\lambda) + \bar{m}_q(\lambda))/\chi_q \tag{9.14}$$

When all functions are expressed in terms of energy, factors $w_{e,L/M}$ and χ_e should be calculated to weight the cone fundamentals in terms of energy.

$$V_F(\lambda) = (w_{e,L/M}\bar{l}(\lambda) + \bar{m}(\lambda))/\chi_e \tag{9.15}$$

In the case of the 2° field, the spectral luminous efficiency function $V_M(\lambda)$ modified by Judd and Vos, and recommended by the CIE in 1988,[45] encompasses the same historic defects as the spectral luminous efficiency function $V(\lambda)$ agreed by the CIE for the standard photometric observer in 1924. For the fundamental observer, the CIE is inclined to prefer $V_F(\lambda)$ as the best fit to an experimentally determined photopic sensitivity curve using heterochromatic minimum flicker photometry with 40 observers.[46] Then, when all functions are expressed in quantum units, the fit yields

$$w_{q,L2/M2} = 1.55 \quad \text{and} \quad \chi_q = 2.476985$$

$$w_{e,L2/M2} = 1.624340 \quad \text{and} \quad \chi_e = 2.525598$$

The advantage of relating the L- and M-cone fundamentals to a spectral luminosity function is that the luminance can be divided between L- and M-cone contributions and can be expressed in luminance units or retinal illuminance units.

In the case of the 10° field, the $\bar{y}(\lambda)$ color-matching function of the standard observer was essentially based on color matches from the Stiles and Burch observers. It could well be approximated by a weighted sum of 10° cone fundamentals. Then the linear model $F_{10}(\lambda)$, not $\bar{y}_{10}(\lambda)$, would constitute the luminous efficiency function associated to the cone fundamentals. No decision has been taken yet on the best values for $w_{L10/M10}$ and χ_{10}.

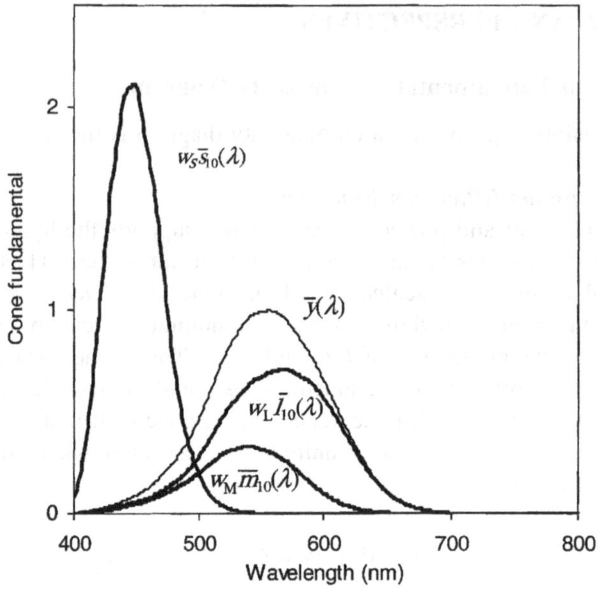

FIGURE 9.7 $\bar{l}(\lambda)$ and $\bar{m}(\lambda)$ fundamentals scaled to luminance and $\bar{s}(\lambda)$ fundamental scaled to equate L plus M tristimulus values for the energy spectrum.

This leaves the scale for S value arbitrary.

In line with the luminance normalization, Boynton and Kambe suggest to set the S-cone fundamental so that one luminance unit of an equal energy white stimulus produces one S-cone fundamental excitation unit.[47] In other words, when $\bar{l}(\lambda)$ and $\bar{m}(\lambda)$ are scaled, as they contribute to the luminance, the integral of $\bar{s}(\lambda)$ would be made equal to the summed integrals of the scaled values of $\bar{l}(\lambda)$ and $\bar{m}(\lambda)$. Such conventions are used to produce Figure 9.7.

The l, s Chromaticity Diagram

By definition, the chromaticity diagram represents the color stimulus in the unit plane, using two relative rather than three absolute tristimulus values. In the fundamental chromaticity diagram, we would use the relative responses of the cones. The fraction of L-cone fundamental response is plotted along the abscissa and the fraction of S-cone fundamental response is plotted along the ordinate, but, contrary to the usual practice, these fractions are relative to the sum of L- and M-cone fundamental responses.

$$\begin{aligned} l &= w_{L/M} \cdot L/(w_{L/M} \cdot L + M) \\ s &= S/(w_{L/M} \cdot L + M) \end{aligned} \tag{9.16}$$

This representation offers the possibility to plot chromaticity in a constant-luminance plane, a feature that is not achieved by the CIE colorimetric system (Figure 9.8). The S-cone fundamental response can vary without changing the

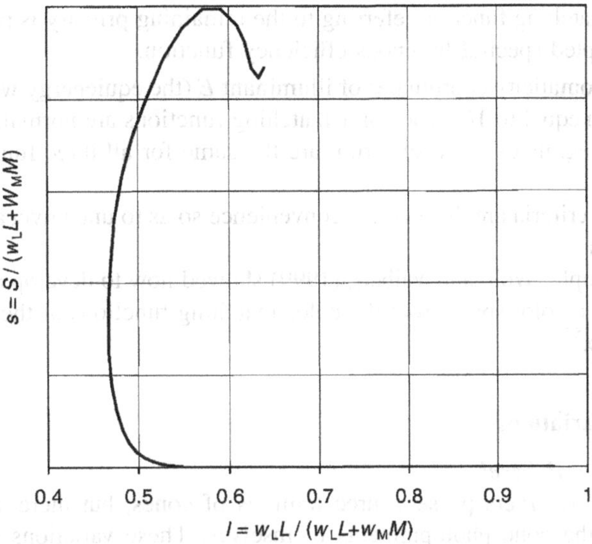

FIGURE 9.8 Isoluminant chromaticity diagram.

luminance, and any increase of the L-cone fundamental contribution to luminance is exactly compensated by a decrease of the M-cone contribution. Such a development has been made by MacLeod and Boynton,[48] who found the precursors of their idea in the papers by Luther[49] and by Le Grand.[50] Their proposal has received an enthusiastic response from the scientific community because the constant-luminance property does not belong to the *x*, *y* chromaticity diagram.

The ordinate scaling proceeds from the choice of units for the S-cone fundamental response. A possible disadvantage is that the white point appears to fall almost on the abscissa. Nevertheless, for colorimetric purpose, CIE would agree to linear scaling.

A CIE-Like Chromaticity Diagram

Users are so familiar with the *x*, *y* chromaticity diagram that it is worthwhile to derive a CIE-like chromaticity diagram from the cone fundamentals. The criteria used by CIE[51] to derive the *XYZ* colorimetric system are given in Chapter 3 of this volume. They are also quoted here.

1. All color stimuli have all nonnegative tristimulus values. The new (virtual) primaries **X**, **Y**, and **Z** are chosen, in the chromaticity diagram of the CIE *RGB* representation referring to red, green, and blue physical primaries, such that their chromaticity coordinates define the vertices of a triangle that fully encompasses the spectrum locus.
2. In the chromaticity diagram, the alychne (the locus of the color stimuli of zero luminance) is represented by a line coinciding with the abscissa. Two of the primaries are represented on the alychne with the consequence that the

color-matching function referring to the remaining primary is proportional to the adopted spectral luminous efficiency function.

3. The chromaticity coordinates of illuminant *E* (the equienergy white stimulus) are each equal to 1/3. The color-matching functions are normalized such that their integrals over the spectrum are the same for all three functions.

Additional criteria are dictated for convenience so as to unequivocally define the representation.

As an example, Wold and Valberg (1999) showed how to develop an *XYZ* representation of the color space from the color-matching functions of the Stiles–Burch 2° pilot group.[52]

Individual Variations

At the Receptoral Level

Color normal observers possess three families of cones, but there are individual variations in the cone photopigments themselves. These variations have become evident from the microspectrophotometric records and from molecular genetics studies. Thus, the recent finding that the longwave sensitive photopigment exists in two varieties, about equally distributed in the normal population, cannot be ignored. It modifies color matches and photometric matches. Estimation from color matching yields a standard deviation in λ_{max} of 1.5 nm, slightly less, for L-cone sensitivity, 0.9 nm for M-cone sensitivity, and 0.8 nm for S-cone sensitivity.[53] The shift in the wavelength of peak sensitivity of the cone photopigments and the variation in effective optical density are major causes of interindividual variations in the Rayleigh matches.[54]

It should be emphasized that cone fundamentals derived by Stockman and Sharpe and agreed by the CIE are representative of the average population.

Postreceptoral Processing: Weighting L-Signals and M-Signals for Luminance

Since the standard photometric observer was established in 1924, a very large photometric interobserver variability has been noticed. It is now explained by variations in the L- versus M-cone numerical ratio. For example, objective *in vivo* count has revealed retinal mosaics with ratios as different as 0.37 and 16.5.[55] A recent psychophysical investigation assumed a range from 0.47 to 15.82.[46]

Examples of Applications: The Future

Color Vision Deficiencies

Because dichromatic vision is a reduced form of normal trichromatic vision, the proposed cone fundamentals should reflect the ability of true congenital dichromats to match color.

The simulation of dichromatic vision proposed by Viénot et al.[56] is based on the LMS colorimetric system, which specifies colors in terms of the relative excitations of the longwave sensitive (L), of the middlewave sensitive (M), and of the shortwave sensitive (S) cones. As dichromats lack one class of cone photopigment, they confuse colors that differ only in the excitation of the missing class of photopigment. In contrast to the case of the trichromatic observer, who requires color specification by three components, two components are sufficient to specify color for the dichromat. One can construct a rule to reduce any set of the confused colors to a single three-component color specification.

Observer Metamerism

Observer metamerism covers the possible failures of a color match due to the variation in normal color vision. Several factors could be responsible. Changes in lens absorbance, macular pigmentation, and subfamily of visual pigments explain interobserver disagreements. Changes in macular pigmentation and effective transmittance optical density of the photopigments explain field size disparity. Changes in the effective optical density that modify the shape of the cone fundamentals explain the discrepancy between maximum saturation color matches and Maxwell color matches (see Chapter 10 of this volume and the plot of the results obtained by Crawford, 1965, and by Lozano and Palmer, 1968, in Wyszecki and Stiles, 1982, p. 385[25]), the departure of the match when photopigments are bleached,[57] and the Stiles–Crawford II effect that occurs when light enters the eye through the periphery of the pupil.[58,59]

Interobserver variability has been measured in variants of the classical Rayleigh match or in pigment surface metameric matches and is possibly related to preretinal absorption and photopigment polymorphism.[54,60]

Color Differences

The discrimination ellipses published by MacAdam[61] were examined by Le Grand[50] and Nagy et al.[62] These authors derived rules to explain the pattern of change that could be attributable to fundamentals. Later, color discrimination was measured in the isoluminant plane under controlled adaptation conditions.[63] When the observer was adapted to the region of color space in which the discriminations were made, the thresholds for detecting changes along the s axis increased linearly with the excitation of the shortwave sensitive cones. Thresholds for detecting changes along the l axis were independent of the locus of adaptation along this axis.

Such rules could allow constructing a uniform chromaticity diagram.

Color Appearance Models

Recent versions of the CIE color appearance model described in Chapter 11 of this volume twice refer to the linear transformation from *XYZ* colorimetric representation to some other colorimetric representation. In the initial stage that deals with chromatic adaptation, the linear transformation yields narrowly tuned functions, with a few negative parts and peaks for the middle- and longwave function slightly apart compared with the L- and M-cone fundamentals. These functions largely reflect cone responses, but the differences with cone fundamentals probably reveal postreceptoral adaptation. Thus, the Von Kries rule applies to the receptor channels rather than to receptors only. The intermediate stage where the cone-opponent signals are constructed simply conceals the transformation from *XYZ* to *LMS* colorimetric representation.

CONCLUSION

By making a clear connection between the color specification and the underlying physiology, the fundamental chromaticity diagram provides a unified framework for various scientific communities. Indeed, the cone responses are the signals entering the visual system and should be known in order to fully analyze a color situation. CIE is very much concerned that the color community should use a fundamental chromaticity diagram with physiological significant axes, not only for scientific and pedagogic purposes but also for specification and as a basis for future developments.

ACKNOWLEDGMENTS

We would like to acknowledge the efforts of all members of CIE TC 1–36, with a special mention to A. Stockman and L. T. Sharpe who provide easily downloadable tables at http://www.cvrl.org/ and useful comments on the manuscript from M. Pointer.

REFERENCES

1. Lee BB (1999) Receptor inputs to primate ganglion cells, in: *Color Vision: From Genes to Perception* (Eds., K. R. Gegenfurtner & L. T. Sharpe, Cambridge University Press, New York, pp. 203–217.
2. König A, Dieterici C (1886) Die Grundempfindungen und ihre Intensitäts-Vertheilung im Spectrum. *Sitz. Akad. Wiss. Berlin*, 805–829.
3. Judd DB (1945) Standard response functions for protanopic and deuteranopic vision. *J. Opt. Soc. Am.*, **35**, 199–121.
4. Vos JJ Walraven PL (1971) On the derivation of the foveal receptor primaries., *Vis. Res.*, **11**, 795–818.
5. Smith VC Pokorny J (1975) Spectral sensitivity of the foveal cone pigments between 400 and 500 nm. *Vision Res.*, **15**, 161–171.
6. Smith VC, Pokorny J, Zaidi Q (1983) How do sets of color-matching functions differ? in: *Color Vision, Physiology and Psychophysics* (Eds., J.D. Mollon, and L.T. Sharpe), Academic Press, London, pp. 93–105.
7. Vos JJ, Estévez O & Walraven PL (1990) Improved color fundamentals offer a new view on photometric additivity. *Vis. Res.*, **30**, 937–943.
8. Stockman A, MacLeod DIA & Johnson NE: (1993) The spectral sensitivities of the human cones. *J. Opt. Soc. Am. A*, **10**, 2491–2521.

9. Stockman A, Sharpe LT (2000) The spectral sensitivities of the middle- and long-wavelength-sensitive cones derived from measurements in observers of known genotype. *Vis. Res.*, **40**, 1711–1737.
10. Stockman A, Sharpe LT (2000) Cone spectral sensitivities and color matching in: *Color Vision: From Genes to Perception* (Eds., K. R. Gegenfurtner, and L. T. Sharpe), Cambridge University Press, New York, pp. 53–87. http://www.cvrl.org/
11. Dartnall HJA, Bowmaker JK, Mollon JD (1983) Human visual pigments: microspectrophotometric results from the eyes of seven persons. *Proc. R. Soc., Lond., Ser. B*, **220**, 115–130.
12. Baylor DA, Nunn BJ, Schnapf JL (1987) Spectral sensitivity of cones of the monkey Macaca fascicularis. *J. Physiol.*, **390**, 145–160.
13. Roorda A, Williams DR (1999) The arrangement of the three cone classes in the living human eye. *Nature*, **397**, 520–522.
14. Williams DR, Roorda A (2000) The trichromatic cone mosaic in the human eye in: *Color Vision: From Genes to Perception* (Eds., K. R. Gegenfurtner, and L. T. Sharpe), Cambridge University Press, New York, pp. 113–122.
15. Naka KI, Rushton WAH (1966) S- potentials from luminosity units in the retina of fish (Cyprinidae). *J. Physiol. Lond.*, **185**, 587–599.
16. Dartnall HJA (1953) The interpretation of spectral sensitivity curves. *Br. Med. Bull.*, **9**, 24–30.
17. Dartnall HJA (1957) *The Visual Pigments*, Methuen, London.
18. Barlow HB (1982) What causes trichromacy? A theoretical analysis using comb-filtered spectra. *Vis. Res.*, **22**, 635–643.
19. Lamb TD (1995) Photoreceptor spectral sensitivities: Common shape in the long-wavelength region. *Vis. Res.*, **35**, 3083–3091.
20. Stockman A, Sharpe LT, Fach CC (1999) The spectral sensitivity of the human short-wavelength sensitive cones derived from thresholds and color matches. *Vis. Res.*, **39**, 2901–2927.
21. Stiles WS (selected papers, 1978) *Mechanisms of Colour Vision*, selected papers of W. S. Stiles, Academic Press, London.
22. Stockman A, MacLeod DIA, Vivien JA (1993) Isolation of the middle- and long-wavelength-sensitive cones in normal trichromats. *J. Opt. Soc. Am. A*, **10**, 2471–2490.
23. Le Grand Y (1972) *Optique physiologique, tome II: lumière et couleurs*, Masson, Paris [English translation (1968) *Light, Colour and Vision*, Chapman & Hall, London].
24. Wright WD (1981) Golden Jubilee of Color in the CIE—The Historical and Experimental Background to the 1931 CIE System of Colorimetry, Bradford.
25. Wyszecki G, Stiles WS (1982) *Color Science: Concepts and Methods, Quantitative Data and Formulae*, Wiley.
26. Fairman HS, Brill MH, Hemmendinger H (1997) How the CIE 1931 color-matching functions were derived from the wright-guild data. *Color Res. Appl.*, **22**, 11–23. [(1998) Erratum. *Color Res. Appl.*, **23**, 259].
27. Selected Papers on Colorimetry—Fundamentals (1993) Ed., D. L. MacAdam), *SPIE Milestone Series*, **MS 77**.
28. Helmholtz H (1860) *Handbuch der Physiologischen Optik*, Verlag von Leopold Voss, Hamburg, part II. English translation (1924–1925) *Treatise on Physiological Optics* (Ed. James P. C. Southall), The Optical Society of America, 3 vols.

29. Maxwell C (1860) On the theory of compound colors, and the relations of the colors of the spectrum. *Phil. Trans.*, **150**, 57–86.

30. Grassmann HG (1853). Zur Theorie der Farbenmischung, *Poggendorf Annalen der Physik und Chemie*, **89**, 69–84 (Theory of Compound Colors. (1993) English translation reprinted in Selected Papers on Colorimetry-Fundamentals .ed : David L. MacAdam DL, *SPIE Milestone Series*, **MS 77**).

31. Ives HE (1915) The transformation of colour-mixture equations from one system to another. *J. Franklin Inst.*, **180**, 673–701.

32. Guild J (1931) The colorimetric properties of the spectrum. *Phil. Trans. R. Soc. Lond.*, **230A**, 149–187.

33. Wright WD (1928-29) A re-determination of the mixture curves of the spectrum. *Trans. Opt. Soc. Lond.*, **30**, 141–164.

34. Judd DB (1951) Report of U.S. Secretariat Committee on Colorimetry and Artificial Daylight, in: *Proceedings of the Twelfth Session of the CIE*, Stockholm, Bureau Central de la CIE, Paris, p. 11.

35. Vos JJ (1978) Colorimetric and photometric properties of a 2° fundamental observer. *Color Res. Appl.*, **3**, 125–128.

36. Trezona PW (1987) Individual observer data for the 1955 Stiles-Burch 2 degree pilot investigation. *J. Opt. Soc. Am. A*, **4**, 769–782.

37. Stiles WS, Burch JM (1955) Interim report to the Commission International de l'Eclairage, Zurich, 1955, on the National Physical Laboratory's investigation of color-matching. *Optica Acta*, **2**, 168–181.

38. Stiles WS, Burch JM (1959) NPL colour-matching investigation: Final report (1958). *Optica Acta*, **6**, 1–26.

39. Speranskaya NI (1959) Determination of spectral color co-ordinates for twenty-seven normal observers. *Optics Spectrosc*, **7**, 424–428.

40. Commission Internationale de l'Eclairage, *Fundamental Chromaticity Diagram with Physiological Axes—Part 1*, CIE **170-1**-2006

41. van Norren D, Vos JJ (1974) Spectral transmission of the human ocular media. *Vis. Res.*, **14**, 1237–1244.

42. Moreland JD, Alexander EC (1997). Effect of macular pigment on color matching with field sizes in the 1° to 10° range. *Doc. Ophthalmol. Proc. Ser.*, **59**, 363–368.

43. Pokorny J, Smith VC (1976) Effect of field size on red-green colour mixture equations. *J. Opt. Soc. Am.*, **66**, 705–708.

44. Pokorny J, Smith VC, Lutze M (1987) Aging of the human lens. *Appl. Opt.*, **26**, 1437–1440.

45. Commission Internationale de l'Eclairage, *CIE 1988 2° spectral luminous efficiency function for photopic vision*, CIE **86**, 1990.

46. Sharpe L T, Stockman A, Jagla W, Jägle H (2005). A luminous efficiency function, $V^*(\lambda)$, for daylight adaptation. *J. Vision*, **5**, 948–968.

47. Boynton RM, Kambe N (1980) Chromatic difference steps of moderate size measured along theoretically critical axes. *Color Res. Appl.*, **5**, 13–23.

48. MacLeod DIA, Boynton R M (1979) Chromaticity diagram showing cone excitation by stimuli of equal luminance. *J. Opt. Soc. Am.*, **69**, 1183–1186.

49. Luther R (1927). Aus dem Gebiet der Farbreizmetrik. *Zeitschrift für technische Physik*, **8**, 540–558.

50. Le Grand Y (1949) Les seuils differentiels de couleurs dans la théorie de Young. *Revue d'Optique* **28** 61–278. English translation (1994) *Color Res. Appl.*, **19**, 296–309.
51. Commission Internationale de l'Eclairage, *Colorimetry*, 3rd ed., **15**-2004.
52. Wold JH, Valberg A (1999) General method for deriving an XYZ tristimulus space exemplified by use of the Stiles–Burch1955 2° color matching data. *J. Opt. Soc. Am. A*, **16**, 2845–2858.
53. Webster MA, MacLeod DIA (1988) Factors underlying individual differences in the color matches of normal observers. *J. Opt. Soc. Am. A*, **5**, 1722–1735.
54. He JC, Shevell SK (1994). Individual differences in cone photopigments of normal trichromats measured by dual Rayleigh-type color matches. *Vis. Res.*, **34**, 367–376.
55. Hofer H, Singer B, Williams DR (2005) Different sensations from cones with the same photopigment. *J. Vis.*, **5**, 44–454.
56. Viénot F, Brettel H, Ott L, Ben M'Barek A, Mollon JD (1995) What do colour-blind people see? *Nature*, **376**, 127–128.
57. Walraven PL (1961) On the Bezold-Brücke phenomenon, *J. Opt. Soc. Am.*, **51**, 1113–1116.
58. Stiles WS (1937) The luminous efficiency of monochromatic rays entering the eye pupil at different points and a new color effect. *Proc. R. Soc. Lond.*, **B123**, 90–118.
59. Walraven PL, Bouman MA (1960) Relation between directional sensitivity and spectral response curves in human cone vision. *J. Opt. Soc. Am.*, **50**, 780–784.
60. Diaz Navas JA, Chiron A, Viénot F (1998) Tracing a metameric match to individual variations of color vision. *Color Res. Appl.*, **23**, 379–389.
61. MacAdam DL (1942) Visual sensitivities to color differences in daylight. *J. Opt. Soc. Am.*, **32**, 247–274.
62. Nagy AL, Eskew RT, Boynton RM (1987) Analysis of color-matching ellipses in a cone-excitation space. *J. Opt. Soc. Am. A*, **4**, 756–758.
63. Krauskopf J, Gegenfurtner K (1992) Colour discrimination and adaptation. *Vis. Res.*, **32**, 2165–2175.

10

OPEN PROBLEMS ON THE VALIDITY OF GRASSMANN'S LAWS

MICHAEL H. BRILL[a] and ALAN R. ROBERTSON[b]

[a] *Datacolor, 5 Princess Road, Lawrenceville, NJ 08648, USA*
[b] *2428 Ogilivie Road, Ottawa K1J 7N5, Canada*

DEFINITION OF THE PROBLEM

Hermann Grassmann[1] disclosed the rules for assessing color sameness more than a century ago, and all of basic color theory depends on these rules. Grassmann's Laws and their corollaries have been formulated in modern terminology by Wyszecki and Stiles.[2] The laws allow us to represent additive color matches by simple equations that can be manipulated by the usual rules of algebra. In particular, we can transform between the different sets of primaries and calculate the tristimulus values of complex stimuli using straightforward linear algebra. The derivation and use of the CIE Standard Observers is firmly based on these principles, which can be formulated succinctly as

$$\text{Symmetry:} \quad \text{If } \mathbf{A} = \mathbf{B} \text{ then } \mathbf{B} = \mathbf{A} \tag{10.1}$$

$$\text{Transitivity:} \quad \text{If } \mathbf{A} = \mathbf{B} \text{ and } \mathbf{B} = \mathbf{C} \text{ then } \mathbf{A} = \mathbf{C} \tag{10.2}$$

$$\text{Proportionality:} \quad \text{If } \mathbf{A} = \mathbf{B} \text{ then } k\mathbf{A} = k\mathbf{B} \tag{10.3}$$

$$\text{Additivity:} \quad \text{If } \mathbf{A} = \mathbf{B} \text{ and } \mathbf{C} = \mathbf{D} \text{ then } \mathbf{A} + \mathbf{C} = \mathbf{B} + \mathbf{D} \tag{10.4}$$

$$\text{If } \mathbf{A} = \mathbf{B} \text{ and } \mathbf{A} + \mathbf{C} = \mathbf{B} + \mathbf{D} \text{ then } \mathbf{C} = \mathbf{D} \tag{10.5}$$

Colorimetry: Understanding the CIE System, Edited by János Schanda

where **A**, **B**, **C**, and **D** represent the color stimuli (i.e., spectral power distributions $A(\lambda)$, $B(\lambda)$, $C(\lambda)$, and $D(\lambda)$, respectively, as functions of visible wavelength λ), and k is any positive factor by which the radiant power of the stimulus is increased or decreased while its relative spectral distribution is kept the same. In describing the color matches, the symbol "=" is pronounced as "matches" and the symbol "+" indicates "additively mixed with." However, by treating **A**, **B**, **C**, and **D** as 3D vectors and k as a scalar, the formulations can be treated as normal mathematical equations and manipulated in the usual way.

Grassmann's laws are tested by what is called a *symmetric-matching* experiment: An observer compares two lights that are presented on identical backgrounds and with a visual system that is adapted in the same way for both sides of the match.

Grassmann's laws are known not to be exactly true in human color matching. Symmetry could be called into question by color difference formulas, such as CIE94,[3] that are asymmetric between batch and standard. Transitivity can be considered to be violated if we take the term "color match" to mean that two colors are within a just-noticeable difference of each other. In this case, adding two subthreshold differences together could produce a combined difference that is above threshold. Proportionality and additivity can also be compromised. Besides the three cone types that herald the trichromacy of vision at high (photopic) light intensities, a fourth photoreceptor type (rods) contributes to vision at low (mesopic and scotopic) light intensities and away from the center of vision (fovea). At very high light intensities, unbleached photopigments deplete and, in aggregate, change their action spectrum. At still higher light intensities, a photopigment molecule can absorb multiple photons but respond as if it absorbed only one photon. All these effects compromise Grassmann's laws, but the successful application of the laws, for example, in photography and television, has led us to believe that the compromises are not serious.

HISTORICAL REVIEW

As early as 1934, Wright[4,5] showed that matches between a monochromatic yellow and a mixture of red and green could break down if the eye was adapted to high levels of luminance (>15,000 td). Then in 1947, Blottiau[6] showed evidence of breakdowns of additivity in blue–green matches. Trezona[7,8] confirmed the general nature of Blottiau's results (at retinal illuminances between 100 nm and 1000 td) but suggested that the deviations from additivity were typically of the order of a few percent in the tristimulus values and not large when compared with just discriminable color differences. She suggested that Blottiau's method of presentation exaggerated the deviations but agreed that deviations do occur.

These investigations and others led the CIE to call for new studies of color matching that eventually led to the establishment of the 1964 (10°) Standard Observer. A major part of these studies was carried out at the NPL in Teddington and, in a paper describing the results, Stiles[9] commented "I think we have to face the fact that in work with colours of high saturation there may occur deviations from additivity which may well repay study from the standpoint of visual theory while having little effect on practical colour measurement." In a subsequent study, Crawford[10] demonstrated differences between color-matching functions (CMFs) obtained by the Maxwell method (all matches made with a white test stimulus) and by the maximum-saturation method (all matches made at high saturation). The differences appeared to be significant, especially in large fields where adaptation effects play a significant role. Lozano and Palmer[11] continued Crawford's work in more detail and concluded that a breakdown of Grassmann's laws was evident. Their studies were performed at a retinal illuminance of about 160 td.

In 1980, Wyszecki and Stiles[12] published a detailed study of the pigment-bleaching hypothesis, comparing Maxwell-type matches at retinal illuminances of 1000 td and 100,000 td. They found strong and predictable bleaching characteristics for the "red" and "green" fundamentals, but the "blue" fundamental exhibited unexpected and unexplained behavior. In 1982, the same authors[2] published the CMFs for a single observer measured by the Maxwell and maximum-saturation methods. The spectral chromaticity coordinates associated with these CMFs are shown in Figure 10.1. The deviations are considerable, and the authors

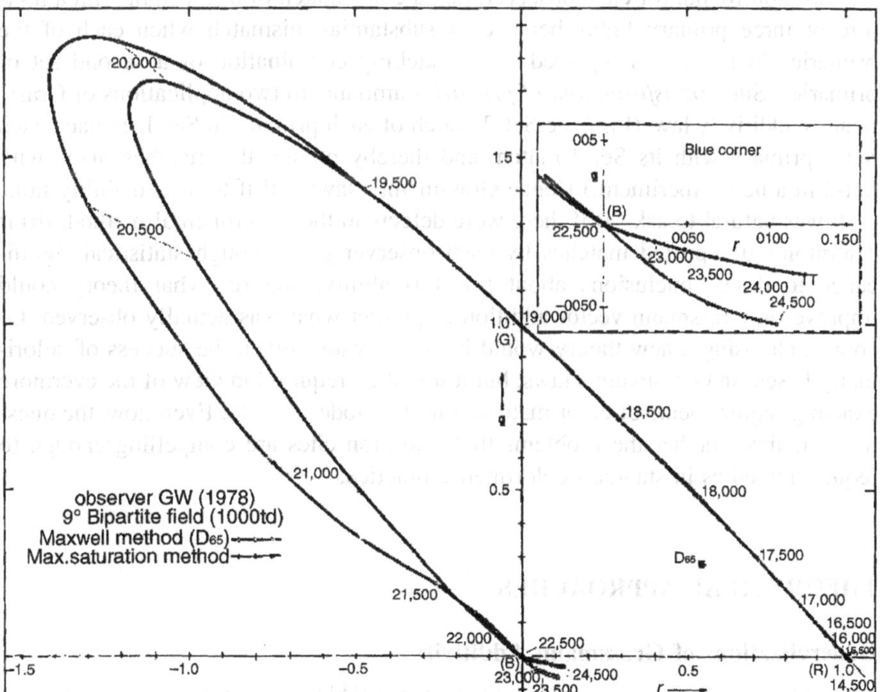

FIGURE 10.1 Spectrum loci derived from color matches in a 9° bipartite field by means of the Maxwell method and the maximum-saturation method. The primary stimuli have wave numbers 15,500 cm^{-1}, 19,000 cm^{-1}, and 22,500 cm^{-1}. (Reprinted from Ref 2, p. 386, Fig. 4(5.6.6) with permission of John Wiley & Sons.)

conclude that they represent failures of the additivity law. They list chromatic adaptation, the Maxwell spot, and interactions or linkages between different cone mechanisms as possible causes of the deviations but conclude that "further work is obviously needed to resolve the conundrum."

Building on the previous work, Zaidi[13] in 1986 made the same maximum-saturation matches with and without the superposition of monochromatic desaturating light. He found additivity failures and showed that they were not caused by computational imprecision, prereceptoral filtering, rod intrusion, variation of cone absorption spectra, and two pigments feeding into the same channel. He concluded that his results were consistent with the hypothesis that additivity failures were introduced by postreceptoral interactions.

Further cause for questioning the practical sufficiency of Grassmann's laws emerged in 1992, when Thornton[14] conducted symmetric color-matching experiments to test the transformability of primaries. Through these experiments, Thornton inferred the CMFs for six observers using three different sets of nearly monochromatic primary lights, and also for a virtual seventh observer whose CMFs are averaged from the other six observers. His observers made many matches, but each observer made each match only once.

Thornton found, for each observer, that a color match of a test light with a mixture of three primary lights becomes a substantial mismatch when each of the primaries in this set is replaced by a matching combination of a second set of primaries. Such *transformation of primaries* amounts to two applications of Grassmann's additivity law. (Find the Set-2 match of each primary in Set-1, replace each Set-1 primary with its Set-2 match, and thereby predict the matches made with Set-2 in a new experiment.) Hence Grassmann's laws fail if transformability fails.

It was natural to ask (a) if there were defects in the experimental method, (b) if Thornton's unrepeated matches by each observer gave enough statistical significance to draw conclusions about transformability, and (c) what theory could improve on Grassmann vector addition to predict what was actually observed. Of course, choosing a new theory would be a last resort, given the success of colorimetry based on Grassmann's laws, but it might be required in view of the evermore exacting requirements of color matching in the modern world. Even now, the question remains whether the problems that Thornton cites are compelling enough to require remedies in standard colorimetric practice.

THEORETICAL APPROACHES

Generalizations of Grassmann Additivity

When Thornton's data were first published, Brill[15,16] pursued the theoretical question (c) by generalizing the usual Grassmann idea that a symmetric color match equates three weighted sums of quantum catches, and hence that matches are additive in the domain of quantum catches. Implicit in Grassmann's additivity rule is the interpretation that "+" means addition at each wavelength of light intensities (or quantum fluxes) per unit wavelength interval. Other additivity domains could be imagined. For example, if the visual system were counting the square root of the number of quanta per unit wavelength interval, then Grassmann's additivity rule would state "If A, B, C, and D are spectral power distributions of light, and A matches C and B matches D, then $(A^{0.5} + B^{0.5})^2$ matches $(C^{0.5} + D^{0.5})^2$." Such an interpretation is difficult to justify physically, but might improve match prediction.

Brill[15,16] tried two theories, each of which contained a parameter whose value could optionally be set to retrieve conventional Grassmann additivity, but covered other alternatives for other parameter values. One such *covering theory* postulated that photon counts in the given-sized wavelength–time bins undergo a power-function transformation before being summed into three "tristimulus-like" numbers; this idea is an extension of the square root law in the previous paragraph. Although contrived, this extension has the virtues of simplicity (only one adjustment parameter—the exponent value) and the ability to change the additivity law while leaving the proportionality law untouched. The extension also has some precedent in colorimetric-like theories of speech perception[17] and texture perception.[18]

The other theory posited a depleted optical density (absorbance) of photopigments under more intense lights; that too produces additivity failure. In this *photopigment-depletion* theory, light bleaches photopigment from each cone species (at a rate depending on the instantaneous light intensity), the retinal metabolism "unbleaches" or restores the photopigment (at a rate that does not depend on the instantaneous light intensity), and the rates of bleaching and unbleaching are assumed to be equal to each other during a color-matching experiment. The photopigment-depletion theory has the advantages of a physical basis and a single parameter of adjustment (ratio of rates of bleaching and unbleaching at one light intensity). However, the theory is more difficult to manipulate than the power-function theory, and affects both the Grassmann additivity and proportionality rules. Neither of the above theories improved the transformability of Thornton's dataset.

Of course, adding more fitting parameters is expected to improve the data fit. Color matches can be cast into a formalism that applies to the spectral power distribution a possibly different nonlinearity f_i at each wavelength before performing three weighted sums to produce "tristimulus-type" values:

$$y_j = \sum_i k_{ij} f_i(S_i) \tag{10.6}$$

Here, $i = 1, \ldots, n$ enumerates the wavelength intervals in which a test light's spectral power S_i (in watts) is gathered, y_j ($j = 1, 2, 3$) are the tristimulus values resulting from the test light, and k_{ji} are spectral-sensitivity weighting factors. The generalization of Grassmann additivity corresponding to Equation (10.6) would be

$$y_{jA} + y_{jB} = y_{jC} + y_{jD} \tag{10.7}$$

A model with the form of Equations (10.6) and (10.7) is the power-function theory above, for which $f_i(S_i) = S_i^a$, and a is the chosen exponent.

Motivated by Wright's[19] formalism of trichromatic units, Oulton[20,21] (and personal communication) used Equations (10.6) and (10.7) and fit a different function f_i to Thornton's data at each wavelength. The fit was close, but questions about the number of parameters and the statistical significance of the data must be answered before the approach is adopted. The question of statistical significance is further discussed in Section "Numeric example."

Theory of Transformation of Primaries

In the remainder of this chapter, it will be important to understand two ways to use color-matching data with one set of the primaries to predict the color matches that would occur when another set of primary lights are used. Both methods rely on Grassmann's laws, but have slightly different computational implications.

The *inverse-matrix method* is the usual method to transform from an old set of CMFs to a new basis defined by a different set of primaries. Denote the old set as $X_j(\lambda)$, and the new set as $Y_j(\lambda)$. Here, $j = 1, 2, 3$, and visible wavelength λ spans N values (say, 341 values from 360 nm to 700 nm at 1 nm increments). The old set of CMFs may be associated with monochromatic primaries at wavelengths λ_j, and the new set with monochromatic primaries at wavelengths μ_j. [Either CMF set could be defined by nonmonochromatic physically realizable primaries, but the test lights needed to define the CMFs by color matches must be nearly monochromatic.] The task is to predict $Y_j(\lambda)$, given $X_j(\lambda)$, λ, j and μ_j. Here are the steps in the inverse-matrix method:

Step 1: Create the $3 \times N$ matrix of CMFs $\mathbf{Q}$ whose rows are the old CMFs $X_j(\lambda)$.

Step 2: Create the 3×3 matrix $\mathbf{A}$ of the old set of CMFs evaluated at the new set of the primary wavelengths: $A_{ij} = X_i(\mu_j)$. [In general, A_{ij} is the ith old-tristimulus value of the jth new primary.]

Step 3: Operate on $\mathbf{Q}$ with $\mathbf{A}^{-1}$, to get $\mathbf{P} = \mathbf{A}^{-1}\mathbf{Q}$, whose rows are the transformed CMFs.

In contrast to the inverse-matrix method, Thornton used a *forward-matrix* method to transform primaries. Step 1 is the same as above, but one additionally knows $Y_i(\lambda_j)$ and Steps 2 and 3 are replaced by the following:

Step 2′: Create the 3×3 matrix $\mathbf{A}$ of the new set of CMFs evaluated at the old set of the primary wavelengths: $A_{ij} = Y_i(\lambda_j)$. [In general, A_{ij} is the ith new-tristimulus value of the jth old primary.]

Step 3′: Operate on $\mathbf{Q}$ with $\mathbf{A}$ to get $\mathbf{P} = \mathbf{AQ}$, whose rows are the transformed CMFs.

The use of Thornton's steps 2' and 3' (instead of the traditional steps 2 and 3) is mathematically correct, but is not usable if the new CMFs are completely unknown (because this method assumes they are known at the old-primary wavelengths). However, in a transformability study, the new CMFs are known because they are measured as part of the test. Because Thornton's method involves no matrix inversions, it avoids the possible error instability that could beset the inversion of a nearly singular matrix. Thus no statistical explanation for Thornton's result can rely on a claim of the near-singularity of a set of primaries. Therefore, in the simulation in the section below, we use Thornton's forward-matrix method of transforming CMFs.

NUMERICAL EXPERIMENT

To what extent could statistical variations account for Thornton's primary-transformation data? In the absence of replicate matches, a numerical simulation—summarized here—was reported in 2001.[22] A particularly relevant part of Thornton's above-cited work is Section IV.B.1, called "Transformation of Primaries."

Summary of the Method

In broad outline, the simulation transforms the CIE 1931 CMFs to Thornton's Prime-Color (PC) and Anti-Prime (AP) primaries, adds Gaussian noise to each set of CMFs, and transforms from each set to estimate the CMFs of the other. The simulation starts by using the first method as described in the Section "Theory of transformation of primaries" to transform the 1931 CMFs to two of the Thornton's primary sets. Two methods suggest themselves for the stochastic (random-variable) analysis: propagate variances and covariances through partial derivative matrices, or perform a Monte Carlo simulation. The Monte Carlo approach was chosen because the partial-derivative approach fails if the relative errors are large—that is, when 2% of the maximum is added to a small "true" value. We report here a Monte Carlo study.

The numerical experiment (in Matlab) starts from input CIE 1931 CMFs (from 360 nm to 700 nm, in 1 nm increments). The program uses the *inverse-matrix method* to transform the CIE CMFs to the expected CMFs for two of Thornton's sets of assumed-monochromatic primaries: Prime-Color (PC—wavelengths 452 nm, 533 nm, and 607 nm), and Anti-Prime (AP—wavelengths 497 nm, 579 nm, and 653 nm). The program then adds random noise at each wavelength to each of these CIE-derived CMF sets. The noise is Gaussian with a mean of zero and a standard deviation of 0.02. The value 0.02 is based on a value quoted by Wyszecki and Stiles (p. 390 of Ref. 2) and is referenced to the value of unity obtained in the match of a monochromatic test light to a primary of the same wavelength.

Finally, the program uses the *forward-matrix method* to transform the noisy CMFs from the PC and AP sets into estimates of each other. The results are plotted in Figures 10.2–10.4. Each plot is an *individual realization of the random process*, and contains: the original (pre-noise) CMFs, the same CMFs with noise added, and the same CMFs predicted by transformation performed on the other noisy CMF set. As the random number generator resets at each wavelength of each function, one can readily distinguish noise on the CMF (high-frequency grass) from noise on the transformation matrix (low-frequency admixtures of CMFs). The high-frequency grass may not be representative of the wavelength-to-wavelength correlation in real measurements.

Results and Discussion

The results of this study, shown in Figures 10.2–10.4, suggest that random errors from a single realization generate artifacts commensurate with Thornton's observed failures of transformability. Random variations in the test-wavelength intensities imposed a low-amplitude artifacts uncorrelated at adjacent wavelengths, and random variations in the primary-wavelength light gave high-amplitude smooth variations over the visible spectrum. The errors are quite asymmetric between predicted PC and AP primaries. When one starts with PC functions and estimates AP functions, the results show only small errors (Figure 10.2). However, starting with AP functions and estimating PC functions gives very large errors (Figures 10.3 and 10.4).

It is instructive to compare the errors with Thornton's Figures 55 and 56.[14] The same comparative magnitudes are observed, but Thornton's are slightly larger. (This difference might be attributed to the fact that several observer functions, each a bit different, were averaged together.)

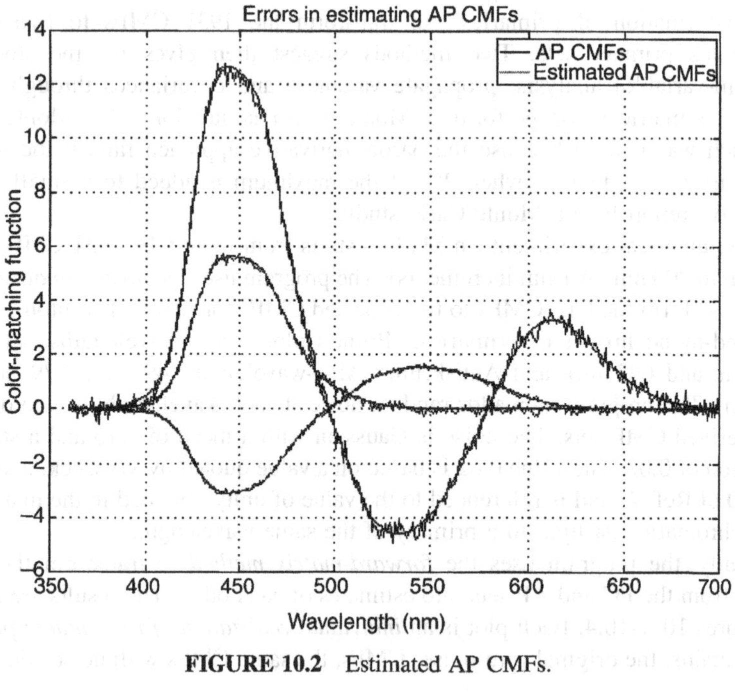

FIGURE 10.2 Estimated AP CMFs.

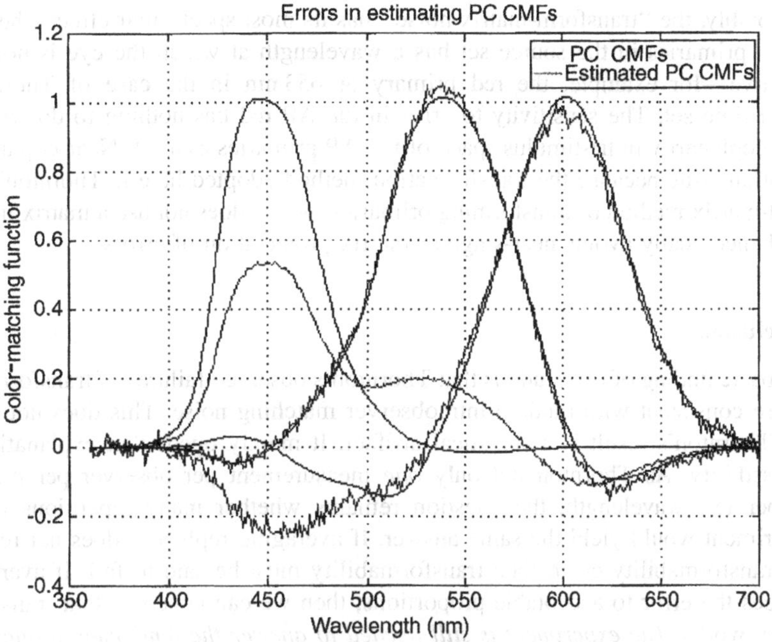

FIGURE 10.3 Estimated PC CMFs, realization 1.

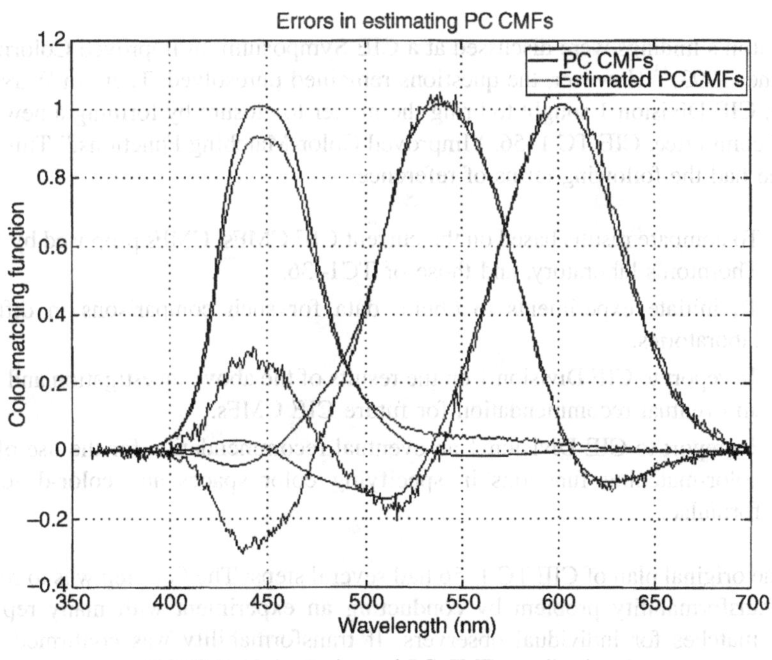

FIGURE 10.4 Estimated PC CMFs, realization 2.

Notably, the "transform-matrix noise" has its most spectacular effect when one of the primaries in the source set has a wavelength at which the eye is not very sensitive—for example, the red primary at 653 nm in the case of Thornton's Anti-Prime set. The sensitivity to error in the AP red has nothing to do with the near-coplanarity in tristimulus space of the AP primaries as a set. Near-coplanarity is not an issue because the transformation method adopted here is Thornton's forward-matrix method of transforming primaries, which does not use a matrix inverse and hence neatly avoids near-singular-matrix propagation of errors.

Conclusion

The basic finding of this study is that Thornton's observed failures of transformability are consistent with random intraobserver matching noise. This does not *prove* that Thornton's result is a statistical artifact. It merely opens that explanation as a possibility. As Thornton did only *one* measurement per observer per primary set per test wavelength, the question remains whether many repetitions of the experiment would yield the same answer. If averaging replicates does not remove the transformability error, then transformability must be said to fail. If averaging reduces the error to acceptable proportions, then we can conclude that transformability works. *The experiment is still needed to answer the fundamental question.*

ACTIVITIES OF CIE TC 1–56

Thornton's findings were discussed at a CIE Symposium on Improved Colorimetry in June, 1993.[23] However, the questions remained unresolved. Then, in Warsaw in 1999, CIE Division 1 sought to bring the matter to closure by forming a new technical committee, CIE TC 1–56, "Improved Color-Matching Functions." This committee had the following terms of reference:

1. To compare results based on the current CIE CMFs, CMFs proposed by W. A. Thornton's laboratory, and those of TC1–36.
2. To initiate experiments to obtain data for such comparisons in different laboratories.
3. To report to CIE Division 1 on the results of the above investigation and make an eventual recommendation for future CIE CMFs.
4. To report to CIE Division 1 an eventual recommendation for the use of new color-matching functions in specifying color spaces and color-difference formulae.

The original plan of CIE TC 1–56 had several steps: The first step was to resolve the transformability problem by conducting an experiment with many replicate color matches for individual observers. If transformability was confirmed to be true for a single observer (after averaging over replicate observations), the next step was to be to look at the differences between observers, and to weigh the statistical significance of the deviations of the average of these observers from the CIE functions. Finally, if improvements could be made in the Standard Observer's fit to the color matches of the average observer, the committee was chartered to suggest improved CMFs, using the findings of other CIE technical committees.

Between its inception in 1999 and its first face-to-face meeting in 2001, TC1–56 discussed details of a proposed color-matching experiment to test transformability of primaries. Some debates ensued about theoretical interpretation, and a Call for Volunteer Labs was issued in several publications to solicit participants in the experimental program. Finally, the numerical study as described in the previous section was performed.

The Call for Volunteer Labs emphasized the first stage in this process, in which the following steps were recommended:

(a) Acquire proven apparatus for measuring color matches: spectroradiometers, monochromators, bipartite 10° field (binocularly viewed). The radiometer should be calibrated to a standard lamp.

(b) Screen from 6 to 10 subjects.

(c) Select two sets of primary-light wavelengths: 452 nm, 533 nm, 607 nm; and 497 nm, 579 nm, 653 nm.

(d) Perform maximum-saturation matches to develop CMFs for each observer and for each primary set.

(e) Repeat Step (d) to obtain a statistically significant estimate of intraobserver variability.

(f) For each subject, compute a transformation from Set-1 CMFs to predicted functions from Set-2 primaries. Compare these predicted functions to the Set-2 functions obtained directly by experiment.

The first face-to-face meeting of CIE TC 1–56 was held June 22, 2001, in Rochester, NY. The nine members and about 40 interested nonmembers discussed the following issues at that meeting:

(a) *Use 2° or 10° CMFs?* It was suggested to construct the apparatus to conduct 10° measurements, but to stop down the aperture to a 2° field for the initial experiments. In this way, the apparatus could be functional for both kinds of match, even though additivity is more likely for a 2° experiment.

(b) *Genetic heterogeneity?* It was thought that genetic heterogeneity is desirable.

(c) *Build CMFs assuming that the spectral luminous efficiency function is one of them?* The spectral luminous efficiency function should not be assumed to be one of the CMFs. However, luminance (or some other intensity variable) is needed to quantitatively specify the reference and background fields in the experiment.

(d) *What should be the surround of the matching colors?* Several suggestions emerged: (i) choose a gray surround with luminance equal to that of the monochromatic light that is matched by either primary set; (ii) choose a gray surround with half the luminance of the match (not of the monochromatic light, but of the entire side of the match); (iii) choose a surround that is identical to the reference field so the test field is just half a circular disk that is made to vanish in the course of the match; (iv) draw a dark line between the two sides of the match to facilitate the discrimination in the blue-yellow direction; and (v) use a gray level as described by Ralph Evans instead of luminance to establish a good level for both the match and the background.

(e) *Maxwell or maximum-saturation matching?* In Maxwell matching, the eye is always adapted to the same color of light (white), and this adaptation state sensitizes the visual system to departures from a match. In maximum-saturation matching, the adapting light has a chromaticity that is on the triangle of the primaries; such lights do not sensitize departures from a color match. Therefore, several attendees favored Maxwell matches over maximum-saturation matches.

(f) *Assume monochromatic lights have been used for the color match, or refine the matrix algebra to include spectrum integrals in place of the samples at a wavelength that now characterize the formalism?* Proper accuracy requires the use of the integrals for any bandwidths greater than 2 nm.

(g) *Measure a full suite of CMFs for each observer using multiple replicates of the match?* Once an apparatus is up and running, it is desirable to measure a full set of CMFs. However, small pilot studies investigating the extent of Grassmann's laws are necessary to start the experimental program. These studies would help decide the luminance and background questions, and to give more empirical weight to the decisions that have to be made in these regards.

After this fairly stringent plan, no apparent committee activity occurred in 2002–2004, other than some tentative indications of interest in performing the critical experiment. A suggestion that the TC be closed, however, brought response from three groups who indicated that they were doing relevant work: Ronnier Luo, Boris Oicherman et al. (University of Leeds, UK) at about 3 cd/m^2, Claudio Oleari at about 30 cd/m^2, and Yasuhisa Nakano at about 300 cd/m^2. These expressions of interest solidified at the next meeting of TC1–56, held May, 16 2005 in León, Spain. At this meeting, all three groups presented their preliminary results; two of the groups had already reported in Granada the preceding week.[24,25]

The main reason for the differences in luminance was that, whereas the Leeds group used narrowband primaries and was hence constrained in light throughput, Oleari used broader band primaries, and Nakano used very broadband functions. Despite the differences in their experimental designs, all three groups reported a marked departure from CIE CMFs. Oicherman et al. described a mesopic condition;

they found that multiple trials of a single observer showed lack of Grassmann additivity, but averaging fewer trials of several observers revealed reduced statistical significance in the Grassmann failure. Results by Nakano indicated greater likelihood of Grassmann additivity at the higher luminances. However, none of the groups had arrived at a definitive conclusion concerning the domain or extent of Grassmann additivity through the transformability of primaries.

As of the León meeting, TC1–56 down-scoped its goals relative to the stated terms of reference: Henceforth, the main goal is not to find better CMFs, but to test the transformability of primaries for many trials on a single observer. Also, rather than try to insist on a single experimental design for the color matching (as was attempted at the 2001 meeting), the TC agreed to receive and synthesize the results of all participating laboratories into a single recommendation. Diversity in the luminance levels of the first three laboratories was already a step forward.

Despite the title of TC1–56, the new main goal does not require the measurement of the CMFs. Only the following are strictly needed for a meaningful result: A least seven lights, comprising an independent test light and two sets of three primaries each. The primary sets need not be even nearly monochromatic. For each of the primaries in a set, statistical robustness requires measuring the match at least 10 times with the opposite set of primaries. Then the seventh light must be repeatedly matched using both sets of primaries. Transformability will be verified if, by averaging the primary matching data (iterates for a single observer), the inferred coordinates of the test light under the two sets of primaries match the coordinates obtained from direct matching. This goal, which agrees with the subsidiary goal agreed upon at the 2001 meeting of TC1–56, is now the main mission of the TC.

A continuation of the effort beyond the newly defined mission would include the following: (a) Extension of the findings to more than one observer; and (b) a side study to determine whether single iterates of the matches produce as much failure of transformability as was observed by Thornton.

THE FUTURE

Within the next four years, the three laboratories who have volunteered to perform replicate color matches to test transformability should have gathered enough data so that CIE TC1–56 can assess the usability of Grassmann's laws in the evermore-demanding environments of the today's world. As many as three additional laboratories may contribute still more data, from which still more might be learned. At that point, either the Grassmann formalism will be vindicated in all practical applications or additional standards will be indicated for particular viewing conditions (high/low light levels, or large/small fields, for example). The new standards would require new theory, and Grassmann covering theories would be an item of continuing research. Such theories are already needed for conditions such as mesopic vision, for which colors are seen, but four receptor types are operative.

REFERENCES

1. Grassmann HG (1853) Zur Theorie der Farbenmischung. *Ann. Phys. Chem. (Poggendorf)*, **89**, 69–84. English translation (anon.) in: *Philosophic Magazine* **4(7)**, 254–264 (1854), reprinted in MacAdam DL (Ed.) (1970), *Sources of Color Science*, MIT Press, Cambridge, MA, 1970.
2. Wyszecki G, Stiles WS (1982) *Color Science: Concepts and Methods, Quantitative Data and Formulae*, 2nd ed., Wiley, New York.
3. CIE (1995) *Industrial Color-Difference Evaluation*, CIE Technical Report 116, Commission Internationale de l'Eclairage, Vienna, Austria.
4. Wright WD (1934) The measurement and analysis of colour adaptation phenomena. *Proc. R. Soc. (Lond.)*, **115B**, 49–87.
5. Wright WD (1936) The breakdown of a colour match with high intensities of adaptation. *J. Physiol.*, **87**, 23–33.
6. Blottiau F (1947) Les défauts d'additivité de la colorimétrie trichromatique. *Rev. d'Opt.*, **26**, 193–201.
7. Trezona PW (1953) Additivity of colour equations. *Proc. Phys. Soc. (Lond.)*, **B66**, 548–556.
8. Trezona PW (1954) Additivity of colour equations: II. *Proc. Phys. Soc. (Lond.)*, **B67**, 513–522.
9. Stiles WS (1955) The basic data of colour-matching. *Phys. Soc. Year Book*, 44–65.
10. Crawford BH (1965) Colour matching and adaptation. *Vision Res.*, **5**, 71–78.
11. Lozano RD, Palmer DA (1968) Large-field color matching and adaptation. *J. Opt. Soc. Am.*, **58**, 1653–1656.
12. Wyszecki G, Stiles WS (1980) High-level trichromatic color matching and the pigment-bleaching hypothesis. *Vision Res.*, **20**, 23–37.
13. Zaidi Q (1986) Adaptation and color matching. *Vision Res.*, **26**, 1925–1938.
14. Thornton WA (1992) Toward a more accurate and extensible colorimetry. Part I. *Color Res. Appl.*, **17**, 79–122; Part II. *Color Res. Appl.*, **17**, 162–186; Part III. *Color Res. Appl.*, **17**, 240–262.
15. Brill MH (1993) *Additivity failures for color matches: toward a covering theory and a new standard observer.* CIE Symposium on Advanced Colorimetry, CIE Publication x007, 32–40. [A version also appears in: R.J. Motta and H.A. Berberian (Eds) (1993) Device-independent color imaging and imaging systems integration. *Proc. SPIE*, 1909, 44–53.]
16. Brill MH (1979) Formalizing Grassmann's laws in a generalized colorimetry. *Sensory Processes*, **3**, 370–372.
17. Yilmaz H (1967) A theory of speech perception. *Bull. Math. Biophys.*, **29**, 793–825.
18. Richards W (1979) Quantifying sensory channels: generalizing colorimetry to orientation and texture, touch and tones. *Sensory Processes*, **3**, 207–229.
19. Wright WD (1969) *The Measurement of Colour*, 4th ed. Adam Hilger, London.
20. Oulton D (2004) The properties of multiple CMF determinations using alternative primary sets. Part I: Evidence and modelling. *Colour Res. Appl.*, **29**, 273–284.
21. Oulton D (2004) The properties of multiple CMF determinations using alternative primary sets. Part II: A data unification methodology. *Color Res. Appl.*, **29**, 438–450.

22. Brill MH (2001) Numerical experiment on transformability of primaries. Presented at *Council for Optical Radiation Measurements Meeting*, April 2001, Gaithersburg, MD.
23. CIE (1993) *Symposium on advanced colorimetry.* CIE Publication x007, Commission Internationale de l'Eclairage, Vienna, Austria.
24. Oicherman B, Luo MR, Robertson AR, Tarrant A (2005) Experimental verification of trichromatic generalisation, in: *Proceedings of the 10th Congress of the International Colour Association, AIC Colour 05*, Granada, Spain, May 8–13, 2005, pp. 29–32.
25. Oleari C, Pavesi M (2005) New measures of the color-matching functions in foveal vision, *Proceedings of the 10th Congress of the International Colour Association, AIC Colour 05*, Granada, Spain, May 8–13, 2005, pp. 599–602.

11

CIE COLOR APPEARANCE MODELS AND ASSOCIATED COLOR SPACES

M. RONNIER LUO and CHANGJUN LI

Department of Color Science, University of Leeds, Leeds LS2 9JT, UK

INTRODUCTION

In 1931, the Commission Internationale de l'Eclairage (CIE) recommended a color specification system.[1] After various additions over the years it now includes a series of colorimetric measures[2] such as the tristimulus values (*XYZ*), chromaticity coordinates, dominant wavelength, and excitation purity for color specification and color matching, CIELAB and CIELUV color spaces for presenting color relationships, and CIELAB, CIELUV, and more recently the CIEDE2000[3] formulas, for evaluating color differences (see Chapters 3 and 4).

Although the CIE system has been successfully applied for over 70 years, it can only be used under quite limited viewing conditions, for example, daylight illuminant, high luminance level, and some standardized viewing/illuminating geometries. However, with recent demands on cross-media color reproduction, for example, to match the appearance of a color or an image on a display to that on hard copy paper, conventional colorimetry is becoming insufficient. It requires a color appearance model capable of predicting color appearance across a wide range of viewing conditions.

A great deal of research has been carried out to understand color appearance phenomena and to model color appearance. In 1997, the CIE recommended a color appearance model designated CIECAM97s,[4,5] in which the "s" represents a simple version and the "97" means the model was considered as an interim model with the expectation that it would be revised as more data and better theoretical

Colorimetry: Understanding the CIE System, Edited by János Schanda

understanding became available. Since then, the model has been extensively evaluated by not only academic researchers but also industrial engineers in the imaging and graphic arts industries. Some shortcomings were identified and the original model was revised. In 2002, a new model: CIECAM02[6,7] was recommended, which is simpler and has a better accuracy than CIECAM97s. Both CIE color appearance models CIECAM97s and CIECAM02 are introduced.

Colorimetry includes three major topics: color specification, color difference evaluation, and color appearance measurement. In the past, they have been separately studied. The only attempt to unify these functions into one model was the LLAB model developed by Luo et al.[8] In this chapter, it will be shown that in addition to quantifying color appearance, the CIECAM02 model can be extended to accurately predict color differences and hence become a universal colorimetric tool.

VIEWING CONDITIONS

Various aspects of the viewing field impact on the color appearance of a stimulus. Hence accurate definitions and descriptions of the components of the viewing field as shown in Figure 11.1 are necessary for the development and correct use of a color appearance model. Here we follow the definitions given by Hunt[9,10] and Fairchild.[11]

Stimulus

A stimulus is a color element for which a measure of color appearance is required. Typically, the stimulus is taken to be a uniform patch of about 2° angular subtense.

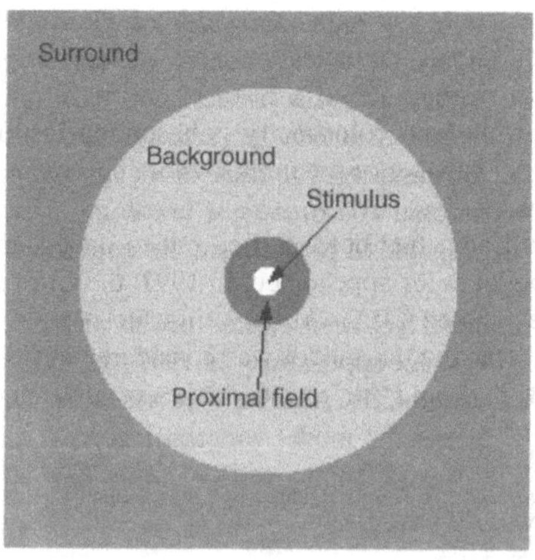

FIGURE 11.1 An illustration of specification of components of viewing field.

Proximal Field

A proximal field is the immediate environment of the color element considered, extending typically for about 2° from the edge of that color element in all or most directions.

Background

The background is defined as the environment of the color element considered, extending typically for about 10° from the edge of the proximal field in all, or most directions. When the proximal field is the same color as the background, the latter is regarded as extending from the edge of the color element considered.

Surround

A surround is a field outside the background. In practical situations, the surround can be considered to be the entire room or the environment in which the image is viewed. For example, printed images are usually viewed in an illuminated (average) surround, projected slides in a dark surround, and domestic television displays in a dim surround.

Adapting Field

An adapting field is the total environment of the color element considered, including the proximal field, the background, and the surround, and extending to the limit of vision in all directions.

COLOR APPEARANCE DATASETS

Color appearance models based on color vision theories have been developed to fit various experimental datasets, which were carefully generated to study particular color appearance phenomena. Over the years, a number of experimental datasets were accumulated to test and develop various color appearance models. Datasets investigated by CIE TC 1-52 *Chromatic Adaptation Transforms* include the following investigators: Mori et al.[12] from the Color Science Association of Japan, McCann et al.[13] and Breneman[14] using a haploscopic matching technique; Helson et al.,[15] Lam and Rigg[16] and Braun and Fairchild[17] using the memory matching technique; and Luo et al.[18,19] and Kuo et al.[20] using the magnitude estimation method. These datasets, however, do not include visual saturation correlates. Hence, Juan and Luo[21,22] investigated a data set of saturation correlates using the magnitude estimation method. The data accumulated played an important role in the evaluation of the performance of different color appearance models and the development of the CIECAM97s and CIECAM02.

CHROMATIC ADAPTATION TRANSFORMS

Chromatic adaptation can be considered as the most important color appearance phenomena and has long been extensively studied. A chromatic adaptation transform (CAT) is capable of predicting corresponding colors, which are defined as pairs of colors that look alike when one is viewed under one illuminant (for example, D65*) and the other under a different illuminant (for example, A). The following is divided into two parts: light and chromatic adaptation, and the development of CAT02.

Light and Chromatic Adaptation

Adaptation can be divided into two: light and chromatic. The former is the adaptation due to the change of light levels. It can be further divided into two: light adaptation and dark adaptation. Light adaptation is the decrease in visual sensitivity upon increase in the overall level of illumination. An example occurs when entering a bright room from a dark cinema. Dark adaptation is opposite to light adaptation and occurs, for example, when entering a dark cinema from a well-lit room.

Physiological Mechanisms

The physiology associated with adaptation mainly includes rod-cone transition, pupil size (dilation and constriction), receptor gain, and offset. It is known that there are two kinds of receptors: cones and rods. The former are less sensitive and respond to high (photopic) levels of illumination (above approximately 10 cd/m^2) whereas the latter are more sensitive and respond to low (scotopic) levels of illumination (below approximately 0.01 cd/m^2). From the example given earlier, when we enter a cinema from a well-lit room, the rods respond to the scotopic level in the cinema and gradually take over from the cones to provide vision. Conversely, when moving from the cinema to the well-lit room, the cone responses take over from the rods. Both adaptation processes will take a finite period of time, sometimes quite a substantial time, to stabilize. In some cases, both rods and cones are functioning in the so-called mesopic region (approximately 0.01 cd/m^2–10 cd/m^2). An example might be when driving along a (poorly) lit road at night.

The pupil size plays an important role in adjusting the amount of light that enters the eye by dilating or constricting the pupil: it is able to adjust the light by a maximum factor of 5. During dark viewing conditions, the pupil size is the largest. Each of the three cones responds to light in a nonlinear manner and is controlled by the gain and inhibitory mechanisms.

Chromatic Adaptation

Light and dark adaptations only consider the change of light level, not the difference of color between two light sources (up to the question of Purkinje

*In this chapter we will use for simplified terms "D_{65}" and "A" instead of the complete official CIE terms: "CIE standard illuminant D_{65}" and "CIE standard illuminant A."

shift due to the difference in the spectral sensitivity of the rods and cones). Under photopic adaptation conditions the difference between the color of two light sources produces chromatic adaptation. This is responsible for the color appearance of objects and leads to the effect known as color constancy (see also Chapter 8, "Color rendering of light sources", where color appearance changes of samples that is discussed can occur if the illumination color is unchanged, only the spectrum of the two lamps is different). The effect can also be divided into two stages: a "chromatic shift" and an "adaptive shift." Consider, for example, what happens when entering a room lit by a tungsten light from outdoor daylight. We experience that all colors in the room instantly become reddish reflecting the relative hue of the tungsten source. This is known as the "colorimetric shift", and it is due to the operation of the sensory mechanisms of color vision, which occur because of the changes in the spectral power distribution of the light sources in question. After a certain short adaptation period, the color appearances of the objects become more normal. This is caused by the fact that most of the colored objects in the real world are more or less color constant (they do not change their color appearance under different illuminants). The most obvious example is the white paper that always appears white regardless of which illuminant it is viewed under. The second stage is called the "adaptive shift", and it is caused by physiological changes and by a cognitive mechanism, which is based upon an observer's knowledge of the colors in the scene content in the viewing field. Judd[23] stated that "the processes by means of which an observer adapts to the illuminant or discounts most of the effect of non-daylight illumination are complicated; they are known to be partly retinal and partly cortical."

The Von Kries coefficient law is widely used to quantify chromatic adaptation. In 1902, von Kries [24] assumed that although the responses of the three cone types (RGB)† are affected differently by chromatic adaptation, the spectral sensitivities of each of the three cone mechanisms remain unchanged. Hence, chromatic adaptation can be considered as a reduction of sensitivity by a constant factor for each of the three cone mechanisms. The magnitude of each factor depends upon the color of the stimulus to which the observer is adapted. The relationship, given in Equation (11.1), is known as the Von Kries coefficient law.

$$\begin{aligned} R_c &= \alpha \cdot R \\ G_c &= \beta \cdot G \\ B_c &= \gamma \cdot B \end{aligned} \tag{11.1}$$

where R_c, G_c, B_c and R, G, B are the cone responses of the same observer, but viewed under reference and test illuminants respectively. α, β, and γ are the Von Kries coefficients corresponding to the reduction in sensitivity of

†In this chapter the RGB symbols will be used for the cone fundamentals, in other chapters the reader will find the LMS symbols. The use of RGB here should not be confused with the *RGB* primaries used in visual color matching.

the three cone mechanisms due to chromatic adaptation. These can be calculated using Equation (11.2).

$$\alpha = \left(\frac{R_{wr}}{R_w}\right); \quad \beta = \left(\frac{G_{wr}}{G_w}\right); \quad \gamma = \left(\frac{B_{wr}}{B_w}\right) \tag{11.2}$$

where

$$\frac{R}{R_w} = \frac{R_c}{R_{wr}}; \quad \frac{G}{G_w} = \frac{G_c}{G_{wr}}; \quad \frac{B}{B_w} = \frac{B_c}{B_{wr}} \tag{11.3}$$

Here R_{wr}, G_{wr}, B_{wr}, and R_w, G_w, B_w are the cone responses for the reference white under the reference and test illuminants, respectively. Over the years, various CATs have been developed but most are based on the Von Kries coefficient law.

Development of the CAT02 Used in CIECAM02

In 1997, Luo and Hunt[25] modified the best available CAT at that time, the Bradford transform[16] derived by Lam and Rigg. The transform, named CMCCAT97, was then recommended by the Colour Measurement Committee (CMC) of the Society of Dyers and Colourists (SDC). This transform is included in the CIECAM97s[4,5] for describing color appearance under different viewing conditions. CMCCAT97 was originally derived by fitting only one data set, Lam and Rigg.[16] Although it gave a reasonably good fit to many other datasets, it predicted badly the McCann data set.[13] In addition, CMCCAT97 includes an exponent p for calculating the blue corresponding spectral response (hence, it can be considered as a modification of the Von Kries type of transform). This causes uncertainty in reversibility and complexity in the reverse mode. Li et al.[26] addressed this problem and provided a solution by including an iterative approximation using the Newton method. However, this is unsatisfactory in imaging applications where the calculations need to be repeated for each pixel. Li et al.[27] gave a linearization version by optimizing the transform to fit all the available datasets, rather than just the Lam and Rigg set.[16] The new transform, named CMCCAT2000, not only overcomes all the problems with respect to reversibility discussed above but also gives a more accurate prediction than other transforms of almost all the available datasets.

At a later stage, CIE TC 8-01 *Colour Appearance Modelling for Colour Management Systems* had to choose a linear chromatic transform for CIECAM02. Multiple candidates such as CMCCAT2000,[27] the sharp chromatic transform[28] developed by Finlayson et al., and CAT02[6,7] were proposed for use as a Von Kries type transform. All had similar levels of performance with respect to the accuracy of predicting various combinations of previously derived sets of corresponding colors. The main difference between these CATs is in the transform from the tristimulus values to the cone responses. Figures 11.2, 11.3, and 11.4 show the spectral sensitivity functions of CMCCAT97, CMCCAT2000, and Finlayson

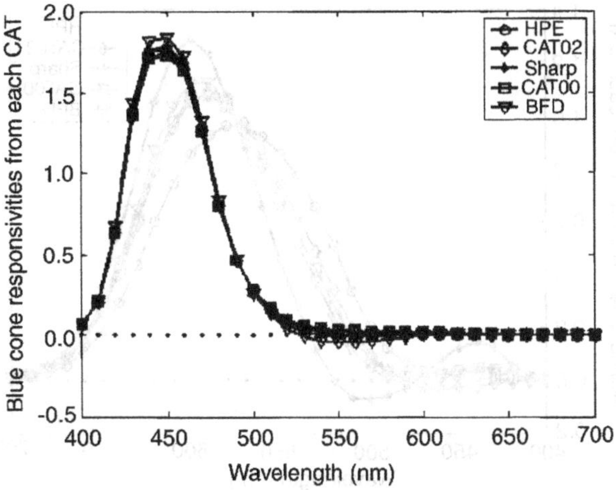

FIGURE 11.2 Blue spectral sensitivity functions for the HPE, CAT02, Finlayson et al., CMCCAT2000, CMCCAT97.

et al. corresponding to the blue, green, and red channels, respectively. In addition, the Hunt–Pointer–Estevez (HPE)[29] spectral sensitivity functions are also plotted, which provide a widely used transform based on the study carried out by Estevez.[30] It can be clearly seen that there are small differences between the functions in the blue channel. However, there are large variations between all the other functions and that of the HPE functions for the red and green channels, that is, all the other functions are much sharper and have negative values compared with the HPE functions. Their peak wavelengths are also very similar and correspond to Thornton's prime-color wavelengths at 448 nm, 537 nm, and 612 nm,[31,32] and they provide the least degree of metamerism if they are used as light sources.

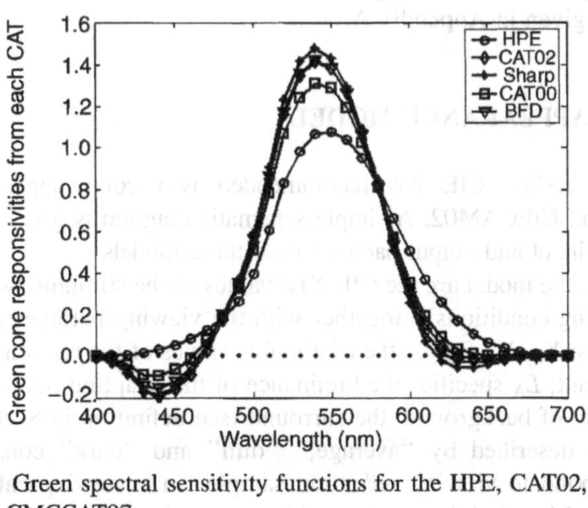

FIGURE 11.3 Green spectral sensitivity functions for the HPE, CAT02, Finlayson et al., CMCCAT2000, CMCCAT97.

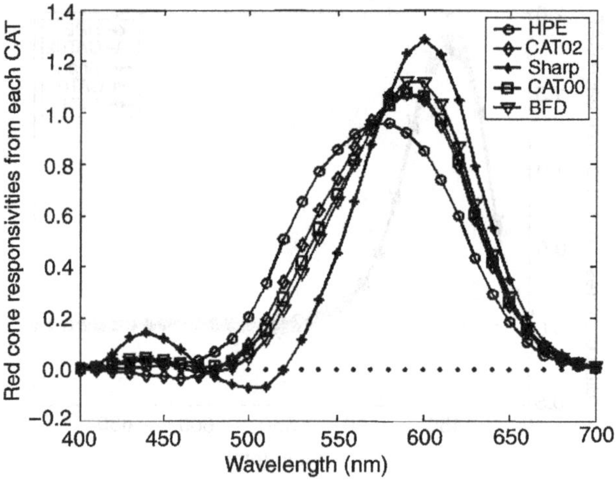

FIGURE 11.4 Red spectral sensitivity functions for the HPE, CAT02, Finlayson et al., CMCCAT2000, CMCCAT97.

In addition to the sharpening of the spectral sensitivity functions, considerations used to select the CIE transform included the degree of backward compatibility with CIECAM97s and error propagation properties by combining the forward and inverse linear CATs, and the datasets which were used during the optimization process. Finally, CAT02 was selected because it is compatible with CMCCAT97 and was optimized using all available datasets except the McCann et al. set, which includes a very chromatic adapting illuminant. It is interesting to note that the primaries in CAT02 are sharper than those used in CMCCAT97. However, the sharper primaries are less backward compatible with CIECAM97s, which was optimized using the Lam and Rigg dataset. The full forward and reverse equations for CAT02 are given in Appendix A.

CIE COLOR APPEARANCE MODELS

As mentioned earlier, CIE has recommended two color appearance models, CIECAM97s and CIECAM02. A simple schematic diagram is given in Figure 11.5 to illustrate the input and output parameters of these models.

The inputs to the model are the CIE *XYZ* values of the stimulus (see definition in Section "Viewing conditions") together with the viewing parameters as shown in the shaded areas: X_w, Y_w, Z_w are the tristimulus values of the reference white under the test illuminant; L_A specifies the luminance of the adapting field; Y_b defines the luminance factor of background; the surround (see definition in Section "Viewing conditions") is described by "average," "dim" and "dark" conditions, which roughly correspond to viewing reflection samples in a viewing cabinet, viewing TV with dim ambient lighting, and watching movie in a cinema, respectively.

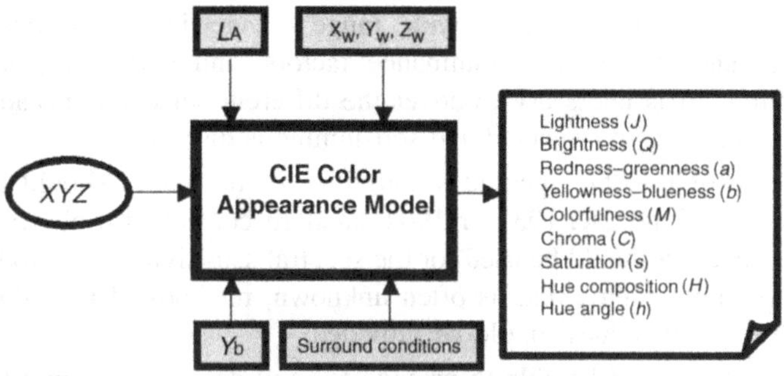

FIGURE 11.5 A schematic diagram of a CIE color appearance model.

There are many output parameters from the model: lightness (*J*), brightness (*Q*), redness–greenness (*a*), yellowness-blueness (*b*), colorfulness (*M*), chroma (*C*), saturation (*s*), hue composition (*H*), and hue angle (*h*). These attributes, defined in the Glossary of Terms section, can be combined to form various spaces according to different applications. They can be divided into two types for evaluating color appearance and color difference, respectively. For example, *JCh* and *JCH* spaces are typically used by the color and imaging industries. The hue angle (*h*), ranges from 0° to 360° in the *a* and *b* plane and is based on the concept of equal perceived difference, and the hue composition (*H*) describes color appearance in terms of four unitary hues, ranged from 0 (pure red), 100 (pure yellow), 200 (pure green), 300 (pure blue), and back to pure red at 400. Note that the 0°, 90°, 180°, 270°, and 360° in the *JCh* space do not correspond to pure hue perceptions of red, yellow, green, blue, and red perceptions, respectively.

CIECAM97s

The CIE held an expert symposium on "Color Standards for Image Technology" in 1996. A decision was made to develop a CIE color appearance model based on the 12 principles outlined by Hunt[33]:

1. The model should be as comprehensive as possible, so that it can be used in a variety of applications; but at this stage, only static states of adaptation should be included because of the great complexity of dynamic effects.
2. The model should cover a wide range of stimulus intensities, from very dark object colors to very bright self-luminous colors. This means that the dynamic response function must have a maximum and cannot be a simple logarithmic or power function.
3. The model should cover a wide range of adapting intensities, from very low scotopic levels, such as that occurs in starlight, to very high photopic levels, such as that occurs in sunlight. This means that rod vision should be included in the model; but because many applications will be such that rod vision is negligible, the model should be usable in a mode that does not include rod vision.

4. The model should cover a wide range of viewing conditions including backgrounds of different luminance factors, and dark, dim, and average surrounds. It is necessary to cover the different surrounds because of their widespread use in projected and self-luminous displays.
5. For ease of use, the spectral sensitivities of the cones should be a linear transform of the CIE 1931 or 1964 standard colorimetric observer, and the $V'(\lambda)$ function should be used for the spectral sensitivity of the rods. Because scotopic photometric data is often unknown, methods of providing approximate scotopic values should be provided.
6. The model should be able to provide for any degree of adaptation between complete and none, for cognitive factors, and for the Helson-Judd effect, as options.
7. The model should give predictions of the perceptual correlates in terms of hue angle, hue composition, brightness, lightness, saturation, chroma, and colorfulness.
8. The model should be capable of being operated in a reverse mode.
9. The model should be no more complicated than is necessary to meet the above requirements.
10. Any simplified version of the model, intended for particular applications, should give the same predictions as the complete model for some specified set of conditions.
11. The model should give predictions of color appearance that are not appreciably worse than those given by the model that is best in each application.
12. A version of the model should be available for application to unrelated colors such as those seen in dark surrounds in isolation from other colors.

Four color appearance models were considered to be the most advanced at that time: Hunt[9,34], Nayatani,[35] RLAB,[36] and LLAB.[8] An agreement was achieved that CIE TC1-34 *Testing Colour Appearance Models* should examine the existing color appearance models and combine their best features into a high performance model for general use, and the model should adequately predict all available datasets. At the meeting held in Kyoto in 1997, CIE TC1-34 agreed to adopt a simplified model, which was named CIECAM97s.[4,5] The comprehensive version was never formulated due to an apparent lack of demand and a lack of suitable data to aid its formulation.

CIECAM02

Soon after the recommendation of CIECAM97s, CIE TC8-01, *Colour Appearance Modelling for Colour Management Systems*, was formed to evaluate CIECAM97s for its predictions of color appearance, and its appropriateness for engineering and implementation requirements for open color management systems. Various trials were conducted and some problems were identified as summarized below:

1. To simplify and improve CMCCAT97 transform by adopting CAT02 as described in Section "Development of the CAT02 used in CIECAM02."
2. To correct the error that the lightness (J) was not equal to zero for a stimulus having a Y tristimulus value of zero, as reported by Li et al.[26]
3. To ensure that the sizes of the gamut volumes from the color appearance model rank from the largest to smallest in the order of average, dim, and dark surround conditions as addressed by Moroney[37] and Li et al.[26]
4. To improve the prediction of chroma for near neutral colors: Newman and Pirrotta[38] had reported that the predictions given by CIECAM97s for colorfulness and chroma are too high for colors close to the neutral axis.
5. To improve the fit to the saturation results accumulated by Juan and Luo,[21,22] which are the only available saturation data to test the color appearance model.

Various methods[26,39,40,41] were proposed for overcoming the above identified shortcomings in the CIECAM97s model and in 2002, CIE TC8-01 recommended a new model: CIECAM02.[6,7] It is not only a refinement of CIECAM97s, removing many shortcomings, but also an improvement giving equivalent or better predictions of color appearance datasets.[42,43] A typical example is given here. Figure 11.6 plots the Munsell chroma data against the chroma predictions from (a) CIECAM97s, (b) CIELAB, and (c) CIECAM02. The results show that the CIECAM02 model outperforms the other two models, that is, it gives the smallest scattering of the data and converging to zero for neutral colors. The full forward and reverse modes of the CIECAM02 model are given in Appendix B. These are different from those given in the CIE publication[7] in some of the computational steps, in that all computations that depend only on the test illuminant and the surround conditions are grouped together as Step 0. As they do not depend on the samples, they only need to be computed once. This is very useful for image processing applications.

Color Appearance Phenomena

This section describes a number of color appearance phenomena studied by various researchers. Examples are given to illustrate how the CIECAM02 model predicts these effects.

Chromatic Adaptation

Chromatic adaptation has been extensively investigated by many researchers. In fact, most of the data described in the previous section were accumulated to study this effect. The results are formulated in the form of corresponding colors for which each pair of colors represents the same color appearance when viewed under different illuminants.

Figure 11.7 illustrates 52 pairs of corresponding colors predicted by CIECAM02 (or its chromatic adaptation transform, CAT02) from illuminant A (open circles of

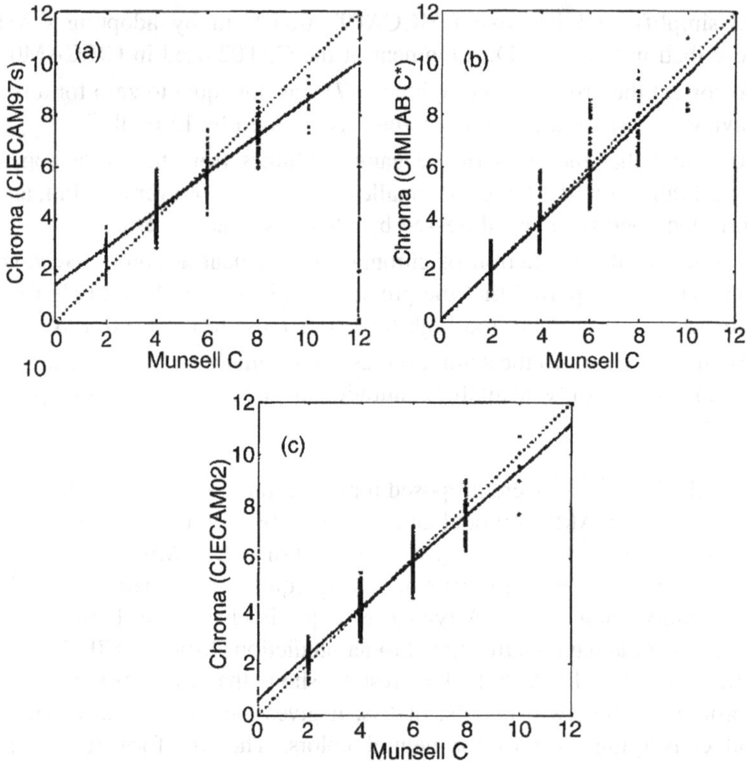

FIGURE 11.6 The predictions from (a) CIECAM97s, (b) CIELAB, and (c) CIECAM02 are plotted against the Munsell Chroma data. Both the 45° line and the best-fit line are plotted. For perfect results, these lines should overlap.

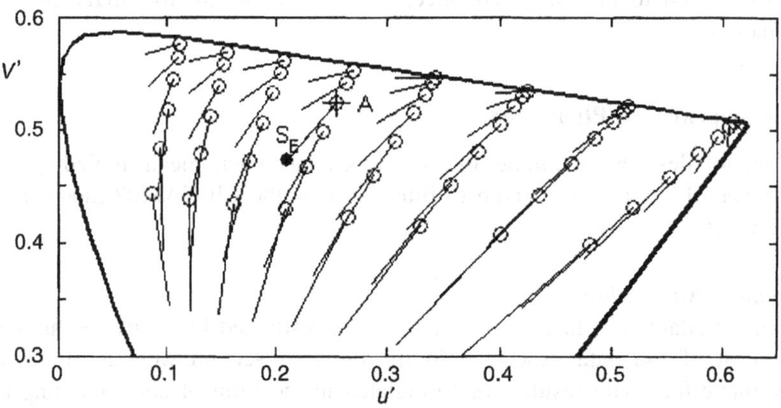

FIGURE 11.7 The corresponding colors predicted by the CIECAM02 from illuminant A (open circles of vectors) to illuminant S_E(open ends of vectors) plotted in CIE $u'v'$ chromaticity diagram for the CIE 1931 standard colorimetric observer. The plus (+) and the dot (•) represent illuminants A and S_E, respectively.

vectors) to S_E (open ends of vectors) plotted in the CIE $u'v'$ chromaticity diagram for the 2° observer. The open circle colors have a value of L^* equal to 50 according to CIELAB under illuminant A. These were then transformed by the model to the corresponding colors under illuminant S_E (the equienergy illuminant). Thus, the ends of each vector represent a pair of corresponding colors under the two illuminants. The input parameters are (the luminance of adapting field) $L_A = 63.7\,cd/m^2$ and average surround. The parameters are defined in the end of Appendix B.

The results show that there is a systematic pattern, that is, for colors below v' equal to 0.48 under illuminant A, the vectors are predicted toward the blue direction under the illuminant S_E. For colors outside the above region, the appearance change is in a counter-clockwise direction, that is, red colors shift to yellow, yellow to green, and green to cyan as the illuminant changes from A to S_E.

Hunt Effect

Hunt[44] studied the effect of light and dark adaptation on color perception and collected data for corresponding colors via a visual colorimeter using the haploscopic matching technique, in which each eye was adapted to different viewing conditions and matches were made between stimuli presented in each eye.

Figure 11.8 illustrates this effect as successfully modeled by the CIECAM02 model. Five colors were selected, having a constant L^* (CIELAB lightness) of 50

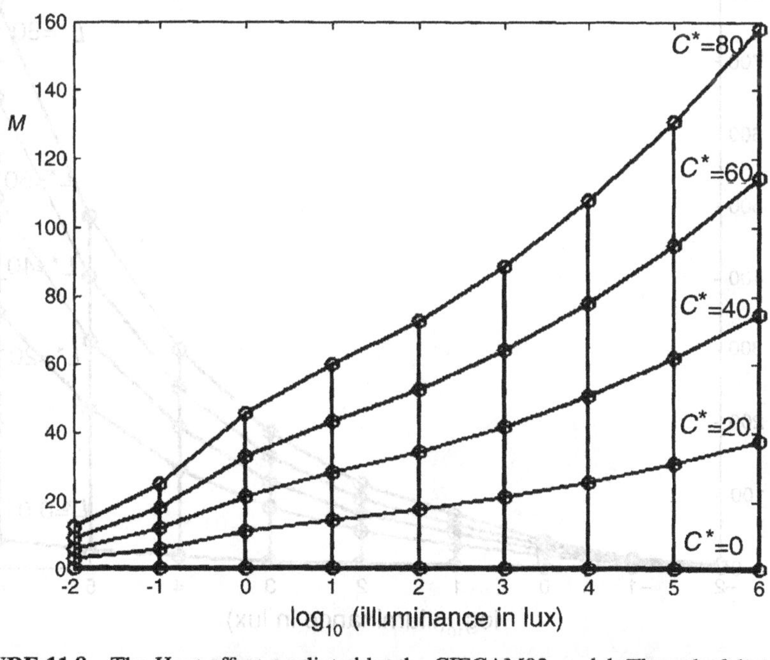

FIGURE 11.8 The Hunt effect predicted by the CIECAM02 model. The colorfulness (M) predictions for five samples of varying CIELAB chroma (C^*) values are plotted against nine illuminance levels on a $\log_{10}$ scale.

and hue angle of 2° (red) with C^*(CIELAB chroma) varying from 0 (neutral color) to 80 (a high chroma color) under illuminant S_E. These colors were predicted by CIECAM02 under 9 illuminance levels ranging from 0.01 lx to 1,000,000 lx.

Each (nearly) horizontal curve represents the change of colorfulness appearance for a particular sample. Each vertical line expresses the degree of colorfulness contrast under a particular illuminance level. The results clearly demonstrate the Hunt effect, that is, each sample represented by each curve increases its colorfulness (M) (except for the neutral colors) when the illuminance of the reference white increases until reaching a value of about 1,000,000 lx. In addition, the colorfulness contrast increases from dark to bright illuminance levels as shown by the lengths of the vertical lines between the dark and bright levels.

Stevens Effect

Stevens and Stevens[45] asked observers to make magnitude estimations of the brightness of stimuli across various adaptation conditions. The results showed that the perceived brightness contrast increased with an increase in the adapting luminance level according to a power relationship.

Five neutral samples having L^* values of 0.01, 20, 40, 60, and 80 under illuminant S_E were selected to demonstrate the Stevens effect as predicted by CIECAM02. Figure 11.9 illustrates the Stevens effect by plotting brightness (Q)

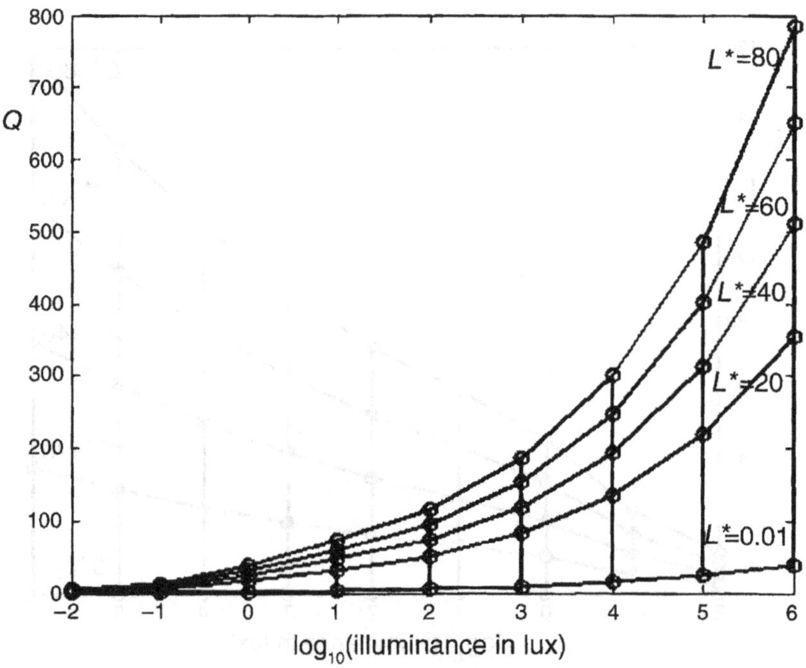

FIGURE 11.9 The Stevens effect predicted by the CIECAM02 model. The brightness (Q) predictions for five neutral samples of varying CIELAB lightness L^* are plotted against nine illuminance levels in $\log_{10}$ scale.

against the base 10 logarithm of E_w (the illuminance of the light source in lux, see the end of Appendix B) of the test illuminant, that is, an increase of brightness contrast with an increase of the illuminance. The incremental increase in brightness is very marked for the lighter samples (above L^* of 20) but the effect is almost zero for the darkest sample ($L^* = 0.01$). This leads to an increase of brightness contrast, that is, the lighter samples appear much brighter.

Surround Effect

Bartleson and Breneman[46] found that the perceived contrast in colorfulness and brightness increased with increasing illuminance level from dark surround, dim surround to average surround. This is an important color appearance phenomenon to be modeled, especially for the imaging and graphic arts industries where, on many occasions, it is required to reproduce images on different media under quite distinct viewing conditions.

Two figures are used to illustrate the surround effect: the colorfulness (M) and lightness (J) predicted by CIECAM02 under the average, dim, and dark surrounds. These are plotted in Figures 11.10 and 11.11, respectively. Figure 11.10 shows the colorfulness (M), with different surrounds, of samples with CIELAB C^* values of 0, 20, 40, 60, and 80 with constant L^* of 50 and h (the CIELAB hue angle) of 2° (red) under the illuminant S_E. (The other parameters were set to: $L_A = 63.7\,\text{cd/m}^2$, $Y_b = 20$, and constants F, c, and N_c were chosen according to Table 11.A1 in Appendix B). Figure 11.11 shows the lightness (J), with different surrounds, of neutral samples with CIELAB L^* of 0.001–80 with the same model parameters as used for Figure 11.10. Note that constant F is a factor for degree of adaptation, c the impact of surround, and N_c the chromatic surround induction factor (see Appendix B).

Figure 11.10 shows that for each of the five test colors having C^* values of 0, 20, 40, 60, and 80, there is a slight decease of colorfulness from average, through dim to dark surround conditions except for C^* of zero. This leads to a reduction of

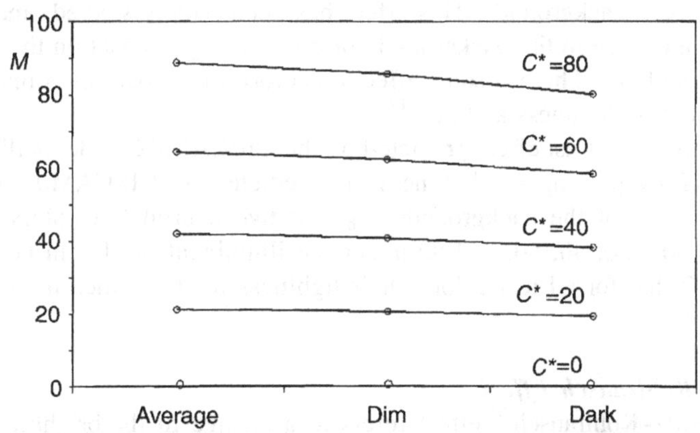

FIGURE 11.10 The surround effect predicted by the colorfulness (M) scale of CIECAM02. The colorfulness (M) predictions for the five samples varying in CIELAB chroma C^* values are plotted against the "average," "dim," and "dark" surround conditions.

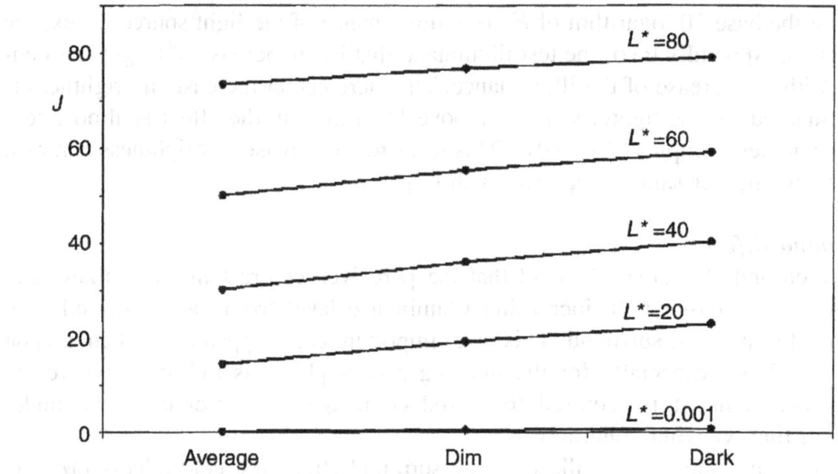

FIGURE 11.11 The surround effect predicted by the lightness (J) scale of the CIECAM02. The lightness (J) predictions for the five neutral samples varying in CIELAB lightness L^* values are plotted against the "average," "dim," and "dark" surround conditions.

colorfulness contrast from average to dark surround conditions. Figure 11.11 shows that for each of the five neutral test colors having L^* of 0.001, 20, 40, 60, and 80, there is a decrease of lightness contrast from average, through dim to dark surround conditions.

Lightness Contrast Effect

The lightness contrast effect[47] reflects that the perceived lightness increases when colors are viewed against a darker background and vice versa. It is a type of simultaneous contrast effect considering the change of color appearance due to different colored backgrounds. This effect has been widely studied, and it is well known that a change in the background color has a large impact on the perception of lightness and hue. There is some effect on colorfulness, but this is much smaller than the effect on lightness and hue.[47]

The lightness contrast effect predicted by the CIECAM02 model is illustrated in Figure 11.12 by plotting the lightness (J) predicted by CIECAM02 against the luminance factor of the backgrounds (Y_b) for five neutral test colors having L^* values of 0.001, 20, 40, 60, and 80 under the illuminant S_E. It can be seen from Figure 11.12 that for all test colors, their lightness reduces when the background becomes lighter.

Helmholtz–Kohlrausch Effect

The Helmholtz–Kohlrausch[48] effect refers to a change in the brightness of color produced by increasing the purity of a color stimulus while keeping its luminance constant within the range of photopic vision. This effect is quite small compared with others and is not modeled by CIECAM02.

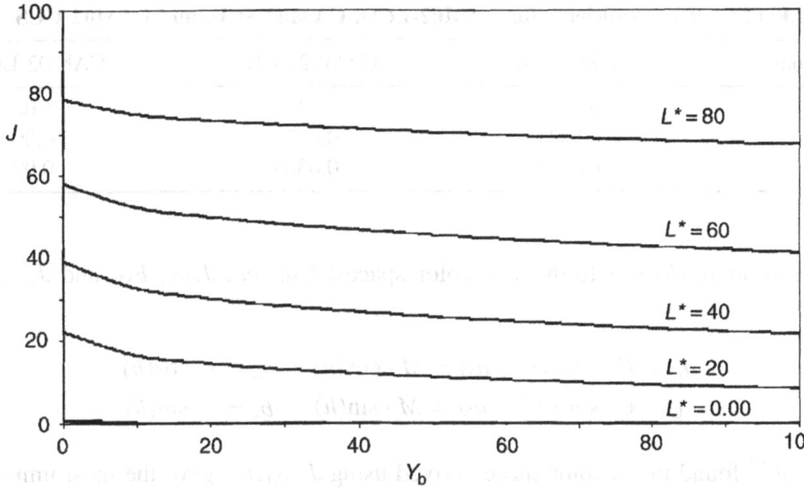

FIGURE 11.12 The lightness contrast effect predicted by the CIECAM02 model. The lightness (J) predictions for the five neutral samples varying in CIELAB lightness L^* values are plotted against neutral background having different luminance factors.

Helson–Judd Effect

When a gray scale is illuminated by a light source, the lighter neutral stimuli will exhibit a certain amount of the hue of the light source and the darker stimuli will show its complementary hue, which is known as the Helson–Judd effect.[49] Thus for tungsten light, which is much yellower than daylight, the lighter stimuli will appear yellowish, and the darker stimuli bluish. This effect is not modeled by CIECAM02.

UNIFORM COLOR SPACES BASED ON CIECAM02

As mentioned in the previous section, CIECAM02 gives an accurate prediction[42] of all the available color appearance data described. Attempts have been made by the authors to extend CIECAM02 for predicting available color discrimination datasets, which include two types, for Large and Small magnitude Color Differences, designated by LCD and SCD, respectively. The former includes six datasets: Zhu et al.,[50] OSA,[51] Guan and Luo,[52] BADB-Textile,[53] Pointer and Attridge,[54] and Munsell.[55,56] They have 144, 128, 292, 238, 1308, and 844 pairs respectively, having an average 10 ΔE^*_{ab} units over all the sets. The SCD data, having an average 2.5 ΔE^*_{ab} units, are a combined data set used to develop the CIE 2000 color difference formula: CIEDE2000.[3]

CIECAM02-Based Color Spaces

CIECAM02[6,7] includes three attributes in relation to the chromatic content: chroma (C), colorfulness (M), and saturation (s). These attributes together with lightness (J)

TABLE 11.1 The coefficients for CAM02-LCD, CAM02-SCD, and CAM02-UCS

Versions	CAM02-LCD	CAM02-SCD	CAM02-UCS
K_L	0.77	1.24	1.00
c_1	0.007	0.007	0.007
c_2	0.0053	0.0363	0.0228

and hue angle (h) can form three color spaces: J, a_C, b_C, J, a_M, b_M, and J, a_s, b_s where

$$a_C = C \cdot \cos(h) \quad a_M = M \cdot \cos(h) \quad a_s = s \cdot \cos(h)$$
$$b_C = C \cdot \sin(h)\,, \quad b_M = M \cdot \sin(h), \quad b_s = s \cdot \sin(h)$$

Li et al.[57] found that a color space derived using J, a_M, b_M gave the most uniform result when analyzed using the large and small color difference datasets. Hence, various attempts[57,58] were made to modify this version of CIECAM02 to fit all available datasets. Finally, a simple, generic form, Equation (11.4) was found that adequately fitted all available data.

$$J' = \frac{(1 + 100 \cdot c_1) \cdot J}{1 + c_1 \cdot J} \qquad (11.4)$$
$$M' = (1/c_2) \cdot \ln(1 + c_2 \cdot M)$$

where c_1 and c_2 are constants given in Table 11.1.

The corresponding color space is J', a'_M, b'_M where $a'_M = M' \cdot \cos(h)$, and $b'_M = M' \cdot \sin(h)$. The color difference between two samples can be calculated in J', a'_M, b'_M space using Equation (11.5).

$$\Delta E' = \sqrt{(\Delta J'/K_L)^2 + \Delta a'^2_M + \Delta b'^2_M} \qquad (11.5)$$

where $\Delta J'$, $\Delta a'_M$, and $\Delta b'_M$ are the differences of J', a'_M, and b'_M between the "standard" and "sample" in a pair. Here K_L is a lightness parameter and is given in Table 11.1.

Three color spaces named CAM02-LCD, CAM02-SCD, and CAM02-UCS were developed for large, small, and combined large and small differences, respectively. The corresponding parameters in Equations (11.4) and (11.5) are listed in Table 11.1.

Comparing the Performance of the New UCSs With Some Selected Color Models

The three new CIECAM02- based color spaces, CAM02-LCD, CAM02-SCD, and CAM02-UCS, together with the best available color difference formulas including

TABLE 11.2 Testing uniform color spaces and color difference formulas using the combined LCD and SCD datasets

Combined LCD dataset	PF/3	Combined SCD dataset	PF/3
CIELAB	26	CIELAB	52
IPT	26	IPT	52
OSA	24	CIEDE2000	**33**
GLAB	24	DIN99d	35
CIECAM02	25	CIECAM02	47
CAM02-LCD	**23**	CAM02-LCD	41
CAM02-SCD	27	CAM02-SCD	34
CAM02-UCS	25	CAM02-UCS	35

CIEDE2000[3] and DIN99d,[59] and uniform color spaces such as CIELAB,[2] IPT,[60] OSA,[51] and GLAB[52] were also tested by Luo et al.[58] using the available small and large color difference datasets. It was found that CAM02-LCD and CAM02-SCD performed either better than or equal to the other best available color spaces for the LCD and SCD data, respectively. The performance results are summarized in Table 11.2 in terms of PF/3 measure.[61] For a perfect agreement between the visual results and a formula's or space's predictions, PF/3 should equal zero. A larger PF/3 value means a larger prediction error. A PF/3 of 30 can roughly be considered as 30% disagreement between the visual data and a formula prediction. It was also very encouraging that CAM02-UCS, developed to fit both the large and small color difference datasets, also gave an excellent performance in predicting the datasets. When selecting one UCS to evaluate color differences across a wide range, CAM02-UCS can be considered a suitable candidate.

The experimental color discrimination ellipses used in the previous studies[62,63] were also used for comparing different color spaces. Figures 11.13, and 11.14 show

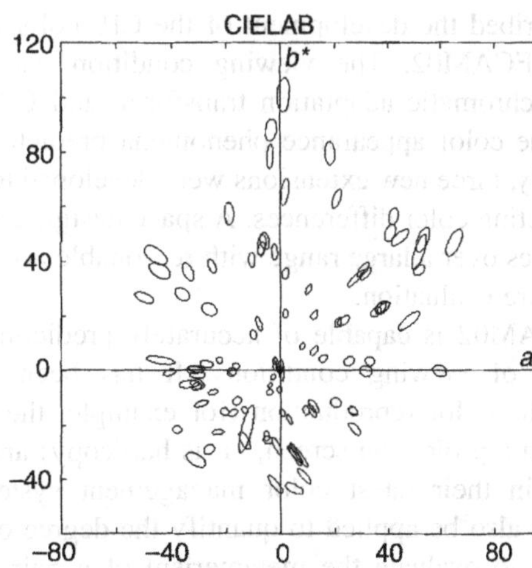

FIGURE 11.13 Experimental chromatic discrimination ellipses plotted in CIELAB.

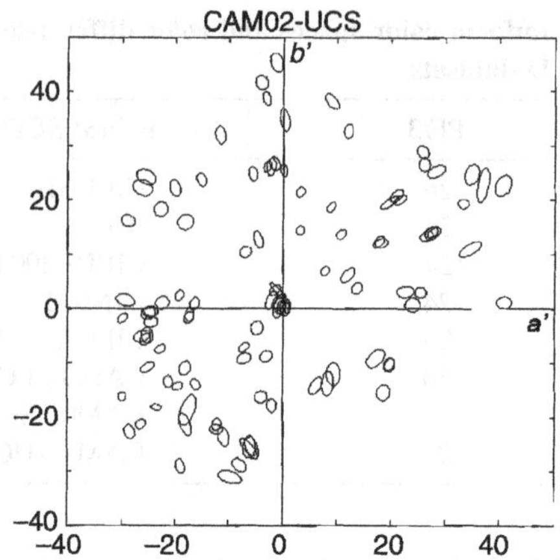

FIGURE 11.14 Experimental chromatic discrimination ellipses plotted in CAM02-UCS.

the ellipses plotted in CIELAB and CAM02-UCS spaces, respectively. The size of the ellipse was adjusted by a single factor in each space to ease visual comparison. For perfect agreement between the experimental results and a uniform color space, all ellipses should be constant radius circles. Overall, it can be seen that the ellipses in CIELAB (Figure 11.13) are smaller in the neutral region and gradually increase in size as chroma increases. In addition, the ellipses are orientated approximately toward the origin except for those in the blue region in CIELAB space. All ellipses in CAM02-UCS (Figure 11.14) are approximately equal-sized circles. In other words, the newly developed CAM02-UCS is much more uniform than CIELAB.

CONCLUSIONS

This chapter has described the development of the CIE color appearance models, CIECAM97s and CIECAM02. The viewing condition parameters are clearly defined. The CAT02 chromatic adaptation transform, and CIECAM02 are given in the appendixes. The color appearance phenomena predicted by the model are also introduced. Finally, three new extensions were developed to form new uniform color spaces for predicting color differences. A space designated CAM02-UCS can predict color differences over a large range with reasonable accuracy and should be recommended for future evaluation.

Overall, the CIECAM02 is capable of accurately predicting color appearance under a wide range of viewing conditions. It has been proved to achieve successfully crossmedia color reproduction (for example, the reproduction of an image on a display, on a projection screen, or as hardcopy) and is adopted by the Microsoft Company in their latest color management system, Window Color System (WCS). It can also be applied to quantify the degree of color inconstancy of a single specimen, to evaluate the metamerism of a pair of samples, and to estimate the color-rendering properties of light sources. Furthermore, it can be used to specify the color appearance of each stimulus in terms of a comprehensive set of clearly defined color appearance attributes. With the extension to include new color spaces as described in the previous section, it can accurately evaluate color differences under various viewing conditions. (In contrast, all existing color difference equations can only be used under daylight illuminants.) Hence, CIECAM02 performed satisfactorily for all three major colorimetric tasks: color specification, color difference evaluation, and color appearance prediction. It can be considered a universal color model.

REFERENCES

1. CIE (1932) Colorimétrie, resolutions 1 – 4, *in recueil des travaux et compte rendu des séances*, Hutième Session Cambridge – Septembre 1931, Bureau Central de la Commission, The National Physical Laboratory Teddington, Cambridge at the University Press, pp. 19–29.
2. CIE Technical Report (2004) *Colorimetry*, 3rd ed., Publication **15**:2004, CIE Central Bureau, Vienna.
3. Luo MR, Cui GH, Rigg B (2001) The development of the CIE 2000 colour difference formula. *Color Res. Appl.*, **26**, 340–350.
4. Luo MR, Hunt RWG (1998) The structure of the CIE 1997 colour appearance model (CIECAM97s), *Color Res. Appl.*, **23**, 138–146.
5. CIE (1998) *The CIE 1997 Interim colour appearance model (simple version)*, CIECAM97s. CIE Publication **131**, CIE Central Bureau, Vienna, Austria.
6. Moroney N (2002) The CIECAM02 color appearance model, in: *Proceeding of the Tenth Color Imaging Conference: Color Science, Systems, and Applications*, pp. 23–27.
7. CIE (2004) *A colour appearance model for colour management systems: CIECAM02*, CIE Publication **159**, CIE Central Bureau, Vienna, Austria.
8. Luo MR., Lo MC, Kuo WG (1996) The LLAB(l:c) colour model, *Colour Res. Appl.*, **21**, 412, 1996.
9. Hunt RWG (1991) Revised colour appearance model for related and unrelated colours. *Color Res. Appl.*, **16**, 146–165.
10. Hunt RWG (1998) *Measuring Colour*, 3rd ed., Fountain Press, Kingston-upon-Thames, England.
11. Fairchild MD (2005) *Color Appearance Models*, 2nd ed., J. Wiley, Chichester.
12. Mori L, Sobagaki H, Komatsubara H, Ikeda K (1991) Field trials on CIE chromatic adaptation formula, in: Proceedings of the CIE 22nd Session, pp. 55–58.
13. McCann JJ, McKee SP, Taylor TH (1976) Quantitative studies in Retinex theory: a comparison between theoretical predictions and observer responses to the 'color mondrian' experiments, *Vision Res.*, **16**, 445–458.
14. Breneman EJ (1987) Corresponding chromaticities for different states of adaptation to complex visual fields. *J. Opt. Soc. Am. A*, **4**, 1115–1129.
15. Helson H, Judd DB, Warren MH (1952) Object-color changes from daylight to incandescent filament illumination. *Illum. Eng.*, **47**, 221–233.
16. Lam KM (1985) *Metamerism and colour constancy*, PhD thesis, University of Bradford, UK.

17. Braun KM Fairchild MD (1996) Psychophysical generation of matching images for cross-media colour reproduction, in: *Proceedings of the 4th Color Imaging Conference*, IS&T, Springfield, VA, pp. 214–220.
18. Luo MR, Clarke AA, Rhodes PA, Schappo A, Scrivener SAR, Tait C (1991) Quantifying colour appearance. Part I. LUTCHI colour appearance data. *Color Res. Appl.*, **16**, 166–180.
19. Luo MR, Gao XW, Rhodes PA, Xin HJ, Clarke AA, Scrivener SAR (1993) Quantifying colour appearance, Part IV: Transmissive media. *Color Res. Appl.*, **18**, 191–209.
20. Kuo WG, Luo MR, Bez HE (1995) Various chromatic adaptation transforms tested using new colour appearance data in textiles. *Color Res. Appl.*, **20**, 313–327.
21. Juan LY, Luo MR (2000) New Magnitude estimation data for evaluating color appearance models. Colour and Visual Scales 2000, NPL, April 3–5, UK.
22. Juan LY, Luo MR (2002) Magnitude estimation for scaling saturation. Proc. 9th Session of the Association Internationale de la Couleur (AIC Color 2001), Rochester, USA, (June 2001), *Proc. SPIE*, **4421**, 575–578.
23. Judd DB (1940), Hue, saturation, and lightness of surface colors with chromatic illumination, *J. Opt. Soc. Am.*, **30**, 2–32.
24. Von Kries (1902), *Chromatic Adaptation*, Festschrift der Albrecht-Ludwig-Universitat (Fribourg), [Translation: MacAdam DL, *Sources of Color Science*, MIT Press, Cambridge, Mass. (1970)].
25. Luo MR, Hunt RWG (1998) A chromatic adaptation transform and a colour inconstancy index. *Color Res. Appl.*, **23**, 154–158.
26. Li CJ, Luo MR, Hunt RWG (2000), A revision of the CIECAM97s Model. *Color Res. Appl.*, **25**, 260–266.
27. Li CJ, Luo MR, Rigg B, Hunt RWG (2002) CMC 2000 chromatic adaptation transform: CMCCAT2000. *Color Res. Appl.*, **27**, 49–58.
28. Finlayson GD, Süsstrunk S (2000) Performance of a chromatic adaptation transform based on spectral sharpening, in: Proceedings of the IS&T/SID 8th Color Imaging Conference, pp. 49–55.
29. Hunt RWG, Pointer MR (1985), A color appearance transform for 1931 standard colorimetric observer. *Color Res. Appl.*, **10**, 165–179.
30. Estevez O (1979) *On the fundamental data base of normal and dichromatic vision*, Ph.D. thesis, University of Amsterdam.
31. Thornton WA (1973), Matching lights, metamers, and human visual response, *J. Color Appearance*, **II**, 23–29.
32. Brill MH, Finlayson GD, Hubel PM, Thornton WA (1998), Prime colours and colour imaging, in: *Proceedings of the IS&T/SID 6th Color Imaging Conference*, pp. 33–42.
33. CIE Publication (1996) *Proceedings of the CIE Expert Symposium '96 Colour Standards for Image Technology*, CIE Publ. No. **x010**-1996, Central Bureau of the CIE, Vienna, Austria.
34. Hunt RWG, Luo MR (1994) An improved predictor of colorfulness in a model of colour vision. *Color Res. Appl.*, **19**, 23–26.
35. Nayatany Y, Sobagaki H, Hashimoto K, Yano T (1997) Field trials of a nonlinear color-appearance model. *Color Res. Appl.*, **22**, 240–258.
36. Fairchild MD (1996) Refinement of the RLAB color space. *Color Res. Appl.*, **21**, 338–346.

37. Moroney N (1998) A comparison of CIELAB and CIECAM97s, in: *Proceeding of the Sixth Color Imaging Conference: Color Science, Systems, and Applications*, pp. 17–21.

38. Newman T, Pirrotta E (2000) *The darker side of color appearance models and gamut mapping*. Colour Image Science 2000, University of Derby, pp. 215–223.

39. Fairchild MD (2001) A revision of CIECAM97s for practical applications. *Color Res. Appl.*, **26**, 418–427.

40. Hunt RWG, Li CJ, Juan LY, Luo MR (2002) Further improvements to CIECAM97s. *Color Res. Appl.*, **27**, 164–170.

41. Hunt RWG, Li CJ, Luo MR (2002) Dynamic cone response functions for models of colour appearance. *Color Res. Appl.*, **28**, 82–88.

42. Li CJ, Luo MR, Hunt RWG, Moroney N, Fairchild MD, Newman T (2002) The Performance of CIECAM02, in: *Proceedings of the IS&T/SID 10th Color Imaging Conference*, pp. 28–32.

43. Li CJ, Luo MR (2005) Testing the robustness of CIECAM02. *Color Res. Appl.*, **30**, 99–106.

44. Hunt RWG (1952) Light and dark adaptation and perception of color. *J Opt. Soc. Am.*, **42**, 190–199.

45. Stevens JC, Stevens SS (1963) Brightness functions: Effects of adaptation. *J. Opt. Soc. Am.*, **53**, 375–385.

46. Bartleson CJ, Breneman EJ (1967) Brightness perception in complex fields. *J. Opt. Soc. Am.*, **57**, 953–957.

47. Luo MR, Gao XW, Sciviner SAR (1995) Quantifying color appearance, Part V, Simultaneous contrast. *Color Res. Appl.*, **20**, 18–28.

48. Wyszecki G, Stiles WS (1982) *Color Science: Concepts and Methods, Quantitative Data and Formulae*, John Wiley & Sons, New York.

49. Helson H (1938) Fundamental problems in color vision. I. The principle governing changes in hue, saturation, and lightness of non-selective samples in chromatic illumination. *J. Exp. Psychol.*, **23**, 439–477.

50. Zhu SY, Luo MR, Cui G (2001) New experimental data for investigating uniform colour spaces, in: *The 9th Congress of the International Colour Association* (*AIC Color 2001*), June 24–29, Rochester, USA, pp. 626–629.

51. MacAdam DL (1974) Uniform color scales. *J. Opt. Soc. Am.*, **55**, 1619–1702.

52. Guan SS, Luo MR (1999) A colour-difference formula for assessing large colour differences. *Color Res. Appl.*, **24**, 344–355.

53. Badu S (1986) *Large colour differences between surface colours*, PhD thesis, University of Bradford, UK.

54. Pointer MR, Attridge GG (1997) Some aspects of the visual scaling of large colour differences. *Color Res. Appl.*, **22**, 298–307.

55. Newhall SM (1940) Preliminary report of the O.S.A. subcommittee on the spacing of the Munsell colors. *J. Opt. Soc. Am.*, **30**, 617–645.

56. Newhall SM, Nickerson D, Judd D (1943) Final report of the O.S.A. subcommittee on spacing of the Munsell colors. *J. Opt. Soc. Am.*, **33**, 385–418.

57. Li CJ, Luo MR, Cui GH (2003) Colour-difference evaluation using colour appearance models, in: Proceedings of the 11th Colour Imaging Conference, IS&T and SID, Scottsdale, Arizona, November, pp. 127–131.

58. Luo MR, Cui GH, Li CJ, Rigg B (2006), Uniform colour spaces based on CIECAM02 colour appearance model. *Color Res. Appl.,* **31**, 320–330.
59. Cui G, Luo MR, Rigg B, Roesler G, Witt K (2002), Uniform colour spaces based on the DIN99 colour-difference formula, *Color Res. Appl.*, **27**, 282–290.
60. Ebner F, Fairchild MD (1998), Development and testing of a color space (IPT) with improved hue uniformity, in: *Proceedings of the Sixth Color Imaging Conference*, pp. 8–13.
61. Guan SS, Luo MR (1999), Investigation of parametric effects using small colour differences. *Color Res. Appl.*, **24**, 331–343.
62. Luo MR, Rigg B (1986) Chromaticity-discrimination ellipses for surface colours. *Col Res. Appl.*, **11**, 25–42.
63. Berns RS, Alman DH, Reniff L, Snyder GD, Balonon-Rosen MR (1991) Visual determination of suprathreshold color-difference tolerances using probit analysis. *Col Res. Appl.*, **16**, 297–316.

APPENDIX A: CHROMATIC ADAPTATION TRANSFORM: CAT02

Part 1: Forward Mode

Input data:
Sample in test illuminant: X, Y, Z
Adopted white in test illuminant: X_{w}, Y_{w}, Z_{w}
Reference white in reference illuminant: X_{wr}, Y_{wr}, Z_{wr}
Luminance of test and reference adapting fields (cd/m^2): L_{A}
(Note that for the calculation of L_{A} see the note in the end of Appendix B.) In addition, when applying CAT, the other viewing conditions such as surround, luminance factor of background, luminance level of reference, and test fields should be fixed.

Transformed data to be obtained
Sample corresponding color in reference illuminant: X_{c}, Y_{c}, Z_{c}

Step 1: Calculate cone responses

$$\begin{pmatrix} R \\ G \\ B \end{pmatrix} = M_{\mathrm{CAT02}} \cdot \begin{pmatrix} X \\ Y \\ Z \end{pmatrix}, \begin{pmatrix} R_{\mathrm{w}} \\ G_{\mathrm{w}} \\ B_{\mathrm{w}} \end{pmatrix} = M_{\mathrm{CAT02}} \cdot \begin{pmatrix} X_{\mathrm{w}} \\ Y_{\mathrm{w}} \\ Z_{\mathrm{w}} \end{pmatrix}, \begin{pmatrix} R_{\mathrm{wr}} \\ G_{\mathrm{wr}} \\ B_{\mathrm{wr}} \end{pmatrix} = M_{\mathrm{CAT02}} \cdot \begin{pmatrix} X_{\mathrm{wr}} \\ Y_{\mathrm{wr}} \\ Z_{\mathrm{wr}} \end{pmatrix}$$

where $$M_{\mathrm{CAT02}} = \begin{pmatrix} 0.7328 & 0.4296 & -0.1624 \\ -0.7036 & 1.6975 & 0.0061 \\ 0.0030 & 0.0136 & 0.9834 \end{pmatrix}$$

Step 2: Calculate the degree of adaptation, D

$$D = F \cdot \left[1 - \left(\frac{1}{3.6}\right) \cdot e^{-\left(\frac{L_A + 42}{92}\right)}\right]$$

where F equals 1.0, 0.9, and 0.8 for average, dim, and dark surround viewing conditions, respectively, and where L_A is the luminance of adapting field (reference and testing). If D is greater than one or less than zero, set it to one or zero accordingly. Note that for the selection of surround parameters please see the note at the end of Appendix B.

Step 3: Calculate the corresponding response

$$R_c = R \cdot \left(\alpha \cdot \frac{R_{wr}}{R_w} + 1 - D\right)$$

$$G_c = G \cdot \left(\alpha \cdot \frac{G_{wr}}{G_w} + 1 - D\right)$$

$$B_c = B \cdot \left(\alpha \cdot \frac{B_{wr}}{B_w} + 1 - D\right)$$

where $\alpha = D \cdot \frac{Y_w}{Y_{wr}}$

Step 4: Calculate the corresponding tristimulus values

$$\begin{pmatrix} X_c \\ Y_c \\ Z_c \end{pmatrix} = M_{CAT02}^{-1} \cdot \begin{pmatrix} R_c \\ G_c \\ B_c \end{pmatrix}$$

Note that for the coefficients in the inverse matrix are given in the note at the end of Appendix B.

Part 2: Reverse Mode

Input data:
Corresponding color in reference illuminant: X_c, Y_c, Z_c
Others are the same as the forward.
Output data:
Sample color in test illuminant: X, Y, Z

Step 1: Calculate cone responses

$$\begin{pmatrix} R_w \\ G_w \\ B_w \end{pmatrix} = M_{CAT02} \cdot \begin{pmatrix} X_w \\ Y_w \\ Z_w \end{pmatrix}, \begin{pmatrix} R_{wr} \\ G_{wr} \\ B_{wr} \end{pmatrix} = M_{CAT02} \cdot \begin{pmatrix} X_{wr} \\ Y_{wr} \\ Z_{wr} \end{pmatrix}, \begin{pmatrix} R_c \\ G_c \\ B_c \end{pmatrix} = M_{CAT02} \cdot \begin{pmatrix} X_c \\ Y_c \\ Z_c \end{pmatrix}$$

Step 2: Calculate the D using Step 2 of the Forward mode.

Step 3: Calculate the cone response

$$R = \frac{R_c}{\left(\alpha \cdot \frac{R_{wr}}{R_w} + 1 - D\right)}$$

$$G = \frac{G_c}{\left(\alpha \cdot \frac{G_{wr}}{G_w} + 1 - D\right)}$$

$$B = \frac{R_c}{\left(\alpha \cdot \frac{B_{wr}}{B_w} + 1 - D\right)}$$

where $$\alpha = D \cdot \frac{Y_w}{Y_{wr}}$$

Step 4: Calculate the original tristimulus values

$$\begin{pmatrix} X \\ Y \\ Z \end{pmatrix} = M_{CAT02}^{-1} \cdot \begin{pmatrix} R \\ G \\ B \end{pmatrix}$$

Note that for the coefficients in the inverse matrix are given in the note at the end of Appendix B.

APPENDIX B: CIE COLOR APPEARANCE MODEL: CIECAM02

Part 1: The Forward Mode

Input: X, Y, Z (under test illuminant X_w, Y_w, Z_w)

Output: Correlates of lightness J, chroma C, hue composition H, hue angle h, colorfulness M, saturation s, and brightness Q

Illuminants, viewing surrounds setup, and background parameters

(See the note at the end of this appendix for determining all parameters.)

Adopted white in test illuminant: X_w, Y_w, Z_w

Background in test conditions: Y_b

(Reference white in reference illuminant: $X_{wr} = Y_{wr} = Z_{wr} = 100$, which are fixed in the model.)

Luminance of test adapting field (cd/m^2): L_A

All surround parameters are given in Table 11.A1

TABLE 11.A1 Surround parameters

	F	c	N_c
Average	1.0	0.69	1.0
Dim	0.9	0.59	0.9
Dark	0.8	0.535	0.8

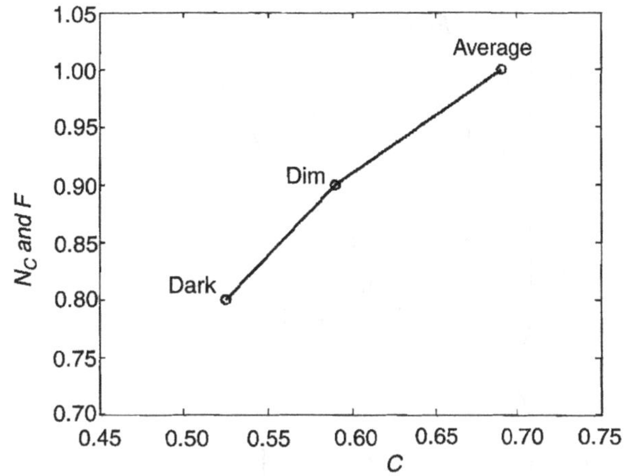

FIGURE 11.A1 N_c and F varies with c.

Note that for determining the surround conditions see the note at the end of this appendix. N_c and F are modeled as a function of c and can be linearly interpolated as shown in Figure 11.A1, using the above points.

Step 0: Calculate all values/parameters that are independent of input samples

$$\begin{pmatrix} R_w \\ G_w \\ B_w \end{pmatrix} = M_{CAT02} \cdot \begin{pmatrix} X_w \\ Y_w \\ Z_w \end{pmatrix}, D = F \cdot \left[1 - \left(\frac{1}{3.6}\right) \cdot e^{\left(\frac{-L_A - 42}{92}\right)}\right]$$

Note if D is greater than one or less than zero, set it to one or zero, respectively.

$$D_R = D \cdot \frac{Y_w}{R_w} + 1 - D,\ D_G = D \cdot \frac{Y_w}{G_w} + 1 - D,\ D_B = D \cdot \frac{Y_w}{B_w} + 1 - D$$

$$F_L = 0.2k^4 \cdot (5L_A) + 0.1(1 - k^4)^2 \cdot (5L_A)^{1/3}$$

$$\text{where } k = \frac{1}{5 \cdot L_A + 1}$$

$$n = \frac{Y_b}{Y_w},\ z = 1.48 + \sqrt{n},\ N_{bb} = 0.725 \cdot \left(\frac{1}{n}\right)^{0.2},\ N_{cb} = N_{bb}$$

$$\begin{pmatrix} R_{wc} \\ G_{wc} \\ B_{wc} \end{pmatrix} = \begin{pmatrix} D_R \cdot R_w \\ D_G \cdot G_w \\ D_B \cdot B_w \end{pmatrix}, \begin{pmatrix} R'_w \\ G'_w \\ B'_w \end{pmatrix} = M_{HPE} \cdot M_{CAT02}^{-1} \cdot \begin{pmatrix} R_{wc} \\ G_{wc} \\ B_{wc} \end{pmatrix}$$

$$M_{HPE} = \begin{pmatrix} 0.38971 & 0.68898 & -0.07868 \\ -0.22981 & 1.18340 & 0.04641 \\ 0.00000 & 0.00000 & 1.00000 \end{pmatrix}$$

$$R'_{\mathrm{aw}} = 400 \cdot \left(\frac{\left(\frac{F_{\mathrm{L}} \cdot R'_{\mathrm{w}}}{100}\right)^{0.42}}{\left(\frac{F_{\mathrm{L}} \cdot R'_{\mathrm{w}}}{100}\right)^{0.42} + 27.13} \right) + 0.1$$

$$G'_{\mathrm{aw}} = 400 \cdot \left(\frac{\left(\frac{F_{\mathrm{L}} \cdot G'_{\mathrm{w}}}{100}\right)^{0.42}}{\left(\frac{F_{\mathrm{L}} \cdot G'_{\mathrm{w}}}{100}\right)^{0.42} + 27.13} \right) + 0.1$$

$$B'_{\mathrm{aw}} = 400 \cdot \left(\frac{\left(\frac{F_{\mathrm{L}} \cdot B'_{\mathrm{w}}}{100}\right)^{0.42}}{\left(\frac{F_{\mathrm{L}} \cdot B'_{\mathrm{w}}}{100}\right)^{0.42} + 27.13} \right) + 0.1$$

$$A_{\mathrm{w}} = \left[2 \cdot R'_{\mathrm{aw}} + G'_{\mathrm{aw}} + \frac{B'_{\mathrm{aw}}}{20} - 0.305 \right] \cdot N_{\mathrm{bb}}$$

Note that all parameters computed in this step are needed for the following calculations. However, they depend only on surround and viewing conditions, hence when processing pixels of image, they are computed once for all. The following computing steps are sample dependant.

Step 1: Calculate (sharpened) cone responses (transfer color-matching functions to sharper sensors)

$$\begin{pmatrix} R \\ G \\ B \end{pmatrix} = M_{\mathrm{CAT02}} \cdot \begin{pmatrix} X \\ Y \\ Z \end{pmatrix},$$

Step 2: Calculate the corresponding (sharpened) cone response (considering various luminance level and surround conditions included in D, hence in D_{R}, D_{G}, and D_{B})

$$\begin{pmatrix} R_{\mathrm{c}} \\ G_{\mathrm{c}} \\ B_{\mathrm{c}} \end{pmatrix} = \begin{pmatrix} D_{\mathrm{R}} \cdot R \\ D_{\mathrm{G}} \cdot G \\ D_{\mathrm{B}} \cdot B \end{pmatrix}$$

Step 3: Calculate the HPE response

$$\begin{pmatrix} R' \\ G' \\ B' \end{pmatrix} = M_{\mathrm{HPE}} \cdot M_{\mathrm{CAT02}}^{-1} \cdot \begin{pmatrix} R_{\mathrm{c}} \\ G_{\mathrm{c}} \\ B_{\mathrm{c}} \end{pmatrix}$$

Step 4: Calculate the postadaptation cone response (resulting in dynamic range compression)

$$R'_a = 400 \cdot \left(\frac{\left(\frac{F_L \cdot R'}{100}\right)^{0.42}}{\left(\frac{F_L \cdot R'}{100}\right)^{0.42} + 27.13} \right) + 0.1$$

If R' is negative, then $$R'_a = -400 \cdot \left(\frac{\left(\frac{-F_L \cdot R'}{100}\right)^{0.42}}{\left(\frac{-F_L \cdot R'}{100}\right)^{0.42} + 27.13} \right) + 0.1$$

and similarly for the computations of G'_a, and B'_a, respectively.

Step 5: Calculate redness–greenness (a), yellowness–blueness (b) components, and hue angle (h):

$$a = R'_a - \frac{12 \cdot G'_a}{11} + \frac{B'_a}{11}$$

$$b = \frac{(R'_a + G'_a - 2 \cdot B'_a)}{9}$$

$$h = \tan^{-1}\left(\frac{b}{a}\right)$$

make sure h is between 0 and 360°.

Step 6: Calculate eccentricity (e_t) and hue composition (H), using the unique hue data given in Table 11.A2; set $h' = h + 360$ if $h < h_1$, otherwise $h' = h$. Choose a proper i ($i = 1$, 2, 3, or 4) so that $h_i \leq h' < h_{i+1}$. Calculate

$$e_t = \frac{1}{4} \cdot \left[\cos\left(\frac{h' \cdot \pi}{180} + 2\right) + 3.8 \right]$$

TABLE 11.A2 Unique hue data for calculation of hue quadrature

	Red	Yellow	Green	Blue	Red
i	1	2	3	4	5
h_i	20.14	90.00	164.25	237.53	380.14
e_i	0.8	0.7	1.0	1.2	0.8
H_i	0.0	100.0	200.0	300.0	400.0

which is close to, but not exactly the same as, the eccentricity factor given in Table 11.A2.

$$H = H_i + \frac{100 \cdot \frac{h' - h_i}{e_i}}{\frac{h' - h_i}{e_i} + \frac{h_{i+1} - h'}{e_{i+1}}}$$

Step 7: Calculate achromatic response A

$$A = [2 \cdot R'_{\mathrm{a}} + G'_{\mathrm{a}} + \frac{B'_{\mathrm{a}}}{20} - 0.305] \cdot N_{\mathrm{bb}}$$

Step 8: Calculate the correlate of lightness

$$J = 100 \cdot \left(\frac{A}{A_{\mathrm{w}}}\right)^{c \cdot z}$$

Step 9: Calculate the correlate of brightness

$$Q = \left(\frac{4}{c}\right) \cdot \left(\frac{J}{100}\right)^{0.5} \cdot (A_{\mathrm{w}} + 4) \cdot F_{\mathrm{L}}^{0.25}$$

Step 10: Calculate the correlates of chroma (C), colorfulness (M), and saturation (s)

$$t = \frac{\left(\frac{50,000}{13} \cdot N_{\mathrm{c}} \cdot N_{\mathrm{cb}}\right) \cdot e_t \cdot (a^2 + b^2)^{1/2}}{R'_{\mathrm{a}} + G'_{\mathrm{a}} + \left(\frac{21}{20}\right) \cdot B'_{\mathrm{a}}}$$

$$C = t^{0.9} \cdot \left(\frac{J}{100}\right)^{0.5} \cdot (1.64 - 0.29^n)^{0.73}$$

$$M = C \cdot F_L^{0.25}$$

$$s = 100 \cdot \left(\frac{M}{Q}\right)^{0.5}$$

Part 2: The Reverse Mode

Input: J or Q; C, M, or s; H or h

Output: X, Y, Z (under test illuminant X_w, Y_w, Z_w)
Illuminants, viewing surrounds, and background parameters are the same as those given in the Forward mode. See notes at the end of this appendix calculating/defining the luminance of the adapting field and surround conditions.

Step 0: Calculate viewing parameters

Compute all $F_L, n, z, N_{bb} = N_{bc}, R_w, G_w, B_w, D, D_R, D_G, D_B, R_{wc}, G_{wc}, B_{wc}, R'_w, G'_w, B'_w\, R'_{aw}, G'_{aw}, B'_{aw}$, *and* A_w using the same formulas as in Step 0 of the Forward model. They are needed in the following steps. Note that all data computed in this step can be used for all samples (for example, all pixels for an image) under the viewing conditions. Hence, they are computed once for all. The following computing steps are sample dependant.

Step 1: Obtain J, C, and h from H, Q, M, s

The entering data can be in different combination of perceived correlates, that is, J or Q; C, M, or s; and H or h. Hence the following are needed to convert the others to J, C, and h.

Step 1.1: Compute J from Q (if start from Q)

$$J = 6.25 \cdot \left[\frac{c \cdot Q}{(A_w + 4) \cdot F_L^{0.25}}\right]^2$$

Step 1.2: Calculate C from M or s

$$C = \frac{M}{F_L^{0.25}} \text{ (if start from } M)$$

$$\mathrm{Q} = \left(\frac{4}{c}\right) \cdot \left(\frac{\mathrm{J}}{100}\right)^{0.5} \cdot ((\mathrm{A_w} + 4.0) \cdot F_L^{0.25}$$

$$\text{and } C = \left(\frac{s}{100}\right)^2 \cdot \left(\frac{Q}{F_L^{0.25}}\right) \text{ (if start from } s)$$

Step 1.3: Calculate h from H (if start from H)

The correlate of hue (h) can be computed by using data in Table 11.A2 in the Forward mode.

Choose a proper i (i = 1, 2, 3, or 4) so that $H_i \leq H < H_{i+1}$.

$$h' = \frac{(H - H_i) \cdot (e_{i+1} h_i - e_i \cdot h_{i+1}) - 100 \cdot h_i \cdot e_{i+1}}{(H - H_i) \cdot (e_{i+1} - e_i) - 100 \cdot e_{i+1}}$$

Set $h = h' - 360$ if $h' > 360$, otherwise $h = h'$.

Step 2: Calculate t, e_t, p_1, p_2, and p_3

$$t = \left[\frac{C}{\sqrt{\frac{J}{100}}\cdot(1.64-0.29^{n})^{0.73}}\right]^{\frac{1}{0.9}}$$

$$e_t = \frac{1}{4}\cdot\left[\cos\left(h\cdot\frac{\pi}{180}+2\right)+3.8\right]$$

$$A = A_w\cdot\left(\frac{J}{100}\right)^{\frac{1}{c\cdot z}}$$

$$p_1 = \left(\frac{50,000}{13}\cdot N_c\cdot N_{cb}\right)\cdot e_t\cdot\left(\frac{1}{t}\right), \quad \text{if t} \neq 0$$

$$p_2 = \frac{A}{N_{bb}}+0.305$$

$$p_3 = \frac{21}{20}$$

Step 3: Calculate a and b

If $t=0$, then $a=b=0$ and *go to Step 4*

(be sure transferring h from degree to radian before calculating $\sin(h)$ and $\cos(h)$)

If $|\sin(h)| \geq |\cos(h)|$ then

$$p_4 = \frac{p_1}{\sin(h)}$$

$$b = \frac{p_2\cdot(2+p_3)\cdot\left(\frac{460}{1403}\right)}{p_4+(2+p_3)\cdot\left(\frac{220}{1403}\right)\cdot\left(\frac{\cos(h)}{\sin(h)}\right)-\left(\frac{27}{1403}\right)+p_3\cdot\left(\frac{6300}{1403}\right)}$$

$$a = b\cdot\left(\frac{\cos(h)}{\sin(h)}\right)$$

If $|\cos(h)| > |\sin(h)|$, then

$$p_5 = \frac{p_1}{\cos(h)}$$

$$a = \frac{p_2\cdot(2+p_3)\cdot\left(\frac{460}{1403}\right)}{p_5+(2+p_3)\cdot\left(\frac{220}{1403}\right)-\left[\left(\frac{27}{1403}\right)-p_3\cdot\left(\frac{6300}{1403}\right)\right]\cdot\left(\frac{\sin(h)}{\cos(h)}\right)}$$

$$b = a\cdot\left(\frac{\sin(h)}{\cos(h)}\right)$$

Step 4: Calculate R'_a, G'_a, and B'_a

$$R'_a = \frac{460}{1403} \cdot p_2 + \frac{451}{1403} \cdot a + \frac{288}{1403} \cdot b$$

$$G'_a = \frac{460}{1403} \cdot p_2 - \frac{891}{1403} \cdot a - \frac{261}{1403} \cdot b$$

$$B'_a = \frac{460}{1403} \cdot p_2 - \frac{220}{1403} \cdot a - \frac{6300}{1403} \cdot b$$

Step 5: Calculate R', G' and B'

$$R' = \mathrm{sign}(R'_a - 0.1) \cdot \frac{100}{F_L} \cdot \left[\frac{27.13 \cdot |R'_a - 0.1|}{400 - |R'_a - 0.1|}\right]^{\frac{1}{0.42}}$$

Here $\mathrm{sign}(x) = \begin{cases} 1 & \text{if } x > 0 \\ 0 & \text{if } x = 0 \\ -1 & \text{if } x < 0 \end{cases}$, and similarly computing G', and B' from G'_a, and B'_a.

Step 6: Calculate R_C, G_C, and B_C (for the inverse matrix, see the note at the end of the appendix)

$$\begin{pmatrix} R_c \\ G_c \\ B_c \end{pmatrix} = M_{\mathrm{CAT02}} \cdot M_{\mathrm{HPE}}^{-1} \cdot \begin{pmatrix} R' \\ G' \\ B' \end{pmatrix}$$

Step 7: Calculate R, G, and B

$$\begin{pmatrix} R \\ G \\ B \end{pmatrix} = \begin{pmatrix} \frac{R_c}{D_R} \\ \frac{G_c}{D_G} \\ \frac{B_c}{D_B} \end{pmatrix}$$

Step 8: Calculate X, Y, and Z (for the coefficients of the inverse matrix, see the note at the end of the appendix)

$$\begin{pmatrix} X \\ Y \\ Z \end{pmatrix} = M_{\mathrm{CAT02}}^{-1} \cdot \begin{pmatrix} R \\ G \\ B \end{pmatrix}$$

Notes to Appendixes A and B

1. It is recommended to use the matrix coefficients given below for the inverse matrix M_{CAT02}^{-1} and M_{HPE}^{-1}:

$$M_{CAT02}^{-1} = \begin{pmatrix} 1.096124 & -0.278869 & 0.182745 \\ 0.454369 & 0.473533 & 0.072098 \\ -0.009628 & -0.005698 & 1.015326 \end{pmatrix}$$

$$M_{HPE}^{-1} = \begin{pmatrix} 1.910197 & -1.112124 & 0.201908 \\ 0.370950 & 0.629054 & -0.000008 \\ 0.000000 & 0.000000 & 1.000000 \end{pmatrix}$$

2. For implementing the CIECAM02, the testing data and the corresponding results from the Forward and Reverse modes can be found from Ref. 7.
3. The L_A is computed using Equation (11.A1)

$$L_A = \left(\frac{E_W}{\pi}\right) \cdot \left(\frac{Y_b}{Y_W}\right) = \frac{L_W \cdot Y_b}{Y_W}, \tag{11.A1}$$

where $E_w = \pi \cdot L_w$ is the illuminance of reference white in lux unit; L_w the luminance of reference white in cd/m^2 unit, Y_b the luminance factor of the background, and Y_w the luminance factor of the reference white.

4. Surround conditions (average, dim, and dark) are determined by the surround ratio S_R given by Equation (11.A2):

$$S_R = \frac{L_{SW}}{L_{DW}} \tag{11.A2}$$

where L_{SW} is the luminance of the reference white measured in the surround field and L_{DW} is the luminance of the reference white measured in the display area. If S_R is 0, then the surround condition is "dark"; if $0 < S_R < 0.2$, then the surround is "dim"; and if $S_R \geq 0.2$, then the surround is "average."

12

IMAGE APPEARANCE MODELING

Garrett M. Johnson[1,2] and Mark D. Fairchild[1]
[1]*Rochester Institute of Technology, Munsell Color Science Laboratory, 54 Lomb Memorial Drive, Rochester, NY 14623-5604, USA*
[2]*Apple Inc., 1632 5th St., Santa Monica, CA 90401*

INTRODUCTION

Color appearance modeling has made some significant advances in recent years, culminating in the creation of the CIECAM02 color appearance model. Historically the general approach of most color difference and appearance models is to isolate the color stimuli from other dimensions of visual performance as much as possible. This typically means predicting the appearance of a single stimulus on a uniform background. Viewing conditions such as the surround lighting and the overall viewing luminance are certainly included, but only so far as their influences on the single color patch. These models have been used successfully in an imaging modality, though the color interactions between individual pixels are mostly ignored. Although traditional color appearance models are very useful, it is possible that such models have progressed about as far as they can and that further advances will require different types of models. These models will need to take into account both spatial and color properties of the stimuli and viewing conditions. Recently, Fairchild and Johnson have proposed a different approach to color appearance modeling specifically for use with complex images, referred to as an image appearance model.[1–3] An image appearance model builds upon color appearance models to incorporate properties of spatial, and perhaps temporal, vision allowing prediction of appearance in complex stimuli. In addition to predicting overall image appearance these models can be used to calculate image differences and from there can be applied as a first step toward an image quality

Colorimetry: Understanding the CIE System, Edited by János Schanda

metrics. This chapter reviews the concept of and need for image appearance modeling, a question still under discussion in CIE TC 8-08, Spatial appearance models, and presents one such model, known as iCAM. Two applications of iCAM for rendering high-dynamic-range (HDR) images and calculating image differences will also be discussed.

FROM SIMPLE TO COMPLEX COLOR APPEARANCE

The earlier chapters of this book give a thorough accounting for the historical progression of color and image measurement, and the role the CIE has played in developing these techniques. A brief review of these contributions in the context of image measurement helps outline the development of, need for, and application of image appearance models. Although image appearance modeling may be represented as a new approach, it can actually be considered a natural evolution and combination of color appearance, color difference metrics, and spatial vision. These individual tools have been used for years in the development of many imaging systems, although often without specific knowledge or intent. Early imaging systems were often not scientifically measured at all, or measured with systems designed to specify the variables of the imaging system itself. These measurements provided little information about what the resulting image actually looked like, though they did provide useful information about the imaging process. For example, densitometers were developed for measuring photographic materials with the intent of specifying the amounts of dye or silver produced in the film, though these measurements could not be easily transferred to a nonphotographic imaging system. Similarly for printing presses, measurements could be made for the specific inks in the process as well as measures of the dot area coverage for halftone systems. Again, these measurements provided little information as to what the resulting prints actually looked like. In electronic systems like television, system measurements such as signal voltages were used to colorimetrically quantify the imaging system.[4] Non-device-specific measurements of imaging systems for image quality do have a long history as illustrated by the example of Schade's vision-based pioneering work.[5] As imaging systems have evolved in complexity and more importantly openness, the need for device-independent image measurements became obvious.

Electronic imaging systems, specifically the development of color television, prompted the first application of device-independent color measurements of images. As discussed in the earlier chapters of this book and by Wright it appears quite plausible that color television could not have been invented without colorimetry. The CIE system was used very successfully in the design and standardization of color television systems and is once again being called upon for the more recent digital and high-definition television systems.

As digital imaging systems have evolved, the need for device-independent color has grown significantly. Interchange of images between different devices, computing platforms, and modalities is now impossible to avoid. Application of CIE colorimetry to these many imaging systems became much more prevalent especially with the use of computer systems to generate and proof content ultimately destined for other media. The use of CIE colorimetry to specify images across the various devices, and manifested in such ways as ICC color management, have promised to solve some of the many new color reproduction problems that were created with open, digital systems. The power and flexibility of these digital systems has also made it possible, and finally practical, to perform colorimetric transformations on massive amounts image data to generate matching images across disparate devices and media.

Research on imaging device calibration and characterization has spanned the range from fundamental color measurement techniques to the specification of a variety of devices including CRT, LCD, and projection displays, scanners and digital cameras, and various film recording and print media. In recent time these techniques have been extended into everyday use, with the use of cellular phone cameras and the Internet. Some of the concepts and results of this research have been summarized by Berns, Day, and Wyble.[6–8] Such measurement capabilities are a fundamental requirement for research and development in color and image appearance. Research on device characterization and calibration provides a means to tackle more fundamental problems in device-independent color imaging. This need will continue, as new imaging devices and modalities are being created at an incredible pace. For example, precise color measurement has lead to extensive research gamut mapping algorithms (see Chapter 7) to deal with the reproduction of desired colors that fall outside the range that can be obtained with a given imaging device.[9,10] As new types of imaging devices are created, such as HDR displays, wide-gamut displays, and multiprimary printers research on color measurement and gamut-expansion measurement will be necessary. Research on device-independent color measurement built upon, and contributed to, research on the development and testing of color appearance models for crossmedia image reproduction.

Following a similar track as color appearance modeling, color difference research has culminated with the recently published CIEDE2000 color difference formula.[11] The history of this research has been described in Chapter 4 of this book. At the heart of all color difference equations, and color appearance models lies some form of uniform color space. As described in Chapters 3 and 4, the CIE initially recommended two uniform color spaces in 1976, CIELAB and CIELUV. Like CIECAM97s, both spaces were initially described as interim color spaces, with the knowledge that they were far from complete. Amazingly some 30 years later, these spaces are still going strong. They are not without problems, though. With a true perceptually uniform color space, perceived color differences can be taken to be a simple measure of distance between two colors in the space, such as CIE ΔE^*_{ab}. The CIE recognized the nonuniformity of the CIELAB color space, and much research has resulted in the formulation of more advanced color difference equations such as ΔE^*_{94} and CIEDE2000. These more complicated equations are very capable of predicting perceived color differences, though like traditional colorimetry they are designed for use with simple color patches on a uniform background.

The experiments used to develop the CIE color difference formulas were performed using simple color patches in precisely controlled viewing conditions. Although quite successful, there is no reason to believe that they are adequate for predicting color difference for spatially complex image stimuli. The S-CIELAB model was designed as a spatial preprocessor to the standard CIE color difference equations, to account for complex color stimuli such as halftone patterns.[12] The spatial preprocessing uses separable convolution kernels to approximate the contrast sensitivity functions (CSF) of the human visual system. These kernels behave as bandpass functions on the "luminance" channels and low-pass functions for the chromatic opponent channels. It is important to stress that the CSF kernels chosen are tied heavily into the color space in which they are applied. For S-CIELAB the CSF serves to remove information that is imperceptible to the visual system and normalize color differences at spatial frequencies that are perceptible. For instance, when viewing halftone dots at a certain distance the dots tend to blur and integrate into a single color. A pixel-by-pixel color difference calculation between a continuous image and a halftone image would result in very large errors, although the perceived difference might in fact be small. The spatial preprocessing would blur the halftone image so that it more closely resembles the continuous tone image.

S-CIELAB represents the first incarnation of an image difference model based upon a simple extension of the CIELAB color space and color difference equations. Recently this model has been refined and extended into a modular framework for image color difference calculations.[13,14] This framework refined the CSF equations from the S-CIELAB model and adds modules for spatial frequency adaptation, spatial localization, and local and global contrast detection. The choice of opponent color space in which the spatial filtering is performed has also been examined and refined.[15,16]

As described in detail in the previous chapter, fundamental CIE colorimetry does not provide a complete solution for color specification. Rather, it described a thorough means for describing color matches. By their very nature colors and images produced or captured by various systems are examined in widely disparate viewing conditions, from the original captured scene to a computer display in a dim room, to printed media under a variety of light sources and to projection displays in dark rooms. Thus color appearance models were developed to extend traditional CIE colorimetry to the prediction of overall color appearance, not just color matches, specifically across changes in media and viewing conditions. Color appearance modeling research applied to simple stimuli and digital imaging systems was very active throughout the 1990s culminating with the recommendation by the CIE of the CIECAM97s model in 1997 and its revision, CIECAM02, in 2002. Luo describes the progression of these models in detail in the previous chapter. The development of these models for use with complex images was, in part, enabled by visual experiments performed to test the performance of published color appearance models in realistic image reproduction situations.[17] Such research on color appearance modeling in imaging applications naturally highlighted the areas that are not adequately addressed for spatially complex image appearance and image quality problems.

As with color difference equations, the heart of a color appearance model is a perceptually uniform color space. In addition to traditional colorimetry, color appearance models strive to account for additional changes in viewing conditions, but are mainly focused on changes in the color of the illumination through chromatic adaptation transforms, as well as changes in the overall illumination level, and the background and surround relative luminances. Such models do not attempt to directly incorporate any of the spatial properties of human vision and the perception of images. They essentially treat each pixel of an image as completely independent stimuli.

Visual adaptation to scenes and images is not just a result of the overall color and luminance of the light source. Indeed, visual adaptation is in a constant state of flux, essentially determined by observers' eye movements and fixations in a scene. The overall behavior tends toward a combination of overall lighting as well as some spatially localized adaptation according to some low-pass characteristics of the scene itself. Likewise, perception of overall contrast is not just a function of a single surround and background condition but also a function of the scene itself. This can be readily imagined by the concept of simultaneous contrast, whereas two identical color stimuli can appear quite different depending on their location in a scene, or likewise two differing stimuli can be made to look identical. An example of this behavior is shown in Figure 12.1.

Although color appearance modeling has been successful in facilitating device-independent color imaging across a variety of viewing conditions and is incorporated into many modern color management systems, there remains significant room for improvement and extension of capabilities. To address these issues with respect

FIGURE 12.1 An example of real-world simultaneous contrast. The colors in the circles are identical, but appear very different. See color insert.

to spatial properties of vision and image perception and image quality, the concept of image appearance models has been recently introduced and a framework implemented.[1–3] These image appearance models evolved to combine attributes of color appearance models with attributes of spatial vision models, along with attributes of color difference equations. The spatial vision models have been historically used for image quality metrics and have traditionally been focused on black-and-white "quality." As described above and in the previous chapter, color appearance models have largely ignored spatial vision (e.g., CIECAM97s, CIECAM02) while spatial vision models for image quality largely ignored color. Example of these type of models are the Daly visual differences predictor (VDP) and Lubin's Sarnoff model.[18,19] Some exceptions that contain both color and spatial properties of stimuli include the retinex model and its various derivatives as described by Land and McCann.[20–25] Other examples include the spatial ATD model and the S-CIELAB model as described above.[26] Another unique approach to extending color appearance with spatial vision models was the CVDM metric, which attempted to meld S-CIELAB with Daly's VDP.[27] The retinex model was never designed as a complete model of image appearance and quality, though its spatially variable mechanisms of chromatic adaptation and color constancy serve some of the same purposes in image rendering and certainly provide some of the critical groundwork for current image appearance modeling.

The goal in developing an image appearance model has been to bring these research areas together to create a single model applicable to image appearance, image rendering, and image quality specifications and evaluations. One framework for such model for still images, referred to as iCAM, is detailed in this chapter. In addition to traditional CIE colorimetry and color appearance modeling, this framework was built upon a wide variety of previous research in the perception of images. This includes uniform color spaces, the importance of image surround, algorithms for image difference and image quality measurement, insights into observers eye movements while performing various visual imaging tasks, and adaptation to natural scenes, and earlier model of spatial and color vision applied to color appearance problems and HDR imaging.

IMAGE APPEARANCE MODELING

Just as a color appearance model is necessary to extend CIE colorimetry to fully describe the appearance of color stimuli across a variety of viewing conditions, an image appearance model is necessary to describe spatially complex color stimuli. In this situation, we use the term "image" to describe any complex scene that has been discreetly sampled or generated by some digital imaging device. Color appearance models, in essence, are tools that allow for the description of attributes such as lightness, brightness, colorfulness, chroma, and hue based upon physical measurements of the stimuli and viewing conditions. Image appearance models extend upon these traditional color descriptors to also predict such perceptual attributes as sharpness, graininess, and contrast. In addition to predicting these general attributes, image appearance models should inherently be able to predict the overall perceived color difference between complex image stimuli.

At the heart of all CIE methods for describing color matches, difference, and appearance lies a uniform color space. A uniform color space also lies in the heart of an image appearance model. The modular image difference framework of Johnson and Fairchild described briefly above allows for great flexibility in the choice of color spaces.[13,14] Such color spaces can include the CIELAB space, in which case the predictions are similar to S-CIELAB, as well as a more complex space such as that from the CIECAM02 color appearance model, or a computationally easier space such as Ebner and Fairchild's IPT color space. By providing a flexible foundation, the modular image difference framework can be extended to create a full image appearance and image difference model. This extension has been coined iCAM, and two such implementations are described in this chapter. It is important to stress that the modular nature of iCAM could also allow a similar image appearance model to be implemented in another uniform color space, as need dictates.

Theoretically, models of image appearance can be used to formulate multidimensional models of image quality. Historically many image quality models have been described as weighted sums of individual appearance attributes such as noisiness, graininess, and sharpness to determine a metric of overall image quality. These techniques have been well described by Keelen, as well as Engledrum's "Image Quality Circle."[29,30] Typically, psychophysical experiments are performed to scale each individual image appearance attribute. The goal of an image appearance model would then be to augment, or eventually replace these experiments in order to generate scales of individual attributes. For instance, a traditional model of image quality might involve weighted sums of tonal balance, contrast, sharpness, and graininess. These attributes, as well as others, may be generated through human observations or through an image appearance model.

THE GENERAL ICAM FRAMEWORK FOR IMAGE APPEARANCE

A flowchart of the framework for using the iCAM model for predicting the overall appearance of still images is shown in Figure 12.2. In a typical use the model requires as input colorimetric data for the image and surround in absolute luminance units. Relative colorimetry can be used, with an absolute luminance scaling or approximation required. Images are typically specified in terms of relative CIE *XYZ* tristimulus values, or device-independent *RGB* such as sRGB. The adapting stimulus is then calculated to be a low-pass filtered version of the CIE *XYZ* image that should also tagged with absolute luminance information necessary to predict the degree of chromatic adaptation. A second low-passed version of the absolute luminances (CIE *Y* value) of the image data is used to control various luminance-dependant aspects of the model intended to predict the Hunt effect (increase in perceived colorfulness with luminance) and the Stevens effect (increase in perceived image contrast with luminance). This version can be identical to the low-passed image used for chromatic adaptation, but is typically an image that is

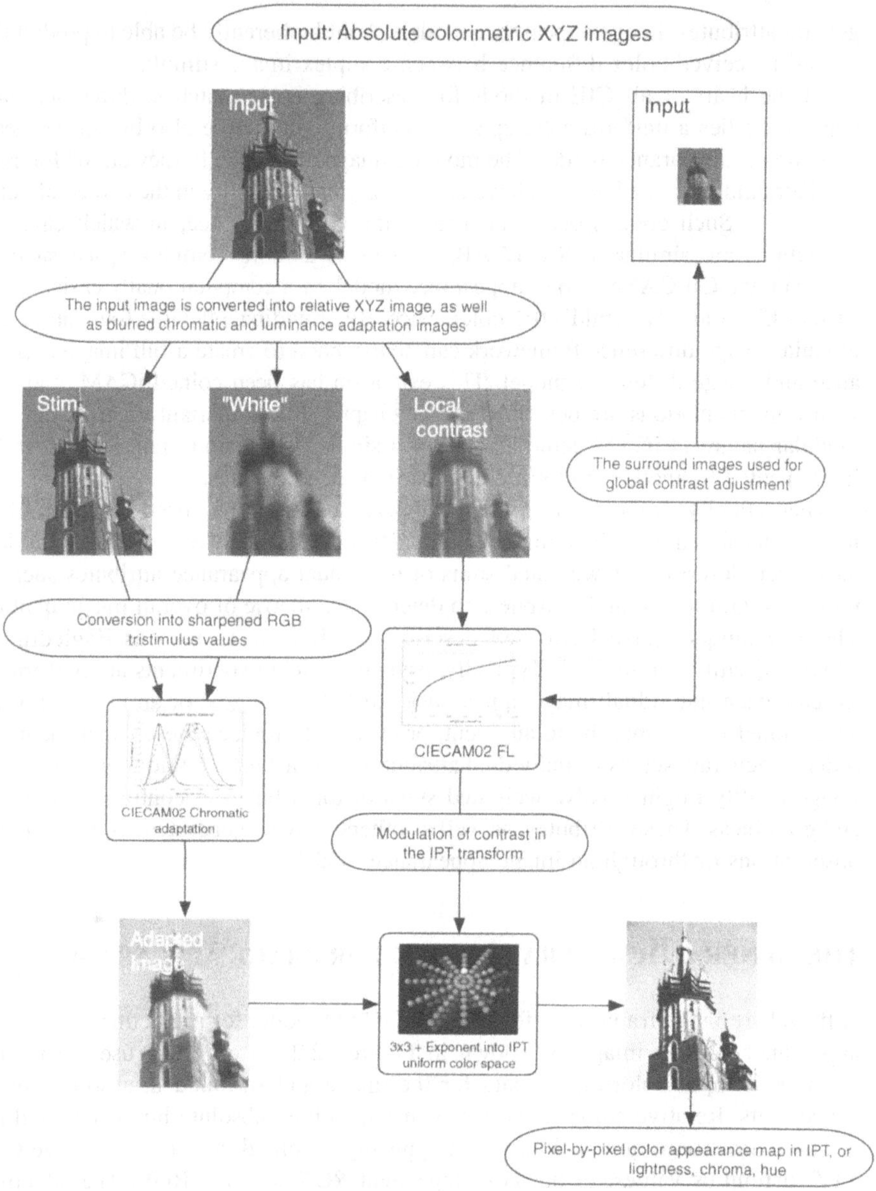

FIGURE 12.2 Flowchart for using the iCAM framework to predict image appearance. See color insert.

less blurred. Ideally another low-passed luminance (CIE Y value) image of significantly greater spatial extent is used to control the prediction of global image contrast. In practice, this image is generally taken to be a single number indicating the overall luminance of the surrounding viewing conditions. In essence, this can be considered a global contrast exponent (or completely low-passed image) that follows the well-established behavior of CIECAM02 for predicting the Bartleson and Breneman equations.[31] The specific low-pass filters used for adapting images depend on viewing distance and application. A typical example might be to specify a chromatic adaptation image as a Gaussian blur that spans 20 cycles per degree of visual angle (cpd), while the local surround may be a Gaussian blur that span 10 cpd.[32] Additionally, in some image-rendering circumstances, such as for HDR images it might be desirable to have different low-pass adapting images for luminance and chromatic information to avoid an overall desaturation of the rendered images due to local chromatic adaptation. Recent research in HDR rendering has shown that the local contrast adaptation blurring may be better served with an edge-preserving low-pass function such as the bilateral filter described by Durand and Dorsey.[33–35] Research in HDR rendering has shown to be one example of application dependence in image appearance modeling. A strong local chromatic adaptation might be generally appropriate for predicting actual perceived image differences or image quality measurements, but inappropriate for image-rendering situations.

The first stage of processing in iCAM, after transforming any input images into CIE *XYZ* tristimulus values (typically using CIE 1931 color-matching functions) is to account for chromatic adaptation. This is identical to the first stage of general color appearance modeling. The chromatic adaptation transform embedded in CIECAM02 has been adopted in iCAM because it was well researched and established to have excellent performance with all available visual data. This transform is a relatively simple and easily invertible, linear chromatic adaptation model amenable to image-processing applications. More details can be found in the previous chapters, though the full equations are given here for ease of reference. The general CIECAM02 chromatic adaptation model, given in Equations 12.1 through 12.6, is a linear "von Kries" normalization of sharpened RGB image signals from a sharpened RGB adaptation "white point." In traditional color appearance modeling, this white point taken to be a single value is typically taken to be the brightest signal in the image, or the *XYZ* tristimulus values of the scene measured off a perfect reflecting diffuser.

It is here that image appearance deviates from traditional color appearance models. The "white point" of the adapting signal is taken to be the low-passed adaptation image at each pixel location ($R_W G_W B_W$). These white point signals can also be modulated by the overall white of the scene, if that is known. The sharpened *RGB* signals are computed using a linear transformation from *XYZ* to *RGB* derived by CIE TC8-01 in the formulation of CIECAM02. The von Kries normalization is further modulated with a degree-of-adaptation factor, D, which can vary from 0.0 for no adaptation to 1.0 for complete chromatic adaptation. Equation 12.3 is provided in the CIECAM02 formulation and can also be used in iCAM for computation of D as a function of adapting luminance, L_A, for various viewing conditions. Alternatively, like in CIECAM02, the D factor can be established manually. The chromatic adaptation model is used to compute pixelwise corresponding colors for CIE Illuminant D65 that are then used in the later stages of the iCAM model. This is accomplished by taking the adapted signals for the viewing condition, $R_C\ G_C\ B_C$, and then inverting Equations (12.1) through (12.6) for a single (nonspatially localized) illuminant D65 adapting white point and with complete chromatic adaptation ($D = 1.0$). It should be noted that the chromatic although adaptation transformation is identical to that in CIECAM02, the iCAM model is already significantly different because it uses the blurred image data itself to spatially modulate the adaptation white point.

$$\begin{bmatrix} R \\ G \\ B \end{bmatrix} = \mathbf{M}_{\mathrm{CAT02}} \begin{bmatrix} X \\ Y \\ Z \end{bmatrix} \tag{12.1}$$

$$\mathbf{M}_{\mathrm{CAT02}} = \begin{bmatrix} 0.7328 & 0.4296 & -0.1624 \\ -0.7036 & 1.6975 & 0.0061 \\ 0.0030 & 0.0136 & 0.9834 \end{bmatrix} \tag{12.2}$$

$$D = F\left[1 - \left(\frac{1}{3.6}\right)e^{\left(\frac{-L_A - 42}{92}\right)}\right] \tag{12.3}$$

$$R_C = \left[\left(100\frac{D}{R_W}\right) + (1 - D)\right]R \tag{12.4}$$

$$G_C = \left[\left(100\frac{D}{G_W}\right) + (1 - D)\right]G \tag{12.5}$$

$$B_C = \left[\left(100\frac{D}{B_W}\right) + (1 - D)\right]B \tag{12.6}$$

The use of the blurred *XYZ* image as a spatially modulated adapting white point implies that the content of an image itself, as well as the overall illumination, controls our state of chromatic adaptation. In this manner, iCAM behaves similar in regard to color constancy as the spatial modulations of the Retinex approach to color vision.[20] This behavior can result in a decrease in overall colorfulness or chroma, especially for large uniform areas such as the blue sky shown in Figure 12.2. Although this may be the correct prediction for the overall image appearance, it may produce undesirable results when using an image appearance model for image-rendering applications, where pleasing colors are the desired outcome.

Another example of the localized spatial behavior inherent in an image appearance model is the modulation of local and global contrast using the absolute luminance image and surround luminance image. This is accomplished by borrowing the F_L function from CIECAM02, as given in Equation (12.7). This

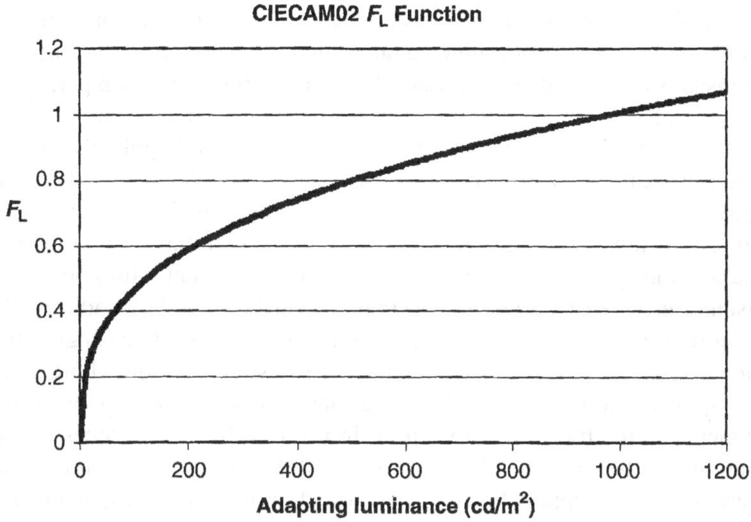

FIGURE 12.3 The luminance level adaptation function, F_L, from CIECAM02, and used in the iCAM framework.

function, shown in Figure 12.3, slowly varies with luminance and has been established to predict a variety of luminance-dependent appearance effects in CIECAM02 and earlier models. Because the function has been established and well understood, it was also adopted for the early stages of iCAM. However, the global manner in which the F_L factor is used in CIECAM02 and the spatially localized manner used in iCAM is quite different.

$$F_L = 0.2\left(\frac{1}{(5L_A + 1)}\right)^4 (5L_A) + 0.1\left(1 - \left(\frac{1}{(5L_A + 1)}\right)^4\right)^2 (5L_A)^{\frac{1}{3}} \qquad (12.7)$$

As mentioned above, the two most important elements of a color appearance model are the chromatic adaptation transform and the uniform color space. Having completed the spatially variant chromatic adaptation, the next stage of the iCAM framework is the conversion into a uniform opponent color space. The chromatic adaptation was performed on the sharpened RGB signals, which are roughly analogous to cone signals in the human visual system. The *RGB* signals, which have been converted to corresponding colors for CIE Illuminant D65 are transformed into *LMS* cone responses using Equation (12.8).

$$\begin{bmatrix} L \\ M \\ S \end{bmatrix} = \begin{bmatrix} 0.4002 & 0.7075 & -0.0807 \\ -0.2280 & 1.1500 & 0.0612 \\ 0.0 & 0.0 & 0.9184 \end{bmatrix} \cdot \begin{bmatrix} X_{D65} \\ Y_{D65} \\ Z_{D65} \end{bmatrix} \qquad (12.8)$$

These *LMS* cone signals are then converted into opponent color signals (light–dark, red–green, and yellow–blue; analogous to higher level encoding in the human visual system) that are necessary for constructing a uniform perceptual color space and correlates of various appearance attributes. In choosing this transformation for the iCAM framework, simplicity, accuracy, and applicability to image processing were the main considerations. The uniform color space chosen was the *IPT* space previously published by Ebner and Fairchild.[36] The *IPT* space was derived specifically for image-processing applications to have a relatively simple formulation and specifically to have a hue-angle component with good prediction of constant perceived hue. Predicting lines of constant hue has traditionally been very important in gamut-mapping applications and will be increasingly important with any gamut-expansion algorithms that are desired for new HDR and wide-gamut displays. The mathematical transformation into the *IPT* opponent space is far simpler than the transformations used in CIECAM02. The process, expressed in Equations (12.9) through (12.12), involves a linear transformation to a different cone-response space, application of power-function nonlinearities, and then a final linear transformation to the *IPT* opponent space (*I*: light–dark; *P*: red–green, *T*: yellow–blue). Although it seems counter-intuitive to have negative *LMS* cone responses because of the chromatic adaptation on sharpened *RGB* values, as well as the linear transformation from CIE *XYZ*, it is possible to have negative *LMS* values in the *IPT* transform. For this reason, it is necessary to use the absolute values when applying the exponent to the *LMS* cone signals, as shown in Equations (12.9) through (12.11).

$$\begin{aligned} L' &= L^{0.43}; \quad L \geq 0 \\ L' &= -|L|^{0.43}; \quad L \leq 0 \end{aligned} \tag{12.9}$$

$$\begin{aligned} M' &= M^{0.43}; \quad M \geq 0 \\ M' &= -|M|^{0.43}; \quad M \leq 0 \end{aligned} \tag{12.10}$$

$$\begin{aligned} S' &= S^{0.43}; \quad S \geq 0 \\ S' &= -|S|^{0.43}; \quad S \leq 0 \end{aligned} \tag{12.11}$$

$$\begin{bmatrix} I \\ P \\ T \end{bmatrix} = \begin{bmatrix} 0.4000 & 0.4000 & 0.2000 \\ 4.4550 & -4.8510 & 0.3960 \\ 0.8056 & 0.3572 & -1.1628 \end{bmatrix} \cdot \begin{bmatrix} L' \\ M' \\ S' \end{bmatrix} \tag{12.12}$$

The power-function nonlinearities in the *IPT* transformation are a critical aspect of the iCAM model. First, they are necessary to predict response compression that is prevalent in most human sensory systems. This response compression helps to convert from signals that are linear in physical metrics (e.g., luminance) to signals that are linear in perceptual dimensions (e.g., lightness). The CIECAM02 model uses a hyperbolic nonlinearity, whereas the CIELAB model uses a cube root for this purpose. For most luminance levels encountered in practice the compression of the visual system can be well represented by a power function. A key component of iCAM is for these exponents to be spatially modulated according to the overall luminance of the image itself (low-pass filtered) as well as that of the surround. In practice, this is accomplished by multiplying the base exponent (0.43) in the *IPT* formulation by the computed F_L factors with appropriate normalization. The F_L factors are calculated on a pixel-by-pixel basis from the blurred luminance image, resulting in a unique "gamma" curve for every pixel. This allows for a local-contrast adjustment based upon the information contained in the image itself. These modulations of the *IPT* exponents allow the iCAM model to be used for predictions of many luminance-based global color appearance phenomena such as the Hunt and Stevens effects, as well as global contrast behavior as described by Bartleson and Breneman.[31] The spatially modulated F_L exponents, as expressed in Equation (12.13), also happen to enable the tone mapping of HDR images into low-dynamic range display systems in a visually meaningful way. In essence they dynamically adapt to compress the large luminance range into the lower dynamic range of the visual system.

$$\begin{aligned} L' &= L^{0.43 \cdot F_L}; \quad L \geq 0 \\ L' &= -|L|^{0.43 \cdot F_L}; \quad L \leq 0 \end{aligned} \tag{12.13}$$

Once the *IPT* coordinates are computed for the image data, a simple coordinate transformation from rectangular to cylindrical coordinates is applied to obtain imagewise predictors of the traditional color appearance attributes such as lightness (*J*), chroma (*C*), and hue angle (*h*) as shown in Equations (12.14) through (12.16). Differences in these dimensions can be used to compute image difference statistics and those used to derive image quality metrics, as discussed in more detail below. In some instances, correlates of the absolute appearance attributes of brightness (*Q*) and colorfulness (*M*) are required. These are obtained by scaling the relative attributes of lightness and chroma with the appropriate function of F_L again derived from the imagewise luminance map as shown in Equations (12.17) and (12.18).

$$J = I \tag{12.14}$$

$$C = \sqrt{P^2 + T^2} \tag{12.15}$$

$$h = \tan^{-1}\left(\frac{P}{T}\right) \tag{12.16}$$

$$Q = \sqrt[4]{F_L} J \tag{12.17}$$

$$M = \sqrt[4]{F_L} C \tag{12.18}$$

The general iCAM model as described above has been successfully applied to prediction of a variety of color appearance phenomena such as chromatic adaptation (corresponding colors), color appearance scales, constant hue perceptions, simultaneous contrast, crispening, spreading, and image rendering.[1]

Because iCAM uses the same chromatic adaptation transform as CIECAM02, it performs identically for situations in which only a change in state of chromatic adaptation is present (i.e., change in white point only). CIE TC 8-01 has worked very hard to arrive at this adaptation transform, and it is clear that no other model currently exists with better performance (although there are several with equivalent performance). Thus the chromatic adaptation performance of iCAM is as good as possible at this juncture.

The appearance scales of iCAM are identical to the *IPT* scales for the reference viewing conditions. The *IPT* space has the best available performance for constant hue contours and thus this feature is retained in iCAM.[37] This feature makes accurate implementation of gamut-mapping algorithms far easier in iCAM than in other appearance spaces. In addition, the predictions of lightness and chroma in iCAM are very good and comparable with the best color appearance models in typical viewing conditions. The brightness and colorfulness scales will also perform as well as any other model for typical conditions. In more extreme viewing conditions, such as for HDR images, the performance of iCAM and other models will begin to deviate. It is in these conditions that the potential strengths of iCAM will become evident and is discussed in more detail below.

Specific Implementations of Image Appearance Models: High-Dynamic Range Tone-Mapping

The general framework for applying iCAM to predict image-wide appearance attributes was described in the section above. This can be considered equivalent to the forward model predictions of traditional color appearance models, such as CIECAM02. The output is an image appearance map where each pixel is now represented with color appearance attributes such as lightness, chroma, and hue. Although these image-appearance maps are very useful and can provide insight into the overall color perception of complex image stimuli, often we are interested in using image appearance to generate "appearance matches" across widely disparate viewing conditions. One such application is for tone mapping, or rendering HDR images.

In everyday life we encounter a huge range of absolute luminance levels, most of which the visual system handles with ease. Perhaps more impressive is the visual systems, ability to instantaneously and seamlessly adapt to scenes with a large dynamic range, scenes that can exceed 10,000 to 1 between sunlight and shadows. Typical examples of the range of luminances can be found in Figure 12.4.

Recent advances in color imaging have lead to systems that are capable of capturing these HDR scenes. These systems can be based upon multiple photographic exposures, as described by Debevec[38] and Xiao et al.[39] or sensor improvements that make it capable to capture HDR information with a single exposure.[40,41,42] Likewise, these systems might be high-contrast computer graphics renderings as

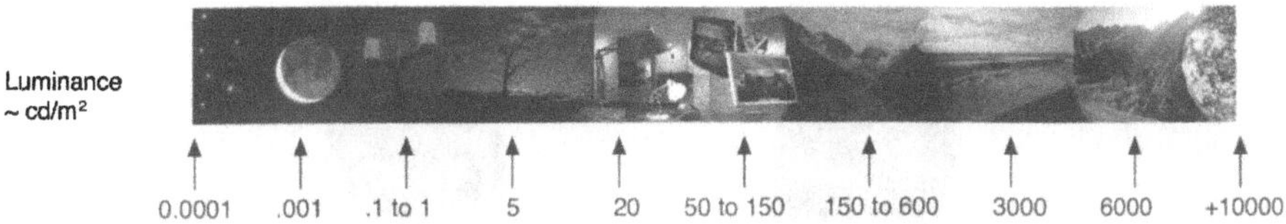

FIGURE 12.4 Typical luminance ranges we encounter in everyday life. See color insert.

described by Ward et al.[43,44] An excellent overview of HDR imaging and rendering can be found in Reinhard et al.[45]

Although the systems for capturing high dynamic range images have improved over the years, the systems for displaying these images have not kept up. A typical desktop display is capable of displaying only one or two orders of magnitude of dynamic range. Although this is slowly changing, with the introduction of HDR displays and wide-gamut displays, currently most desktop displays and all printers are still only capable of reproducing a limited dynamic range.

Image appearance models attempt to predict the perceptual response toward spatially complex stimuli. As such, they can provide a unique framework for the prediction of the appearance of HDR images. Because the encoding in our visual system is of a rather low-dynamic-range, this is essentially a replication of the image appearance processing that goes on in the human observer and is being modeled by iCAM. It is important to stress that these image appearance models are not designed specifically as tone-mapping algorithms but rather as predictors of overall color appearance. However, the general iCAM framework does not need to be changed to be useful for the rendering of HDR scenes. Several of the parameters of the iCAM framework, as discussed above, can be specifically tuned for this application. It is also necessary to "invert" the model, to get an appearance image that can be displayed on another device. The flowchart for using iCAM for HDR rendering is shown in Figure 12.5.

Once again, the input is an absolute *XYZ* image, though for this application the image happens to be of HDR. For image-rendering application, it is very important to stress that the overall luminance range will have a large impact on the output appearance. This is because both the chromatic adaptation and the local contrast, as controlled by the CIECAM02 F_L function relies on absolute luminance. Often this information is not actually known, and for HDR-rendering applications an approximate "scaling" factor can be calculated. This calculation as described in Equation (12.19) has no physical meaning and should be avoided if the actual luminance values are known. In essence, it is used to find the "key" of the image, based upon image statistics of the 0.99 quantile or percentile (where the median of the image is the 0.50 quantile), and scale the relative dynamic range based upon that.

$$\text{key} = \text{Quantile}\left\{\frac{\text{image}}{\max(\text{image})}, 0.99\right\}$$
$$\text{scale} = \frac{50}{\text{key}} \tag{12.19}$$

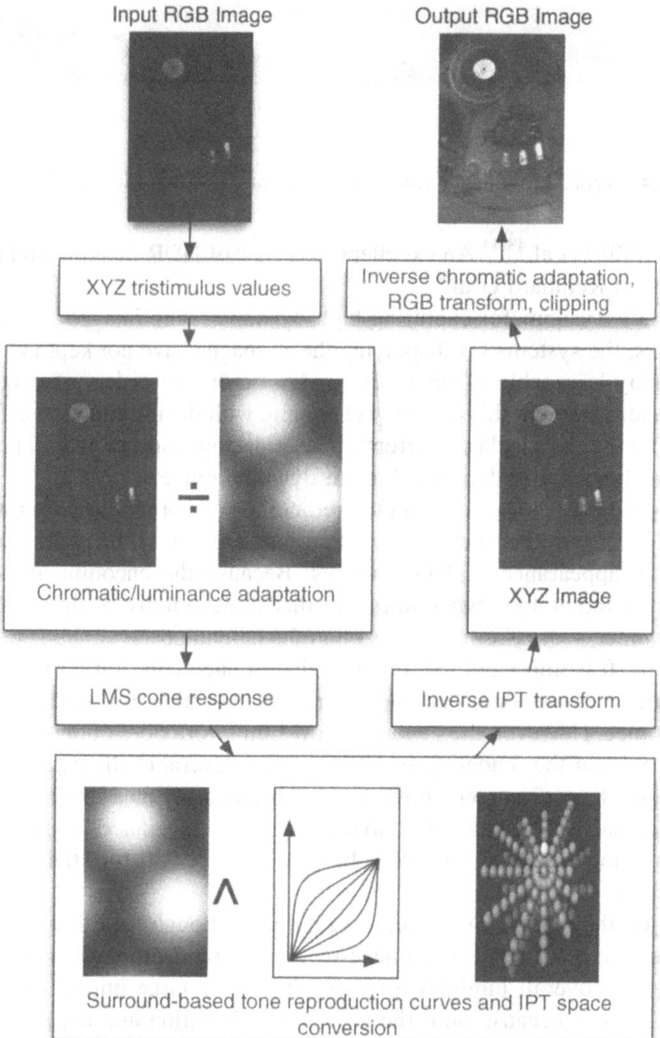

FIGURE 12.5 Flowchart for using iCAM as a predictor of HDR images. See color insert.

The scaled *XYZ* data are then processed with the same chromatic adaptation transformed as described in the general iCAM framework above. In this case, the choice of low-pass function as well as degree of adaptation, *D*, has a strong influence on the rendered image. An example of this is shown in Figure 12.6. Typically the extent of the Gaussian blur is taken to be a function of image size itself, rather than viewing distance. This is because for most HDR applications, the ultimate viewing conditions are unknown. The chromatic adaptation transform itself will result in a decrease in chroma of the rendered image, which may be appropriate

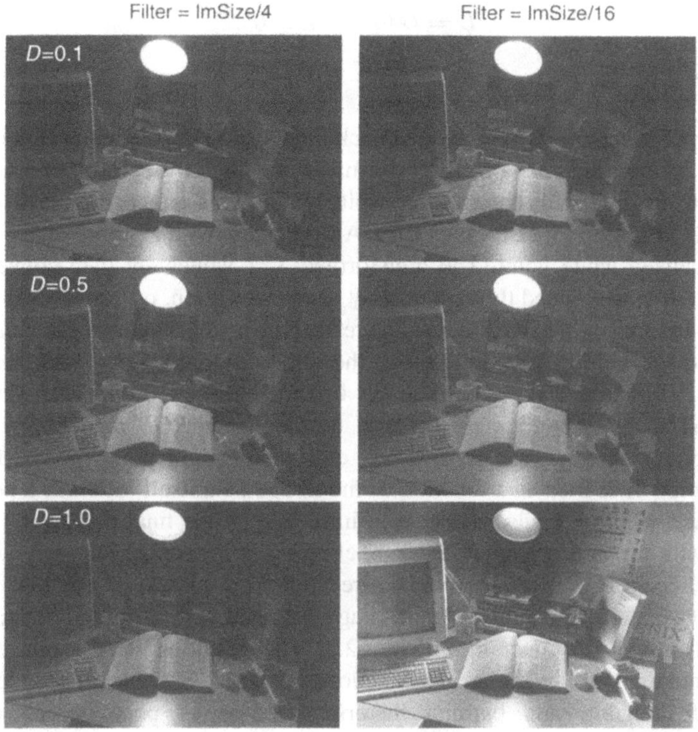

FIGURE 12.6 Influence of Gaussian blur and degree of adaptation on rendered HDR images. See color insert.

when predicting actual appearance. This decrease in chromatic content may not be desirable for rendering HDR images, and so it may be necessary to use only luminance adaptation. This is an application specific example where changing the iCAM framework may be beneficial.

Following the chromatic, or luminance adaptation, the adapted image is once again transformed into *LMS* cone signals. These signals are transformed into the *IPT* color space using the spatially modulated F_L function from Equations (12.7) and (12.13), which is calculated using the low-passed luminance channel. Again, the choice of low-pass function has a large influence on the rendered output. Whereas the original iCAM framework discussed above suggests using a Gaussian filter of approximately 10 degrees of visual angle, this choice may not be appropriate for rendering HDR images. Recent research has suggested that using an edge preserving low-pass function, such as the bilateral filter may be beneficial.[34,35]

The IPT color appearance attributes are then inverted for display. This is accomplished using the standard inverse *IPT* transform, assuming a single uniform contrast condition (not spatially modulated). This is accomplished using Equation (12.20) with similar equations for the *M* and *S* cones.

$$L' = L^{\frac{1}{0.43}}; \qquad L \geq 0$$
$$L' = -|L|^{\frac{1}{0.43}}; \qquad L \leq 0 \tag{12.20}$$

The *LMS* cones are then converted back into CIE *XYZ* tristimulus values. To display the *XYZ* images on a monitor we must first invert the chromatic adaptation transform, from D65 to the monitor white point. This is accomplished using the chromatic adaptation transform of CIECAM02, again with a global rather than spatially localized white point. The transformed *XYZ* values are then converted into device *RGB* values found through display characterization, or through using a standard device such as sRGB. The results are still linear *RGB* values. The final images can be displayed by accounting for the display nonlinearity with an inverse "gamma" function and by scaling the images between 0 and 255. Often it is beneficial to apply a clipping function to the linear *RGB* data before scaling. This clipping function can remove any extremely bright pixels prior to display. The clipping is defined as a function of the image data itself, often as a percentile. The choice of clipping also has a large influence on the final displayed image, as shown with the three different levels of clipping in Figure 12.7.

The clipped *RGB* is then gamma-corrected and displayed on the low-dynamic-range device. An example of an HDR image rendered with a "global" gamma function and iCAM is shown in Figure 12.8. The resulting iCAM image can be considered quite acceptable as reproductions of the HDR scenes (equivalent to the result of dodging* and burning historically done in photographic printing). Discussions on testing HDR-rendering techniques are described in the section below.

Testing High-Dynamic Range Rendering Algorithms

As described in great detail in the previous chapters of this book, the CIE is dedicated to providing discussion, information, and guidance in the science and art of light and lighting. The terms of reference of Division 8 of the CIE is "to study procedures and prepare guides and standards for the optical, visual and metrological aspects of the communication, processing, and reproduction of images, using all types of analogue and digital imaging devices, storage media and imaging media."

Along those lines, CIE Technical Committee (TC) 8-08 is tasked with developing guidelines and testing methods for using spatial or image appearance models, specifically for use with HDR images. The goal of TC 8-08 is not to create a CIE recommended image appearance model, but rather to design and conduct experimental techniques for evaluating these models. Details on the goals of CIE TC 8-08 can be found in Johnson.[46]

Several experiments have already been performed to test the performance of many existing tone-mapping operators, including iCAM. Details of scaling preference and accuracy of rendered images can be found in Kuang et al. as well as Ledda.[34,35,47,48] These experiments have shown that image appearance models

*Manipulation of the light projected through a negative by an enlarger to lighten or darken selected part of the resultant print (*Chambers Dictionary of Science and Technology*).

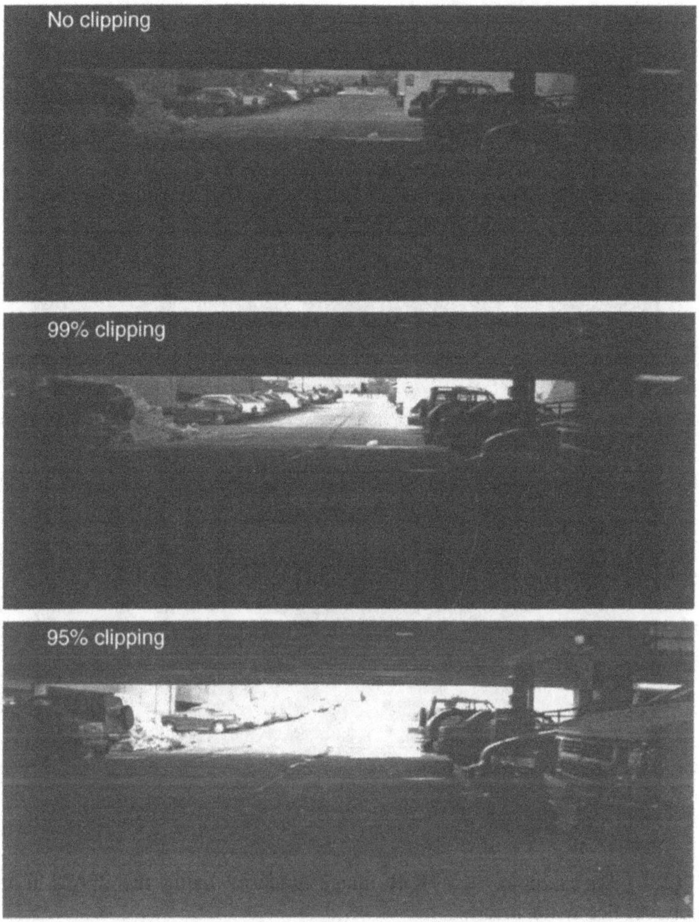

FIGURE 12.7 The choice of clipping the *RGB* image prior to display has a large influence on the final appearance of the rendered image. See color insert.

perform fairly well for producing pleasing images, though specialized HDR tone-mapping algorithms such as the bilateral filter technique of Durand and Dorsey[33] performs significantly better.[34,35] When testing for accuracy of appearance, the color appearance heritage of iCAM results in accurate appearance predictions.[47,48]

These experiments indicate that a generic image appearance model can be used as a form of automatic HDR rendering. By comparing a wide variety of algorithms, these experiments have also indicated the potential benefit of a more application specific algorithm. An example of this may be replacing the Gaussian low-pass filter used in iCAM with an edge preserving low-pass filter, such as the bilateral filter. The flexible nature of the iCAM framework should easily allow for this type of enhancements.

FIGURE 12.8 An example of a HDR image-rendered using the iCAM framework as described above. See color insert.

AN IMPLEMENTATION OF IMAGE APPEARANCE FOR CALCULATING IMAGE DIFFERENCES

One of the goals in creating an image appearance model, such as iCAM, is to combine the historical research in color appearance as well as color difference metrics and spatial vision. The general iCAM framework presented in Figure 12.2 is used to predict overall appearance attributes for an image, but does not take into account some of the spatial vision properties necessary for calculating color image differences. An extension to the general iCAM framework, inspired by the S-CIELAB spatial extension to the CIELAB color space has been developed. This was adapted from the modular color image difference metric described by Johnson and Fairchild.[13,14]

The behavior of the human visual system in regard to spatially complex stimuli has been well studied over the years dating back to the seminal work of Campbell

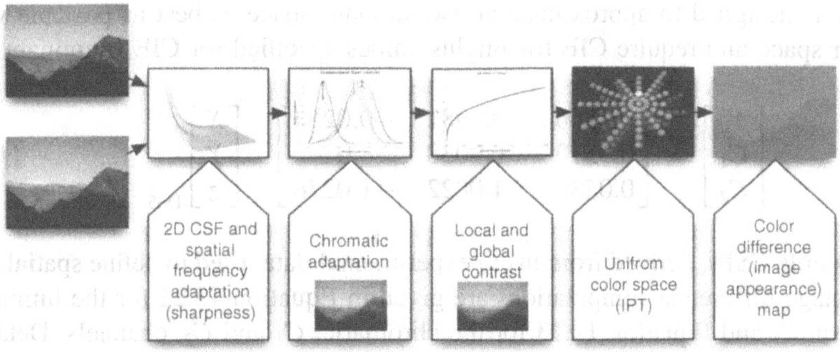

FIGURE 12.9 Using iCAM as an image difference metric. See color insert.

and Robson[49] and Mullen[50]. The CSF describes this behavior in relation to spatial frequency. Essentially the CSF is described in a postretinal opponent color space, generally with a bandpass nature for the luminance channel and low-pass nature for the chrominance channels. S-CIELAB uses separable convolution kernels to approximate the CSF for use in image processing, and these kernels serve to eliminate details that are imperceptible.[12] Along similar lines, image processing with CSFs in the frequency domain, that include both modulation and frequency enhancement, were discussed in detail by Johnson and Fairchild.[13] Other models with similar features include the previously mentioned Lubin's Sarnoff, Daly's VDP, and the spatial ATD model.[18,19,26] Another interesting approach, and probably the first general image appearance model is the Multi-scale Observer Model, as described by Pattanaik et al.[51]

For image difference and image quality predictions within the iCAM framework, it is also necessary to apply spatial filtering to the image data to eliminate any image variations at spatial frequencies too high to be perceived. This is performed, as in S-CIELAB, as a preprocessing stage. The flowchart for using iCAM for calculating image differences is shown in Figure 12.9.

Again, the spatial preprocessing serves to eliminate information that is imperceptible to the visual system and normalize color differences at spatial frequencies that are visible. For example, the dots in a printed halftone image are not visible if the viewing distance is sufficiently large, and so the spatial filtering would blur the dots into a continuous representation. This computation is highly dependent on viewing distance and is based on filters derived from human CSFs. Because the human CSFs vary for luminance (bandpass with sensitivity to high frequencies) and chromatic (low-pass) information, it is necessary to apply these filters in an opponent color space. The choice of opponent color space is crucial for this step as it is necessary that the luminance and chromatic channels be mathematically orthogonal. Problems such as chromatic fringing can arise when this is not the case. As such, the iCAM framework performs spatial filtering in a specially designed orthogonal color space called $Y'C_1C_2$. More details on the development of this space can be found in Song et al. and Johnson.[15,16] Equation (12.21) shows the linear transform from CIE XYZ into $Y'C_1C_2$. It is important to note that this space is designed to approximate an isoluminant space as best as possible with a linear space and require CIE tristimulus values specified for CIE Illuminant D65.

$$\begin{bmatrix} Y' \\ C_1 \\ C_2 \end{bmatrix} = \begin{bmatrix} 0.0556 & 0.9981 & -0.0254 \\ 0.9510 & -0.9038 & 0 \\ 0.0386 & 1.0822 & -1.0276 \end{bmatrix} \cdot \begin{bmatrix} X \\ Y \\ z \end{bmatrix}_{D65} \tag{12.21}$$

Sample CSFs, derived from fits to experimental data, used to define spatial filters for image difference computations are given in Equation 12.22 for the luminance, Y', channel and Equation 12.23 for the chromatic, C_1 and C_2, channels. Details of the general formulation of these equations can be found in Johnson and Fairchild.[14] The specific details of optimizing these spatial filters for use with the $Y'C_1C_2$ color space can be found in Song.[52]

$$\mathrm{csf}_{\mathrm{lum}}(f) = a \cdot f^{c} \cdot e^{-b \cdot f} \tag{12.22}$$

$$\mathrm{csf}_{\mathrm{chrom}}(f) = a_1 \cdot e^{-b_1 \cdot f^{c_1}} + a_2 \cdot e^{-b_2 \cdot f^{c_2}} \tag{12.23}$$

The parameters, a, b, and c, in Equation (12.22) are set to 0.63, 0.085, and 0.616, respectively for the luminance CSF, as applied to the Y' channel. In Equations (12.22) and (12.23), spatial frequency, f, is defined in terms of cpd. To apply these functions as image-processing filters, f is described as a two-dimensional map of spatial frequencies of identical size to the image itself. For the red–green chromatic CSF, applied to the C_1 dimension, the parameters (a_1, b_1, c_1, a_2, b_2, c_2) in Equation (12.23) are set to (91.228, 0.0003, 2.803, 74.907, 0.0038, 2.601). For the blue yellow chromatic CSF, applied to the C_2 dimension, the parameters are set to (5.623, 0.00001, 3.4066, 41.9363, 0.083, 1.3684). Figure 12.10 shows the one-dimensional projection of the CSFs.

The bandpass nature of the luminance CSF, as well as the low-pass nature of the two chromatic channels can be seen in Figure 12.10. Two other important features can be seen in regards to the luminance CSF: its behavior at 0 cpd (the DC component) and that the response goes above 1.0. Special care must be taken with regards to the DC component when performing spatial filtering in the frequency domain. The DC component contains the mean value of the image, for that particular channel. Since we would like the image difference metric to collapse down into a traditional color difference metric for solid patches, it is important that the mean value does not change. The luminance spatial filter described by Equation (12.22), and shown in Figure 12.10 goes to zero at the DC component. Therefore it is necessary to first subtract the mean value of the luminance channel, apply the spatial filter, and then add the mean value back to the image. The entire filtering process, for the luminance channel, is shown in Equation (12.24).

$$\mathrm{Image}_{\mathrm{filt}} = \mathrm{FFT}^{-1}\{(\mathrm{FFT}\{\mathrm{Image} - \mathrm{mean}\ (\mathrm{Image})\}) \cdot \mathrm{Lum\ CSF}\} + \mathrm{mean}\ (\mathrm{Image}) \tag{12.24}$$

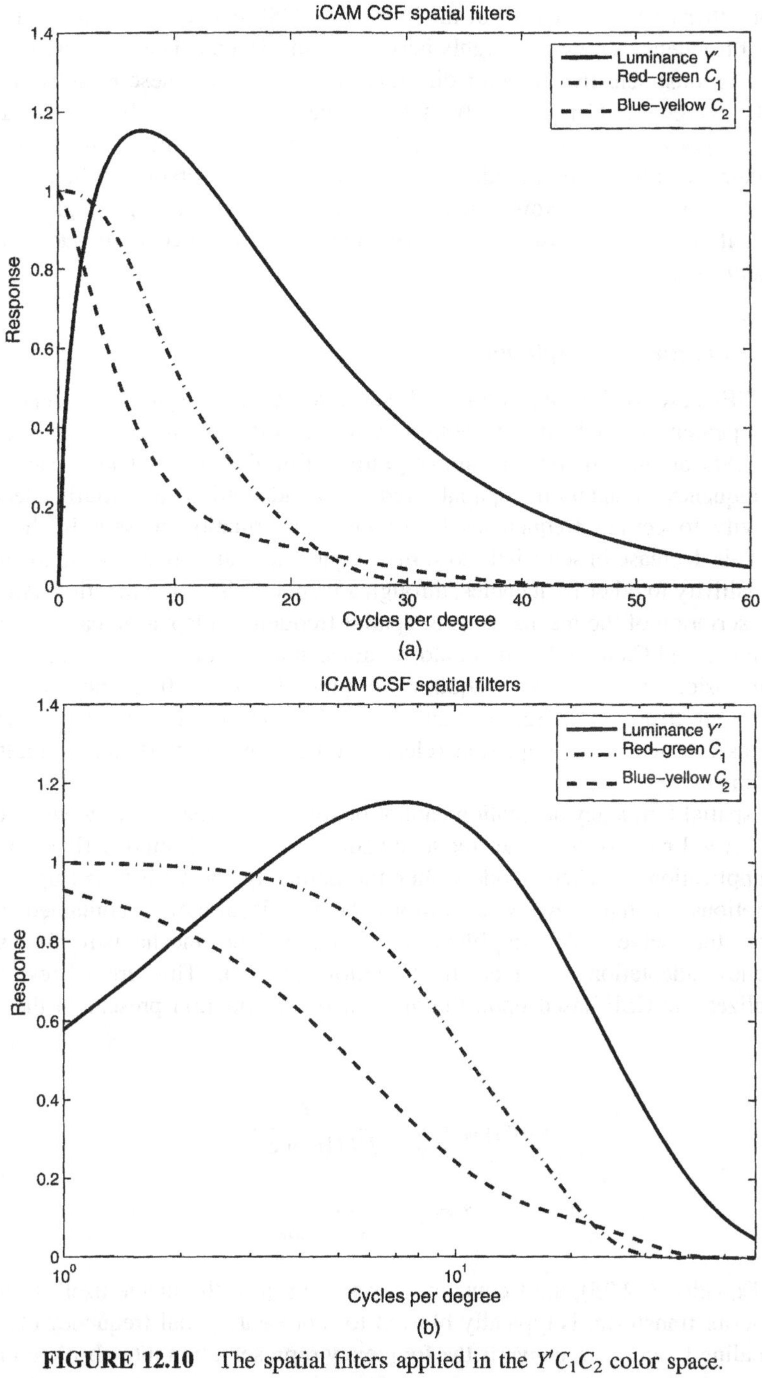

FIGURE 12.10 The spatial filters applied in the $Y'C_1C_2$ color space.

The other important feature of the luminance CSF is that it goes above 1.0 for a band of frequencies ranging roughly between 3 and 15 cpd. This is where the visual system is most sensitive to color differences, and as such these regions are more heavily weighted. Care must be taken when applying a frequency domain filter that goes above 1.0, as this can often lead to severe ringing artefacts. When the filter is sufficiently broad, this is often not a problem. When the filter itself becomes very narrow, such as when applied to a large high-resolution image, it may be necessary to renormalize the luminance CSF such that the maximum is at 1.0.

Spatial Frequency Adaptation

The CSFs described in this framework serve to modulate spatial frequencies that are not perceptible and enhance certain frequencies that are most perceptible. Generally CSFs are measured using simple grating stimuli with care taken to avoid spatial frequency adaptation. Spatial frequency adaptation essentially decreases sensitivity to certain frequencies based upon information present in the visual field. This decrease in sensitivity to some frequencies can actually serve to enhance the sensitivity to other frequencies, through a form of CSF normalization. An excellent description of the mechanisms of spatial frequency adaptation can be found in Blakemore and Campbell.[53] It should be noted that a multiscale, or multichannel, spatial vision model is not required to predict spatial frequency adaptation. Instead, all that is required is that the CSF functions be allowed to change shape as a function of adaptation (clearly indicating the existence of multiscale mechanisms).

As spatial frequency adaptation cannot be avoided in real-world viewing conditions, several models of spatial frequency adaptation have been described for practical applications.[13] These models alter the nature of the CSF based upon either assumptions of the viewing conditions or the information contained in the images themselves. A simplified image-dependent mechanism for spatial frequency adaptation is given in Equation (12.25). This model essentially normalizes the CSF based upon the amount of information present in the image itself.

$$\begin{aligned} CSF_{\text{adapt}} &= \frac{CSF}{\alpha \cdot FFT(\text{Image}) + 1} \\ \alpha &= \frac{1}{D \cdot X_{\text{size}} \cdot Y_{\text{size}}} \end{aligned} \tag{12.25}$$

In Equation (12.25), the frequency representation of the image itself, found via the Fourier transform, is typically blurred to represent spatial frequency channels. The scaling function, α, converts the frequency representation into absolute units of contrast at each spatial frequency. The D factor is similar to the degree of chromatic adaptation factor found in CIECAM02, and is traditionally set to 1.0 for complete

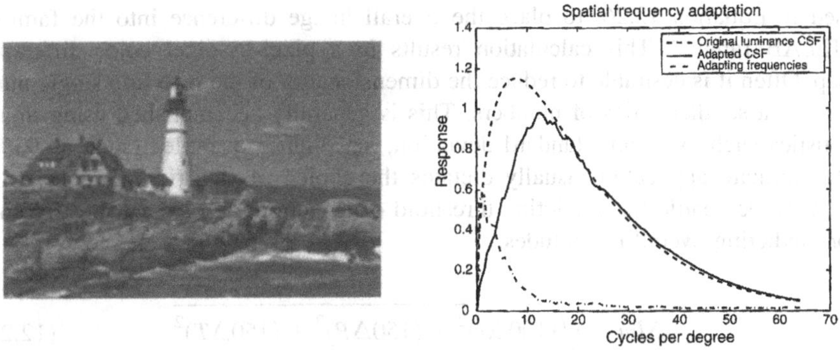

FIGURE 12.11 An example of spatial frequency adaptation to a halftoned image. The halftone pattern in this image was at 22 cpd, resulting in a slight decrease in sensitivity at that spatial frequency.

spatial frequency adaptation. Spatial frequency adaptation is important when calculating image differences between images that may have regular periodic patterns, such as a stochastic halftone pattern, or a jpeg-compressed image that has an 8-pixel blocking pattern. The regular period of these patterns actually reduces the visual sensitivity to the pattern itself and makes it less visible. An example of spatial frequency adaptation for a halftoned image is shown in Figure 12.11. The degree of adaptation factor used to generate Figure 12.11 was chosen to be 3.0 to exaggerate the overall spatial frequency adaptation. Another potential benefit of spatial frequency adaptation is the ability to predict visual masking without the need for multiscale approaches. If a masking frequency is present in an image, the CSF for that particular frequency region (depending on the extent of the blur) will become less sensitive.

Calculating Image Differences

To calculate image differences using the iCAM image appearance model two images are first processed with the spatial filtering, allowing for spatial frequency adaptation. This is illustrated in the flowchart in Figure 12.9. The two spatial filtered images are then processed using the general iCAM framework discussed above. This results in two pixel-by-pixel color appearance maps. These color appearance maps are in a uniform color space and as such can be used to calculate perceived color differences through simple subtraction. Differences in these dimensions can be used to compute image difference statistics and can also be used to derive image quality metrics. An overall image color difference can be found by taking Euclidean distances in the uniform color space. The overall Euclidean difference in *IPT* is referred to as ΔIm (Equation 12.26), for image difference, to distinguish it from a traditional color difference metrics, such as CIELAB ΔE_{ab}, that include no spatial filtering. A scaling factor of 100 for ΔI and 150 for ΔP and ΔT is used in Equation 12.26, to place the overall image difference into the familiar CIELAB range.[28] This calculation results in a pixel-by-pixel color difference map. Often it is desirable to reduce the dimensionality of the map into single numbers, or a smaller series of numbers. This is generally accomplished using image statistics such as mean, standard deviation, and higher quantiles (such as 95%). The ultimate application usually dictates the choice of statistics, such as using the 95% percentile for predicting threshold of detection, and the mean difference for predicting overall magnitudes.

$$\Delta Im = \sqrt{(100\Delta I)^2 + (150\Delta P)^2 + (150\Delta T)^2} \tag{12.26}$$

CONCLUSIONS AND FUTURE CONSIDERATIONS

Advances in imaging and computing technologies along with increased knowledge of the function and performance of the human visual system have allowed for the integration of models of color, spatial, and temporal vision to create a new type of color appearance model, referred to as an image appearance model. Just as color appearance models were born to extend CIE colorimetry to predict changes in appearance across disparate viewing conditions, image appearance models aim to further extend this ability across spatially complex images. The traditional color appearance attributes of lightness, chroma, hue, brightness, and colorfulness need to be augmented with image attributes such as sharpness, graininess, and contrast.

This chapter described the general framework of one example of an image appearance model referred to as iCAM. At the heart of this image appearance model, just as in CIECAM02, is a chromatic adaptation transform and uniform color space. The actual implementation differs greatly from CIECAM02, for example, relying on the spatial properties of the image itself to determine chromatic adaptation. Two specific implementations for HDR image tone mapping and image difference calculations were discussed. The CIE has recognized the potential for using image appearance in HDR applications and has created CIE TC 8-08 to develop guidelines for testing such models. The image difference calculations follow the same general spatial filter preprocessing that S-CIELAB applied to the CIELAB color space.

The model presented in this chapter is not intended as the only solution to image appearance, but rather a framework for future image appearance research. Future efforts could be directed for adding spatiotemporal filters and time-course chromatic adaptation that would be required for using image appearance models for video difference metrics. The collection of more psychophysical data on image and video appearance and differences is required to tune the individual parameters. Psychophysical testing has already shown that future improvements in HDR-rendering are possible, such as the use of edge-preserving low-pass spatial filters. The formulation of specific iCAM algorithms for two applications has been discussed, but those are certainly not the only applications. The iCAM model is not proprietary, and source code and updates are freely available at ⟨www.cis.rit.edu/mcsl/iCAM⟩ for those interested in evaluating the model and potentially suggesting improvements.

REFERENCES

1. Fairchild MD, Johnson GM (2002) Meet iCAM: A next-generation color appearance model, in: *IS&T/SID 10th Color Imaging Conference*, Scottsdale, pp. 33–38.
2. Fairchild MD, Johnson GM (2003) Image appearance modeling, in: *Proceedings of the SPIE/IS&T Electronic Imaging Conference*, SPIE Vol. **5007**, Santa Clara, pp. 149–160.
3. Fairchild MD, Johnson GM (2004) The iCAM framework for image appearance, image difference, and image quality, *J. Electronic Imaging*, **13**, 126–138.
4. Hunt RGW (2005) *The Reproduction of Colour*, 6th ed., John Wiley & Sons, NJ.
5. Schade OH (1956) Optical and photoelectric analog of the eye, *J. Opt. Soc. Am.*, **46**, 721–739.
6. Berns RS (1997) A generic approach to color modeling, *Color Res. Appl.*, **22**, 318–325.
7. Day EA Taplin LA, Berns RS (2004) Colorimetric characterization of a computer-controlled liquid crystal display, *Color Res. Appl.*, **29**, 365–373.
8. Wyble DR, Zhang H (2003) Colorimetric characterization model for DLP projectors, in: *IS&T/SID 11th Color Imaging Conference*, Scottsdale, pp. 346–350.
9. Braun GJ, Fairchild MD (2000) General-purpose gamut-mapping algorithms: Evaluation of contrast-preserving rescaling functions for color gamut mapping, *J. Imaging Sci. Technol.*, **44**, 343–350.
10. Morovic J (1998) *To develop a universal gamut mapping algorithm*, University of Derby Ph.D. thesis.
11. CIE (2001) *Improvement to industrial colour-difference evaluation*, CIE Publication **142**-2001, Bureau of the CIE, Vienna.
12. Zhang XM, Wandell BA (1996) A spatial extension to CIELAB for digital color image reproduction, in: *Proceedings of the SID Symposiums*, pp. 731–734.
13. Johnson GM, Fairchild MD (2001) On contrast sensitivity in an image difference model, *Proceedings of the IS&T PICS Conference*, 18–23.
14. Johnson GM, Fairchild MD (2003) Measuring images: Differences, quality, and appearance, in: *SPIE/IS&T Electronic Imaging Conference*, SPIE Vol. **5007**, Santa Clara, pp. 51–60.
15. Song X, Johnson GM, Fairchild MD (2004) Minimizing the perception of chromatic noise in digital images, in: *IS&T/SID 12th Color Imaging Conference*, Scottsdale, pp. 340–346.
16. Johnson GM, Fairchild MD (2005) The effect of opponent noise on image quality, in: *SPIE/IS&T Electronic Imaging Conference*, SPIE Vol. **5668**, Santa Clara, pp. 82–89.
17. Braun KM, Fairchild MD (1997) Testing five color appearance models for changes in viewing conditions, *Color Res. Appl.*, **22**, 165–174.

18. Daly S (1993) The visible differences predictor: An algorithm for the assessment of image fidelity, in: *Digital Images and Human Vision* (Ed., A. Watson), MIT, Cambridge, pp. 179–206.
19. Lubin J (1995) A visual discrimination model for imaging system design and evaluation, in: Vision Models for target Detection and Recognition (Ed., E. Peli), World Scientific, Singapore, pp. 245–283.
20. Land EH (1986) Recent advances in retinex theory, *Vision Res.*, **26**, 7–21.
21. Land EH, McCann JJ (1971) Lightness and the retinex theory, *J. Opt. Soc. Am.*, **61**, 1–11.
22. McCann JJ, McKee S, Taylor T (1976) Quantitative studies in retinex theory: A comparison between theoretical predictions and observer responses to 'Color Mondrian' experiments, *Vis. Res.*, **16**, 445–458.
23. Funt B, Ciurea F, McCann JJ (2000) Retinex in Matlab, in: *Proceedings of the IS&T/SID 8th Color Imaging Conference*, pp. 112–121.
24. Barnard K, Funt K (1997) Analysis and improvement of multi-scale retinex, in: *Proceedings of the 5th IS&T/SID Color Imaging Conference*, Scottsdale, pp. 221–226.
25. Brainard DH, Wandell BA (1986) Analysis of the retinex theory of color vision. *J. Opt. Soc. Am. A*, **3**, 1651–1661.
26. Granger EM(1993) Uniform color space as a function of spatial frequency, in: *SPIE/IS&T Electronic Imaging Conference*, SPIE Vol. **1913**, San Jose, pp. 449–457.
27. Jin EW Feng XF, Newell J (1998) The development of a color visual difference model (CVDM), in: *Proc IS&T PICS Conference*, Portland, pp. 154–158.
28. Ebner F, Fairchild MD (1998) Development and testing of a color space (IPT) with improved hue uniformity, in: *IS&T/SID 6th Color Imaging Conference*, Scottsdale, pp. 8–13.
29. Keelan BW (2002) *Handbook of Image Quality: Characterization and Prediction*, Marcel Dekker, New York, NY.
30. Engledrum PG (2002) Extending image quality models, in: *Proceedings of the IS&T PICS Conference*, pp. 65–69.
31. Bartleson CJ, Breneman EJ (1967) Brightness perception in complex fields, *J. Opt. Soc. Am.*, **57**, 953–957.
32. Yamaguchi H, Fairchild MD (2004) A study of simultaneous lightness perception for stimuli with multiple illumination levels, in: *IS&T/SID 12th Color Imaging Conference*, Scottsdale, pp. 22–28.
33. Durand F, Dorsey J (2002) Fast bilateral filtering for the display of high-dynamic-range image, in: *Proceedings of ACM SIGGRAPH 02*, pp. 257–266.
34. Kuang J, Yamaguchi H, Johnson GM, Fairchild MD (2004) Testing HDR image rendering algorithms, in: *IS&T/SID 12th Color Imaging Conference*, Scottsdale, pp. 315–320.
35. Kuang J, Johnson GM, Fairchild MD (2004) Image preference scaling for HDR image rendering, in: *IS&T/SID 13th Color Imaging Conference*, Scottsdale, pp. 8–13.
36. Ebner F, Fairchild MD (1998), Development and testing of a color space (IPT) with improved hue uniformity, in: *IS&T/SID 6th Color Imaging Conference*, Scottsdale, pp. 8–13.
37. Fairchild MD (2005) *Color Appearance Models, 2nd ed.*, John Wiley & Sons, NJ.
38. Debevec PE, Malik J (1997) Recovering high dynamic range radiance maps from photographs, in: *Proceedings of the SIGGRAPH'97*, pp. 369–378.

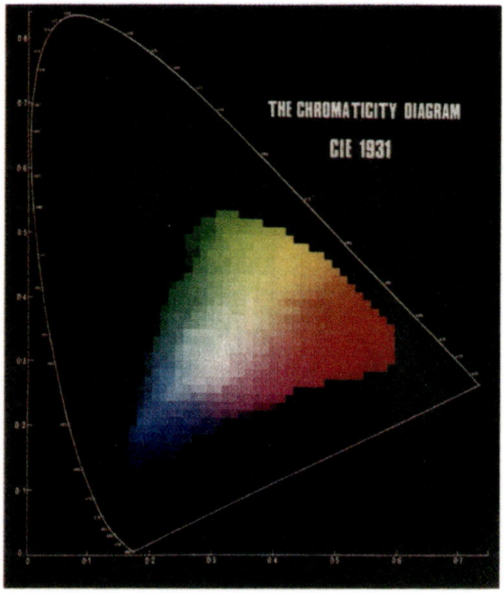

Figure 2.1 The CIE 1931 xy chromaticity diagram. Each square shows the brightest surface colour which can be achieved using non-fluorescent dyes or pigments

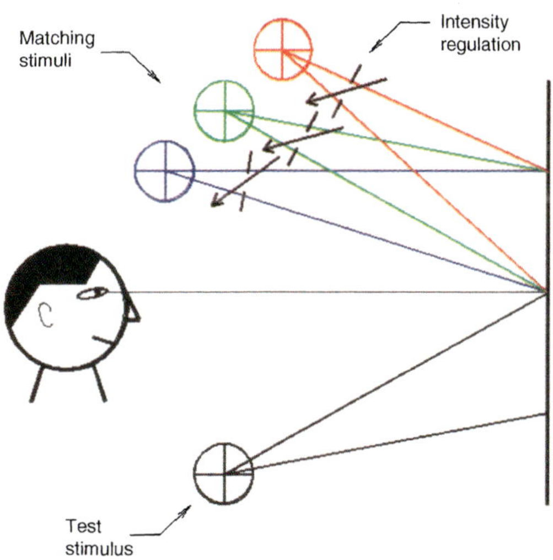

Figure 3.1 Basic experiment of color matching.

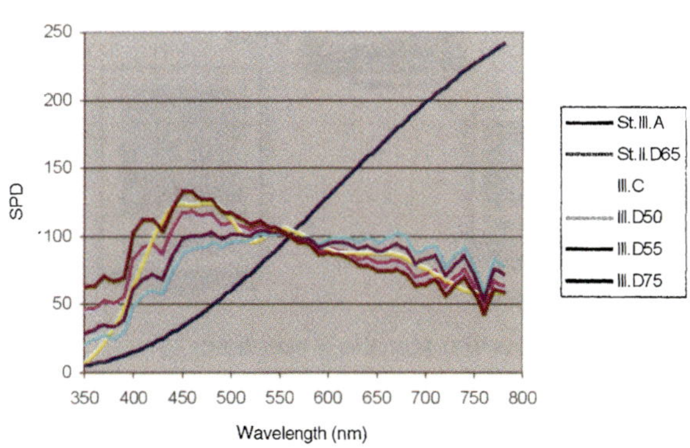

Figure 3.5 Relative spectral power distribution of the CIE standard illuminants and a further three daylight illuminants and illuminant C.

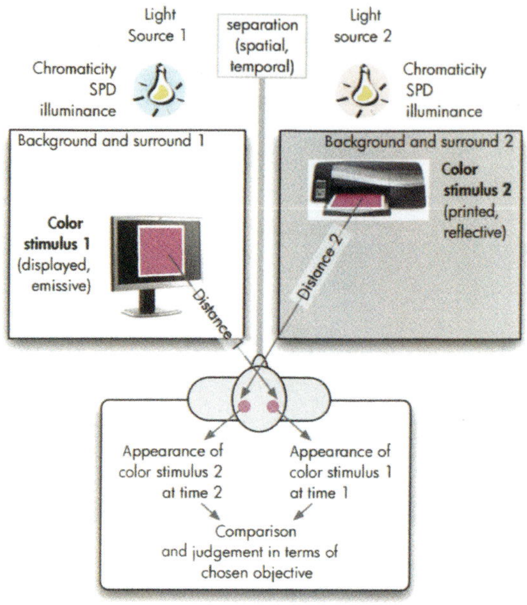

Figure 7.1 Viewing a pair of colors generated using different imaging devices.

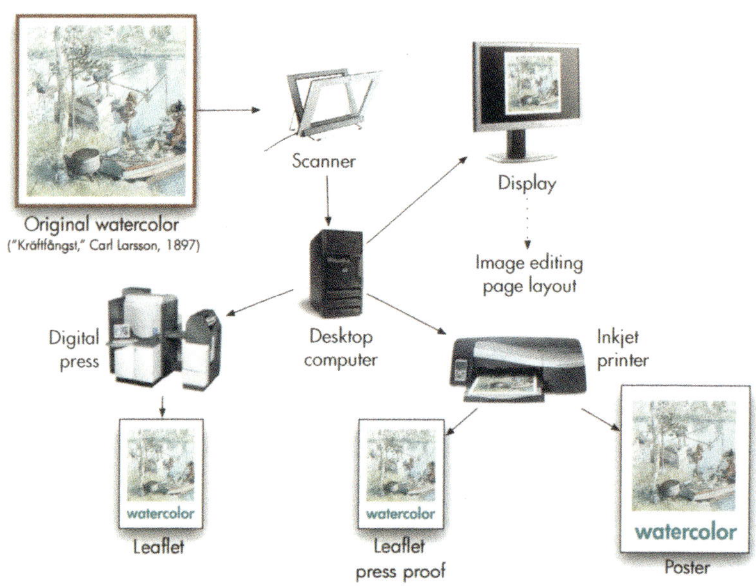

Figure 7.6 A watercolor reproduction scenario's workflow.

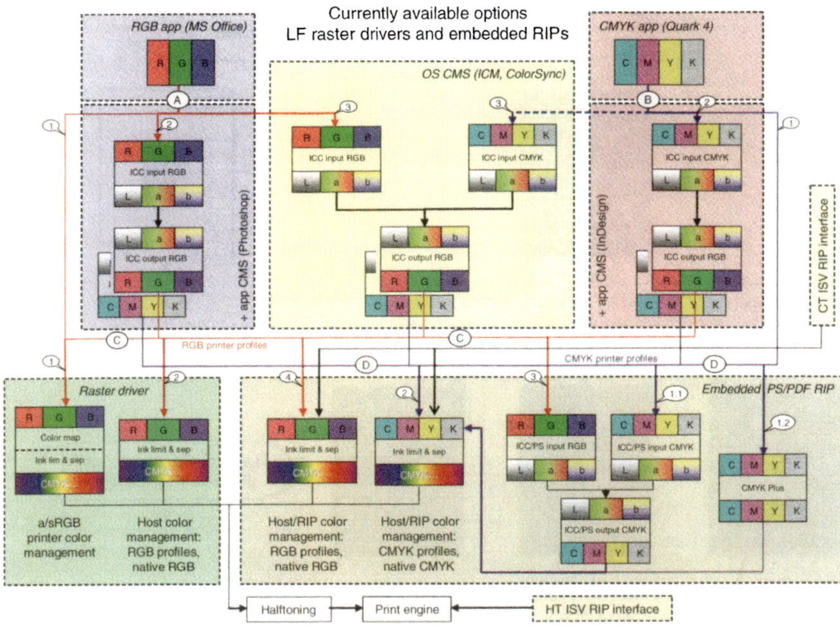

Figure 7.9 Proliferation and redundancy of color management functionality in heterogeneous color managed systems. The large shaded boxes represent RGB-based applications, operating system functionality, CMYK-based applications, raster printer drivers, and printer embedded PS/PDF RIPs respectively, in English reading order. Smaller boxes inside the former represent color profile and color data types, and lines among them the different ways in which they can in principle be connected. Although in principle there is only one color transformation from input to output color space involved, a system as depicted allows 27 different paths to be constructed. Almost half of those result in wrong output, and about 75% of them are redundant, resulting in the same output as some other one.

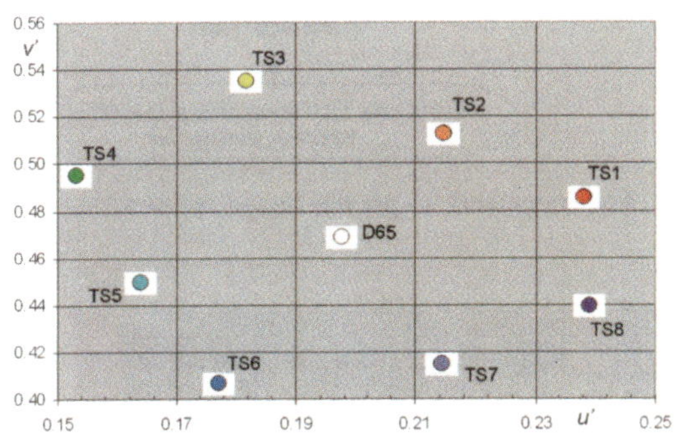

Figure 8.1 The CIE 13.2 test samples for D_{65} illumination and the CIE 2° observer in the u', v' diagram.

Figure 12.1 An example of real-world simultaneous contrast. The colors in the circles are identical, but appear very different.

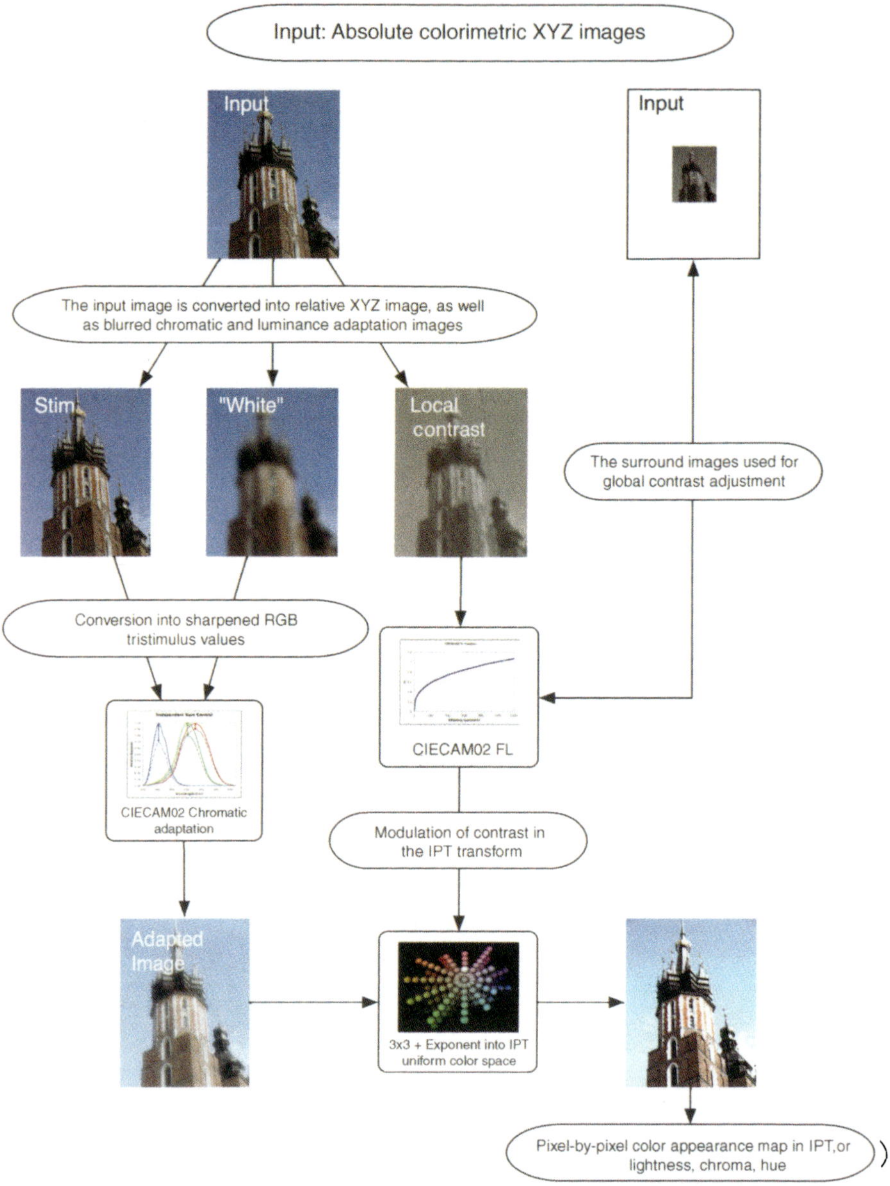

Figure 12.2 Flowchart for using the iCAM framework to predict image appearance.

Figure 12.4 Typical luminance ranges we encounter in everyday life.

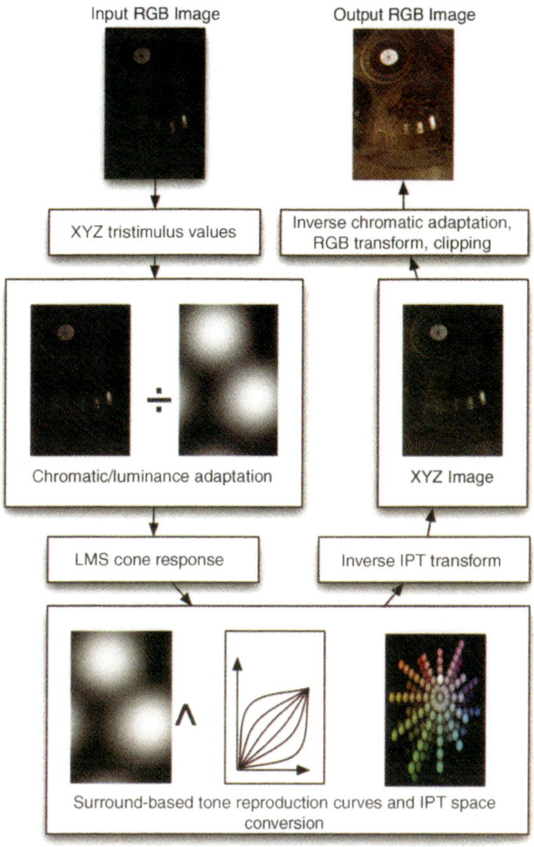

Figure 12.5 Flowchart for using iCAM as a predictor of HDR images.

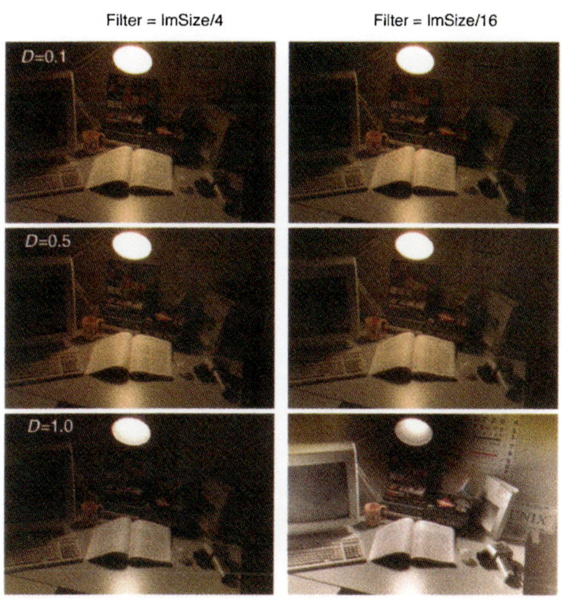

Figure 12.6 Influence of Gaussian blur and degree of adaptation on rendered HDR images.

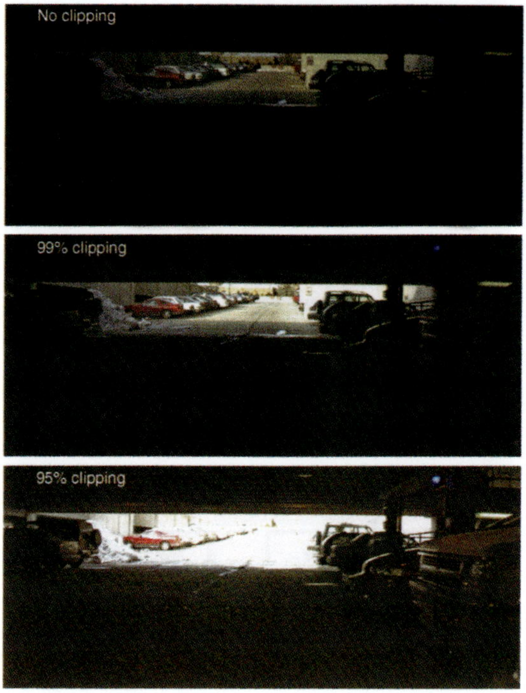

Figure 12.7 The choice of clipping the *RGB* image prior to display has a large influence on the final appearance of the rendered image.

Figure 12.8 An example of a HDR image-rendered using the iCAM framework as described above.

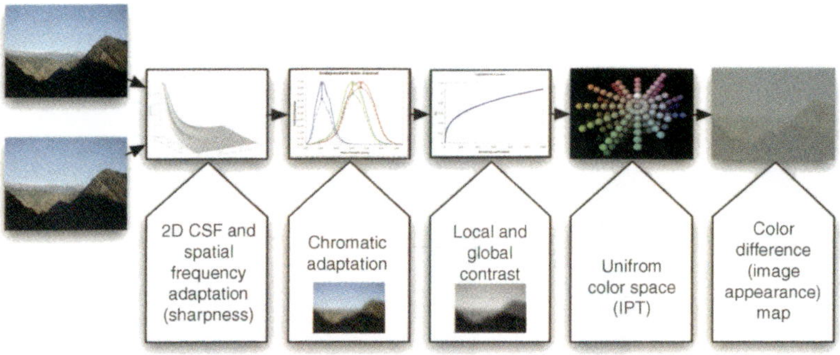

Figure 12.9 Using iCAM as an image difference metric.

Figure 13.2 Dali's painting "Dawn, Noon, Evening and Twilight" as seen from three distances: (A) From 420 cm; (B) From 180 cm; and (C) From 60 cm. (C) corresponds to the region within the white rectangle in A and B. The human figure is one of Dali's versions of Millet's "Angelus." (Dali Theatre-Museum. Figures, Spain. Photos taken by author with permission.)

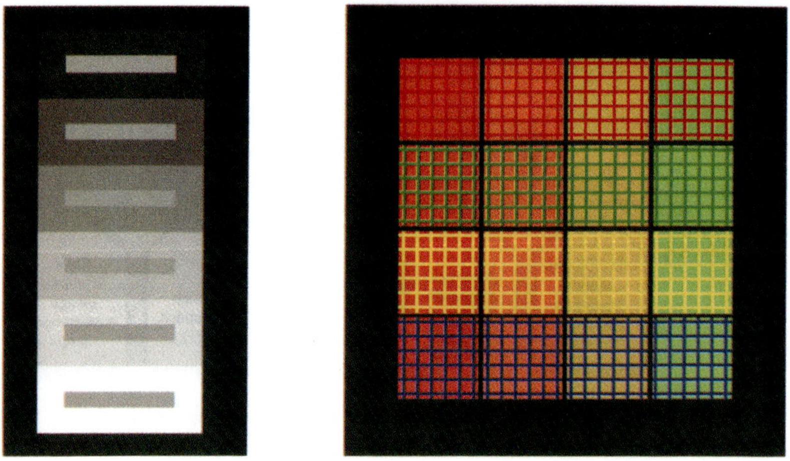

Figure 13.3 Examples of achromatic and chromatic contrast effects.[17] Left: All small gray bars within rectangular backgrounds have the same luminance. Right: All backgrounds behind the grids in each of four columns have the same chromaticity and luminance. Appearance changes considerably at different viewing distances.

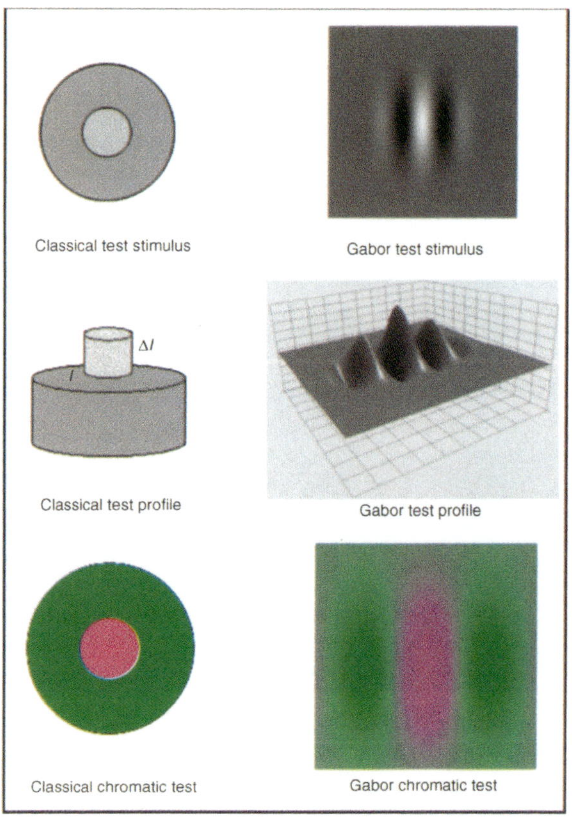

Figure 13.4 Classical spatial stimulus properties (left) and Gabor spatial profiles (right) for chromatic (bottom two) and achromatic (top four) visual stimuli.

39. Xiao F, DiCarlo JM, Catrysse PB, Wandell BA (2002) High dynamic range imaging of natural scenes, in: *Proceedings of the IS&T/SID 10th Color Imaging Conference*, pp. 337–342.

40. Yang D, Fowler B, El Gamal A, Tian H (1999) A 640×512 CMOS image sensor with ultrawide dynamic range floating-point pixel-level ADC. *IEEE J. Solid State Circuits*, **34**, 1821–1834.

41. Nayar SK, Mitsunaga T (2000) High dynamic range imaging: spatially varying pixel exposures, in: *Proceedings of the IEEE CVPR.*

42. Takemura K (2003) Challenge for improving image quality of a digital still camera. *in: Proceedings of the SPIE Electronic Imaging Conference* Santa Clara.

43. Ward G (2001) High dynamic range imaging, in: *Proceedings of the IS&T/SID 9th Color Imaging Conference*, 9–16.

44. Larson GW, Rushmeier H, Piatko C (1997) A visibility matching tone reproduction operator for high dynamic range scenes. *IEEE Trans. Visualization Computer Graphics*, **3**, 291–306.

45. Reinhard E, Ward G, Pattanaik S, Debeve P (2005) *High Dynamic Range Imaging: Acquisition, Display, and Image-Based Lighting*, Morgan Kaufman, San Francisco.

46. Johnson GM (2005) Cares and concerns of CIE TC 8–08: Spatial appearance modeling and HDR rendering, in: *SPIE/IS&T Electronic Imaging Conference*, SPIE Vol. **5668**, Santa Clara, pp. 148–156.

47. Ledda P, Chalmers A, Troscianko T, Seetzen H. (2005) Evaluation of tone mapping operators using a high dynamic range display, in: *Proceedings of ACM SIGGRAPH 05*, pp. 640–648.

48. Kuang J, Liu C, Johnson GM, Fairchild MD (2006) Evaluation of HDR image rendering algorithms using real-world scenes, in: *Proceedings of ICIS 06*, Rochester.

49. Campbell FW, Robson JG (1968) Application of Fourier analysis to the visibility of gratings, *J. Physiol.*, **197**, 551–566.

50. Mullen KT (1985) The contrast sensitivity of human color vision to red-green and blue-yellow chromatic gratings, *J. Physiol.*, **359**, 381–400.

51. Pattanaik SN, Ferwerda JA, Fairchild MD, Greenberg DP (1998) A multiscale model of adaptation and spatial vision for image display, in: *Proceedings of SIGGRAPH 98*, pp. 287–298.

52. Song X (2004) *Chromatic noise perception in digital photography*, RIT M.S. thesis.

53. Blakemore C, Campbell FW (1969) On the existence of neurons in the human visual system selectively sensitive to the orientation and size of retinal images, *J. Physiol.*, **203**, 237–260.

13

SPATIAL AND TEMPORAL PROBLEMS OF COLORIMETRY

Eugenio Martinez–Uriegas

Hewlett–Packard, S.L. (LFP), Avda Graells 501, 08174 Sant Cugat del Valles, Barcelona, Spain

INTRODUCTION

This chapter intends to provide some basic concepts for the reader not familiar with color science in the first three sections and some practical material on contrast sensitivity function (CSF) and multiscale concepts in the last two sections. Readers with color expertise may prefer to leap to the latter sections for discussion on CSF and scalable colorimetry. Unlike the first three, the last two sections present preliminary and tentative proposals to foster further discussion and development.

RADIOMETRY, PHOTOMETRY, COLORIMETRY, AND HUMAN VISION

One thing most useful when starting a conversation about color is to settle a fundamental but often dismissed fact: color is a biosensory percept, not a physical property of illuminated objects nor a physical property of light. That is the substance of the dictum "The rays are not coloured" coined by Isaac Newton in 1730[1] and brought back by Wright in 1967.[2] That is, color is a complex biosensory construct initiated by visible light* reaching retinal photoreceptors, whether that light is reflected from objects or originated in any other way.

*To be exact one should speak about visible radiation, but we will use—as is usual both in vision science texts and in colloquial English—the term visible light instead of visible radiation.

Colorimetry: Understanding the CIE System, Edited by János Schanda

This is a good place to include a philosophical thought along the lines of Helier J. Robinson on reality and perception.[3] The misleading notion that color is a physical property of light and objects (i.e., independent of perception) probably arises, at least in part, from the extreme efficiency of human vision to make us aware of what is out there, to make us think that seeing is believing, and other conventional inferences like the color of objects. These common sense notions, although essential for the paramount biological evolution of human vision and species survival, are misleading and have at times been the most difficult obstacles for understanding color science. Similarly, everyday sensations and experiences of friction and gravity, essential for the evolution of our successful locomotion in arbitrary environments, are still major obstacles for elementary school students to understand Newton's first law of classical mechanics; it is "evident," according to such conventional wisdom ingrained by perception, that to maintain any object in constant motion we need to spend energy.

The theory and measurement of physical properties of all electromagnetic radiation, including the relatively narrow spectrum of visible light, correspond to physics, especially optics and radiometry. The theory and measurement of the basic properties of the biosensory percept that we call human vision, color included, corresponds to the interdisciplinary field of vision science, specially photometry and colorimetry where the metrics and units are inextricably linked to human visual responses. The important point is not where these scientific areas formally classify, but that the building blocks of photometry and colorimetry, namely the CIE standards of luminous efficiency functions and color-matching functions, define efficiency and matching in terms of human visual responses under very specific, spatially and temporally controlled conditions.

Standards of Color: The Role of Biology and Psychophysics

It is useful to think of the following global description to place more specific observations in this context. Human color vision is trichromatic under photopic conditions. That has been supported by abundant biological and psychophysical evidence.[4–6] Under the presentation of a uniform and steady patch of light, trichromacy starts at the retinal photoreceptor stage with three types of photosensitive visual pigments transducing light power into neural signals and, through processing at different neural stages (still an area of intense research), translates into three "sensory" dimensions, one "intensity" dimension related to percepts of luminosity and two "quality" dimensions related to percepts of hue and chroma.

The basic standard of photometry is the luminous efficiency function, a unimodal curve that plots a behavioral measurement of "efficiency"† versus the physical wavelength of light used as visual stimulus; that is, a typical case of the classical

†Several visual responses were used for the standard: direct side-by-side comparison, step-by-step brightness match, and more importantly heterochromatic flicker photometry (HFP). The latter one, together with the minimally distinct border (MDB) technique, is currently preferred by vision researchers over the others because they hold Abney's additivity law.[7] Some details on HFP and MDB are reviewed in the section on "Classical separation of spatial, temporal, and color vision" of this chapter.

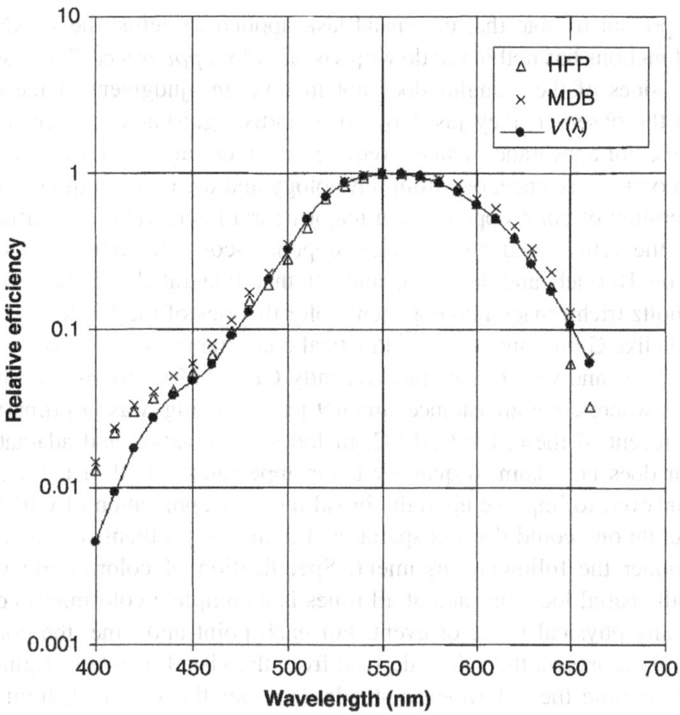

FIGURE 13.1 The line through circles shows the standard $V(\lambda)$ curve (with Judd's 1951 revision). Triangles are data measured with visual responses of HFP and crosses are measured under visual responses of MDB. HFP and MDB data from Wagner and Boynton.[9]

paradigm of psychophysics. Figure 13.1 shows the standard $V(\lambda)$ curve (with Judd's 1951 revision) and two other experimental datasets also obtained using visual responses. The behavioral efficiency axis represents the inverse of the light power required to produce certain visual response. The less light power required, the more efficient is the visual response. As wavelength (λ) is varied through the visible spectrum, say from 380 nm to 780 nm, efficiency varies in a unimodal fashion, low efficiency at the spectrum extremes and maximum efficiency around the middle. The point emphasized here is how the visual response confers a psychophysical nature to the standard, but the reader is encouraged to look into all specifications of the actual family of standard $V(\lambda)$ functions established by the CIE.[8]

Similarly to photometry, the basis of colorimetry is grounded on psychophysical measurements. It is a three-function set known as CIE color-matching functions (see Figure 3.2 in Chapter 3). These are three functions of wavelength, derived as discussed in Section "CIE standard colorimetric observer" of Chapter 3. The fact that a mixture of three well-chosen lights is sufficient to visually match a monochromatic light of any wavelength in the visible spectrum is the evidence that human color vision is trichromatic. Strong evidence from electrophysiology and retinal densitometry of visual pigments from retinal photoreceptors is also consistent with human trichromacy.[4,5]

It is important to note that the visual task applied to define the standard color-matching functions has nothing to do with visual color *appearance*. The visual match of the two sides of the stimulus does not involve any judgment of the color that appears to the observer; they just have to be indistinguishable. Understanding and quantifying color appearance actually required the merging of two fundamental lines of study into vision science, one from physiology and the other from psychophysics: (a) consideration of color opponency, a major neural interaction from different cone signals at the retina, and (b) Hering's opponent-color hypothesis, updated and advanced by Hurvich and Jameson, and all that integrated together with Young and Helmholtz trichromacy into opponent-color theories of the Muller type[10] producing models like Guth's and others.[30] Practical colorimetry evolved into color spaces like CIE $L^*u^*v^*$ and CIE $L^*a^*b^*$ (and recently CIECAM02) to provide metrics for applications where color appearance, and not just matching, was of primary interest. The more recent of these, CIECAM02, includes illumination and adaptation parameters, but does not claim to quantify color appearance in all visual scenarios. It intends, however, to improve upon the broad industry application of CIE $L^*a^*b^*$.

In principle one could dismiss spatial and temporal problems of colorimetry as ill-posed under the following argument: Specification of colorimetric values for each and all spatial locations and at all times is a complete colorimetric characterization of any physical scene or event. For each point and time, the colorimetric values merely represent the values derived from the visual matches originally measured to determine the colorimetric standards under the specified, limited conditions. No claim can be made about color matches made under other spatial and temporal conditions, or about any other visual response to parts or the totality of the scene space or about the presented sequence of temporal events. In other words, the argument is that visual color experiences, including color matching, produced by realistic, dynamic scenes and events are not colorimetric problems but vision science problems. A different argument, adopted in the rest of this chapter, is that it would be very useful to develop additional standards *a la* colorimetry, where visual responses to spatial and temporal variations in color under specified conditions are the basic measurements of additional colorimetric units. This would be useful because practical applications of colorimetry most often deal with such realistic, dynamic scenes and events where metrics and benchmarks are needed to evaluate and predict complex visual tasks. For example, just consider the huge variety of visual tasks under electronic imaging, both analog and digital. Chapter 12 already provided one approach to deal with the spatial interaction of adjacent color stimuli; in this chapter we will go in further detail regarding temporal effects also.

Three summary points close this introduction: (a) Photopic color vision is trichromatic; it has three degrees of freedom with respect to light wavelength; (b) the characteristics of the photometry and colorimetry fundamental functions ($V(\lambda)$ and the color-matching functions) not only result from trichromacy but also result from visual tasks measured under specific spatial and temporal conditions; changing these conditions changes the shape of the fundamental functions; and (c) there is a practical need for additional metrics for visual responses to spatial and temporal variations in color.

These three points are important to understand the color issues related to spatial and temporal dimensions that are discussed in the rest of this chapter.

SPATIAL AND TEMPORAL CONSTRAINTS OF COLORIMETRY: A SELECTIVE OVERVIEW

It is interesting that only about a dozen pages of Wyzsecki and Stiles's Color Science[10] are explicitly under the heading "Spatial and Temporal Factors." This is probably because of the classical separation of areas of study on vision referred as spatial vision, temporal vision, and color vision. However, the literature about spatial and temporal issues on color is currently so large that even a summary review would be impractical within the limitations of this chapter. Those interested in detailed reviews are directed to supplementary readings.[11,12] For the issues herein explored, only a selective overview is necessary.

Spectral, Spatial, and Temporal Dimensions of Visible Light

In physics, electromagnetic radiation power can be formally described in three fundamental dimensions: spectral, spatial, and temporal. In operational terms, one can think of measuring the spectral composition, distribution in space, and distribution in time of a given light. Therefore, these are three independent dimensions of variability, that is, visible light can separately vary in all three domains. Correspondingly, we can measure visible light separately in each dimension. We can think of a function $f(x, y, z, t, \lambda)$ describing the power of a light measured at each point of 3D space represented by coordinates x, y, z, at each time t, and for each wavelength λ of the spectrum.

When visible light is under consideration, there is a first constraint that is implied but not always made explicit in each of those three fundamental dimensions. What does visible light mean? Most commonly, the answer is given in terms of the wavelength range (or frequency range) of light that can be visually detected, say, light wavelengths between 380 nm and 780 nm (and that is usually explained in terms of photosensitive pigments in human retinal photoreceptors). The criterion of visibility is then defined by a specific visual task. An example, in very general terms, is visual detection. If the observer's eye is completely dark adapted, visibility refers to scotopic visibility; if the eye is light adapted, say above 5 cd/m^2, it refers to photopic visibility. In both cases, the relevant measurement is the minimum light power required at each wavelength in the visible range to change an observer's response from "I don't see the light" to "I see the light."[‡] The point emphasized here is that the spatial and temporal dimensions of the light stimulus presented to the observer are fixed and predetermined; they are not defined as a range. For example, a 10° uniform and steady adapting field of 10 cd/m^2 upon which a test field of 1° is presented on the fovea during 200 ms. Or for color matching, another visual task, there

[‡]The example here is brief and simple for clarity; the reader is directed to various psychophysical methods to determine visual thresholds in a formal fashion.[13]

are spatial and temporal constraints of aperture colors in a circular, bipartite patch having uniform light on each half, spanning 2° of visual angle and presented as a 500 ms flash on the fovea. Under those constraints, the only dimension that is varied for each data point is the wavelength of the test flash for detection, or the wavelength of primary lights presented on each side for color matching, measuring its power for each setting. So, from an operational point of view, photometry and colorimetry have been founded on the spectral dimension as the fundamental variable to establish units and standards, maintaining spatial and temporal dimensions at predetermined sets of constant values. Fundamental constraints and standards on visibility in the spatial and temporal dimensions of light need to be clearly defined as they have been for the spectral dimension, in terms of a visibility range within a continuum of variability, and that seems a fundamental source of many spatial and temporal problems of colorimetry.

Classical Separation of Spatial, Temporal, and Color Vision

Classical studies of spatial vision focus on luminance detection and discrimination thresholds of stimuli of varying size and geometrical shape and under different backgrounds. Thus, a typical paradigm involves an observer's eye adapted to a stimulus of luminance level L_1, subtending an area A_1 having a certain shape. After a time T_1, the luminance of an area A_2 (being all or part of A_1) is changed to a second value L_2, and after T_2 seconds the luminance L_2 is changed back to L_1. Under a visual stimulus physically determined by those parameters, the observer follows a psychophysical procedure to measure a visual response (e.g., brightness detection or discrimination).

The temporal profile of luminance changes is also specified, usually by different types of ramps and steps between the initial and final stages of the stimulus, but the main variable under study are shapes, areas, and locations of stimuli and backgrounds. It is clear that there is an endless variety of shapes and temporal profiles that could be tested for visibility.

A similar paradigm is related to classical studies on temporal vision. Here, the temporal profiles of stimulation are the main focus and the spatial parameters of the stimulus are kept restricted to a few shapes of light spots. For example, test spots under a variety of conditions and with a variety of flicker profiles also produced a vast collection of published data.[14]

It was in the temporal domain where we could see first the emergence of a range of visibility within a temporal continuum, the many studies where the frequency of light intensity oscillations was the main variable, flicker visibility was the visual response, and flicker fusion frequency marked the higher limit of a visibility range in the time domain, in similar fashion to a visibility range in the light wavelength domain.

One of the views that drove significant weight in the analysis of results of spatial studies of this type was that of spatial and temporal interactions, where the shapes of test field and background and their temporal profiles were analyzed as producing different types of converging (summation) or competing (inhibition) effects on visual responses. There is a wealth of results in the literature, but the large diversity of test stimuli parameters produced such a diversity of results on visibility (detection and discrimination) that it has not been possible to organize them into a unified, practical description. There are optical standards and benchmarks for specific applications, like acuity or resolution (e.g., detection of Landlot C, Snellen letters, Ronchi rulings or gratings),[15] as well as temporal standards for flicker detection and discrimination, but systematic relationships between spatial and temporal variables of visibility with colorimetry are still in the initial stages of development.[16]

Nonetheless, many spatial and temporal visual phenomena have been studied in terms of chromatic or achromatic conditions, sometimes with surprising results. For example, it is interesting that there are two conventional matching tests, one based on a temporal visual response and the other based on a spatial visual response, such that both produce practically the same visual sensitivity curve, and such curve practically matches the photopic visual efficiency function $V(\lambda)$. The temporal criterion test is known as heterochromatic minimum-flicker match and the spatial criterion test is known as MDB match. In simple terms, both tests compare a fixed-intensity light with a variable-intensity light of different wavelength. Importantly, neither test is based on a color match of the two lights. The flicker test presents both lights alternating in time exactly on the same location; the intensity of the variable light is changed up and down until the observer sees that *flicker visibility is minimal* and that value of intensity is registered as a match. The test is repeated for any number of lights of different wavelengths, each tested against the same comparison light, and the result is a collection of intensity values required for the flicker match at each of those wavelengths. To keep this example short and simple (but see references for important details), imagine that the same collection of light pairs is now presented on a steady bipartite field, the comparison light on one side and the test light on the other. The minimum border test also varies the intensity of the test light, but in this case the visual task is a spatial criterion: The intensity should be such that *the border between the lights is minimally visible.* As mentioned, it is quite interesting that both tests produce practically the same matches, and that the values closely fall on the spectral luminous efficiency, $V(\lambda)$ curve, that is, all matched light pairs are isoluminant whether measured by a minimum-flicker or minimum-border visual task. These curves were previously shown in Figure 13.1. Another interesting observation is that if the same pairs of lights are now presented side by side on separate circular patches, one constant and the other set by the observer's response under the criterion of equal brightness, the resulting curve is significantly different from the $V(\lambda)$ curve; it is also unimodal, with the maximum about the same wavelength (550 nm), but broader than $V(\lambda)$.[8,9] Therefore, in general, changing spatial and temporal parameters of a stimulus changes visibility results, although there are remarkable exceptions where different spatiotemporal parameters together with radically different visual tasks like heterochromatic flicker and MDB, produce the same $V(\lambda)$ curve.

Two Examples of Spatial Limitations of Colorimetry

There are two CIE standard colorimetric observers (1931 and 1964, see Chapter 3). The important difference is the size of the field of view (2° for 1931, and 10° for

1964). Such large difference in field size has implications for applicability of one or the other standard. For example, at lower luminance levels, one should not apply the 1964 standard, among other reasons, because the larger stimulus area, unlike the 2° field of the 1931 standard, involves retinal areas where rods are present and their contribution to visual detection is significant at low luminance levels. In specific applications, it is sometimes possible to introduce correction factors to discount these effects when they manifest in the form of systematic errors.

At least two contributing factors originate the most practical limitations of colorimetry. The first is that colorimetry units are defined in terms of elementary visual responses obtained in highly specialized laboratory environments; and the second is that spatial and temporal constraints necessary in those definitions imply that those metrics are consistent only with those specialized visual responses under those constraints unless more complex visual responses are adequately represented by combinations of the elementary visual responses, which is rarely the case. Thus, when colorimetry is applied to realistic light scenarios, and to realistic visual behaviors, a colorimetric characterization is very useful as a quantitative standard representation of the scenarios, but should not be expected to accurately correlate with actual visual tasks that take place in such scenarios.

For example, consider the scenario illustrated in Figure 13.2 (See color insert). It shows three views of the painting entitled "Dawn, Noon, Evening and Twilight" by Salvador Dali as seen from three viewing distances.

The three images in this figure are from the same physical painting but show increasing detail (both, shapes and color) when seen at shorter distances. A colorimetric characterization could be made, at least in principle, by measuring luminance and chromaticity "point by point" on the painting, as densely as physically possible, and under the same illumination. For clarity, let us ignore (e.g., by masking out) everything but the area within the small rectangle in A and B to compare with panel C. The colorimetric values of that region are the same independently of the three viewing distances, but any visual evaluation would produce different results at such different scales.

One way to think about this problem is to consider that within its spatial resolution, vision differentiates lights that arise from different places; however, beyond its limits on spatial resolution, vision integrates lights that physically arise from different places, and in the latter case, the resulting color perceived at a distance is merely a combination, a kind of sum of the colors that are perceived as separate, differentiated colors at closer distances. A difference dominates when the two lights are separate enough, and a sum dominates when the lights are closer to each other. Consider a grating made of two lights, alternating green and red bars, for example. It turns out that if the two lights are different in both achromatic and chromatic variables (say different luminance and different chromaticity), then the bars can be thinner and spatially closer to each other and still be visually distinguishable, but if they have the same luminance and are only different in chromaticity, then the bars need to be coarser, with boundaries farther apart to be visually distinguishable. The MDB test exploits precisely this phenomenon: the border between two color hemifields, being equivalent to the border between two bars, is minimally perceived when the two lights are equated in luminance.

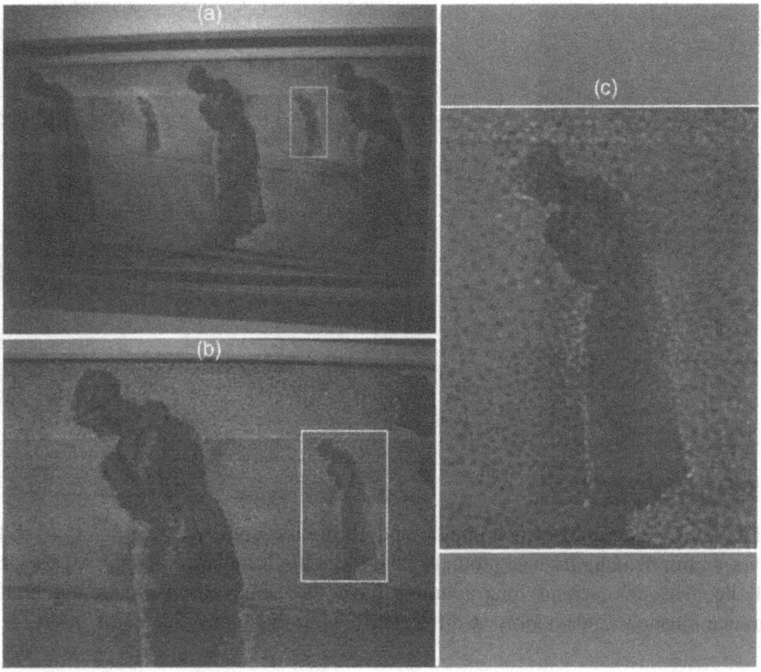

FIGURE 13.2 Dali's painting "Dawn, Noon, Evening and Twilight" as seen from three distances: (A) From 420 cm; (B) From 180 cm; and (C) From 60 cm. (C) corresponds to the region within the white rectangle in A and B. The human figure is one of Dali's versions of Millet's "Angelus." (Dali Theatre-Museum. Figeres, Spain. Photos taken by author with permission.) See color insert.

Similarly, for temporal alternation of two lights shined on the same area, if they have different luminance, they can be seen flickering until they reach a relatively high frequency of alternation, but if they have equal luminance, it does not matter how different they are in chromaticity; they will visually fuse into a steady spot at a much lower frequency of alternation. This is the basis of heterochromatic flicker being minimal when alternating lights have the same luminance. Again, vision differentiates lights that alternate at low frequencies and integrates lights alternating at higher rates. When these lights have different luminance, the crossover point, the alternation frequency where differentiation turns into integration is higher than the crossover point found when the lights are different in chromaticity but equal in luminance. In Section "Multiscale colorimetry," we shall revisit the notions of differentiation and integration of spectral, spatial, and temporal visual information.

Going back to Dali's painting to apply this reasoning, there are some spots of different colors in the framed area shown in the figure; some of these spots are closer in luminance to the background, and they are visually fused (integrated) with it at the intermediate viewing distance, while others have a larger difference in luminance and are fused only at the longer viewing distance.

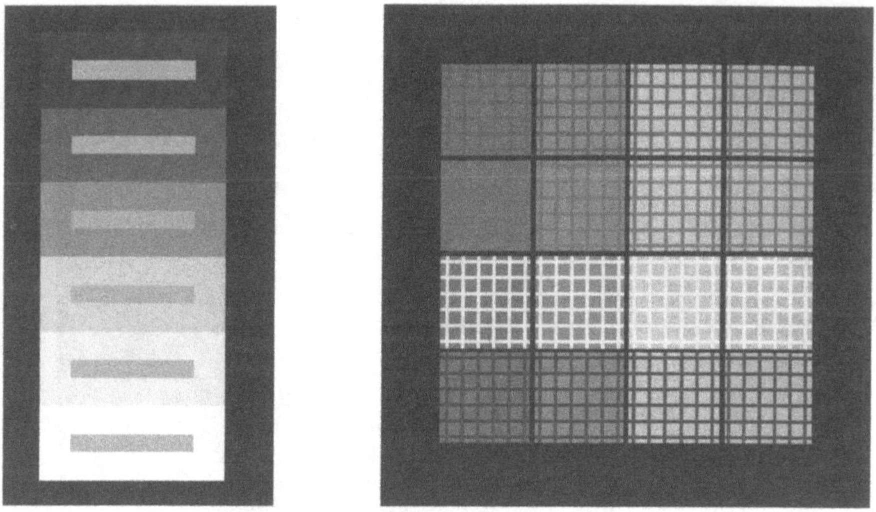

FIGURE 13.3 Examples of achromatic and chromatic contrast effects.[17] Left: All small gray bars within rectangular backgrounds have the same luminance. Right: All backgrounds behind the grids in each of four columns have the same chromaticity and luminance. Appearance changes considerably at different viewing distances. See color insert.

Another illustration of colorimetry limitations, this time using a simple geometrical drawing, is shown in Figure 13.3 (See color insert), reconstructed by this author from an illustration by Steve Shevell.[17] The left side is the well-known example of how the spatial surround (the increasing luminance backgrounds) of a constant test (the smaller, constant luminance rectangles) produces the strong visual perception of increasing luminance of the test in the opposite direction of the background change. Now, the right side on the same figure shows an example of a similar spatial phenomenon but in color. All backgrounds behind square grids in each column have the same chromaticity and luminance values, but they are seen as different colors. The background and the grids are clearly distinguishable, but some kind of color "mixing" is undoubtedly seen. The reader can experience that there is less perceived mixing at closer viewing distances. And attention seems to play a role too. One can easily imagine realistic situations where patterned materials or objects may present practical colorimetric problems.

What happens is that the visual response is different at locations and times where colorimetry is the same. The question is, should colorimetry be developed to deal with this type of situations, or these types of problems belong to a different area of vision science? In other words, if colorimetry is to deal with these problems, how should it be developed to be practically useful when the relevant visual tasks involve spatial and temporal variables?

The perspective adopted by an increasing number of specialists is that it would be very useful and of practical interest to develop colorimetry into spatial and temporal dimensions. An initial step in that direction is to define suitable representations of spatial and temporal changes of light to be used as visual stimulus, and choose appropriate visual responses to define units in those dimensions; all this while maintaining consistency with already defined colorimetric units.

REPRESENTATION OF SPATIAL AND TEMPORAL PROPERTIES OF VISIBLE LIGHT

Considering that light can be manipulated in its wavelength composition, in the way that is distributed in physical space, and in the way it varies over time, three fundamental questions come to mind: (a) How should one vary the spatial and/or temporal distributions of light in a systematic way, and what constraints should be applied to the spectral distribution during those variations? (b) What type of visual responses should be used to determine visibility under variations in spatial and temporal dimensions? (c) Are changes in visual response related across the three domains (spatial, temporal, spectral)? Although we can independently vary light power in three domains that does not necessarily mean that visual mechanisms process information independently in those domains.

Let us provide short answers to each question followed by related comments.

Spatial and Temporal Distributions of Visible Light

Q: How should one vary the spatial and/or temporal distributions of light in a systematic way, and what constraints should be applied to the spectral distribution during those variations?

A: A useful family of space–time distributions of light to be applied as visual stimuli is the family of representations in the spatiotemporal frequency domain. A practical subdivision in spectral properties of these distributions is in chromatic and achromatic variations.

This is probably one of the more visited topics in current vision science because it represents a change of paradigm from a conventional object/shape/event definition of a visual stimulus to a more abstract representation based on frequency-content analysis. An excellent introduction to frequency representation of visual stimuli is given by Cornsweet.[18] This change of paradigm is linked to a different view of the initial stages of the visual system from an object/shape/event detector to a space–time pattern analyzer.[12] This more recent view of initial visual processes does not preclude the proposition of theories of higher neural processes that build more complex internal representations in terms of objects and scenarios. Such initial stage is based on the notion of a "Primal Sketch" representation of all visually acquired information; a bottom-up representation that is rich in information and very efficient as input for higher visual processes that build internal constructs in terms of objects and scenarios.[19]

In a spatiotemporal pattern approach, light power distributions to be visually tested are represented in terms of their spatiotemporal frequency components instead of their shapes, areas, and temporal profiles. Figure 13.4 illustrates a

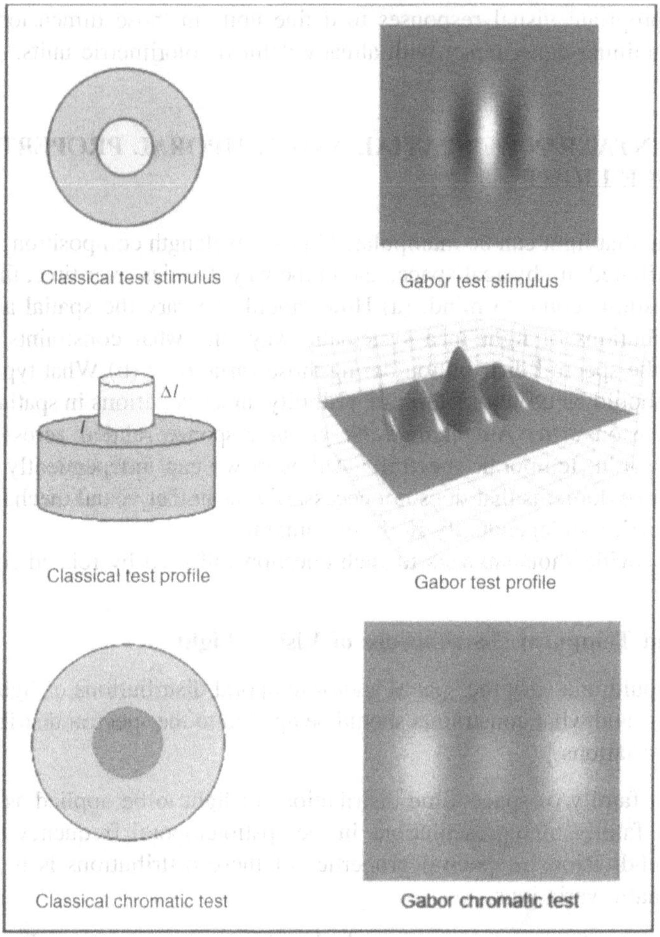

FIGURE 13.4 Classical spatial stimulus properties (left) and Gabor spatial profiles (right) for chromatic (bottom two) and achromatic (top four) visual stimuli. See color insert.

schematic example of two types of visual stimulus; the conventional target based on the shapes of a luminance increment on a background, on the left side of the figure, and a more current type based on spatial patterns of luminance or chromatic modulation on the right. The illustrations on the right are called "Gabor test stimuli" and are the product of a two-dimensional Gaussian times a cosine grating. For the chromatic Gabor patch, it is the product of a two-dimensional Gaussian times the sum of two cosine gratings of the same spatial frequency but 180° out of phase, and made of light of different spectral composition, red and green, for example. In the next section, we shall see how the two cosine gratings can be modulated such that average luminance across the whole patch can be maintained constant while the chromaticity modulation is varied to test visual response dependence on such variation.

The use of conventional targets of increment on background leads to the characterization of visual responses in terms of "$\Delta I/I$ contrast" where ΔI is the luminance increment over the background's luminance I. The ratio $\Delta I/I$ is also known as the Weber fraction, and there is a wealth of descriptions on how that fraction changes with spatial variations (area, shape), temporal profiles of background and target presentation, as well as adaptation to selected settings of luminance and chromaticity of the surroundings. In fact, $\Delta I/I$ is a metric with long tradition in other sensory dimensions aside of vision.[20]

In comparison, spatial-pattern targets lead to analyses of responses in terms of "Michelson contrast," $(I_{max} - I_{min})/(I_{max} + I_{min})$, where I is the luminance and maxima and minima are taken over the entire stimulus pattern. The spatial variable is defined by the spatial frequency of the pattern. In a sinusoidal grating, it is the spatial frequency of the sinusoid. In the Gabor profile, it is the spatial frequency of the cosine function used in its construction. A "fine" pattern has a high spatial frequency while a low spatial frequency corresponds to a "coarse" pattern. If the pattern is set in motion, say at constant speed, with respect to the observer's eye, the result is a spatiotemporal pattern characterized by the same contrast value but with independent means to change spatial and temporal frequency (from coarse to fine pattern and from slow to fast motion). This approach leads to the operational concept of visual contrast sensitivity. In general, higher contrast sensitivity results from less contrast required in a given pattern to perform a predetermined visual task. Let us see in more detail the achromatic and chromatic pattern stimuli.

Visualizing the achromatic contrast of a sinusoidal grating is straightforward: luminance of a light with fixed spectral composition sinusoidally oscillates around an average value. Zero contrast corresponds to no oscillation, just the uniform average value. One hundred percent contrast corresponds to the magnitude of the oscillation that goes from a minimum of zero to a maximum luminance of twice the average. A practical implication for display specifications to produce these patterns is that it should be capable of delivering twice the desired average luminance, and should vary its luminance linearly across the full range to accurately produce the desired profiles.

Contrast of a chromatic sinusoidal grating is not so straightforward. Let us start with two sinusoidal gratings like that just described, both with the same modulation frequency and each with separate independent controls for average luminance and modulation amplitude. And let us make them such that they have different spectral composition, say one looks red and the other looks green. Next, shift them laterally such that they are 180° out of phase, that is, all maxima of one coincide with all minima of the other. Next, shift the average luminance level of one such that it has the same luminance as the average luminance of the other and keep them constant[§]. Finally, lock the modulation controls of both gratings into a single one, such that

[§]The equiluminance condition can be implemented by a standard luminance criterion (standard luminance values for the two lights), or by a observer-dependent, visual-task-dependent criterion, for example, by each observer making a heterochromatic flicker match of the two lights being used. Commonly, the first is preferred in color engineering, while the second criterion is common in color vision research.

modulation changes are in percent of the average units for both gratings. That is, if one turns the common contrast knob up from zero contrast, both gratings increase their modulation amplitude in the same percent with respect to their corresponding average values. The result is a composite grating that, independently of the contrast value set with the unified control, has constant luminance across the whole field of presentation because being in opposite phase, and both locked to change amplitude by the same percent, the red and green gratings add up to the same luminance value everywhere. That constant luminance is actually the sum of the two average values. There is no luminance modulation anywhere; turning the contrast knob produces changes exclusively in chromaticity. This is called an isoluminant, chromatic, sinusoidal grating; it is the type of stimulus one needs to study visual responses due to chromatic contrast, without intermixed effects possibly due to the presence of luminance contrast. Appearance of this grating is not uniform at all. As the contrast knob is turned up from its zero position, the appearance of the grating goes from a uniform yellow field (no grating), through barely visible bars slightly alternating as greenish and reddish, up to a full 100% chromatic contrast grating that appears as saturated red and green alternating bars joined by a saturated yellow bar between them, all of the same luminance. As contrast is decreased the vividness of the red and green maxima gradually decreases until all that is seen is a uniform yellow field. Detection threshold of these isoluminant chromatic gratings is the chromaticity contrast required for visual detection of the grating. Gabor chromatic patches, like those illustrated in Figure 13.4 are simply the product of a Gaussian times the two counter phase sinusoids just described, holding the constant luminance condition. Now let us consider the second of our three questions.

Detection and Discrimination Thresholds

Q: What type of visual responses should be used to determine visibility under variations in spatial and temporal dimensions?

A: Detection and discrimination thresholds are basic visual responses that are in current use to measure visual efficiency with different spatiotemporal distributions of light.

A simple example of the method of adjustments is useful to illustrate contrast detection thresholds with gratings. Using as visual stimulus a luminance achromatic grating of a given average luminance, an observer is asked to manipulate a contrast knob up and down, straddling around a setting where he sees the grating appear and disappear, with the task of choosing such transition point as his setting. The procedure is repeated for each of a set of spatial frequency values, that is, the observer obtains a set of detection threshold values corresponding to a set of coarse-to-fine gratings. Sensitivity is defined as the inverse of threshold. The smaller the contrast threshold setting, the higher the contrast sensitivity. Typical results produce a unimodal curve with a maximum around 3 cpd (cycles per degree), and with lower sensitivity for higher or lower spatial frequencies; the cut-off frequency is around 60–80 cpd. The shape of this curve varies depending on the stimulus parameters like average luminance and the size of the field of view, among many others.[11]

A more reliable procedure than that of adjustments is the two-alternative forced-choice staircase procedure, where the observer is shown two stimulus fields (usually in sequence at the same location) of the same average luminance but with only one of them having a grating of a given contrast. To maintain adaptation, the two uniform fields are not turned on and off, but are present all the time; the contrast grating is introduced in either the first or the second field. Sound cues are provided to warn the observer of each time frame. The observer must decide which of the two had the grating, or guess if in doubt. Observer response and patterns of response determine the subsequent changes in presentation following preestablished protocols of contrast steps in ascending or descending staircases. The objective is to obtain a final contrast value statistically justified as having a 75% chance of being detected, and all that in a reasonable amount of time. This is just one example from a variety of forced-choice procedures commonly applied in psychophysical experiments, visual or otherwise.[21] Independently of the psychophysical method used, the resulting curve is called CSF for visual detection.

Compared to the achromatic thresholds, detection thresholds for isoluminant chromatic gratings show two main differences. First, they have a low-pass shape, that is, they have maximum sensitivity at low spatial frequencies, monotonically decreasing as spatial frequency increases. Second, the cutoff frequency is much lower, 8–18 cpd. Similar to achromatic CSFs, they show significant variability depending on average luminance and size of the field of view.[22] Figure 13.5 shows examples of chromatic and achromatic CSFs for visual detection. One important point should be clarified. The isoluminant requirement imposed on the gratings or patterns used to obtain chromatic CSFs intends to avoid any influence from

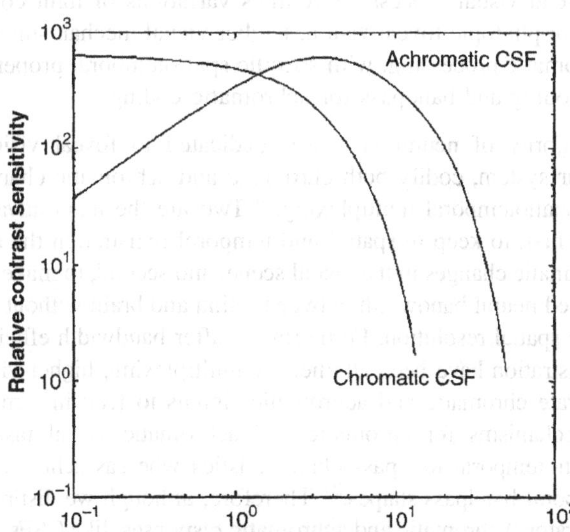

FIGURE 13.5 Chromatic and achromatic CSFs for visual detection to illustrate the main differences in shape.

achromatic mechanisms and study chromatic mechanisms per se, but that does not mean that chromatic and achromatic mechanisms do not interact when both chromatic and achromatic contrasts are present, which is the case in most natural visual scenes. The discussion is still very much alive on the nature and characterization of such chromatic–achromatic interactions.[23]

When both comparison and test stimuli contain a grating (with contrast > 0), the visual task becomes one of distinguishing between them. Contrast discrimination thresholds involve the visual task of deciding if two gratings presented in sequence (e.g., with the forced-choice procedure) have enough contrast difference or enough spatial frequency difference to visually tell them apart. Therefore, unlike detection thresholds, grating discrimination thresholds can be measured with respect to differences in spatial frequency. Classical literature on discrimination measurements include the paradigm of masking, where the first two of the gratings presented to be compared is said to mask the second grating.[24]

In terms of explanation and interpretation of detection and discrimination, it is currently thought that detection data correspond to visual mechanisms closer to the front end of the visual system, whereas discrimination data correspond to visual mechanisms at a higher level, built upon the early ones for detection.[25] Let us now consider our third question.

Visual Multiplexing of Spatiotemporal Chromatic and Achromatic Information

Q: Are changes in visual response related across the three domains (spatial, temporal, and spectral)?

A: Major front-end visual processing codifies variations of light conjointly in all three domains for photopic foveal vision. Higher visual mechanisms separate chromatic and achromatic processing, with specific spatiotemporal properties, low-pass for chromatic coding and bandpass for achromatic coding.

The vast majority of neural processes dedicated to foveal vision, known as the parvocellular system, codify both chromatic and achromatic changes through a process called spatiotemporal multiplexing.[26] Two are the main advantages of this neural strategy: first, to keep in spatial and temporal registration the code for chromatic and achromatic changes in the visual scene, and second, to make more efficient use of the reduced neural bandwidth between retina and brain without sacrificing the highest possible spatial resolution. Furthermore, after bandwidth efficiency and spatiotemporal registration have been attained by multiplexing, higher visual processes efficiently separate chromatic and achromatic signals to feed in turn the relatively independent mechanisms for chromatic and achromatic visual tasks. Chromatic CSFs show spatiotemporal low-pass characteristics whereas achromatic CSFs produce spatiotemporal bandpass shapes.[27] Therefore, although we distinguish and isolate (in the laboratory) chromatic and achromatic responses, like CSFs, it is important to keep in mind that most likely they share a common neural mechanism in their early processing. In fact, there is quantitative evidence of this possibility: Chromatic low-pass and achromatic bandpass CSF data have both been explained in terms of the same canonical retinal receptive field; that is, a model receptive field applicable to all members of a physiological class of receptive fields; in this case, parvocellular receptive fields of foveal photopic vision.[28]

It makes sense to consider that the same differentiation and integration principles underlying chromatic and achromatic vision underlie its spatial and temporal properties. That is not a mere speculation because fundamental neural processing involves two main modes of neural signal interaction, excitation, and inhibition.[4]

From three types of cone photopigments, photopic vision builds two very different types of processing: Adding spectral sensitivities (i.e., excitatory cone signals interaction) produces an achromatic, unimodal function. For example, $V(\lambda)$ is obtained by adding $1.65L(\lambda) + M(\lambda)$ and normalizing the result, where $L(\lambda)$ and $M(\lambda)$ are Smith and Pokorny cone fundamentals.[29] Subtracting spectral sensitivities (i.e., inhibitory cone signals interaction) produces a bipolar, chromatic function, for example, $L(\lambda) - M(\lambda)$ yields a so-called r–g color opponent function adopted in color opponent models.[30] Therefore, in the spectral domain, addition of different cone signals leads to achromatic processing whereas subtraction of those signals leads to chromatic processing.

Similar to the spectral domain, in the spatiotemporal domain we also have functions that sample their domain with different sensitivity curves. We have at least two general functions of space and time; let us call them $N(x,y,t)$ and $W(x,y,t)$, respectively, for narrow-fast and wide-slow properties (say narrow and wide spatial Gaussians with exponential decay in their time responses, but many other specific models are feasible). These functions are commonly used to represent sensitivity variations in each of two concentric, circular regions at the retina, one smaller (narrow) and with faster response than the other (wide and slow). This is known as the center-surround receptive field organization. Applying the same neural principles to this case, addition, $N(x,y,t) + W(x,y,t)$, leads to a low-pass space–time CSF and subtraction, $N(x,y,t) - W(x,y,t)$, produces a bandpass space–time CSF.

Therefore, there is a somewhat puzzling situation when one tries to put it all together: addition would produce achromatic, spatiotemporal low-pass characteristics, whereas subtraction would produce chromatic, spatiotemporal bandpass characteristics. But that inference is the opposite of most electrophysiological and psychophysical evidence: achromatic vision is spatiotemporally tuned, that is, it shows bandpass properties, whereas chromatic vision is not tuned, showing spatiotemporal low-pass properties.

Multiplexing solves the apparent puzzle.[27] Assigning different spectral sensitivity functions to center (narrow) and surround (wide) functions one can represent, for example, a "center" function as the product of sensitivities: $L(\lambda) \cdot N(x,y,t)$ and a "surround" function as the product $M(\lambda) \cdot W(x,y,t)$. The multiplexing identity shows that the resulting sensitivity of "center minus surround" is identically represented as:

$$
\begin{aligned}
L(\lambda) \cdot N(x,y,t) - M(\lambda) \cdot W(x,y,t) = \\
(1/2)\{[L(\lambda) + M(\lambda)] \cdot [N(x,y,t) - W(x,y,t)] \\
+ [L(\lambda) - M(\lambda)] \cdot [N(x,y,t) + W(x,y,t)]\}
\end{aligned}
$$

The implication revealed by that algebraic identity is consistent with the evidence that achromatic $[L(\lambda) + M(\lambda)]$ visual encoding has spatiotemporal bandpass characteristics $[N(x,y,t) - W(x,y,t)]$ (first half of the expression in braces) whereas chromatic $[L(\lambda) - M(\lambda)]$ visual encoding has spatiotemporal low-pass characteristics $[N(x,y,t) + W(x,y,t)]$ (second half of the expression in braces). Furthermore, according to this theory, higher level neural processes separate multiplexed terms to produce relatively independent achromatic and chromatic mechanisms with three fundamental properties: (a) They are already in spatiotemporal register (because of their multiplexed locked-in origin); (b) they have gone through the optic nerve bandwidth bottleneck; and (c) they have acquired (by the separation process) an emergent property not manifested in the early stages, the property of orientation selectivity for achromatic mechanisms.[27]

DEVELOPING CSF STANDARDS

The essential standards for colorimetry established by the CIE are the luminous efficiency function $V(\lambda)$, and the color-matching functions $\bar{x}(\lambda), \bar{y}(\lambda), \bar{z}(\lambda)$. There is an ongoing initiative for the CIE to establish similar standards for spatial vision. A natural candidate to be the focal point of this effort is the CSF; therefore, a CIE Technical Committee, TC 1–60, has been established to study the issues and present a technical report. The general approach under discussion is summarized here.

General Approach: Data-Based or Theory-Based Standard

In similarity to the usefulness of $V(\lambda)$ and $\bar{x}(\lambda), \bar{y}(\lambda), \bar{z}(\lambda)$ to characterize the infinite variety of light spectral compositions that may enter the eye, a CSF standard would be useful to characterize the infinite variety of spatial distributions of lights in the visual scene. The possible characterization of the spatial structure of any image into its chromatic and achromatic CSF components is a concept analog to the characterization of any light into its luminous efficiency and colorimetric components. Such a representation of the visual scene in terms of light colorimetry and its CSFs does not necessarily imply that the visual system operates as a spectrophotometer or as a spatial frequency analyzer. That kind of conceptualization requires theories and models of how the visual system works. In the same way that colorimetry is not expected to account for many color aspects of human vision, an extension of colorimetry in terms of chromatic and achromatic CSFs is not expected to account for most aspects of spatiotemporal vision. However, it is also reasonable to expect that a spatiotemporal extension of colorimetry will provide useful tools and benchmarks applicable to practical scenarios, in the same way that current colorimetry has provided the basis to develop such industrial applications. One possible explanation of the success of current colorimetry is that it was built upon the fundamental property of trichromacy at the early stages of neural visual processes, even if photoreceptor signals interact and go through many stages before visual behaviors take place. With that in mind, it seems reasonable to consider the CSF as a feasible candidate because physiological and psychophysical evidence point to an early visual stage as its origin.

There are interesting analogies between $V(\lambda)$ and the spatial CSF. In very general terms, visual response given by $V(\lambda)$ for each wavelength is conceptually analogous to visual response given by the CSF for different spatial frequencies. A standard CSF will quantify the limited visibility range in the spatial frequency domain in a similar way as $V(\lambda)$ quantifies a limited visibility range in the light-wavelength domain. CSF describes visibility of spatial contrast while $V(\lambda)$ describes the visibility of radiance. A major task for TC 1-60 is to gather and unify criteria from the vision laboratories working in this area, in order to propose an appropriate definition of visibility of spatial contrast, that is, an operational definition that implies a set of specified conditions and a set of specified measurement techniques. The resulting technical proposition of a CSF, from an appropriate population of subjects and appropriate statistics, will then be proposed as part of a standard observer.

The conceptual analogy between CSF and $V(\lambda)$ continues at higher levels. According to the vast literature on the visual mechanisms that produce CSF, where there are competing, sometimes antagonistic models, the CSF could be a manifestation of two or more submechanisms, in a similar fashion as the photopic luminous efficiency function $V(\lambda)$ can be conceived as a result from the combination of at least two cone mechanisms with different spectral sensitivities. But this is where TC 1-60 needs to keep in sight the distinction between CSF as a set of data that characterizes visual responses under established conditions and any theory, model, or formula that intends to explain CSF and aims to predict visual responses under stimuli and conditions that are different from the reference conditions specified for a proposed CSF.

The initial step is the ongoing work on a candidate for a CSF baseline standard for achromatic, steady conditions and is based on detection thresholds. Additional levels of complexity will then be studied to deal with extensions of the standard to cover chromatic and temporal variables of spatial contrast, as well as contrast sensitivity in terms of discrimination thresholds. TC 1-60 plans to continue work on those extensions, but only after a standard CSF baseline is established for achromatic steady conditions, to be used as a basis for more complex metrics.

Initial Results

The plot in Figure 13.6 is a simple comparison of the well-known CSF laboratory data, but it has an interesting feature: Data obtained with Gabor patches (triangles) are in the same plot as data obtained with sinusoidal gratings (all the other data points).

Data are normalized to 1 at each set's maximum (as gain equalizing factors); the shape of the curve obtained from each set is not altered with this manipulation. That is, spatial frequencies where maxima for each set occur, and the (extrapolated) cutoff frequency are not changed, and they are very similar. The reasonable agreement obtained in spite of significant differences in conditions, methods, and stimulus parameters is interesting. Also, apart from normalization, no special effort was

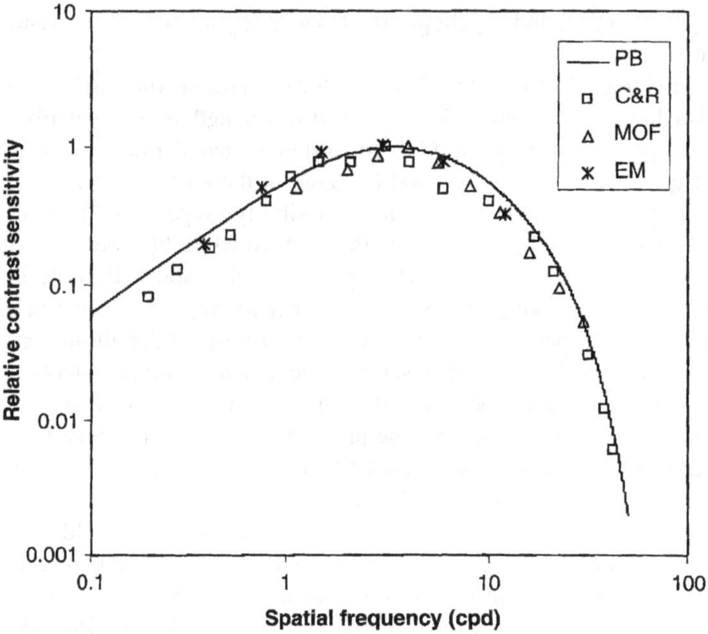

FIGURE 13.6 Squares represent Campbell and Robson's data[31]; triangles represent Modelfest 2001 average data from 16 observers,[32] and crosses are average data from seven observers.[33] The smooth line is a plot from the simplified version of Barten's formula.[34]

made to vary Barten's formula parameters to optimize the fit between data and the calculated curve. A field size of 8° and an average luminance of 100 cd/m^2 were used in that formula.

The curve obtained from Barten's simplified formula[34] is included. This formula defines sensitivity S as a function of spatial frequency u as follows:

$$S(u) = \frac{5200 e^{-0.0016u^2(1+100/L)^{0.08}}}{\sqrt{\left(1 + \frac{144}{X_o^2} + 0.64u^2\right)\left(\frac{63}{L^{0.83}} + \frac{1}{1 - e^{-0.02u^2}}\right)}}$$

In this formula, the luminance L is expressed in cd/m^2 and the field size X_o in degrees of the visual angle. A square field is assumed. For a rectangular field, $X_o{}^2$ has to be replaced by X_oY_o, and for a circular field by $\pi/4\ D^2$. Binocular vision is assumed with equal luminance surround. For monocular vision, the same formula can be used, but multiplied by 0.7. Typical values of the constants are given in ref. 34. The constant 5200 in the numerator corresponds with the nominal signal-to-noise ratio 3. If this value appears to be larger in practice, this constant has to be taken smaller.

Following the historical procedures of colorimetric standards, in TC1-60 it is being discussed that a candidate standard should be a table derived from the actual laboratory data plus a detailed definition of a standard CSF observer. Formulas like Barten's together with some other models under consideration could be included as a technical annex of tools for further research and applications.

MULTISCALE COLORIMETRY: A SPATIOTEMPORAL PATH FORWARD

Example of Multiscale Image Decomposition

As mentioned in the previous section, there are psychophysical and physiological data indicating that CSF is actually the envelope of several component mechanisms operating at different spatial frequency ranges,[10,11] that is, at different scales. From this general view of scaling of the chromatic and achromatic spatiotemporal processing, one can sketch a possible path for the development of spatiotemporal colorimetry; a path that could be called scalable colorimetry. The following is just an example to illustrate the general idea. Let us start only in the spatial domain for simplicity and include the temporal domain afterwards. Let us sketch what would be a colorimetric characterization of a digital image if we start with CIE $L^*a^*b^*$ values for each pixel (or *Jab* values from CIECAM02 to take into account illumination and adaptation parameters), and suppose that we have established standards for chromatic and achromatic CSFs of the type illustrated previously in Figure 13.5.

Figure 13.7 shows four formulas applied to the L^* pixel values of a digital image using a sampling cell of 2 × 2 pixel values. The four input values a_{11}, a_{12}, a_{21}, and

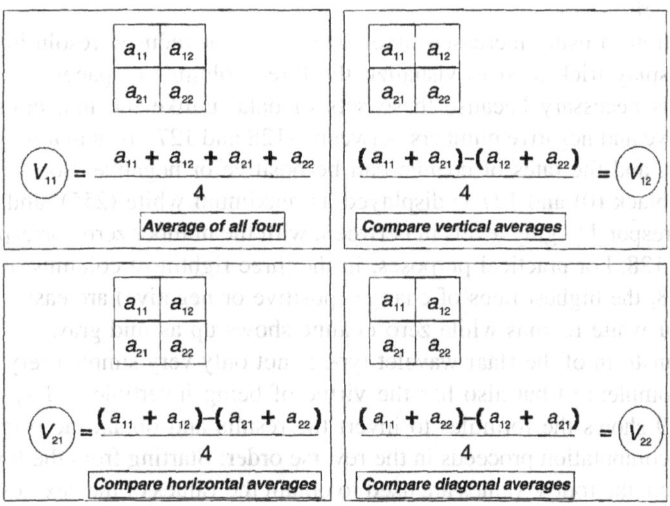

FIGURE 13.7 Four image transformations to build a scalable image representation. Four pixel values in a 2 × 2 cell are combined using the four formulas to obtain four outputs V_{11}, V_{12}, V_{21}, and V_{22}. The process is repeated for the next nonoverlapping 2 × 2 cell of pixels, and so on, until the whole image is represented this way. In fact, this is a simple case of a 2D Haar wavelet transform.[35]

a_{22} of that cell produce four output values V_{11}, V_{12}, V_{21}, and V_{22}. The transform is applied to all non-overlapping cells of 2×2 pixels of the image. The result is four sets of numbers corresponding to the four formulas applied to all 2×2 cells. So, all the V_{11} values produce version V_{11} of the original image. V_{11} has the same aspect proportion but only one-fourth the number of original pixels, and each of these is the average of the values of each 2×2 cell of original pixels. The other three versions, V_{12}, V_{21}, and V_{22} also keep the same proportions and each has one-fourth the number of original pixels. However, these are not just averages; they are local derivatives in three orientations—vertical, horizontal, and diagonal. And, unlike version V_{11}, they can have negative values. It is not by coincidence but by design inspired by neural multiplexing of early vision that the selected operations include both addition (averages) and subtraction (local derivatives) of values.

The process described in Figure 13.7 completes Level 1 of this image representation. Scalability comes to play when version V_{11}, made of 2×2 averages, is used as input to repeat the whole process and obtain Level 2 of the representation, with its own four versions V_{11}, V_{12}, V_{21}, and V_{22} which keep the same proportions but have now one sixteenth of the original number of pixels. The process is recursively applied to compute successive levels, each with one-fourth the resolution of the previous one.

Figure 13.8 illustrates four levels showing the reduction in resolution as increased pixel size. The reason to show it this way is to facilitate visual inspection and comparison of the panels in the figure. So, each row shows the input pixels on the left panel, and the computed values are shown in the other four panels with pixels four times in size. Given that Level 2 takes as input version V_{11} of Level 1, the last four panels of Level 2 show pixels 16 times the size of the original pixels in the top left panel.

In addition to using increased pixel size in lieu of reduced resolution, there is another display trick used to visualize the three columns of panels on the right; this trick is necessary because these sets of data, unlike the first column, have both positive and negative numbers between -128 and 127 given that they are local derivatives, and the rates of change can be positive or negative. So, -128 is displayed as black (0) and 127 is displayed as maximum white (255), and all others values correspond to gray levels in between, with the number zero corresponding to gray level 128. For practical purposes, in the three rightmost columns of panels in Figure 13.8, the highest rates of change (positive or negative) are easily identified by black or white regions while zero change shows up as mid gray.

This transform of the Haar wavelet type is not only very simple (very low computation complexity) but also has the virtue of being invertible and symmetrical. Figure 13.9 shows the formulas to invert the results and obtain the original pixel cells. The computation proceeds in the reverse order: Starting from the lowest level of resolution, the four V values are used to obtain the values of the next level, and so on. The most important characteristics of this transform are (a) there is the same number of input and output values, (b) there is a procedure to obtain all original pixel values, and (c) this procedure has the same simplicity as the forward transformation formulas. In practical terms, you can always get the original image from the transformed image.

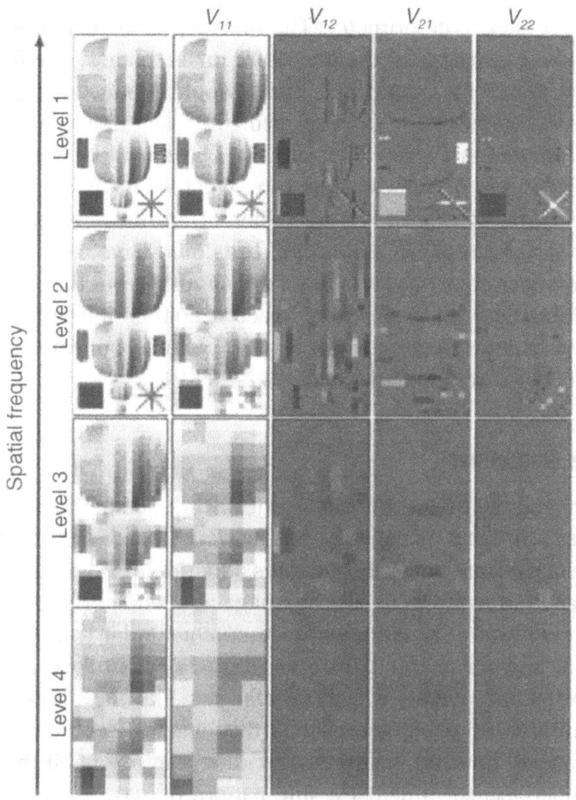

FIGURE 13.8 Illustration of a scaling strategy applied to an image. Top row: In Level 1 of scalability, versions V_{11}, V_{12}, V_{21}, and V_{22} are computed from the original on the left. Second row: Versions V_{11}, V_{12}, V_{21}, and V_{22} for Level 2 of scalability are computed from version V_{11} of Level 1; and so on. See text for explanation of the spatial frequency axis.

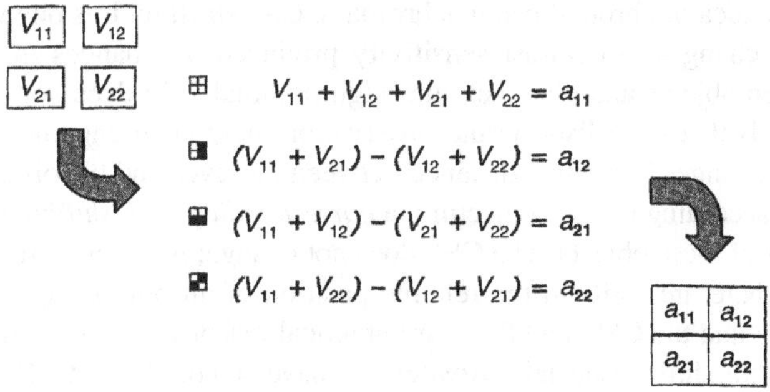

FIGURE 13.9 Top left illustrates forward transformed values V used as input for the inverse transform formulas. (center) which produce the original values a (bottom right). Small squares on the left of each formula illustrate the property of the *inverse* transform of having essentially the same operations as the forward formulas: integral for the first, and vertical, horizontal, and diagonal local derivatives for the other three.

Now let us get back to colorimetry. The first step is to keep in mind that similar transforms can be computed for the two other image planes, one for a^* values and the other for b^* values for each pixel. These would produce corresponding image versions as those computed for L^*. The result for each image would be three sets like the one illustrated in Figure 13.8; one for L^* values like the one in the figure, and one each for a^* and b^* values. Therefore, what we have is merely an invertible spatial representation of all the $L^*a^*b^*$ values of an image through a two-dimensional wavelet transform of the Haar type. We could have used RGB, XYZ, CIECAM02, or any other color space values for each pixel, but the example intends to be tutorial and practical, and the $L^*a^*b^*$ space is commonplace for color industrial applications and belongs to the general class of color-opponent spaces which have been used to model chromatic and achromatic CSFs.

Scale-Shifting Conjecture

There is a very common behavior observed when a person is asked to examine a picture: holding it in front of the eyes and moving it back and forth depending on the image features that are under the person's attention. It is brought closer to examine fine detail and farther away to see the image as a whole. If the picture is a painting hanging on the wall, it is common to observe a similar behavior, getting closer to the canvas to appreciate small features, even brush strokes, and then stepping away to see better the overall visual effects, just like the different views of the painting of Dali used as example earlier in this chapter (see Figure 13.2). In practical terms, suppose someone measures one data point of a CSF by measuring the minimum contrast needed to detect a sine pattern of, say, 8 cycles per centimeter from a distance of, say, 40 cm. Then the viewing distance is halved to 20 cm but the pattern remains *unchanged* at 8 cycles per cm. The two main consequences are (a) The spatial frequency of the pattern projected *on the person's retina* is halved, and (b) the new setting of minimum contrast for detection of that pattern should correspond to the CSF for the *new* spatial frequency projected on the eye. Figure 13.10 illustrates the idea of chromatic and achromatic CSF shifting. It is one way to represent relative changes in contrast sensitivity produced by changes in viewing distance between objects and observer, although nominal CSF does not change itself.

The point is that in realistic visual observation there is an ongoing scale shifting produced by changes in relative distance between our eyes and the observed objects, and there is, according to this conjecture, a *corresponding CSF shifting* when applied to the images of these objects. The CSF does not change; it is a property of the visual system; but there are shifts in the relative position of an object's spatial frequency representation and the CSF, and there are practical reasons to account for such shifts in terms of CSF shifts, especially if/when we have standards for CSFs.

Multiscale Colorimetry: A Spatiotemporal Path Forward

The idea is then to develop colorimetry that is consistent with the CSF changes brought about by such scale shifting. A tentative name would be multiscale

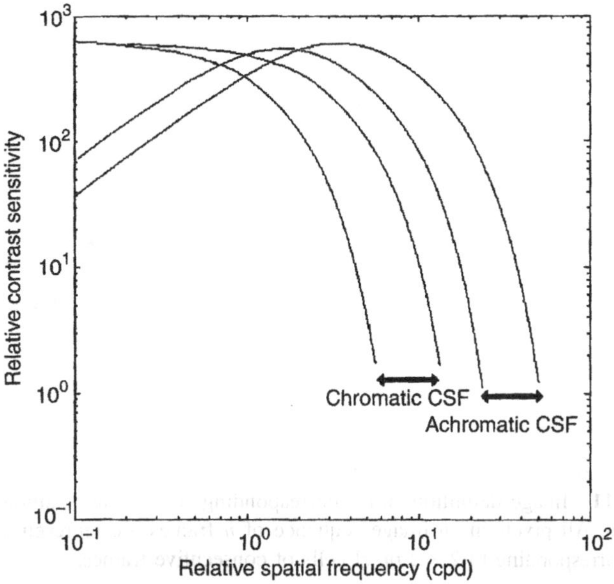

FIGURE 13.10 Illustration of the scale-shifting conjecture for chromatic and achromatic CSFs. Relative changes in viewing distance between objects and observer can be represented by CSF shifts even though CSFs do not change themselves. Note the horizontal axis in relative units.

colorimetry. This approach is rooted in the so-called pyramid procedures of information processing that have been successful in several practical areas.[36,37]

These are just initial ideas for a possible development. Many details await study. From the practical point of view, consider the vertical axis in Figure 13.8 labeled "spatial frequency." It is consistent with low spatial frequencies corresponding to lower resolution levels of the transform, and higher spatial frequencies corresponding to finer resolution levels, in this example by frequency doubling from one level to the next. However, there are no specific values on that axis. Let us say that all possible levels of wavelet scaling are computed (in the example, it is possible to compute Levels 5 and 6). Therefore, image scaling is complete and it will not change if the image is viewed from far or close. But CSF sensitivities will shift up and down depending on the viewing distance. However, to apply any shifting, it would be convenient to have a benchmark value for each and all images, to use it as reference. For example, to specify which of the image levels of resolution has a 2×2 pixel area such that when viewed from 40 cm spans 0.01 degrees of visual angle. After that, one of the tasks to do is to assign *appropriate* CSF values to each of these levels of image resolution (the easiest solution would be to apply the sampling theorem limit to the 2×2 cell size in visual angle units, e.g., 0.01 degrees would represent a spatial frequency of 50 cpd). The next problem would be to establish appropriate operations to *weight* the wavelet-transformed L^*, a^*, b^* values with assigned CSF values, and from that, it should be possible

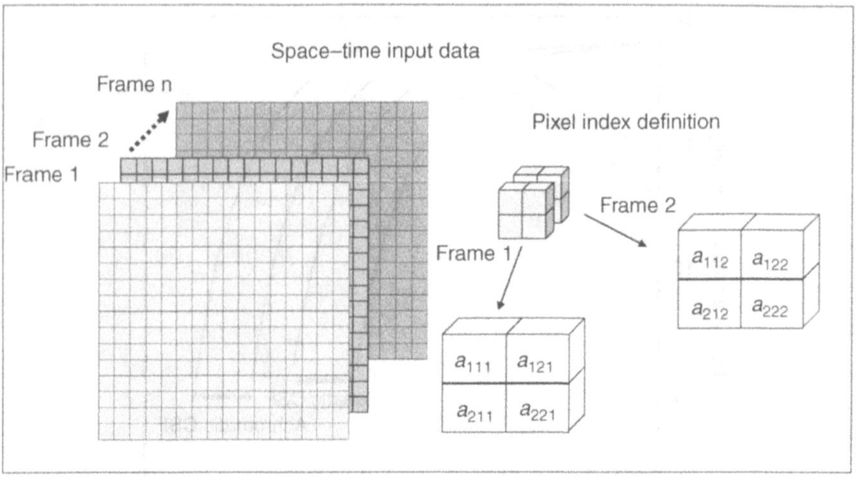

FIGURE 13.11 Image definitions for a corresponding space–time example of multiscale transformation. All pixels in an image sequence of n frames are represented as $2 \times 2 \times 2$ pixel cubes corresponding to 2×2 pixel cells of consecutive frames.

to define some *adequate statistics* to summarize in practical terms the multiscale colorimetry of any image. It is easy to realize that there are many important details pending to be worked out within this approach.

For completeness, it should be mentioned that a complementary reasoning can be followed to include temporal variations. Consider Figure 13.11 illustrating successive, changing frames of a digital color image. Taking the first two frames (two shots at the highest temporal frequency), one can define cubes of $2 \times 2 \times 2$ pixels that represent spatial and temporal information. Then, as shown in Figure 13.12, a wavelet of the Haar type can be defined, such that for a single frame it is reducible to the one applied above, before time was introduced.

Gray cubes in Figure 13.13 represent the set of transformed values obtained by applying the first forward transform formula (average) which simply produces one half of the frames containing a total of one eighth of the original pixels. This set is used as input to transform the next level, and so on. At each level, the other seven transform formulas produce the values represented by the clear space inside the largest cubes, which are computed only once. After all possible levels are computed, the total of transformed values V is equal to the original number of values a in the input.

Following the iteration procedure illustrated in Figure 13.13, several levels of spatiotemporal resolution can be computed. These can be applied in turn to shifting spatiotemporal CSFs to define useful metrics and statistics for specific applications of multiscale colorimetry.

Therefore, the conceptual frame outlined for multiscale colorimetry is relatively easy to extrapolate from spatial to spatiotemporal scaling, but the complexity grows fast, and difficult problems can be foreseen ahead; so, this is work in its very early stages. After spatiotemporal CSFs for chromatic and achromatic contrast

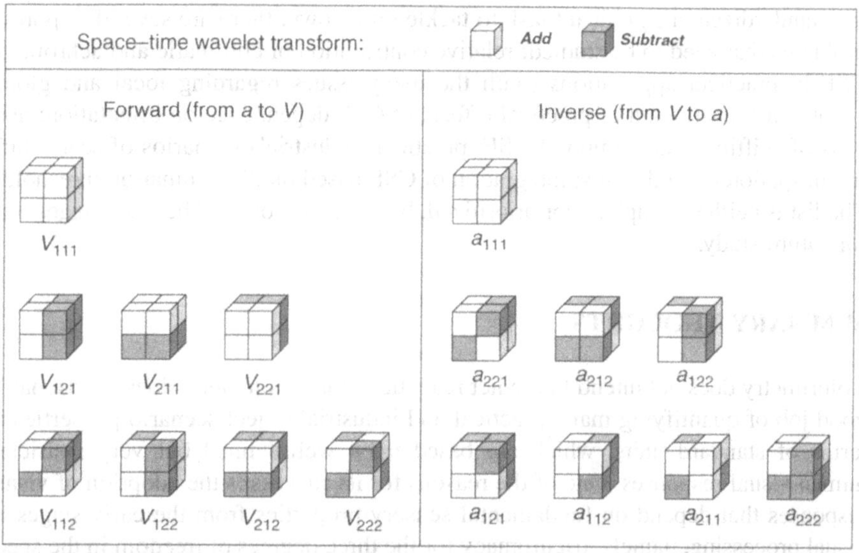

FIGURE 13.12 Forward and inverse transform formulas for each space–time cube. The eight formulas are represented by eight cubes with gray or white pixels within each cube to indicate subtraction or addition. For clarity, it was omitted that all forward transforms are divided by 8. Pixel subindices for each forward transform formula follow definition as in Figure 13.11, and similarly for inverse transform formulas, with the same index definition applied to V values used as input.

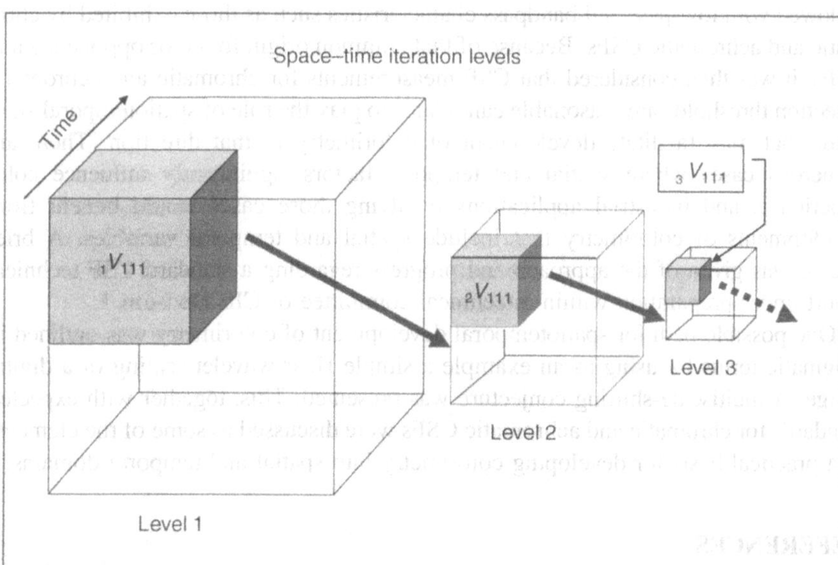

FIGURE 13.13 Illustration of space–time scaling produced by iteration of the Haar wavelet transform.

are standardized, a significant task to tackle on its own, there are several important problems that need to be studied: relative contribution of chromatic and achromatic CSF in practical applications (with the many issues regarding local and global adaptation); evaluation of practical effects of CSF dependence on orientation; analysis of shifting spatiotemporal CSF; practical (industrial) scenarios of scale shifting in spatiotemporal terms; integration of CSF based on discrimination thresholds. The list is neither complete nor prioritized, but merely indicated here as suggestions for future study.

SUMMARY THOUGHTS

Colorimetry does not intend to predict realistic color vision, but it does a reasonably good job of quantifying many practical and industrial object-scenario properties in terms of standard units, which are based upon well-defined but very restricted human visual responses. One of the reasons for its success is the adoption of visual responses that depend on fundamental sensory properties from the early stages of visual processing, namely, trichromacy for the three degrees of freedom in the spectral dimension, and early color-opponent, center-surround neural interactions for the spatial and temporal dependence of the shape and properties of the luminance and color-matching functions.

A brief discussion was outlined about the relationship between spatiotemporal and color opponency in terms of neural multiplexing that occurs at the early processing stages of vision, at the same level as color opponency. The main consequences are that chromatic and achromatic signals are locked in spatiotemporal register and are endowed with low-pass and bandpass characteristics such as those exhibited by chromatic and achromatic CSFs. Because of that common origin for color opponency and CSFs, it was then considered that CSF measurements for chromatic and achromatic detection thresholds are reasonable candidates to play the role of spatiotemporal standards that may facilitate development of colorimetry in that direction. There are numerous cases where spatial and temporal factors significantly influence color experience, and industrial applications involving those cases would benefit from developments of colorimetry that include spatial and temporal variables. A brief update was given of the approach and progress regarding a standard CSF technical report under preparation within a technical committee of CIE Division 1.

One possible path for spatiotemporal development of colorimetry was outlined in pragmatic terms by using as an example a simple Haar wavelet scaling of a digital image. A multiscale-shifting conjecture was presented. This, together with expected standards for chromatic and achromatic CSFs were discussed as some of the elements of a practical basis for developing colorimetry into spatial and temporal domains.

REFERENCES

1. Newton I (1730) *Opticks, or a Treatise of the Reflections, Refractions, Inflections and Colors of Light. Wm. Innys*, London.

2. Wright WD (1967) *The Rays Are Not Colored. Hilger*, London.

3. Robinson HJ (1984) A theorist's philosophy of science. *Phys. Today*, **84**, 24–32.

4. Lennie P, D'Zmura M (1988) Mechanisms of color vision. *CRC Crit. Rev. Neurobiol.*, **3**, 333–400.

5. Rushton WAH (1975) Visual pigments in man, in: *Handbook of Sensory Physiology* Vol. VII/1 (Ed., H. J. A. Dartnall), Springer Verlag, New York.

6. Pokorny J, Smith VC (1986) Colorimetry and color discrimination. in: *Handbook of Perception and Human Performance* Vol. 8 I, Ch 8, (Eds., K. R. Boff, L. Kaufman, and J. P. Thomas), Wiley & Sons, New York.

7. Abney W. deW (1913) *Researches in Color Vision and the Trichromatic Theory*, Longman, Green & Co.

8. CIE (1988) *Spectral luminous efficiency functions based upon brightness matching for monochromatic point sources, 2° and 10° fields*, CIE Publ. **75**-1988.

9. Wagner G, Boynton RM (1972) Comparison of four methods of heterochromatic photometry, *J. Opt. Soc Am.*, **62**, 1508–1515.

10. Wyszeki G, Stiles WS (1982) *Color Science: Concepts and Methods, Quantitative Data and Formulae*, Wiley, New York.

11. Olzak LA, Thomas JP (1986) Seeing spatial patterns. in: *Handbook of Perception and Human Performance*, Vol. I, Ch 7, (Eds., K. R. Boff, L. Kaufman, and J. P. Thomas), Wiley & Sons, New York.

12. De Valois RL, De Valois KK (1988) in: Sensitivity to color variations, *Spatial Vision*, Ch 7, Oxford University Press, Oxford.

13. Riggs LA (1971) *Vision, Holt, Rinehart and Winston*, New York.

14. Watson AB (1986) Temporal Sensitivity. in: *Handbook of Perception and Human Performance* Vol. I, Ch 6, (Eds., K. R. Boff, L. Kaufman, and J. P. Thomas), Wiley & Sons, New York.

15. Riggs LA (1965) Visual acuity, in: *Vision and Visual Perception* (Ed., C. H. Graham), Ch. 11, Wiley & Sons, New York.

16. Kelly DH (1979), Motion and vision. II. Stabilized spatiotemporal threshold surface, *J. Opt. Soc. Am.*, **69**, 1340–1349.

17. Shevell S (Ed.) (2003) *The Science of Color*, OSA and Elsevier Science, New York.

18. Cornsweet TN (1970) *Visual Perception*, Academic Press, New York.

19. Marr D (1982) *Vision, W. H. Freeman*, San Francisco.

20. Laming DRJ (1997) *Sensory Analysis*, Academic Press, New York.

21. Falmagne JC (1986) Psychophysical measurement and theory, in: *Handbook of Perception and Human Performance* Vol. I, Ch 1, (Eds., K. R. Boff, L. Kaufman, and J. P. Thomas), Wiley & Sons, New York.

22. Mullen KT (1985) The contrast sensitivity of human color vision to red-green and blue-yellow gratings. *J. Physiol.*, **359**, 381–400.

23. Webster MA, Wilson JA (2000) Interactions between chromatic adaptation and contrast adaptation in color appearance. *Vis. Res.*, **40**, 3801–3816.

24. Nachmias J, Sansbury RV (1974) Grating contrast: Discrimination may be better than detection. *Vis. Res.*, **14**, 1039–1042.

25. Daly S (1993) The visible differences predictor: an algorithm for the assessment of image fidelity, in: *Digital Images and Human Vision, (Ed., A. B. Watson),* MIT Press, Cambridge, MA.
26. Ingling CR, Martinez-Uriegas E (1985) The spatiotemporal properties of the r-g X-cell channel. *Vision Res.*, **25**, 33–38.
27. Martinez-Uriegas E (1994) Chromatic-achromatic multiplexing in human color vision, in: *Visual Science and Engineering, Models and Applications*, (Ed., D. H. Kelly), Marcel Dekker, New York.
28. Kelly DH (1989) Opponent-color receptive-field profiles determined from large-area psychophysical measurements. *J. Opt. Soc. Am.*, **A6**, 1784–1790.
29. Boynton RM (1979) *Color Vision*, Holt Rinehart & Winston, New York.
30. Guth SL (1991), Model for color vision and light adaptation. *J. Opt. Soc. Am*, **A8**, 976–993.
31. Campbell FW, Robson JG (1968) Application of Fourier analysis to the visibility of gratings. *J. Physiol.*, **197**, 551–566.
32. Watson AB (2000) Visual detection of spatial contrast patterns: Evaluation of five simple models. *Optics Express*, **6**, 12–33.
33. Martinez-Uriegas E, Larimer JO, Lubin J, Gille J (1995) Evaluation of image compression artifacts with ViDEOS, a CAD system for LCD color display design and testing. *SPIE Proc. Ser.*, **2411**, 74–82.
34. Barten PB (1999) *Contrast Sensitivity of the Human Eye and Its Effects on Image Quality*, SPIE Press, Bellingham, Washington.
35. Iyengar SS, Prasad L (1997) *Wavelet Analysis with Applications to Image Processing*, CRC Press, Boca Raton.
36. Burt PJ (1988) Algorithms and architectures for smart sensing, in: *Proceedings of the DARPA Image Understanding Workshop*, pp. 139–153.
37. Burt PJ (1988) Smart sensing with a pyramid machine, *IEEE Proc.*, **76**(8), 1006–1015.

14

THE FUTURE OF COLORIMETRY IN THE CIE

ROBERT W.G. HUNT
Barrowpoint, 18 Millennium Close, Odstock Road, Salisbury, Wiltshire SP2 8TB, UK

INTRODUCTION

It has long been accepted that the work of the CIE in colorimetry falls into three broad categories: color matching, color difference, and color appearance. Possible activities of the CIE in these areas in the future will now be reviewed.

COLOR MATCHING

The 1931 Standard Colorimetric Observer has had a remarkably long life. That it is still in general use today is a great tribute to the work of David Wright and John Guild all those years ago.[1,2] Indeed the only significant error in the standard is the underestimate of the luminance at wavelengths below 460 nm, which arose not from the work of Wright and Guild but from the incorporation of the then existing $V(\lambda)$ function.[3] It is interesting to ask whether this error is likely to be corrected in the future. When such a correction has been discussed in the past, it has been concluded that the gain in practical applications would not justify the upheaval caused by such a change, and this seems likely to continue to be the view generally held.

The 1964 Standard Colorimetric Observer does not suffer from the above error because the $V(\lambda)$ function plays no part in this observer. However, the CIE has recently recommended[4] the $\bar{y}_{10}(\lambda)$ function as a 10° luminous efficiency function,

Colorimetry: Understanding the CIE System, Edited by János Schanda

$V_{10}(\lambda)$, and it remains to be seen how justified it will be to use it as a weighting function for deriving 10° luminances from spectral power data; this could well be a subject that the CIE should pursue in the future.

The rule that, if the angular subtense of view is greater than 4°, then the 1964 Observer should be used, instead of the 1931 Observer, is clearly arbitrary; there is no sudden change in the visual color-matching properties of the eye at 4°. This problem and many others are being addressed currently by a CIE Technical Committee (CIE TC1–36, Fundamental Chromaticity Diagram with Physiologically Significant Axes). This committee is going to recommend fundamental response curves for 10° and 2° observers, together with the information on photopigment absorption spectra and spectral absorptions in the ocular media for different field sizes and observer age. This information will enable best estimates to be made of the average color-matching functions for a range of field sizes and observer age. A chromaticity diagram with physiologically significant axes is also being developed. The motivation for the work of this committee has come largely from vision scientists who would find its results valuable in their work. But, for industrial applications, the use of multiple sets of color-matching functions would not be practicable, and the proposed chromaticity diagram will be very nonuniform.

The work of Thornton[5] and Oulton[6] and coworkers has cast doubt on the validity of the additivity of the color matches for individual observers. If such additivity were shown to be in serious error, the whole basis of calculating tristimulus values by summations at a series of wavelengths would be undermined. Fortunately, it now appears that, while nonadditivity evidently occurs for individual observers, when the results of groups of observers are averaged, additivity then holds well enough for practical purposes.[7] However, this is an area that merits further study by the CIE, and another CIE Technical Committee (TC 1–56, Improved Colour Matching Functions), is actively studying some of the issues; in particular, it is intended to test the effects of luminance level and the validity of transforming to different sets of primaries (initially by carrying out many trials using a single observer).

The Standard Deviate Observer introduced by the CIE in 1989 to predict the magnitude of observer metamerism, now appears to underestimate the difference between observers by quite a large factor.[8,9] It is a matter of some urgency that a more representative Standard Deviate Observer be made available.

It is remarkable that the 1931 Observer is used so widely, in view of the fact that it was based on only 17 observers, all from the same country. The 1964 Observer was based on about 60 observers, situated in two countries. Very little work has been carried out to see if there are differences in color-matching properties between observers of different gender, or from different races or cultures. There are several facts that suggest that such differences, if they exist, are not large. First, the CIE 1931 and 1964 Observers are used all over the world with apparent success. Second, problems of illuminant and observer metamerism in the colorant industries do not seem to arise differently among these groups. Third, the color rendering in photography, printing, television, and digital imaging is sufficiently acceptable all over the world to sustain huge industries, in spite of all these systems involving large degrees of metamerism. Fourth, the very sensitive Ishihara test for defective color vision was developed in Japan, but is used successfully all over the world. However, in spite of the above facts, it would be better to have some actual data on the color-matching properties of the different groups, and the CIE might regard this as a desirable topic for future study.

Another area that is in need of urgent attention is that of Standard Daylight Illuminants. The D series of illuminants provide good relative spectral power data for outdoor daylight, but they cannot be realized by actual sources. This means that, when tristimulus values have been computed using any of the D illuminants, there is no practicable method for actually viewing the samples under the same illuminant.[10] Another difficulty is that when spectrophotometry is being carried out on samples that fluoresce, if, as is usually the case, a single monochrometer instrument is used, the sample should be illuminated by the standard light source being adopted; but in the case of the D illuminants these do not exist. The problem was exacerbated by the decision in 1963 by the CIE to define the D illuminants at every 10 nm with values at the intermediate wavelengths being arrived at by linear interpolation; the result is a series of sawtooth spectral power distributions which are impossible to replicate as real sources. A solution to these problems is not easy to come by. One possibility is to standardize a source that can be realized, such as a tungsten halogen lamp, run at a specified correlated color temperature and used together with a filter having a specified spectral transmittance; such a source would not match the spectral power distribution of the typical daylight exactly, but it could probably do so well enough for practical purposes, and could be used in viewing cabinets and in spectrophotometers. A further problem that could be addressed at the same time is that indoor daylight differs significantly from outdoor daylight, particularly in the ultraviolet region, and another CIE Technical Committee (TC1–44, Practical Daylight Sources for Colorimetry) is studying this; to cover this point, two filters could be recommended, one for indoor daylight and the other for outdoor daylight.[11] More complicated sources can be envisaged, such as a collection of Light Emitting Diodes, but obtaining enough illuminance might be difficult, and the stability of such sources would have to be evaluated. Another CIE Technical Committee (CIE TC 1-66, Indoor Daylight) is investigating this concept; its terms of reference are to prepare a CIE recommendation on an Indoor Daylight Illuminant and a corresponding Indoor Daylight Source, considering the needs of the partner international standards organizations. The specific industry in mind is the paper industry, which is still using illuminant C.

COLOR DIFFERENCE

The introduction in 1976 by the CIE of the two color difference spaces, CIELUV and CIELAB, with their associated color-difference formulas, represented a very important step in the promotion of practical colorimetry; prior to 1976, there were a dozen or more color-difference formulas being used. It was hoped that only a single space and formula would be acceptable; but it proved impossible to meet both the requirements of those in the television industry (who wanted an associated approximately uniform chromaticity diagram, as provided by the u',v' diagram in the CIELUV system), and those in the colorant industries (who wanted a space and formula similar to the ANLAB formula then currently widely used, as provided by the CIELAB system).

The absence of an associated chromaticity diagram in the CIELAB space results in the absence of a measure in that space that correlates with saturation. In the CIELUV space, the measure $s_{uv} = C^*_{uv}/L^*$ provides a correlate of saturation. Some workers have suggested using an analogous measure $s_{ab} = C^*_{ab}/L^*$ to provide a correlate of saturation in the CIELAB system, but this is not satisfactory. A series of colors of constant saturation (a shadow series) is obtained by multiplying a set of tristimulus values by a constant. When this is done in the CIELUV system, s_{uv} remains constant as required. But in the case of s_{ab} , if, as usual, Y_n is the value of Y for the reference white, k is the constant, and Y_1 is the value of Y when $k = 1$, and C^*_{ab1} is the value of C^*_{ab} when $k = 1$, then

$$L^* = 116(kY_1/Y_n)^{1/3} - 16$$

and

$$s_{ab} = k^{1/3}C^*_{ab1}/[116(kY_1/Y_n)^{1/3} - 16]$$

and this is not independent of k, so that s_{ab} does not remain constant. In the absence of the term -16, this would simplify to

$$s_{ab} = C^*_{ab1}/[116(Y_1/Y_n)^{1/3}]$$

which is independent of k, but of course the constant -16 is necessary in the formula for L^*. The simplest solution to this problem is to use s_{uv} as the correlate of saturation in both spaces.

Since 1976, improvements to the accuracy with which perceived color differences can be predicted in CIELAB-type color-difference formulas have been offered with the CIE94, CMC, and CIEDE2000 formulas. These formulas are increasingly more complicated than that of CIELAB, but, when the best possible predictions are required, they offer progressively better and better results. It seems unlikely that there will be much more accuracy to squeeze out of the CIELAB basis, so that CIEDE2000 may be the last in this series of formulas.

A factor that merits attention is the effect of the size of the color differences being considered on the choice of a uniform space and difference formula.[12–16] In the colorant industries, differences close to the threshold for juxtaposed samples are of importance, whereas in the imaging industries the larger differences tolerable when the samples are separated in space, and frequently also in time, are of concern. Another CIE Technical Committee (CIE TC 8-02, Colour Difference Evaluation in Images) is addressing the topic of color differences in imaging.

When the CIELUV and CIELAB spaces were introduced, it was stated that they were intended to be used in illuminants of near daylight quality. However, color differences also need to be evaluated in illuminants of other colors, and the best way to do this is to use a chromatic adaptation transform (CAT) to convert the tristimulus values to those of the corresponding colors in a daylight illuminant. If this is not done and the formulas are used as they stand, then errors will occur, particularly with the CIELUV space. The latest CAT embodied in CIE recommendations is the one that is used in CIECAM02; it is desirable that this CAT be recommended by the CIE for general use.

Other limitations of the CIELUV and CIELAB spaces, stated in 1976, are that they were intended to be applied to object colors of the same size and shape, viewed in identical white to mid-gray surroundings. The effects of changes in size, shape, and surroundings, and on the apparent size of color differences are an area that the CIE might encourage workers to address. The best way to tackle these problems might well be to use a color appearance model that has means for evaluating color differences embodied in it, as suggested by Luo et al..[12] This would also have the advantage of unifying the spaces used for evaluating color differences and color appearance.

COLOR APPEARANCE

The introduction in 1997 of the CIECAM97s Colour Appearance Model indicated the entry of the CIE into the area of color appearance in earnest. The drive for this activity came from the imaging industry, in which differences in the viewing conditions often occur. CIECAM97s was quite successful, but, as it was used, various shortcomings became apparent. Therefore, in 2002, the CIE introduce an improved version CIECAM02.[17]

There are features in both these models that might with advantage be improved. These include the correlates of yellowness–blueness, redness–greenness, brightness, and colorfulness.

The correlates of yellowness–blueness and redness–greenness are based on the differences between the cone signals[18,19] (after adaptation and the application of the dynamic-response function), R'_a, G'_a, B'_a:

$$C_1 = R'_a - G'_a \qquad C_2 = G'_a - B'_a \qquad C_3 = B'_a - R'_a$$

The correlate of redness–greenness, a, is based on the magnitude of the departure from the criterion for unique yellow ($C_1 = C_2/11$), the departure from the criterion for unique blue not having being included in an average because of the lower precision with which unique blue hues can be identified experimentally.

$$\begin{aligned} a &= [C_1 - C_2/11] \\ &= [(R'_a - G'_a) - (G'_a - B'_a)/11] \\ &= [R'_a - 12G'_a/11 + B'_a/11] \end{aligned}$$

The correlate of yellowness–blueness, b, is based on the average of the magnitudes of the departures from the criterion for unique red ($C_1 = C_2$) and from the criterion for unique green ($C_1 = C_3$), together with a factor of 1/(4.5) to allow for the paucity of short-wavelength cones.

$$\begin{aligned} b &= (1/2)[C_2 - C_1 + C_1 - C_3]/(4.5) \\ &= (1/2)[(G'_a - B'_a) - (R'_a - G'_a) + (R'_a - G'_a) - (B'_a - R'_a)]/(4.5) \\ &= (1/9)[R'_a + G'_a - 2B'_a] \end{aligned}$$

$C_2 - C_1$ is used instead of $C_1 - C_2$ so that yellowness is positive as in the case of $C_1 - C_3$.

However, because the criteria for unique red, unique green, unique yellow, and unique blue are all different, there is a discontinuity as the color considered passes from one to a neighboring hue quadrant. The four hue quadrants are

Orange (reddish and yellowish), when $C_1 \geq C_2/11$ and $C_2 \geq C_1$
Lime (greenish and yellowish), when $C_1 < C_2/11$ and $C_1 \geq C_3$
Cyan (greenish and bluish), when $C_1 < C_2/4$ and $C_1 < C_3$
Magenta (reddish and bluish), when $C_1 \geq C_2/4$ and $C_2 < C_1$

It would, therefore, be more correct to have two correlates of redness–greenness, a_y for yellowish colors, and a_b for bluish colors, thus

$$\begin{aligned} a_y &= (C_1 - C_2/11) = [(R'_a - G'_a) - (G'_a - B'_a)/11] = [R'_a - 12G'_a/11 + B'_a/11] \\ a_b &= (C_1 - C_2/4) = [(R'_a - G'_a) - (G'_a - B'_a)/4] = [R'_a - 5G'_a/4 + B'_a/4] \end{aligned}$$

and two correlates of yellowness–blueness, b_r for reddish colors, and b_g for greenish colors, thus

$$\begin{aligned} b_r &= (C_2 - C_1)/(4.5) = [(G'_a - B'_a) - (R'_a - G'_a)]/(4.5) = [2G'_a - R'_a - B'_a]/(4.5) \\ b_g &= (C_1 - C_3)/(4.5) = [(R'_a - G'_a) - (B'_a - R'_a)]/(4.5) = [2R'_a - G'_a - B'_a]/(4.5) \end{aligned}$$

Then, a_y and b_r would be used for orange colors; a_y and b_g for lime colors; a_b and b_g for cyan colors; and a_b and b_r for magenta colors.

These more elaborate correlates of redness–greenness and yellowness–blueness should result in better correlation with experimental determinations of these perceptions; and, when incorporated in the formulas for the correlates of chroma, colorfulness, and saturation, they might also result in improvements in the predictions for these perceptions. The use of a_y, a_b, b_r, and b_g in computing hue angle would not be expected to make much difference to hue quadrature because this measure is anchored at the four unique hues.

The formula, in CIECAM02, for the correlate of brightness, Q, contains A_w, the achromatic signal for the reference white. As the adapting luminance decreases, A_w, decreases, and this decreases Q, as required; but the decrease is insufficient so the formula includes a power of F_L, the luminance-level adaptation factor. However, if

F_L were altered so as to increase the separation of the dynamic-response function curves along the log I axis, it might be possible to avoid having to use F_L in the formula for Q, so that its decrease with adapting luminance then depended only on the dynamic-response function (as is the case in CIECAM97s). This would be a more physiologically plausible result.

The formulas, in both CIECAM97s and CIECAM02, for the correlate of colorfulness, M, contain the correlate of chroma, C; but C is derived as a ratio of $[(a^2 + b^2)^{0.5}]$ over $[R'_a + G'_a + (21/20)B'_a]$, and hence if R'_a, G'_a, and B'_a are all multiplied by the same constant (as tends to happen when the adapting luminance changes), the value of C is not changed. This necessitates the inclusion of a power of F_L in the formulas for M, to make M decrease as the adapting luminance decreases. However, if M were made to depend on $[(a^2 + b^2)^{0.5}]$ without the $[R'_a + G'_a + (21/20)B'_a]$ divisor, then, as the adapting luminance decreased, M, would decrease as required. It might then be possible to avoid having to use F_L in the formulas for M, so that its decrease with adapting luminance then depended only on the dynamic-response function. This would be a more physiologically plausible result.

The s in CIECAM97s indicated that this was a simple model in that there were various features that were not provided; these included a response from the rods, the Purkinje effect, cone bleach factors, the Helson–Judd effect, a low-luminance tritanopia factor, the Helmholtz–Kohlrausch effect, and the Bezold–Brűcke effect; these features are also absent from CIECAM02. A comprehensive model is required to include these features. There is also a need for a model for unrelated colors, such as signal lights. It is to be hoped that the CIE will pursue these requirements in the future.

Neither CIECAM97s nor CIECAM02 offers means for predicting the effects of simultaneous contrast. A model that can do this would be useful in certain industries, such as fabric design and the compilation of posters.

A color appearance model that includes a rod response, and other effects that occur at low light levels, would have applications in several practical areas, including the recognition of the color of coded goods (such as electrical resistors), the recognition of colored signs in street lighting, the recognition of objects in security surveillance, railway lighting, navigation, aviation, emergency lighting, and the cinema.

As mentioned earlier, it is also to be hoped that, at some time in the future, the CIE will be able to recommend a single space that serves the needs of both color appearance and the evaluation of the color differences.

Some work has been done to extend color appearance modeling to include the effects of spatial factors.[20,21] Temporal factors are also important in the moving images used in motion picture films and in television and camcorders. These are also legitimate areas for involvement by the CIE.

Other important factors affecting the appearance of color are gloss and translucency. Gloss is important in the raiment, paint, automotive, plastic, and paper industries; translucency is particularly important in the food industry. These are also areas that the CIE could usefully address. Another CIE Technical Committee (TC1–65, Visual Appearance Measurement) has gloss in its remit.

SOURCES OF FUNDS

There are clearly many interesting and potentially useful areas in colorimetry for the CIE to pursue. A continuing problem is obtaining funding for the work that needs to be done. In the past, companies in the colorant, lighting, and imaging industries were sufficiently prosperous to be able to support this type of work. A typical example is the derivation of the MacAdam ellipses in the Eastman Kodak research laboratories. Many companies in these areas are now facing intense global competition, which has substantially reduced the amount of money available for fundamental research of this type. Universities are also often severely limited in the resources available for research. It is perhaps from those industries whose profit margins are still high that funding for these projects should be sought. In this context, it is encouraging that Microsoft has developed its Windows Color System in which the requirements of the International Color Consortium are embodied.[22]

REFERENCES

1. Guild J (1931) The colorimetric properties of the spectrum. *Phil. Trans. R. Soc. A*, **230**, 149–187.
2. Wright WD (1928–1929) A re-determination of the trichromatic co-efficients of the spectral colours. *Trans. Opt. Soc.*, **30**, 141–161.
3. Gibson KS, Tyndall EPT (1923) Visibilty of radiant energy. *Bull. Bureau Stand.*, **19**, 131–191.
4. CIE (2005) CIE 10 *Degree photopic photometric observer*, Publication **165**-2005.
5. Thornton WA (1992, 1997, 1998) Toward a more accurate and extensible colorimetry. Introduction. The visual colorimeter–spectroradiometer. Part I. Experimental results *Color Res. Appl.*, **17**, 79–122; Part II Discussion **17**, 162–186; Part III Discussion (continued) **17**, 240–262; Part IV Visual experiments with bright fields and both 10° and 1.3° field sizes **22**, 189–198; with Fairman HS Part V Testing visually matching pairs of lights for possible rod participation on the Aguilar-Stiles model **23**, 92–103; Part VI Improved weighting functions. Preliminary results **23**, 226–233.
6. Oulton DP (2004) The properties of multiple CMF determinations using alternative primary sets Part I: Evidence and modelling *Color Res. Appl.*, **29**, 273–284; Part II: A data unification methodology **29**, 438–450.
7. Borbely A, Schanda J (2004) Colour matching using LEDs as primaries. *Color Res. Appl.*, **29**, 360–364.
8. North AD, Fairchild MD (1993) Measuring colour matching functions. Part II. New data for assessing observer metamerism. *Color Res. Appl.*, **18**, 163–170.
9. Oicherman B, Luo R, Tarrant A (2005) The uncertainty of colour matching data, in: *13th Color Imaging Conference*, IS&T, Springfield, VA, USA, pp. 326–332.
10. Hunt RWG (1992) Standard sources to represent daylight. *Color Res. Appl.*, **17**, 293–294.
11. Clarke FJJ (1980) *Practical standard illuminant representative of interior daylight.* CIE Publication No. 50. Proc. 19th Session, Kyoto, pp. 73–78.

12. Luo MR, Lo M-C, Kuo W-G (1996) The LLAB (*l:c*) colour model. *Color Res. Appl.*, **21**, 412–429.
13. Guan S-S, Luo MR (1999) A colour difference formula for assessing large colour differences. *Color Res. Appl.*, **24**, 344–355.
14. Attridge GG and Pointer MR (1997) Some aspects of the visual scaling of large colour differences. *Color Res. Appl.*, **22**, 298–307.
15. Attridge GG, Pointer MR (2000) Some aspects of the visual scaling of large colour differences—II. *Color Res. Appl.*, **25**, 116–122.
16. Kuehni RG (2000) Threshold color differences compared to supra-threshold color differences. *Color Res. Appl.*, **25**, 226–229; How does the size of the difference affect perceived larger color difference, **25**, 381.
17. CIE (2004) *A colour appearance model for colour management systems: CIECAM02*, Publication **159**-2004.
18. Hunt RWG (1998) *Measuring Colour*, 3rd ed., Fountain Press, p. 218.
19. Hunt RWG (2004) *The Reproduction of Colour*, 6th ed., Wiley, pp. 606 and 674.
20. Zhang X, Wandell BA (1996) A spatial extension of CIELAB for digital color image reproduction. *SID Symp. Technical Digest*, **27**, 731–734.
21. Fairchild MD (2004). *Colour Appearance Models*, 2nd ed., Wiley, p. 334.
22. Bourgoin M (2005) The Windows Color System—evolution in the Microsoft Color Management Ecosystem, in: *13th Color Imaging Conference*, IS&T, Springfield, VA, USA, p. 135.

APPENDIX 1

MEASUREMENT UNCERTAINTY

GEORG SAUTER
Physikalisch-Technische Bundesanstalt, AG 4.12 Photometrie, Postfach 3345, D-38023 Braunschweig, Germany.

INTRODUCTION

Whenever the value of a quantity is determined as the result of a measurement, the associated uncertainty is also questioned. Since 1993, a mathematical theory is presented in the "Guide to the Expression of Uncertainty in Measurement,"[1] abbreviated as GUM, which summarizes the definitions related to the evaluation of measurement uncertainty. To avoid any confusion with terms existing from former methods stating the quality of a measurement, new terms are introduced in this theory and are explained in this section. Today, these terms have to be used exclusively when dealing with the measurement uncertainty.

A stated uncertainty is accepted more generally, when the procedure and related calculation for the evaluation is transparent. Thus, the GUM recommends a procedure in eight steps, summarized at the end of the next section. Most of the steps are only mathematics, and computer programs are available for their calculation. But the first three steps are specific and need detailed knowledge about the quantity and the measurement process:

(1) Formulation of the measurement equation for all significant input quantities
(2) Estimation of their values
(3) Determination of the associated uncertainties.

Colorimetry: Understanding the CIE System, Edited by János Schanda

The measurement equation, also denoted as "model of evaluation," can be constructed from a series of modules, and a selection is presented in the third section. For transparency and comparison, the information is listed in an "uncertainty budget," which shows all contributions and their significance in the determination of the output quantity.

Quantities varying to some extend due to the change of a common third quantity are correlated: For example, luminous intensity and distribution temperature of an incandescent lamp depend strongly on the electric current for operating the lamp; they are correlated with respect to the lamp current. Another example yields for values presenting a spectral distribution: Whenever a distribution can be recognized as "typical" or "characteristic," then the relation between the values for neighbored wavelengths—or even for all wavelengths—are correlated. Simply speaking, the variation of the value of one quantity, for example, the current from the power supply, will change the whole distribution in a more or less predictable way. Such a correlation can strongly modify the uncertainty associated to an output quantity, which often is defined in colorimetry as an integral over a spectral range.

The recommendations given in the GUM are valid *only* for "linear models," which is sufficient for the evaluation of uncertainties associated to most—but not all—output quantities in colorimetry. A "nonlinear model" is a measurement equation with a nonsteady first derivative with respect to one or more of the input quantities and within that interval of values covered by the associated uncertainty. The GUM "Supplement I"[2] deals with these situations and uses the Monte Carlo method to solve the problem. An example is also given in the third section.

DEFINITIONS AND TYPES FOR THE EVALUATION OF UNCERTAINTY

Definitions of Terms

Uncertainty (of measurement) is a parameter associated with the *result of a measurement* that characterizes the dispersion of the values that could reasonably be attributed to the measurand.

Note: The term "uncertainty" must not be used to characterize properties of instruments or of intervals of accepted values, both of which are denoted as tolerance intervals. The individual reading or a mean value of a series of observations has an empirical standard deviation, which often is taken as associated uncertainty.

Type A evaluation (of uncertainty): The method of evaluation of uncertainty by the statistical analysis of a series of observations.

Type B evaluation (of uncertainty): The method of evaluation of uncertainty by means of other than statistical analysis of a series of observations.

Standard uncertainty is the uncertainty of the result of a measurement expressed as a standard deviation—the half-width of a normal probability distribution (NPD).

Combined standard uncertainty is the standard uncertainty of the result of a measurement, when that result is obtained from the values of a number of other quantities, calculated as the positive square root of a sum of terms, the terms being the variances or covariances of these other quantities weighted according to the variation of the measurement result with changes in these quantities.

Expanded uncertainty is a quantity defining an interval about the result of a measurement that may be expected to encompass a large fraction of the distribution of values that could reasonably be attributed to the measurand (this interval requires explicit or implicit assumptions regarding an arbitrary probability distribution (PD); it is not necessarily the half-width of a NPD).

Degrees of freedom (DOF) depend on the type A or B of the evaluation of uncertainty and is the (effective) equivalent of the number of independent repeated observations reduced by one (type B evaluation often claims a "complete knowledge" of the distribution, which is equivalent to an infinite DOF).

Coverage factor is a numerical factor used as a multiplier of the combined standard uncertainty in order to obtain an expanded uncertainty.

Note: In colorimetry, the fraction of the distribution is set as 95.45%, leading to a coverage factor $k = 2$ for a sufficiently large effective DOF $\nu_{\text{eff}} > 30$.

Correlation between two quantities means that their values are varying to some extend due to a common third quantity, which significantly affects the random properties of the two others.

Sensitivity coefficients describe how the output estimate varies with changes in the values of the input estimates. They are determined as the partial derivatives of the model of evaluation with respect to the input quantities.

Types for the Evaluation of Uncertainty

The "type A" evaluation of uncertainty is based on $1 \leq i \leq n$ repeated (independent) observations. The mean value $\bar{x}$, "empirical variance" $s^2(x_i)$ of the individual observations, "empirical standard deviation" $s(x_i)$, and the "empirical standard deviation of the mean" $s(\bar{x})$ are calculated in Equation (A1.1). Provided $n > 30$, then the latter is taken as standard uncertainty and for the DOF yields $\nu = n - 1$.

$$\bar{x} = \frac{1}{n}\sum_{i=1}^{n} x_i, \quad s^2(x_i) = \frac{1}{n-1}\sum_{i=1}^{n}(x_i - \bar{x})^2$$
$$u(\bar{x}) = s(\bar{x}) = s\frac{(x_i)}{\sqrt{n}}, \quad \nu = n - 1 \tag{A1.1}$$

For two or more input quantities x_i, z_i measured simultaneously, the estimated covariance $u(\bar{x}, \bar{z}) \approx s(\bar{x}, \bar{z})$ and the correlation coefficient $-1 \leq r(\bar{x}, \bar{z}) \leq 1$ can be determined from $1 \leq i \leq n$ repeated observations.

$$\bar{z} = \frac{1}{n}\sum_{i=1}^{n} z_i, \quad s(\bar{x}, \bar{z}) = \frac{s(x_i, z_i)}{n} = \frac{1}{n(n-1)}\sum_{i=1}^{n}(x_i - \bar{x})(z_i - \bar{z})$$
$$r(\bar{x}, \bar{z}) = \frac{s(\bar{x}, \bar{z})}{\sqrt{s(\bar{x})\, s(\bar{z})}} \tag{A1.2}$$

Note: The observations for two quantities measured sequentially appear as taken "simultaneously," if the total duration of the measurements is short compared to the time for the variation of a common, third quantity.

Correlation can also be stated from the knowledge about a common origin affecting two or more input quantities and additionally their random behavior. For example, the individual values of a set of transfer standards are *logically correlated*, when calibrated in the same measurement campaign, and so the uncertainty associated to the average of these values has to take the correlation into account.

Input quantities not being determined from repeated observations but from any other available information have to be taken into account, which is referred to as "type B" evaluation of uncertainty. Often an upper limit a_+ and a lower limit a_- of an interval are known without any further specific knowledge, which means a rectangular probability distribution (RPD) with a mean value $\bar{x}$, an associated standard uncertainty $u(\bar{x})$ and an infinite DOF $\nu(u(\bar{x})) \to \infty$ is known.

$$\bar{x} = \frac{a_+ + a_-}{2}, \quad u(\bar{x}) = \frac{1}{\sqrt{3}}\left|\frac{a_+ - a_-}{2}\right| = \frac{a}{\sqrt{3}}, \quad \left[\nu(u(\bar{x})) = \frac{1}{2}\left(\frac{u(\bar{x})}{\Delta u(\bar{x})}\right)^2\right] \tag{A1.3}$$

If more information is available about the shape of the PD within the limits of the interval, a modified uncertainty could be taken (triangular $u(\bar{x}) = a/\sqrt{6}$, sinus-oscillating $u(\bar{x}) = a/\sqrt{2}$, etc.). The knowledge of the PD and the DOF are joined together: Usually the limits of the interval are stated wide enough, with a PD totally known, which leads to the DOF $\nu(u(\bar{x})) = \infty$. In rare situations—untypical for colorimetry—the limits are judged as only partially sufficient and consequently the associated uncertainty as well as the related DOF are reduced.

Model of Evaluation of Uncertainty

Usually the quantity of interest, the measurand $Y = f(X_1, X_2, \ldots, X_N)$, is calculated from the model of evaluation f from several input quantities $X_1, X_2, \ldots, X_N$ with estimates $y = f(x_1, x_2, \ldots, x_N)$ and $x_1, x_2, \ldots, x_N$, respectively. The theoretical difference between quantity and related estimate has to be regarded carefully. In this report, measurand as well as the estimate are denoted mostly by the same character, to shorten the writing in the following equations.

In linear models, the combined uncertainty $u(y)$ as positive square root of the variance $u^2(y)$ associated to the value of the output quantity is calculated from contributions, which are products of two factors, the standard uncertainties $u(x_1), u(x_2), \ldots, u(x_N)$ associated to the (uncorrelated) input quantities and the partial derivatives of the model of evaluation, and higher-order terms $O(2)$ are neglected. The partial derivatives of the model $f(x_1, x_2, \ldots, x_N)$ with respect to the input quantities $(x_1, x_2, \ldots, x_N)$ are also denoted as sensitivity coefficients c_i

and might be determined analytically or—in the case of a linear model—approximated by the numerical quotient of differences.

$$y = f(x_1, x_2, \ldots, x_N), \quad u^2(y) = \sum_{i=1}^{N} [c_i \cdot u(x_i)]^2 + O(2)$$
$$c_i = \frac{\partial f}{\partial x_i} \approx \frac{f(x_1, \ldots, x_i + u(x_i), \ldots, x_N) - f(x_1, \ldots, x_i - u(x_i), \ldots, x_N)}{2\,u(x_i)} \quad \text{(A1.4)}$$

It should be noted that one of the higher-order terms $O(2) = u^2(x_i) \cdot u^2\,(x_j) + \ldots$ has to be regarded additionally in a model containing a product of two quantities x_i, x_j one of the values being small or even zero and both with limited associated uncertainties. Provided the input quantities have symmetrical PDs and the model includes strong gradients, then the following higher-order terms (up to the forth order $O(4)$) have to be added to the combined uncertainty in Equation (A1.4), and the numerical approximation mentioned there is no longer sufficient.

$$O(2) = \sum_{i=1}^{N}\sum_{j=1}^{N}\left[\left(\frac{1}{2}\left(\frac{\partial^2 f}{\partial x_i \partial x_j}\right)^2 + \frac{\partial f}{\partial x_i}\frac{\partial^3 f}{\partial x_i \partial x_j^2}\right)u^2(x_i)u^2(x_j)\right] + O(4) \quad \text{(A1.5)}$$

For correlated input quantities with correlation coefficients $-1 \leq r(x_i, x_j) \leq 1$ and sensitivity coefficients c_i, c_j, an additional term has to be added or *subtracted* depending on the signs of the correlation and the sensitivity coefficients.

$$y = f(x_1, x_2, \ldots, x_N),\ u^2(y) = \sum_{i=1}^{N} [c_i \cdot u(x_i)]^2 + 2\sum_{i=1}^{N-1}\sum_{j=i+1}^{N} [c_i \cdot c_j \cdot u(x_i)u(x_j)r(x_i, x_j)]$$
$$r(x_i, x_j) = 1,\ u^2(y) = \left(\sum_{i=1}^{N} c_i \cdot u(x_i)\right)^2 \quad \text{(A1.6)}$$

If all input quantities are strongly correlated with a positive correlation coefficient $r(x_i, x_j) = 1$, then the combined variance $u^2(y)$ is no longer the "sum of squares" as known from the propagation of uncertainties, but the "squared sum" of the contributions as shown in the second line of Equation (A1.6). Instead of dealing with correlated input quantities, it is recommended—whenever possible—to introduce an additional independent input quantity for the influence, which is common for several input quantities.

Monte Carlo Method

The standard GUM method explained before is valid only for "linear" models. This means, for example, that the signs of the partial derivatives $(\partial y/\partial x_i)$ of the output quantity y determined with respect to all input quantities x_i must not change within the intervals limited by (at minimum) twice the associated standard uncertainties $(x_i \pm 2u(x_i))$. The consequence may be seen from the following example: The standard GUM method is valid for all angles ε for the model $y = \cos\varepsilon$ with derivative $dy/d\varepsilon = -\sin\varepsilon$, except for the values of the input quantity with associated uncertainty connected by the relation $|\varepsilon| < 2u(\varepsilon)$. The model with values of this exception is nonlinear and is explicitly mentioned in Ref. 1 for an input uncertainty $u(\varepsilon) < |a|$ with RPD. The solution stated there is an output value $y = a^2/6$ with associated output uncertainty $u(y) = a^2/\sqrt{45}$. It is important to notice that the output value changes with the uncertainty of the input quantity, which yields often for nonlinear models, but it is totally different from linear models with output values independent of uncertainties.

The output quantity is defined by the measurement equation just as before, and its value is evaluated from the values of the input quantities x_i with associated standard uncertainties $u(x_i)$. The related PD is known and may be normaly distributed $NPD(x_i, u(x_i))$ or rectangularly distributed $RPD(x_i, u(x_i))$, or similar type of distributions. Now, the output quantity is evaluated very often with the values of the input quantities modified according to the associated PD using random generators, and the mean value, standard deviation, and even the PD function are determined from the simulated values. This numerical way to assign a value and the associated uncertainty to an output quantity is called Monte Carlo method.

Provided that the input quantities are calculated independently and very often $m > 10{,}000$ with values forming a normal probability distribution $NPD_i \equiv NPD(x_i, u(x_i))$, the output quantities $y_{i,j} = f(x_1, x_2, \ldots, NPD_{i,j}, \ldots, x_N)$ are simulated for the variation of the ith input from $1 < j < m$ calculations. For this set, the mean value y_i and the variance of the mean u_i^2 can be calculated. The latter shows the contribution to the combined uncertainty of the output quantity.

$$y_i = \frac{1}{m}\sum_{j=1}^{m} f(x_1, x_2, \ldots, ND_{i,j}, \ldots, x_N);$$
$$u_i^2 = \frac{1}{m(m-1)}\sum_{j=1}^{m} [f(x_1, x_2, \ldots, ND_{i,j}, \ldots, x_N) - y_i]^2 \qquad \text{(A1.7)}$$

The combined contribution of the uncertainties associated to the input quantities is determined using simulations for all input quantities, and the mean value y as well as the associated uncertainty $u(y)$ are determined from the distribution found for the output quantity.

$$y = \frac{1}{m}\sum_{j=1}^{m} f(NPD_{1,j}, NPD_{2,j}, \ldots NPD_{i,j}, \ldots, NPD_{N,j})$$
$$u(y) = \sqrt{\frac{1}{m(m-1)}\sum_{j=1}^{m\cdot N} [f(NPD_{1,j}, NPD_{2,j}, \ldots, NPD_{i,j}, \ldots, NPD_{N,j}) - y]^2} \qquad \text{(A1.8)}$$

Model with Two or More Output Quantities

The color of light—either emitted directly or after being reflected—is stated with a set of two or more output quantities, which are originated by the radiation $\Phi_{e,\lambda}(\lambda)$ depending on the wavelength λ. The radiation $\Phi_{e,\lambda}(\lambda) = \Phi_0 \cdot S(\lambda)$ is presented as a product of an absolute factor Φ_0 and a relative function $S(\lambda)$. For the determination of colorimetric values, only the relative function is needed, and this common origin makes the two or more output quantities—to some extend—correlated.

Numerous quantities for the statement of color are explained within this book and a selection of them like tristimulus values, chromaticity coordinates, dominant wavelength and purity, correlated color temperature plus the distance to the Planckian locus, and the distribution temperature for (nearly) Planckian radiation may be listed as a number $1 \leq i \leq ny$ of output quantities $Y_1, Y_2, \ldots, Y_{ny}$ written as a matrix $\boldsymbol{Y} = (Y_1, Y_2, \ldots, Y_{ny})$ with values $\boldsymbol{y} = (y_1, y_2, \ldots, y_{ny})$. The values of these quantities are determined from a number $1 \leq k \leq nx$ of the input quantities $\boldsymbol{X} = (X_1, X_2, \ldots, X_{nx})$ with values $\boldsymbol{x} = (x_1, x_2, \ldots, x_{nx})$ either by the results of the spectrally integrating measurement devices, like tristimulus colorimeters, or by spectrally resolved measurements with a spectrometer and a mathematical calculation. A list with a number of $1 \leq k \leq nm$ functions $\boldsymbol{F}(\boldsymbol{X}, \boldsymbol{Y}) = (F_1, F_2, \ldots, F_{nm})$ built from all input and output quantities is written as a system of equations $\boldsymbol{F}(\boldsymbol{X}, \boldsymbol{Y}) = \boldsymbol{0}$, which has to be solved for the set of values for the output quantities.

Using matrices to solve the system of equations, it is convenient to organize the uncertainties associated to the values of the input quantities and possible correlations also as a symmetric matrix. The entries in the main diagonal of this uncertainty matrix $\boldsymbol{u}_x$ are the variances $u^2(x_k)$ of the uncertainties associated to the input quantities, and the other places hold the covariances $u(x_k, x_l) = u(x_k)u(x_l)r(x_k, x_l)$, which are the products of the correlation coefficients $r(x_k, x_l)$ and the two related standard uncertainties $u(x_k), u(x_l)$.

$$\boldsymbol{u}_x = \begin{pmatrix} u^2(x_1) & u(x_1, x_2) & \cdots & u(x_1, x_{nx}) \\ u(x_2, x_1) & u^2(x_2) & \ldots & u(x_2, x_{nx}) \\ \vdots & \vdots & \ddots & \vdots \\ u(x_{nx}, x_1) & u(x_{nx}, x_2) & \ldots & u^2(x_{nx}) \end{pmatrix} \tag{A1.9}$$

For a number nx of input quantities sufficiently large—$nx > ny$ if compared with the number ny of output quantities—three cases have to be distinguished:

- If $ny > nm$, no solution exists for more output quantities than equations,
- If $ny = nm$, one solution can be found for independent equations, and
- If $ny < nm$, one solution can be found, solved by a best fit approximation (least mean square fit).

In the case $ny = nm$, the former general system of equations is equivalent to a more specific system with the sensitivity matrix $\boldsymbol{F}_X$ calculated for the values $\boldsymbol{x}$ of the input quantities. The exponent "−1" at a matrix means the inverse matrix. The model $\boldsymbol{F}(\boldsymbol{X},\boldsymbol{Y})$ is the difference between the system of linear equations $\boldsymbol{G}(\boldsymbol{X},\boldsymbol{Y})$ and the vector $\boldsymbol{B}$ with the results for the equations. The solution $\boldsymbol{Y}$ for the coefficients and the associated uncertainty matrix $\boldsymbol{u}_y$ are found from the matrix products with the partial derivatives.

$$\boldsymbol{F}(\boldsymbol{X},\boldsymbol{Y}) = \boldsymbol{0} = \boldsymbol{G}(\boldsymbol{X},\boldsymbol{Y}) - \boldsymbol{B};$$

$$\boldsymbol{Q} = (-\boldsymbol{F}_Y^{-1} \bullet \boldsymbol{F}_X)_x; \quad \boldsymbol{F}_X = \left(\frac{\partial F_i}{\partial X_k}\bigg|_x\right); \quad \boldsymbol{F}_Y = \left(\frac{\partial F_i}{\partial Y_k}\bigg|_x\right) \tag{A1.10}$$

$$\boldsymbol{Y} = \boldsymbol{F}_y^{-1} \bullet \boldsymbol{B}; \quad \boldsymbol{u}_Y = \boldsymbol{Q} \bullet \boldsymbol{u}_X \bullet \boldsymbol{Q}^{\mathrm{T}}$$

The uncertainty matrix $\boldsymbol{u}_y$ with covariances associated to the output quantities is found from the multiplication of the uncertainty matrix $\boldsymbol{u}_x$ associated to the input quantities with the sensitivity matrix and the transposed matrix (superscript "T") of that matrix.

The least mean square fit yields for a number of output quantities $ny \leq nm$ smaller than the number of equations, and an improved solution $\boldsymbol{y}$ is found by an iteration with the starting value $\boldsymbol{y}_0$ indicated by the subscript "0" and improvements $\Delta\boldsymbol{y}$. The next run in the iteration uses the improved solution as new start values and so on. The iteration ends when the improvements of all individual output quantities are smaller than an accepted limit ε, which might be tested by the vector product $\sqrt{\Delta\boldsymbol{y} \bullet \Delta\boldsymbol{y}} < \varepsilon$.

A converging iteration process needs good start values $\boldsymbol{y}_0$ and may be achieved by reduced improvements $\beta \cdot \Delta\boldsymbol{y}$ with values $0 < \beta < 1$. The system of equations uses weights introduced by the uncertainty matrix $\boldsymbol{u}_x$ of the input quantities.

$$\boldsymbol{F}_0 = \boldsymbol{F}(\boldsymbol{X},\boldsymbol{Y}_0) \neq \boldsymbol{0}; \qquad \boldsymbol{F}_{x,0} = \boldsymbol{F}_x(\boldsymbol{X},\boldsymbol{Y}_0); \quad \boldsymbol{q} = (\boldsymbol{F}_{x,0} \bullet \boldsymbol{u}_x \bullet \boldsymbol{F}_{x,0}^{\mathrm{T}})^{-1}$$

$$\boldsymbol{F}_{y,0} = \boldsymbol{F}_y(\boldsymbol{X},\boldsymbol{Y}_0);$$

$$\Delta\boldsymbol{y} = (\boldsymbol{F}_{y,0}^{\mathrm{T}} \bullet \boldsymbol{q} \bullet \boldsymbol{F}_{y,0})^{-1} \bullet (\boldsymbol{F}_{y,0}^{\mathrm{T}} \bullet \boldsymbol{q} \bullet \boldsymbol{F}_0) \tag{A1.11}$$

$$\boldsymbol{y} = \boldsymbol{y}_0 - \Delta\boldsymbol{y}; \qquad \boldsymbol{u}_y = (\boldsymbol{F}_y^{\mathrm{T}} \bullet (\boldsymbol{F}_x \bullet \boldsymbol{u}_x \bullet \boldsymbol{F}_x^{\mathrm{T}})^{-1} \bullet \boldsymbol{F}_y)^{-1}$$

Note: The model $\boldsymbol{F}(\boldsymbol{X},\boldsymbol{Y})$ used to describe the set of input quantities may be not appropriate, which will produce deviations between the input values and the values calculated by the fit function. The "goodness of fit" is also included in the values found as uncertainties associated to the output quantities. This fixed contribution can be found by an additional iteration with zero uncertainties associated to the input quantities, which is achieved when the uncertainty matrix $\boldsymbol{u}_x$ of the input quantities is replaced by an identity matrix of the same dimension.

Expanded Uncertainty

The expanded uncertainty is an interval about the measurement result that may be expected to encompass a large fraction of the distribution of values that could reasonably be attributed to the measurand. The expanded uncertainty $U(y)$ of the output quantity y depends on the combined standard uncertainty $u(y)$ and the coverage factor $k(\nu_{\text{eff}}, p)$, which itself is a function of the effective DOF ν_{eff} and the confidence level p.

$$U(y) = k(\nu_{\text{eff}}, p)u(y) \tag{A1.12}$$

The effective DOF is determined by help of the "Welch–Satterthwaite formula" from the standard uncertainties $u_i(x_i) = c_i \cdot u_i(x_i)$ and the related DOF ν_i contributing to the combined uncertainty $u(y)$ of the output quantity.

$$\nu_{\text{eff}} = u^4(y) \bigg/ \sum_{i=1}^{N} \frac{u_i^4(x_i)}{\nu_i} \tag{A1.13}$$

The effective DOF is an entry to the Student's t-distribution for the recommended confidence level needed for the determination of the coverage factor. In colorimetry—as in many other fields of science—a coverage factor $k = 2$ is recommended for a sufficiently large effective DOF $\nu_{\text{eff}} > 30$, which implies a confidence level or fraction of $p = 95.45\%$.

The coverage factor increases only a little—for example, $k = 2.28$—with a reduced effective DOF $\nu_{\text{eff}} = 10$. Therefore, instead of a calculated value, often a fixed value $k = 2$ is used for simplification.

Note 1: Expanded uncertainties are intervals containing a certain fraction of arbitrary probability distributions. This is totally different from standard uncertainties, which are normal distributed, characterized by two parameters and may be combined by a sum of squares. A combination of the expanded uncertainties is generally not possible due to the different coverage factors in Equation (A1.12). Using the simplification stated before, the result seems to be correct, but the operation is not allowed.

Note 2: A characterization of expanded uncertainties by $k = 2$ without a reference to the confidence level is insufficient, for example, a value $k(\nu_{\text{eff}} = 5, p = 90\%) = 2$ yields a different confidence level at reduced DOF.

Note 3: Sometimes, a standard uncertainty is denoted by a symbol "$k = 1$" from expanded uncertainties, which is not allowed due to the explanations given in Note 1.

Steps for Evaluating Uncertainty

The evaluation of uncertainty can be formalized using the following eight steps, and—as already mentioned—only the steps 1–3 need a specific knowledge about the measured quantity and the measurement procedure. The other steps are calculations using the equations given before.

1. Express the measurement equation (model of evaluation) for the determination of the measurand including all the input quantities, which can contribute a significant component of uncertainty to the result of the measurement.
2. Determine for all input quantities the estimated value.
3. Determine for all values of input quantities the associated standard uncertainty using the type A (statistical method) or type B evaluation (any other method).
4. Determine the covariances for all correlated input quantities.
5. Calculate the result of the measurement with the model of evaluation from all input quantities.
6. Determine the combined standard uncertainties from the standard uncertainties and covariances of the input quantities and the related sensitivity coefficients.
7. If it is necessary to state an expanded uncertainty, determine the effective degree of freedom and the related value for the coverage factor for the intended confidence level.
8. Report the result of the measurement together with its expanded uncertainty and the coverage factor using a presentation in a recommended format.

Measurement results have to be stated at minimum with the associated expanded uncertainty and the statement referring to the GUM method for the determination. Additionally, the traceability chain back to a national or international reference standard should be mentioned. The expanded uncertainty is stated by a number with only two valid figures, and it may be presented in "absolute," "relative," or "percentage" presentation.

PRACTICAL EXAMPLES

Lamps are used as secondary standards for many colorimetric calibration tasks. Thus the operation of sources as well as the use of measurement devices for the determination of operational conditions or optical properties will be demonstrated here. As an example for the development of the "model of evaluation" the calibration of a spectroradiometer by an irradiance standard lamp is discussed in more details. Most of the explanations given below are also valid for other sources, like LEDs, if the characteristic values are adjusted. Reduced uncertainties are achieved with minimum effort for traceability, when the reference standard and the calibration object are of the same kind and when they are as similar as possible.

The f_1' value is used to characterize the match of the relative spectral responsivities (originally of a photometer) here for the channel of a tristimulus head to the related weighting function $\bar{x}(\lambda), \bar{y}(\lambda), \bar{z}(\lambda)$. The definition of the f_1' value deals with the absolute value of the differences, which acts as a nonlinear model.

The evaluation of the associated uncertainty is shown as an example using the Monte Carlo method.

Determination of the Spectral Irradiance of a Source

Principle of a Spectral Irradiance Measurement

Spectroradiometers, like monochromator or polychromator, measure optical radiation as a function of wavelengths. Their wavelength scales have to be calibrated first with the radiation emitted by, for example, low-pressure discharge lamps with a series of spectral lines at well-known peak wavelengths.

Note: A correction of the possible broadening of these narrow lines due to the widened bandpass of the spectroradiometer is explained in Chapter 5.

After the wavelength scale is fixed, the spectroradiometer can be used to compare—by substitution method—the relative spectral distributions of the calibration objects with the spectral distribution of a secondary standard lamp, the reference. Incandescent or halogen lamps calibrated as transfer standards for spectral irradiance are used as reference lamps in the visible spectral region.

Today, Si photodiodes are mainly used as detectors in spectroradiometers generating photocurrents as output quantity. In monochromators, the photocurrent of one detector at the single exit slit is converted by a current-to-voltage amplifier with gain setting resistor R_g to a voltage as output quantity $y(\lambda)$ and measured with a DVM for one wavelength after the other. In a polychromator, series of individual pixels of a CCD detector act as exit slits and the photocurrents of these photodiodes are integrated simultaneously over the expose time as charges in individual capacitors. The sequential readout of the signals and their conversion in a digital presentation $y(\lambda)$ is repeated independent of the type of spectroradiometer for signal averaging $\bar{y}(\lambda)$[1] and a correction for the "dark signal" $\bar{y}_0(\lambda)$ has to be applied $y(\lambda) = \bar{y}(\lambda) - \bar{y}_0(\lambda)$.

The spectral irradiance $E_\lambda(\lambda) = E_{\lambda,0} \cdot S(\lambda)$ of a source is preferably written as a product of an "absolute factor" $E_{\lambda,0}$ independent of wavelength and the relative spectral distribution function $S(\lambda)$. In this report, all quantities related to the reference are indicated by an index "Ref". The spectral irradiance $E_{\lambda_0,\mathrm{Ref}}$ produced by such a reference lamp depends on the distance d^2_{Ref} between the filament of the lamp and the limiting aperture of the spectroradiometer input optics and the spectral radiant intensity $I_{\lambda_0,\mathrm{Ref}}$ following the inverse square law $E_{\lambda_0,\mathrm{Ref}} = I_{\lambda_0,\mathrm{Ref}}/d^2_{\mathrm{Ref}}$. The radiant intensity varies with the consumed electrical power, the mechanical alignment, a possible effect of aging, and some more corrections discussed below.

For this example the fundamental equation for the substitution method is modified by the introduction of a calibration factor $C_{\mathrm{spec}}(\lambda)$ as the ratio of the values of the optical quantity and the output signal of the spectroradiometer. The evaluation

[1] $\bar{y}(\lambda)$ represents "average" in this chapter and should not be confused with the $\bar{y}(\lambda)$ symbol for the "Y color-matching function" used in the rest of the book.

of the uncertainty associated to this calibration factor is shown in the following example.

$$E_\lambda(\lambda) = E_{\lambda_0} \cdot S(\lambda) = R_g \cdot y(\lambda) \cdot C_{\text{spec}}(\lambda)$$

$$C_{\text{spec}}(\lambda) = \frac{E_{\lambda_0,\text{Ref}} \cdot S_{\text{Ref}}(\lambda)}{R_{\text{g,Ref}} \cdot y_{\text{Ref}}(\lambda)} \tag{A1.14}$$

$$E_{\lambda_0,\text{Ref}} = \frac{I_{\lambda_0,\text{Ref}}}{d_{\text{Ref}}^2}$$

Operation of a Spectral Irradiance Standard

Halogen lamps are mostly used as reference lamp for the visible spectral region. The relative spectral distribution $S_{\text{Ref}}(\lambda)$ of their radiation is similar to a Planckian radiator with the *relative* spectral distribution function $P(\lambda, T)$ normalized at wavelength λ_0 and characterized by a distribution temperature T—sometimes incorrectly denoted as color temperature. The small deviation of the relative spectral distribution from the Planckian curve can be approximated by a polynomial $f_{\text{Ref}}(\lambda)$, which can be taken as independent of the distribution temperature.

$$S_{\text{Ref}}(\lambda) = P(\lambda, T_{\text{Ref}}) \cdot f_{\text{Ref}}(\lambda)$$

$$P(\lambda, T) = \left[\lambda^5 \left(\exp\left(\frac{c_2}{\lambda \cdot T}\right) - 1\right)\right]^{-1} \tag{A1.15}$$

$$c_2 = 1.3488\,\text{cm} \cdot \text{K}$$

An incandescent lamp operated at a fixed lamp current J_0 has the related values of spectral radiant intensity I'_{λ_0}, lamp voltage U'_0 and distribution temperature T'_0 (in the example, the symbol "J" is used for currents to avoid any confusion with the radiant intensity symbolized by "I"). A relative variation of lamp current J_0 to a nearby value J changes relatively the values of the related quantities I_{λ_0}, U, T with values of the coefficients given, "as a rule of thumb" indicated, as $m_I \cong 6; m_U \cong 2; m_T \cong 0.7$. It should be noted that similar relations are valid for other light sources, but with quite different values of the coefficients, and with spectral distributions significantly different from the Planckian radiator.

Note: For semiconductor sources like LEDs, instead of a change of the distribution temperature, for example, the whole spectral function including the peak wavelength is shifted to smaller wavelengths.

$$I_{\lambda_0} = I'_{\lambda_0}\left(\frac{J}{J_0}\right)^{m_I}; \quad U = U'_0\left(\frac{J}{J_0}\right)^{m_U}; \quad T = T'_0\left(\frac{J}{J_0}\right)^{m_T} \tag{A1.16}$$

The lamp current J is measured with a DVM as mean value from repeated readings of the voltage drop $\overline{U}_J$ across a shunt resistor R. This mean value is influenced by the calibration factor c_J of the DVM and a possible offset voltage $\overline{U}_{J\text{off}}$ created by thermoelectric effects due to temperature differences in the circuit $J \approx c_J(\overline{U}_J - \overline{U}_{J\text{off}})$. The value R_0 of the shunt resistor at the rated ambient temperature varies with a relative coefficient α_{R_0} and a deviation ΔT_{R_0} from the rated ambient temperature or due to self-heating with a temperature rise $wR_0J_0^2$ by the consumed electrical power and the thermal resistance w. The equation is organized with higher-order terms $O(2)$ omitted and divided in two factors, the second in square brackets is close to unity. For the ratio of lamp currents in Equation (A1.16)

$$\begin{aligned}\frac{J}{J_0} &= \frac{c_J\,\overline{U}_J}{J_0\,R_0}\frac{1-\overline{U}_{J\,\text{off}}/\overline{U}_J}{1+\alpha_{R_0}(\Delta T_{R_0}+w\cdot R_0\cdot J_0^2)}\\ &= \frac{c_J\,\overline{U}_J}{J_0\,R_0}\cdot\left[1-\frac{\overline{U}_{J\,\text{off}}}{\overline{U}_J}-\alpha_{R_0}(\Delta T_{R_0}+w\,R_0\,J_0^2)+O(2)\right]\end{aligned} \tag{A1.17}$$

The distribution temperature in Equation (A1.16) depends on this ratio of lamp currents. As no further quantities affect the distribution temperature value, it yields $T_0 = T_0'$ and the ratio can be calculated for the reference lamp.

$$T_{\text{Ref}} = T_{0\text{Ref}}\left(\frac{c_J}{J_{0\text{Ref}}}\frac{\overline{U}_{J\text{Ref}}}{R_0}\right)^{m_T}\cdot\left[1-m_T\left(\frac{\overline{U}_{J\text{off}}}{\overline{U}_{J\text{Ref}}}+\alpha_{R_0}(\Delta T_{R_0}+w\,R_0\,J_{0\text{Ref}}^2)\right)+O(2)\right] \tag{A1.18}$$

The spectral radiant intensity I''_{λ_0} value of the reference is reduced by aging, with a relative aging factor β_L due to repeated operations with a total duration Δt_L. This aging factor has to be determined individually for each reference lamp similar as the alignment factor γ_L explained below. Thus, the radiant intensity is found by the combination of the Equations (A1.16) and (A1.17) with the correction factor $I'_{\lambda_0} = I''_{\lambda_0}[1 - \beta_L \cdot \Delta t_L - \gamma_L + O(2)]$ and written for the reference lamp. The combination of the effects mentioned above gives the model for the spectral radiant intensity and thus for the spectral irradiance. It should be remembered that the spectral radiant intensity $I'_{\lambda_0,\text{Ref}}$ refers to the rated values in the certificates whereas $I_{\lambda_0,\text{Ref}}$ means the effective radiant intensity during the measurement.

$$I_{\lambda_0,\text{Ref}} = I'_{\lambda_0;\text{Ref}}\left(\frac{c_J\,\overline{U}_J}{J_0\,R_0}\right)^{m_I}\cdot\left[1-m_T\left(\frac{\overline{U}_{J\text{off}}}{\overline{U}_J}+\alpha_{R_0}(\Delta T_{R_0}+wR_0J_0^2)\right)+\beta_L\Delta t_L+\gamma_L+O(2)\right] \tag{A1.19}$$

The combination of the Equation (A1.14) and Equation (A1.19) leads to the equations for the substitution method with an incandescent lamp as reference. The contributions are grouped and show in the first line the principle equation with the ratio of the two output signals corrected for "dark signals," the relative spectral distribution of the reference as product of the Planckian curvature, and the deviation function, all of these depending on the wavelength. The factor "corr" is independent of the wavelength and an important factor for the determination of the absolute irradiance of all calibration objects. The Planckian distribution depends on the distribution temperature, which is given in the second line and mainly depends on the lamp current.

$$C_{\text{spec}}(\lambda) = \frac{P(\lambda, T_{\text{Ref}}) \cdot f_{\text{Ref}}(\lambda)}{y_{\text{Ref}}(\lambda)} \cdot \text{corr} \tag{A1.20}$$

where

$$T_{\text{Ref}} = T_{0\text{Ref}} \left(\frac{c_J}{J_{0\text{Ref}}} \frac{\overline{U}_{J\text{Ref}}}{R_0} \right)^{m_T} \cdot \left[1 - m_T \left(\frac{\overline{U}_{J\,\text{off}}}{\overline{U}_J} + \alpha_{R_0} (\Delta T_{R_0} + w R_0 J_0^2) \right) + O(2) \right]$$

and

$$\text{corr} = \frac{I'_{\lambda_0,\text{Ref}}}{R_{\text{g,Ref}} \cdot d_{\text{Ref}}^2} \left(\frac{c_J}{J_{0\text{Ref}}} \frac{\overline{U}_{J\text{Ref}}}{R_0} \right)^{m_I}$$

$$\cdot \left[1 - m_T \left(\frac{\overline{U}_{J\,\text{off}}}{\overline{U}_J} + \alpha_{R_0} (\Delta T_{R_0} + w R_0 J_{0\text{Ref}}^2) \right) + \beta_{\text{L}} \Delta t_{\text{Ref}} + \gamma_{\text{Ref}} + O(2) \right]$$

Mechanical Alignments

The reference lamp is operated in the spectroradiometer setup with a specified optical axis, needed for the mechanical alignment of six degrees of freedom for each of the two devices, lamp and spectroradiometer entrance aperture. The areas of filament and aperture of the spectroradiometer can be aligned centrally to the optical axis of the setup with negligible deviations. Similarly, a negligible effect can be expected for small misalignments due to the rotations of these devices about the optical axis of the setup.

A possible angle ε_{S} between the optical axis and the direction of normal incidence for the spectroradiometer aperture can be corrected by $\cos \varepsilon_{\text{S}}$ as a factor of the related output signal. Provided that the spectroradiometer is not moved between the measurements of the reference and the test lamp, the substitution method cancels out this correction factor.

Due to the structured filament, the angular intensity distribution of a lamp is often much more complicated than the cosine function of a Lambertian distribution. Therefore, an alignment factor $(1 - \gamma_{\text{L}})$ depending on rotations about horizontal and vertical axes might be determined individually for each reference lamp, with the relative effect γ_{L} determined from the small variations of the direction with respect to the burning position and the direction of emittance as average of repeated alignments and estimated by a rectangular probability distribution (RPD). The factor $(1-\gamma_L)$ was already used in the equations above.

Often, the spectral irradiance of a lamp is certified for a specified distance measured from a certain area—a distance d_L away from the filament—which can increase the reproducibility for the alignment of the distance. In the spectroradiometer setup, two reference planes in the specified distance d_{B_0} are used for the alignment of source and spectroradiometer, respectively, and the distance is measured by a length meter with the calibration factor c_d. A deviation ΔT_d of the ambient temperature will modify the distance $d_B = c_d d_{B_0}(1+\alpha_d \cdot \Delta T_d)$ due to a relative temperature coefficient α_d. The effective distance d is determined from the distance d_B between the reference planes and from the possible offsets d_S and d_L due to (distance) misalignment of each of the two devices.

$$d = (c_d\, d_{B_0}(1+\alpha_d \cdot \Delta T_d) + d_S + d_L) \qquad \text{(A1.21)}$$

Distances squared in Equation (A1.20) are calculated from Equation (A1.21) and given in a simplified form with higher-order terms $O(2)$ omitted and presented as a product of two factors. The factor within square brackets is close to unity.

$$d^2 = c_d^2\, d_{B_0}^2 * \left[1 + 2 \cdot \left(\frac{d_S + d_L}{d_{B_0}} + \alpha_d \cdot \Delta T_d\right) + O(2)\right] \qquad \text{(A1.22)}$$

As typical for the substitution method in Equation (A1.20), not the effective distances for the reference and for the calibration object but the ratio of these values are included, which compensates for most of the contributions. This compensation is also valid for the contribution from the current measurement and other similar factors if the calibration objects have similar properties than the reference standard. The final presentation of the model of evaluation is the basis for the uncertainty budget explained in the next paragraph.

Uncertainty Budget

The uncertainty budget uses the model Equation (A1.23) for the evaluation of the uncertainty, and it presents the list of all components contributing to the combined uncertainty. In the first line the ratio $P(\lambda, T_{\text{Ref}}) \cdot f_{\text{Ref}}(\lambda)/y_{\text{Ref}}(\lambda)$ of the relative spectral distribution of the reference standard and the spectroradiometer, reading is multiplied with a correction factor corr, which is constant for all wavelengths. The relative spectral distribution is a Planckian radiator $P(\lambda, T_{\text{Ref}})$, which varies with the distribution temperature T_{Ref} stated in the second line. For colorimetric calculations, the spectral distribution function of the test device has to be normalized to unity at a certain wavelength. Therefore, the value of the "absolute" factor corr has no meaning. (If for other reasons the absolute spectral distribution function of the irradiance $E_\lambda(\lambda)$ of the test device would be needed, then the correction

TABLE A1.1 Uncertainty budget: Planck function for the wavelength λ= 560 nm

No	name	symbol	Value, x_i	uncertainty $u(x_i)$	DOF	sensitivity coeff. c_i	uncert. contrib. $u_i = c_i\, u(x_i)$
1	DVM voltage drop	$\bar{U}_{J\text{Ref}}$	0.80023 V	0.00085 V	30	0.19767	0.0001680
2	DVM voltage offset	$\bar{U}_{J\text{off}}$	0.00017 V	0.00052 V	30	−0.19776	−0.0001028
3	Exponent temp.	m_T	0.7	0.2	∞	−0.000422	−0.0000843
4	Temp. coefficient	α_{R_0}	0.00005 1/K	0.00002 1/K	∞	−2.34459	−0.0000469
5	DVM cal. factor	c_J	1.00	0.00025	∞	0.158149	0.0000395
6	Shunt resistor	R_0	0.10012 Ω	0.00002 Ω	∞	−1.5806	−0.0000316
7	Rel. therm. resistance	w	2.0 K/W	0.5 K/W	∞	−0.0000507	−0.0000254
8	Amb. temp. deviation	ΔT_{R_0}	2.0 K	1.0 K	∞	−0.000007913	−0.0000079
9	Rated current	$J_{0\text{Ref}}$	8.00 A				
10	Rated distrib. temp.	$T_{0\text{Ref}}$	3000 K				
	Planck function	$P(\lambda, T_{\text{Ref}})$	0.225897		>80	$u(P(\lambda, T_{\text{Ref}}))$	0.0002237

factor corr could be determined using the relations given in the third and forth line of Equation (A.23)).

$$C_{\text{spec}}(\lambda) = \frac{P(\lambda, T_{\text{Ref}}) \cdot f_{\text{Ref}}(\lambda)}{y_{\text{Ref}}(\lambda)} \cdot \text{corr}$$

$$T_{\text{Ref}} = T_{0\text{Ref}} \left(\frac{c_J}{J_{0\text{Ref}}} \frac{\overline{U}_{J\text{Ref}}}{R_0} \right)^{m_T} \cdot \left[1 - m_T \left(\frac{\overline{U}_{J\text{off}}}{\overline{U}_{J\text{Ref}}} + \alpha_{R_0} (\Delta T_{R_0} + w R_0 J_{0\text{Ref}}^2) \right) + O(2) \right]$$

$$\text{corr} = \frac{I_{\lambda_0,\text{Ref}}}{R_{\text{g,Ref}} \cdot d_{\text{B}_0,\text{Ref}}^2} \left(\frac{c_J}{J_{0\text{Ref}}} \frac{\overline{U}_{J\text{Ref}}}{R_0} \right)^{m_I}$$

$$\cdot \left[1 - m_I \left(\frac{\overline{U}_{J\text{off}}}{\overline{U}_{J\text{Ref}}} + \alpha_{R_0} (\Delta T_{R_0} + w R_0 J_{0\text{Ref}}^2) \right) - \beta_{\text{L}} \Delta t_{\text{L}} - \gamma_{\text{L}} - 2 \frac{d_{\text{S}} + d_{\text{Ref}}}{d_{\text{B}_0,\text{Ref}}} + O(2) \right] \tag{A1.23}$$

Replacing in Equation (A1.15) the relative spectral distribution $S_{\text{Ref}}(\lambda) = P(\lambda, T_{\text{Ref}}) \cdot f_{\text{Ref}}(\lambda)$ by a product of the relative Planck function $P(\lambda, T_{\text{Ref}})$ and the polynomial $f_{\text{Ref}}(\lambda)$ with the relation $u_{\text{rel}}^2(S_{\text{Ref}}(\lambda)) = u_{\text{rel}}^2(P(\lambda, T_{\text{Ref}})) + u_{\text{rel}}^2(f_{\text{Ref}}(\lambda))$ of their relative variances opens the possibility to set the relative variance of the Planck function to zero, which leads to the equation $u_{\text{rel}}(S_{\text{Ref}}(\lambda)) = u_{\text{rel}}(f_{\text{Ref}}(\lambda))$. The uncertainty contributions due to the variation of the distribution temperature from the operation of the reference lamp is an additional effect and is shown in the uncertainty budget below. The values of the input quantities, the associated uncertainties, and their degrees of freedom are listed together with the related sensitivity coefficients and the contributions to the combined uncertainty of the output value. This example shows the evaluation of uncertainties originated by the calibration of a spectroradiometer, but it is neither a template for uncertainty evaluation procedures nor a complete list of contributions because in main chapters of this book important other effects are already explained, for example, stray light, bandpass, and sampling. Due to the incomplete budget only fictitious numbers are taken.

$$\begin{aligned} C'_{\text{spec}}(\lambda) &= f_{\text{Ref}}(\lambda) \cdot P(\lambda, T_{\text{Ref}}) \cdot y_{\text{Ref}}^{-1}(\lambda) \\ u_{\text{rel}}(C'_{\text{spec}}(\lambda)) &= \sqrt{u_{\text{rel}}^2(f_{\text{Ref}}(\lambda)) + u_{\text{rel}}^2(y_{\text{Ref}}^{-1}(\lambda)) + u_{\text{rel}}^2(P(\lambda, T_{\text{Ref}}))} \end{aligned} \tag{A1.24}$$

$C_{\text{spec}}(\lambda)$ Calibration factor of a spectroradiometer found from the operation of a halogen lamp as reference with a certified spectral irradiance in the direction of the optical axis of the spectroradiometer setup. The function values of this quantity are result of the spectral calibration procedure, and the associated combined relative uncertainty is the square root of the sum of squares of relative variances.

$f_{\text{Ref}}(\lambda)$ Spectral distribution function $f_{\text{Ref}}(\lambda) = S_{\text{Ref}}(\lambda)/P(\lambda, T_{0\text{Ref}})$ found from the spectral distribution $S_{\text{Ref}}(\lambda)$ and the Planck function $P(\lambda, T_{0\text{Ref}})$ by the adjustment of the distribution temperature $T_{0\text{Ref}}$. The relative spectral distribution $S_{\text{Ref}}(\lambda) = E_{\lambda,\text{Ref}}(\lambda)/E_{\lambda_0,\text{Ref}}$ is the original spectral irradiance function $E_{\lambda,\text{Ref}}(\lambda)$ of the reference lamp normalized to unity at wavelength λ_0. The expanded uncertainty $U(E_{\lambda,\text{Ref}}(\lambda))$ stated in the certificate of the reference for $k = 2$ converted to a relative standard uncertainty $u_{\text{rel}}(E_{\lambda,\text{Ref}}(\lambda)) = \frac{1}{2}U(E_{\lambda,\text{Ref}}(\lambda))/E_{\lambda,\text{Ref}}(\lambda) = u_{\text{rel}}(f_{\text{Ref}}(\lambda))$ is taken as the relative uncertainty of the product $P(\lambda, T_{\text{Ref}}) \cdot f_{\text{Ref}}(\lambda)$ in the model Equation (A1.23). The *relative* sensitivity coefficient is found as unity as shown in Equation (A1.24).

$y_{\text{Ref}}^{-1}(\lambda)$ The output signal of the spectroradiometer determined from series of repeated n readings and corrected for "dark currents." Provided that the resolution $\delta y_{\text{Ref}} < s(y_{\text{Ref}})$ of the DVM is smaller than the standard deviation of the mean, the latter is taken as the uncertainty associated to the output signal at a certain wavelength λ, and the degrees of freedom are DOF $= n - 1$. The *relative* sensitivity coefficient is -1 as shown in Equation (A1.24).

$P(\lambda, T_{\text{Ref}})$ Relative spectral distribution of a reference lamp operated at an effective distribution temperature T_{Ref} determined from the Planckian radiator at value $T_{0\text{Ref}}$, which was found from the adjustment to the original spectral irradiance function of the reference lamp. The *relative* sensitivity coefficient is found as unity as shown in Equation (A1.24). The variation of the uncertainty due to nonperfect operation is shown in the following uncertainty budget.

corr The correction factor is of no need in colorimetry and not regarded any more.

$J_{0\text{Ref}}$ The rated DC current with fixed polarity for the reference lamp stated in the certificate to produce the irradiance; it has a nominal value with no uncertainty. $J_{0\text{Ref}} = 8.000$ A.

$T_{0\text{Ref}}$ The distribution temperature of a Planck function adjusted to fit the original spectral irradiance distribution of the reference lamp. The uncertainty of this adjustment will not affect the output quantity because the polynomial $f_{\text{Ref}}(\lambda)$ takes over the differences completely, $T_{0\text{Ref}} = 3000$ K.

c_J Calibration factor of the DVM used to measure the voltage drop across the shunt resistor. The value and associated expanded uncertainty ($k = 2$) is stated in the certificate of the DVM. $c_J = 1.000; u(c_J) = U(c_J)/2 = 2.5 * 10^{-4}, DOF = \infty$.

$\overline{U}_{J\text{Ref}}$ Mean value of the repeated readings of the voltage drop across the shunt resistor with standard deviation taken as standard uncertainty because the resolution of the DVM was significantly smaller than the standard deviation. $\overline{U}_{J\text{Ref}} = 0.80023$ V, $u(\overline{U}_{J\text{Ref}}) = 0.00085$ V, DOF = 30.]

$\overline{U}_{J\text{off}}$ Mean value of the repeated readings of the voltage drop across the shunt resistor with standard deviation taken as standard uncertainty; the resolution of the DVM was significantly smaller than the standard deviation at zero lamp current. $\overline{U}_{J\text{off}} = 0.00017$ V, $u(\overline{U}_{J\text{off}}) = 0.00052$ V, DOF = 30.

R_0 Resistance of the shunt resistor stated in the related certificate for an ambient temperature of 22°C and a negligible current with relative expanded uncertainty for $k = 2$. $R_0 = 0.10012\,\Omega$, $U_{\text{rel}}(R_0) = 4*10^{-4}$, which is converted to a standard uncertainty $u(R_0) = R_0 \cdot U_{\text{rel}}(R_0)/2 = 2*10^{-5}\,\Omega$, DOF = ∞.

α_{R_0} Relative temperature coefficient of the shunt resistor, the value within an interval with RPD is taken from the technical information of the resistor. $\alpha_{R_0} = (5 \pm 3)^*10^{-5}$ 1/K, therefore value $\alpha_{R_0} = 0.00005$ and uncertainty are determined as $u(\alpha_{R_0}) = 3/\sqrt{3} \times 10^{-5}$ 1/K, DOF = ∞.

ΔT_{R_0} Difference of the ambient temperature near the shunt resistor to the certified ambient temperature of the shunt. This difference is varying during the measurement campaign within the interval (24 ± 1.5)°C with RPD. Therefore, the difference is just $\Delta T_{R_0} = 2$ K and the standard uncertainty associated to the stated interval is $u(\Delta T_{R_0}) = 1.5/\sqrt{3}$ K with DOF = ∞.

w Thermal resistance by the convection of the shunt resistor to the ambient air when cooling the heat of the self-heating. The value was determined for several currents and averaged to get an interval with RPD of $w = (2 \pm 0.9)$K/W: that is, a value $w = 2.0$K/W, an associated standard uncertainty $u(w) = 0.9/\sqrt{3}$ K/W, and a DOF $= \infty$.

m_T Exponent describing the variation of the distribution temperature with the current of the reference lamp. As a rule of thumb, the interval $m_T = 0.7 \pm 0.35$ with RPD is known. Therefore, the value $m_T = 0.7$ and the associated standard uncertainty $u(m_T) = 0.2/\sqrt{3}$ are known with a DOF $= \infty$.

Model:

$$C'(\lambda_i) = P(\lambda, T_{\text{Ref}})$$

$$T_{\text{Ref}} = T_{0\text{Ref}}\left(\frac{c_J}{J_{0\text{Ref}}}\frac{\overline{U}_{J\text{Ref}}}{R_0}\right)^{m_T} \cdot \left[1 - m_T\left(\frac{\overline{U}_{J\text{off}}}{\overline{U}_{J\text{Ref}}} + \alpha_{R_0}(\Delta T_{R_0} + wR_0 J_{0\text{Ref}}^2)\right)\right] \quad \text{(A1.25)}$$

The relative uncertainty of the Planck function due to the operation of the reference lamp and evaluated at wavelength 560 nm is found to be $u_{\text{rel}}(P(\lambda, T_{\text{Ref}})) = u(P(\lambda, T_{\text{Ref}}))/P(\lambda, T_{\text{Ref}}) = 0.0010$. The variation of the relative uncertainty with wavelength is negligible. Therefore, in Equation (A1.24), this result can be used independent of the wavelength.

Chapter 6 contains the example of uncertainties for the calibration constants of a tristimulus colorimeter based on these principles. A further example, that of the distribution temperature, is available elsewhere.[3]

Determination of f_1' Values

The match of the relative spectral responsivity to a weighting function like $V(\lambda)$, or the color-matching functions, is the most important property of a spectrally integrating photometer or a colorimeter channel and is characterized by an f_1' value. A smaller value indicates a better match, and (in theory) the perfect match will be indicated by a zero value of f_1'.

In practice, the values of the relative responsivity—even when matching the weighting function perfectly—are determined with a nonzero associated uncertainty. It is demonstrated below how this will affect the f_1' value and the associated uncertainty.

The relative spectral (irradiance) responsivity function $s_{x,\text{rel}}(\lambda)$ of a colorimeter channel (x stands for $\bar{x}(\lambda), \bar{y}(\lambda), \bar{z}(\lambda)$) is corrected by layers of glass filters to match the related tristimulus function, for example, $\bar{x}(\lambda)$ and is determined for a specified spectral distribution $S_A(\lambda)$, the CIE illuminant A. Consequently, the characteristic $f_{1,x}'$ of the "quality of the match" is normalized for that illuminant, too, as explained in Chapter 6.

$$f_{1,x}' = \frac{\int_{\lambda 1}^{\lambda 2} |s_{x,\text{rel}}^*(\lambda) - \bar{x}(\lambda)|\mathrm{d}\lambda}{\int_{\lambda 1}^{\lambda 2} \bar{x}(\lambda)\mathrm{d}\lambda}; \quad s_{x,\text{rel}}^*(\lambda) = s_{x,\text{rel}}(\lambda) \cdot \frac{\int_{\lambda 1}^{\lambda 2} S_A(\lambda)\,\bar{x}(\lambda)\mathrm{d}\lambda}{\int_{\lambda 1}^{\lambda 2} S_A(\lambda) s_{x,\text{rel}}(\lambda)\mathrm{d}\lambda} \quad \text{(A1.26)}$$

Note: The wavelength range $\lambda 1 \leq \lambda, (\Delta\lambda) \leq \lambda 2$ and the spacing $\Delta\lambda$ are affecting the f_1' value and the associated uncertainty, but these numbers are not well defined in the CIE document and will not be regarded here.

Obviously, the responsivity function $s_{\text{rel}}(\lambda)$ is normalized by a ratio of two integrals, which is already known from the Equation (A1.26) and which cancels out any factor for $s(\lambda)$. The numerator is constant and can be substituted by a value b. The difference between the normalized responsivity function $s_{\text{rel}}^*(\lambda)$ and $V(\lambda)$ is divided by the constant integral over the $V(\lambda)$ function, which is substituted by a constant value a. The integrals in Equation (A1.26) can be approximated by the sum of the values related to a number $1 < i < n$ of the equally spaced wavelengths and with the two substitutions explained before; the model for the evaluation of f_1' reads

$$f_1' = \frac{1}{a}\sum_{i=1}^{n}\left|\frac{b \cdot s(\lambda_i)}{\sum_{i=1}^{n} S_{\text{A}}(\lambda_i)s(\lambda_i)} - V(\lambda_i)\right|; \quad a = \sum_{i=1}^{n} V(\lambda_i); \quad b = \sum_{i=1}^{n} S_{\text{A}}(\lambda_i)V(\lambda_i) \tag{A1.27}$$

The values of CIE illuminant A and $V(\lambda)$ are standardized and defined without any uncertainty. The combination of glass filters gives a variety of spectral responsivity functions close to $V(\lambda)$, and the measurement of this function $s(\lambda)$ for each wavelength adds noise and possible offsets, which are summarized in the associated uncertainties $u(s(\lambda))$. Usually, the value of f_1' characterizing a measured spectral responsivity function is of interest, but here, the evaluation and the procedure as to how to get that value and the associated uncertainty will be explained. Therefore, instead of the measured values for a responsivity function, the simulated values and uncertainties will be used, which allows for everybody to repeat the calculations without a specific set of data.

The values of the relative spectral responsivity function $s(\lambda)$ are simulated with the $V(\lambda)$ function shifted by an adjustable offset $\Delta\lambda$ in the wavelength range. The (absolute) uncertainty is assumed as a fixed value $u(s(\lambda)) = \Delta u$ independent of the wavelength, and the wavelength range as defined for the $V(\lambda)$ function with a 5-nm spacing is used.

$$s(\lambda) = V(\lambda + \Delta\lambda); \quad u(s(\lambda)) = \Delta u; \quad 380 \leq \lambda/\text{nm}, (\Delta\lambda = 5\,\text{nm}) \leq 780 \tag{A1.28}$$

Figure A1.1 shows the $V(\lambda)$ function and the responsivity function $s(\lambda)$ (dotted line) simulated as explained in Equation (A1.27) with a wavelength shift of $\Delta\lambda = 1$ nm. The difference between these two functions is the major criteria for the evaluation of the f_1' value. This difference, magnified by a factor 10, is drawn by a dashed line. The same factor of 10 is used to show the magnitude of the simulated (constant) standard uncertainty $\Delta u = 0.005$ associated to the values of the responsivity function.

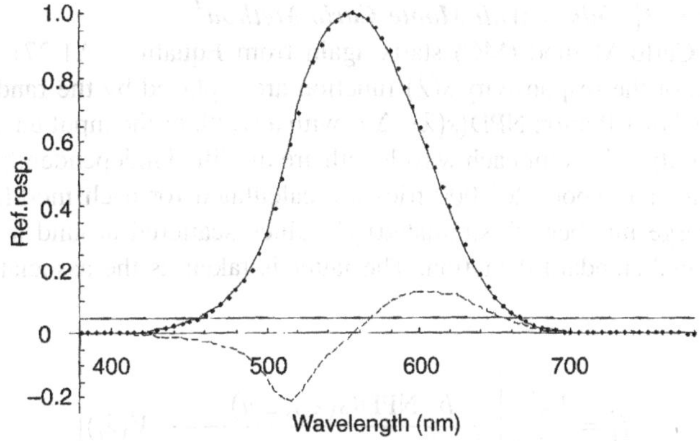

FIGURE A1.1 Four functions are shown: The $V(\lambda)$ function is drawn near the simulated relative spectral responsivity $s(\lambda)$ (dotted line) and magnified by a factor of 10 the difference $V(\lambda) - s(\lambda)$ (dashed line) and the line for the constant uncertainty.

The model in Equation (A1.27) defines the f_1' value. Assuming that this model would be linear, the sensitivity coefficients are the derivatives with respect to the responsivities for each wavelength. For uncorrelated input values, the square root of the sum of squares of the products built from a sensitivity coefficient c_i multiplied with the associated standard uncertainty Δu would be taken as the uncertainty of the output quantity. The value and a (nonvalid) solution for the uncertainty are given in Table A1.2 in the next section.

$$u(f_1') = \sqrt{\sum_{i=1}^{n} \left[\frac{\partial f_1'}{\partial s(\lambda_i)} \Delta u\right]^2} \qquad \text{(A1.29)}$$

Figure A1.1 shows that the difference between the responsivity and $V(\lambda)$ has negative values, which are turned to positive by the absolute sign in the definition of f_1'. This is the reason why the model acts nonlinear for all values smaller than the input uncertainty Δu.

TABLE A1.2 Values and associated standard uncertainties for f_1'. (The results in the part headed by "GUM" are not valid and presented only for comparison.)

Method	GUM			MC			
Uncertainty		0.01	0.005	0.01		0.005	
Result	$\Delta u = f_1'$	$u(f_1')$	$u(f_1')$	f_1'	$u(f_1')$	f_1'	$u(f_1')$
$\Delta\lambda = 2.0$	3.861	0.495	0.248	5.447	0.372	4.508	0.197
$\Delta\lambda = 1.0$	1.922	0.484	0.242	3.889	0.337	2.710	0.182
$\Delta\lambda = 0.5$	0.958	0.479	0.239	3.317	0.304	1.937	0.166
$\Delta\lambda = 0.1$	0.191	0.474	0.237	3.093	0.282	1.561	0.142

***Uncertainty of f_1' Values With Monte Carlo Method*[4]**

The Monte Carlo Method (MC) starts again from Equation (A1.27). The (simulated) values of the responsivity $s(\lambda)$ function are replaced by the random generation of normal distribution $\mathrm{NPD}(s(\lambda), \Delta u)$ with a width of the input uncertainty Δu. The responsivity values for each wavelength are modified independently by the random generator and about 200,000 tries are calculated for each modification. The result is a large number of (simulated) f_1' values scattered around a mean value with the related standard deviation. The latter is taken as the associated standard uncertainty.

$$f_1' = \frac{1}{a}\sum_{i=1}^{n}\left|\frac{b \cdot \mathrm{NPD}(s(\lambda_i), \Delta u)}{\sum_{i=1}^{n} S_A(\lambda_i)\ \mathrm{NPD}(s(\lambda_i), \Delta u)} - V(\lambda_i)\right| \tag{A1.30}$$

$$a = \sum_{i=1}^{n} V(\lambda_i); \quad b = \sum_{i=1}^{n} S_A(\lambda_i)\ V(\lambda_i)$$

In Table A1.2, results are summarized for the values and the associated uncertainties evaluated for the two methods shown in the parts entitled as "GUM" and "MC." The results in the "GUM" part of the table are not valid because they are determined with the standard GUM procedure for linear models as explained earlier—only the part headed by "MC" shows valid results. The calculation was done with two constant values for the simulated uncertainty Δu of the input quantities shown as the header of the related rows. The table has four lines with values and standard uncertainties determined for different spectral responsivity functions simulated by a variation of the wavelength shift $\Delta\lambda$ as defined in Equation (A1.30).

The f_1' values in the "GUM" part are listed in one row because they are independent of the uncertainties associated to the input quantities—just as expected. The similar values in the "MC" part are shown in two rows and are found to be significantly larger and strongly depending on the uncertainty Δu associated to the input quantities. The f_1' values in the "GUM" part tend to zero for a perfect match $\Delta\lambda = 0$, whereas the values in the "MC" part are depending on both the quality of the match as indicated by the value of $\Delta\lambda$ and the input uncertainty Δu.

The uncertainties $u(f_1')$ associated to the output quantity show significant differences between the methods, depending on the input uncertainty. The uncertainties in the "GUM" part are always larger than the related uncertainties in the "MC" part, but for both methods the ratio of the output uncertainties for one value of $\Delta\lambda$ is nearly the same ratio as that of the two fixed uncertainties of input quantities.

The histogram of the results found by the large number of tries from the MC method represents the probability distribution function of the f_1' values. Examples are drawn in Figure A1.2 for the four simulated responsivity functions and a fixed uncertainty of the input.

It is important to understand, that the GUM method is definitely limited to linear models. The use of the GUM method for nonlinear models can give nonvalid values

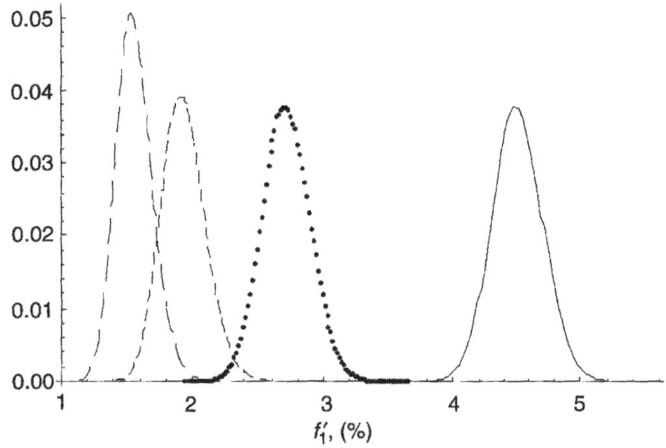

FIGURE A1.2 Four probability distribution functions are shown as result of MC simulation of the f_1' values with spectral responsivity functions simulated by $V(\lambda)$ function calculated with a wavelength scale shifted by $\Delta\lambda = 2\,\text{nm}$ (continuous line), $\Delta\lambda = 1\,\text{nm}$ (dotted line), $\Delta\lambda = 0.5\,\text{nm}$ dashed line, and $\Delta\lambda = 0, 1\,\text{nm}$ (long dashed line).

and associated uncertainties, whereas the procedure of the MC method ensures valid values and associated uncertainties including the related probability distributions for linear and nonlinear models.

REFERENCES

1. International Standards Organisation (1993) *Guide to the Expression of Uncertainty in Measurement, for Standardization*, ISBN 92-67-10188-9.
2. ISO GUM Suppl. 1 (DGUIDE 9998), *Guide to the Expression of Uncertainty in Measurement, Supplement 1: Numerical methods for the propagation of distributions* (2004).
3. Gardner JL (2006) Uncertainties in distribution temperature and correlated color temperature, in: *2nd CIE Expert Symposium on Measurement Uncertainty*, June 2006, Braunschweig.
4. Krüger U, Sauter G (2006) Comparison of methods for indicating the measurement uncertainty of integral parameters on the basis of spectral data by means of the measurement uncertainty of the f1' value, in: *2nd CIE Expert Symposium on Measurement Uncertainty*, June 2006, Braunschweig.

APPENDIX 2

UNCERTAINTIES IN SPECTRAL COLOR MEASUREMENT

James L. Gardner
5 Adamson Ave, Thornleigh 2120 Australia

INTRODUCTION

Color is an important part of everyday life. Although color is strictly a perception that will differ from observer to observer, many fields in industry, commerce, and science require color to be specified by a set of numbers that can be reproduced, that is, for color to be measured in some objective and reproducible manner. Such specification is all based on the determination of the tristimulus values, using CIE specified and agreed color-matching functions, which we have seen in Chapter 3. Random and systematic effects affect any physical measurement, which is hence not exact but has a reasonable probability of occurring in a range about the specified value. It is this range, or uncertainty, for numbers specifying a color measurement, that we are concerned with here.

The tristimulus values are most accurately determined in a spectral measurement, that of relative spectral irradiance or power for source color and of reflectance for surface color. The measured values at different wavelengths are then combined with the color-matching functions to form the tristimulus values, in turn combined to form various color values. Uncertainty in each spectral value then contributes to the uncertainty of the color values, and the principles of the ISO *Guide to the Expression of Uncertainty in Measurement* (GUM)[1] are used to propagate uncertainty from the spectral measurements to the values of the final desired color quantities.

Colorimetry: Understanding the CIE System, Edited by János Schanda

The basic principles of uncertainty propagation are given in Appendix 1. Color measurements are complicated because each of the tristimulus values depends on a common set of spectral values, that is, the tristimulus values are not independent but are correlated. This means that as one of the spectral values moves within its probable range, the tristimulus values do not move randomly but move in a concerted way, even for random effects in the spectral measurement. This correlation flows through to the color values, and it must be considered when propagating uncertainties. Similarly, systematic effects in the spectral measurement correlate those measurements—values do not move independently in their probable ranges—and this in turn means that the tristimulus values are correlated. We shall see below that we can handle these correlations by treating each independent uncertainty component of the spectral measurement separately and propagating its uncertainty through to its contribution to the uncertainty in color values.

In this chapter, we consider the spectral measurement process and its uncertainty components that influence the calculated tristimulus values. Uncertainties in the spectral values are propagated to those in the tristimulus values and thence to uncertainties in the various color quantities. Both source color and surface color are handled in the same manner. Guidelines are given for estimating the uncertainties of systematic and random effects in the measurement process.

Only the measurement process is considered here. Color measurements on one sample can vary because of different illumination and viewing conditions and nonuniformity. Geometric conditions and bandwidth are parameters of the measurement, and they should be clearly stated when quoting results. Corrections can be applied to the spectra for bandwidth effects.[2,3] Nonuniform samples can be measured many times in different positions or orientations, and a further contribution to the uncertainty that would apply to a single, randomly positioned and orientated measurement be estimated from those measurements.

This chapter concentrates on color determined from spectral measurements; uncertainties in the calibration values of a tristimulus colorimeter are covered in Chapter 6.

TRISTIMULUS VALUES

Tristimulus values are integrals representing a product of the CIE color-matching functions,[4] shown in Figure 3.3, and the spectral power distribution of the light reaching the detector. The detector is hence a simulated eye with an agreed response function. The set of color-matching functions chosen (2° or 10° field of view) depends on the application.

The tristimulus values are integrals. Spectral data are recorded at discrete wavelengths, not necessarily at regularly spaced wavelength intervals. This is particularly true for array spectrometers, where the wavelength scale may be nonlinear. For sampled data, the X, Y, Z tristimulus integrals are given by

$$X = \sum_{i=1}^{N} \Delta_i \bar{x}_i S_i$$

$$Y = \sum_{i=1}^{N} \Delta_i \bar{y}_i S_i \tag{A2.1}$$

$$Z = \sum_{i=1}^{N} \Delta_i \bar{z}_i S_i$$

where S_i is the spectral power distribution reaching the eye (at the ith of N wavelengths), $\bar{x}_i, \bar{y}_i, \bar{z}_i$ are the CIE color-matching functions defined over the wavelength range from 360 nm to 830 nm, and Δ_i is the weight of the contribution at the ith point of the integral. For trapezoidal integration,

$$\begin{aligned} \Delta_1 &= (\lambda_2 - \lambda_1)/2 \\ \Delta_N &= (\lambda_N - \lambda_{N-1})/2 \\ \Delta_i &= (\lambda_{i+1} - \lambda_{i-1})/2, \quad i = 2, \ldots, N-1 \end{aligned} \tag{A2.2}$$

Three variables are required to describe color, two representing position in a two-dimensional color plane and the third representing a correlate of lightness or brightness. In the case of the simple chromaticity values (x, y) or (u', v'), the third variable is the Y tristimulus value itself (luminance), often quoted as a ratio to that of an illuminant. Only relative spectral power distributions (relative spectral irradiance at the detector) are important for the chromaticity coordinates. Absolute values will affect luminance.

For source color, the measured quantity S_i is relative spectral irradiance itself. If we are measuring surface color, the spectral irradiance at the detector is a product of the source spectral irradiance at the surface and the surface spectral reflectance. Surface colors are usually specified for a particular CIE Standard Illuminant,[5,6] often D65, and the measured spectral quantity is spectral reflectance for a specified illumination and viewing geometry. The illuminant distributions are agreed values tabulated over the visible spectral range and their values carry no uncertainty. All the color quantities for surfaces can be derived using the same expressions as those for sources provided the color-matching functions are replaced by their product with the reference illuminant spectrum:

$$\bar{x}'_i = S_i^{\text{Ill}} \bar{x}_i, \bar{y}'_i = S_i^{\text{Ill}} \bar{y}_i, \bar{z}'_i = S_i^{\text{Ill}} \bar{z}_i \tag{A2.3}$$

where S_i^{Ill} is the value of the illuminant spectral power distribution at the ith wavelength. Values of the spectral power distribution S_i in Equations (A2.1) are then replaced by the values of R_i, the reflectance at the ith wavelength. In the sections below, we take S_i to refer to the measured spectrum (source or reflectance) and the color-matching functions to be amended as in Equations (A2.3) if we are referring to a surface measurement (with the prime notation dropped).

UNCERTAINTY PROPAGATION

Here, we apply the basic principles described in Appendix 1 and formulate a convenient method for dealing with uncertainties in the tristimulus values, including the correlations between these related quantities.

The uncertainty $u(X)$ (or more correctly, the variance $u^2(X)$) of a quantity X formed by combining measured quantities x_i through the relationship $X = f(x_1, x_2, \ldots x_N)$ is most commonly expressed in the form

$$u^2(X) = \sum_{i=1}^{N} \left(\frac{\partial f}{\partial x_i}\right)^2 u^2(x_i) + 2\sum_{i=1}^{N-1} \sum_{j=i+1}^{N} \frac{\partial f}{\partial x_i}\frac{\partial f}{\partial x_j} u(x_i, x_j) \tag{A2.4}$$

where $u(x_i)$ is the uncertainty in x_i and $u(x_i, x_j)$ is the covariance of x_i and x_j. The derivatives $\partial f/\partial x_i$ are sensitivity coefficients or rates of change for the dependence of X on the various measured quantities x_i. Equation (A2.4) is a convenient form for dealing with uncorrelated (random) input quantities because then the covariance of pairs of variables is zero and Equation (A2.4) reduces to

$$u^2(X) = \sum_{i=1}^{N} \left(\frac{\partial f}{\partial x_i}\right)^2 u^2(x_i) \tag{A2.5}$$

This is the "sum of squares" which is applied for the uncertainty when combining uncorrelated quantities.

Given that $u^2(x_i) \equiv u(x_i, x_i)$, a more general form of Equation (A2.4) is

$$u^2(X) = \sum_{i=1}^{N} \sum_{j=1}^{N} \frac{\partial f}{\partial x_i}\frac{\partial f}{\partial x_j} u(x_i, x_j) \tag{A2.6}$$

If we form another quantity Y by combining the measured quantities x_i through the relationship $Y = g(x_1, x_2, \ldots, x_N)$, the uncertainty in Y is given by an expression similar to that of Equation (A2.6), but now the quantities X and Y are correlated through dependence on the common set x_i. The covariance between X and Y is given by

$$u(X, Y) = \sum_{i=1}^{N} \sum_{j=1}^{N} \frac{\partial f}{\partial x_i}\frac{\partial g}{\partial x_j} u(x_i, x_j) \tag{A2.7}$$

These last two relationships are conveniently expressed in matrix form

$$u^2(X) = \mathbf{f}_x \mathbf{U}_x \mathbf{f}_x^{\mathrm{T}} \tag{A2.8}$$

$$u(X, Y) = \mathbf{f}_x \mathbf{U}_x \mathbf{g}_x^{\mathrm{T}} \tag{A2.9}$$

where

$$\mathbf{f}_x = \left(\frac{\partial f}{\partial x_1} \quad \frac{\partial f}{\partial x_2} \quad \cdots \quad \frac{\partial f}{\partial x_n} \right) \tag{A2.10}$$

$$\mathbf{g}_x = \left(\frac{\partial g}{\partial x_1} \quad \frac{\partial g}{\partial x_2} \quad \cdots \quad \frac{\partial g}{\partial x_n} \right) \tag{A2.11}$$

are row vectors of sensitivity coefficients, and

$$\mathbf{U}_x = (u(x_i, x_j)) \tag{A2.12}$$

is the $N \times N$ variance–covariance matrix of the squares of uncertainty (variance) in diagonal elements, covariance values elsewhere.

Equations (A2.6) and (A2.7) are the most convenient forms when estimating the variances and covariances of the tristimulus values. The matrix forms are extremely useful for propagating uncertainties and correlations from those of the tristimulus values to the various color quantities, as we shall see below.

Correlations are often described in terms of correlation coefficients r. Ranging in value from -1 to $+1$ with a value 0 for uncorrelated quantities, these are normalized covariance values defined by

$$r(x_i, x_j) = \frac{u(x_i, x_j)}{u(x_i)u(x_j)} \tag{A2.13}$$

TRISTIMULUS UNCERTAINTIES BY COMPONENT

A spectrum of measured quantities such as relative spectral irradiance or spectral reflectance will contain a number of sources of uncertainty. Some of these will be random at different wavelengths, such as amplifier or source noise, but others such as scattered light in the background will be systematic or correlated between wavelengths—if scattered light increases at one wavelength because of a misplaced baffle, it will increase at all wavelengths. The mixture of random and correlated components means that the measured spectral values will in general be partly correlated. To propagate uncertainty from spectral measurement to tristimulus values, we then need to know both the total uncertainty at each spectral point and the covariance between values. However, the individual effects contributing to the uncertainties of the spectral values are independent. To find the total variance of each of the tristimulus values, we simply add the variances due to each component effect—this in effect is Equation (A2.5). Similarly, we can also add the covariance due to each effect to find the total covariance between any two tristimulus values. Propagating each measurement uncertainty component through to the tristimulus values means that simple forms particularly can be used, depending on whether the effect is random or systematic between wavelengths.

For each effect, we calculate the uncertainty $u(S_i)$ due to that effect alone in the measured spectral irradiance or spectral reflectance quantity S_i at the ith wavelength. We then propagate those uncertainties through to the tristimulus values. From Equation (A2.1), the sensitivity coefficients for the dependence of the tristimulus on the measured spectral values, $\frac{\partial X}{\partial S_i}, \frac{\partial Y}{\partial S_i}, \frac{\partial Z}{\partial S_i}$, are the color-matching functions (augmented by the chosen illuminance distribution for surface reflectance spectra).

Random Component Effects

For an effect that is random from wavelength to wavelength, the uncertainty in the X tristimulus value is propagated through Equation (A2.5), with the contribution to the variance given by

$$u^2(X) = \sum_{i=1}^{N} \bar{x}_i^2 u^2(S_i) \tag{A2.14}$$

with similar expressions for the variance of the Y, Z tristimulus values. Even though the spectral values are uncorrelated, the tristimulus values are correlated because they depend on the same set of spectral values. We have $u(S_i, S_j) = 0$ for $i \neq j$, $u(S_i, S_i) = u^2(S_i)$, and from Equation (A2.7), the contribution to the covariance between X and Y is

$$u(X, Y) = \sum_{i=1}^{N} \bar{x}_i \bar{y}_i u^2(S_i) \tag{A2.15}$$

with similar expressions for covariances between the other combinations of tristimulus values.

Systematic Component Effects

Systematic components are fully correlated between wavelengths, that is,

$$u(x_i, x_j) = \pm\, u(x_i)u(x_j) \tag{A2.16}$$

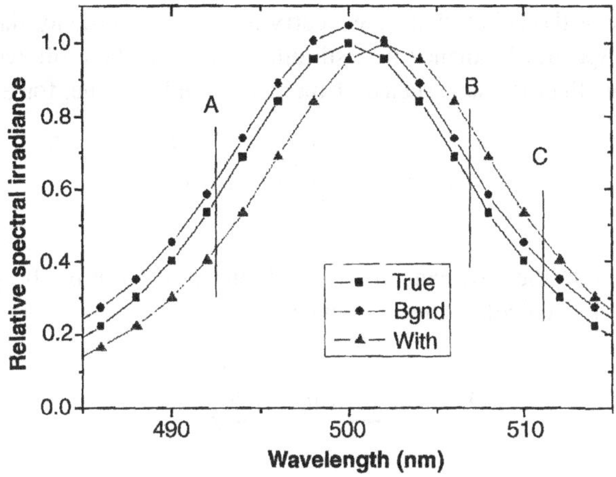

Figure A2.1 Portion of a representative LED relative spectral irradiance distribution (true curve). Curves bgnd and wlth represent shifts from the true curve in the presence of systematic background and wavelength offsets, respectively.

If the correlation has the same sign for all pairs of wavelengths, Equation (A2.4) reduces to

$$u^2(X) = \left(\sum_{i=1}^{N} \frac{\partial f}{\partial x_i} u(x_i) \right)^2 \tag{A2.17}$$

However, the correlation between the spectral values can be positive or negative, as demonstrated in Figure A2.1.

Figure A2.1 uses the measurement of relative spectral irradiance of a LED to illustrate correlations that arise from a systematic offset in the background level and a systematic offset in the wavelength setting, both at particular values in their uncertainty ranges. The true spectrum and its possible measured distribution after these effects are shown schematically. Spectral values at all wavelengths increase if the background offset is negative; hence, the correlations between all the pairs of wavelengths are all positive for this effect. This is not true for a systematic wavelength offset, as shown by comparing the change in measured spectral irradiance at points A and B. At point A, the measured value decreases for a positive offset in wavelength, whereas it increases at point B. Hence, the correlation between points A and B is negative for this effect. It is however positive between points B and C. The magnitude of a given effect at a particular wavelength is the product of the sensitivity coefficient of the spectral value for the effect and the uncertainty of the effect. The sign of the correlation between two wavelengths is the product of the signs of the sensitivity coefficients.

Although uncertainties are always taken as positive, it is shown elsewhere[7] that the correlations are properly handled if we attach the sign of the sensitivity coefficient to the uncertainty of the effect, that is, we carry a signed uncertainty through the propagation from spectral irradiance to tristimulus values. It then follows that the contribution of the effect to the variance of the X tristimulus value, for example, is

$$u^2(X) = \left(\sum_{i=1}^{N} \bar{x}_i u_s(S_i) \right)^2 \tag{A2.18}$$

The contribution to the covariance between X and Y is given by the product of the same sums used to calculate their variances

$$u(X, Y) = \sum_{i=1}^{N} \bar{x}_i u_s(S_i) \sum_{i=1}^{N} \bar{y}_i u_s(S_i) \tag{A2.19}$$

By comparing these last two equations, we also see that

$$u(X, Y) = \pm u(X) u(Y) \tag{A2.20}$$

That is, the tristimulus values are also fully correlated for the systematic effect being considered. The individual sums in Equation (A2.19) can be positive or negative, as can the covariance between the tristimulus values for a given systematic effect.

The complete variance–covariance matrix for the tristimulus values, for our effect (random or systematic) being considered, is

$$\mathbf{U}_{XYZ} = \begin{bmatrix} u^2(X) & u(X,Y) & u(X,Z) \\ u(X,Y) & u^2(Y) & u(Y,Z) \\ u(X,Z) & u(Y,Z) & u^2(Z) \end{bmatrix} \tag{A2.21}$$

PROPAGATION FROM TRISTIMULUS UNCERTAINTIES TO COLOR-VALUE UNCERTAINTIES

Color quantities can all be expressed in terms of the tristimulus values. Uncertainties in the derived quantities are then found by propagating uncertainties through sensitivity coefficients for those quantities in terms of the tristimulus values and the variance–covariance matrix for the tristimulus values, for each effect in turn. The variance and covariance of all effects are then summed to find the total variance–covariance matrix and hence uncertainties and correlations for the color values. While the summation over effects can be at the tristimulus variance–covariance matrix stage, propagating the uncertainty of each effect through to the desired color values shows which components are more significant than others. This can then lead to a review and improvement of the experimental techniques.

Except for specialized values such as correlated color temperature or dominant wavelength, useful forms of color representation for particular types of sources (not covered here), three variables are required to describe color. In the simplest representation, these are the tristimulus values. The tristimulus values may be combined to form other color quantities, and we then need to know not only the uncertainty (or variance) of the tristimulus values but also the relation between them (covariance), which may be quoted as a matrix of correlation coefficients. Instead of calculating individual variance or covariance values using Equations (A2.8) and (A2.9), we form a 3 × 3 matrix of sensitivity coefficients for each desired quantity in terms of the X,Y, and Z tristimulus values. If **Q** is such a matrix with color quantities ordered by rows and tristimulus values by columns, the variance–covariance matrix for the color quantities is given by

$$\mathbf{U} = \mathbf{Q}\mathbf{U}_{XYZ}\mathbf{Q}^{\mathrm{T}} \tag{A2.22}$$

Using x, y, Y as an example of a color triplet, the matrix $\mathbf{Q}_{xyY}$ is given by

$$\begin{bmatrix} \dfrac{\partial x}{\partial X} & \dfrac{\partial x}{\partial Y} & \dfrac{\partial x}{\partial Z} \\ \dfrac{\partial y}{\partial X} & \dfrac{\partial y}{\partial Y} & \dfrac{\partial y}{\partial Z} \\ \dfrac{\partial Y}{\partial X} & \dfrac{\partial Y}{\partial Y} & \dfrac{\partial Y}{\partial Z} \end{bmatrix} \tag{A2.23}$$

Methods of Calculation for Color Triplets

In the following sections, we derive the sensitivity matrices for the various color triplets. Some of these require a second stage of propagation because they in turn depend on other color quantities.

(x,y,Y) Color Coordinates

The (x, y) chromaticity values are given simply as

$$x = \frac{X}{T_{xy}} \qquad y = \frac{Y}{T_{xy}} \tag{A2.24}$$

with $T_{xy} = X + Y + Z$

Y values for a source are a measure of luminance, relative to the reference standard used to calibrate instrument response. Standard illuminants used for surface color calculations have their spectral distributions normalized to a luminance value of 100. Then for surface reflectance values in the range 0–1, the calculated value of Y is luminous reflectance, expressed as a percentage of the source luminance in the range 0–100. The sensitivity matrix is for the dependence of (x, y, Y) color values on the tristimulus values is

$$\mathbf{Q}_{xyY} = \begin{bmatrix} \frac{Y+Z}{T_{xy}^2} & \frac{-X}{T_{xy}^2} & \frac{-X}{T_{xy}^2} \\ \frac{-Y}{T_{xy}^2} & \frac{X+Z}{T_{xy}^2} & \frac{-Y}{T_{xy}^2} \\ 0 & 1 & 0 \end{bmatrix} \tag{A2.25}$$

A general conclusion can be drawn for red LED x, y chromaticities. Here, the Z tristimulus value is effectively zero and the chromaticity is close to the monochromatic boundary. Hence, $x + y = 1$, and any sensitivity coefficients for x are the negative of those for y and any uncertainties for x and y will be equal.

(u,v,Y) Color Coordinates

In a manner similar to the treatment for (x, y, Y) above

$$u = \frac{4X}{T_{uv}} \qquad v = \frac{6Y}{T_{uv}} \tag{A2.26}$$

with $T_{uv} = X + 15Y + 3Z$

The sensitivity matrix for the dependence of (u, v, Y) values on the tristimulus values is

$$\mathbf{Q}_{uvY} = \begin{bmatrix} \frac{60Y+12Z}{T_{uv}^2} & \frac{-60X}{T_{uv}^2} & \frac{-12X}{T_{uv}^2} \\ \frac{-6Y}{T_{uv}^2} & \frac{6X+18Z}{T_{uv}^2} & \frac{-18Y}{T_{uv}^2} \\ 0 & 1 & 0 \end{bmatrix} \tag{A2.27}$$

(u′,v′,Y) Color Coordinates

These are given as a simple scaling of (u, v), $u' = u, v' = 3v/2$. This scaling is applied to the middle row of Equation (A2.27).

(L^*, a^*, b^*) Color Coordinates

L^*, a^*, and b^* are calculated as

$$L^* = 116\left(\frac{Y}{Y_N}\right)^{1/3} - 16$$

$$a^* = 500\left[\left(\frac{X}{X_N}\right)^{1/3} - \left(\frac{Y}{Y_N}\right)^{1/3}\right] \tag{A2.28}$$

$$b^* = 200\left[\left(\frac{Y}{Y_N}\right)^{1/3} - \left(\frac{Z}{Z_N}\right)^{1/3}\right]$$

where X_n, Y_n, and Z_n are the tristimulus values for a perfect reflector (i.e., the illuminant distribution alone—these carry no uncertainty). The sensitivity matrix is

$$\mathbf{Q}_{L^*a^*b^*} = \begin{bmatrix} 0 & \frac{116}{3}Y_N^{-1/3}Y^{-2/3} & 0 \\ \frac{500}{3}X_N^{-1/3}X^{-2/3} & -\frac{500}{3}Y_N^{-1/3}Y^{-2/3} & 0 \\ 0 & \frac{200}{3}Y_N^{-1/3}Y^{-2/3} & -\frac{200}{3}Z_N^{-1/3}Z^{-2/3} \end{bmatrix} \tag{A2.29}$$

($L,^*C^*,h^*$) Color Coordinates (based on a^*, b^*)

The quantities hue angle and chroma may be calculated from (a^*, b^*) chromaticity values as

$$C^* = \sqrt{a^{*2} + b^{*2}}$$

$$h = \tan^{-1}\left(\frac{b^*}{a^*}\right) \tag{A2.30}$$

We first calculate the variance–covariance matrix $\mathbf{U}_{L^*a^*b^*}$ for L^*, a^*, b^* from that of the tristimulus values using the sensitivity matrix Equation (A2.29):

$$\mathbf{U}_{L^*a^*b^*} = \mathbf{Q}_{L^*a^*b^*}\mathbf{U}_{XYZ}\mathbf{Q}^{\mathbf{T}}_{L^*a^*b^*} \tag{A2.31}$$

The sensitivity matrix for the quantities L^*, C^*, and h in terms of L^*, a^*, and b^* is

$$\mathbf{Q}_{L^*C^*h} = \begin{bmatrix} 1 & 0 & 0 \\ 0 & \frac{a^*}{C^*} & \frac{b^*}{C^*} \\ 0 & -\frac{b^*}{C^{*2}} & \frac{a^*}{C^{*2}} \end{bmatrix} \tag{A2.32}$$

*See also Section "CIE 1976 uniform color spaces" in Chapter 3.

and the final uncertainties and correlations are carried in the variance–covariance matrix:

$$\mathbf{U}_{L^*C^*h} = \mathbf{Q}_{L^*C^*h}\mathbf{U}_{La^*b^*}\mathbf{Q}^{\mathbf{T}}_{L^*C^*h} \tag{A2.33}$$

(L*, u*, v*) Color Coordinates

These are defined as

$$\begin{aligned} L^* &= 116\left(\frac{Y}{Y_{\mathrm{N}}}\right)^{1/3} - 16 \\ u^* &= 13L^*(u' - u'_{\mathrm{N}}) \\ v^* &= 13L^*(v' - v'_{\mathrm{N}}) \end{aligned} \tag{A2.34}$$

where u', v' are CIE 1976 chromaticity coordinates and $u'_{\mathrm{N}}, v'_{\mathrm{N}}$ are similar quantities for the illuminant alone.

First we calculate the covariance matrix $\mathbf{U}_{Lu'v'}$ for the quantities L^*, u', and v', for which the sensitivity matrix in terms of the tristimulus values is

$$\mathbf{Q}_{L^*u'v'} = \begin{bmatrix} 0 & \dfrac{116}{3} Y_N^{-1/3} Y^{-2/3} & 0 \\ \dfrac{60Y + 12Z}{T_{uv}^2} & \dfrac{-60X}{T_{uv}^2} & \dfrac{-12X}{T_{uv}^2} \\ \dfrac{-9Y}{T_{uv}^2} & \dfrac{9X + 27Z}{T_{uv}^2} & \dfrac{-27Y}{T_{uv}^2} \end{bmatrix} \tag{A2.35}$$

where $T_{uv} = X + 15Y + 3Z$.

The covariance matrix $\mathbf{U}_{L^*u'v'}$ for the quantities L^*, u' and v' is then

$$\mathbf{U}_{L^*u'v'} = \mathbf{Q}_{L^*u'v'}\mathbf{U}_{XYZ}\mathbf{Q}^{\mathbf{T}}_{L^*u'v'} \tag{A2.36}$$

The sensitivity matrix for the quantities L^*, u^*, and v^* in terms of L^*, u', and v' is

$$\mathbf{Q}_{L^*u^*v^{*\prime}} = \begin{bmatrix} 1 & 0 & 0 \\ 13(u' - u'_{\mathrm{N}}) & 13L^* & 0 \\ 13(v' - v'_{\mathrm{N}}) & 0 & 13L^* \end{bmatrix} \tag{A2.37}$$

and the final uncertainties and correlations are carried in the variance–covariance matrix:

$$\mathbf{U}_{L^*u^*v^*} = \mathbf{Q}_{L^*u^*v^*}\mathbf{U}_{Lu'v'}\mathbf{Q}^{\mathbf{T}}_{L^*u^*v^*} \tag{A2.38}$$

(L^*, C^*, h) Color Coordinates (based on u^*, v^*)
Hue angle and chroma are also calculated from (u^*, v^*) color values as

$$C^* = \sqrt{u^{*2} + v^{*2}}$$
$$h = \tan^{-1}\left(\frac{v^*}{u^*}\right) \tag{A2.39}$$

Uncertainties in these values are found by substituting (u^*, v^*) for (a^*, b^*) in Equation (A2.32) for the sensitivity coefficients and propagating uncertainties from those of L^*, u^*, v^*.

(L^*, s, h) Color Coordinates (based on u^*, v^*)
Saturation s may be required in place of chroma:

$$s = C^*/L^* \tag{A2.40}$$

The sensitivity matrix for the quantities L^*, s, and h in terms of L^*, u^*, and v^* is

$$\mathbf{Q}_{L^*sh} = \begin{bmatrix} 1 & 0 & 0 \\ \frac{-s}{L^*} & \frac{u^*}{sL^{*2}} & \frac{v^*}{sL^{*2}} \\ 0 & -\frac{v^*}{sL^{*2}} & \frac{u^*}{sL^{*2}} \end{bmatrix} \tag{A2.41}$$

Uncertainties in L^*, s, and h are then propagated from the variances and covariances of L^*, u^*, and v^* as

$$\mathbf{U}_{L^*sh} = \mathbf{Q}_{L^*sh}\mathbf{U}_{L^*u^*v^*}\mathbf{Q}^{\mathrm{T}}_{L^*sh} \tag{A2.42}$$

SPECTRAL MEASUREMENT AS A TRANSFER

The spectral measurements from which we calculate color values are made as a comparison against a reference standard. For surface colors, we measure spectral reflectance against a reference standard for the required geometric conditions. Sources are compared with a spectral irradiance standard. In both cases, we are transferring values S_i from those of the reference standard S_i^{Ref} at the ith wavelength as

$$S_i = t_i S_i^{\mathrm{Ref}} \tag{A2.43}$$

Uncertainties in the spectral value S_i arise both from those of the reference value and those introduced by the spectral transfer. For uncertainty components of the reference spectrum, from Equation (A2.43) we have

$$u(S_i) = t_i u(S_i^{\mathrm{Ref}}) \tag{A2.44}$$

whereas for components in the transfer we have

$$u(S_i) = S_i^{\text{Ref}} u(t_i) \tag{A2.45}$$

A double-beam spectrophotometer used to measure reflectance typically reports the transfer ratio directly once an instrument zero spectrum has been recorded. Single-beam instruments, used both for surface and source measurements, report individual signals for the device under test and for the reference. The transfer ratio is strictly the ratio of measured signals, but the monochromator efficiency is constant at a given wavelength and we calculate the transfer value as the ratio $t_i = S_i/S_i^{\text{Ref}}$. In this section, we consider the four generic classes of uncertainty sources that apply to the transfer measurement of S_i. Individual sources of uncertainty within these classes may be random or correlated between wavelengths; we separately treat these types within the class.

Uncertainty of the Reference Values

The spectral reference standard is likely to be measured at a limited number of wavelengths, with both random and systematic errors in the process. The reference values will then be at least partially correlated. In this case, the full expressions of Equations (A2.4) and (A2.7) or their matrix forms Equations (A2.8) and (A2.9) must be used to calculate the tristimulus uncertainties and correlations. Complete specification of the reference uncertainties requires the total random components and the individual systematic components at each wavelength—these can then be individually propagated through the transfer to component uncertainties of the tristimulus values. The correlations or at least the total systematic uncertainties as well as the total or total random uncertainties should be provided at each reference wavelength. At the highest levels of accuracy, reference spectra uncertainties are dominated by systematic effects and they are highly correlated. If the correlation coefficient for the reference values is constant, or approximately constant through the visible spectral range, the reference uncertainties can be split into fully correlated and uncorrelated parts and the simpler expressions of Equations (A2.14), (A2.15), (A2.18), and (A2.19) can be used to propagate the reference contribution to the tristimulus variance and covariance.

Reference spectra usually change relatively slowly with wavelength and are provided at wider wavelength spacing than the transfer measurements. The reference data are then interpolated to the measurement wavelengths. Interpolation introduces correlations,[8] and these must be taken into account when propagating the reference uncertainties. A simple technique when the correlation coefficient of the reference data is constant is to interpolate the reference uncertainties to the measurement wavelengths. The systematic reference uncertainty component will yield the correct result, but the variance of the propagated random component will be underestimated by the ratio of the reference data spacing to that of the measurement.

The remaining classes are all treated in terms of the signed uncertainties of the measured spectral values for the various effects. The uncertainties are those that apply after applying corrections for the effects.

Relative Scaling of the Measured Spectral Values

In this case, the transfer value at each wavelength is multiplied by a variable s whose value is 1 after corrections have been applied. The uncertainty of the scaling is usually given in relative terms. It follows that

$$u_s(S) = S_i u_{s,\mathrm{rel}}(s) \tag{A2.46}$$

Random Scaling Components

Source noise and amplifier gain noise fall into this category. Equations (A2.14) and (A2.15) are used to calculate the contribution to the variances and covariances of the tristimulus values, respectively.

Systematic Scaling Components

These may arise from gain differences between amplifiers, baffles obscuring either the reference or test beams, and so on. Nonlinearity of response is also a systematic scaling component. Linear scalings are not important if only simple chromaticity coordinates are required, but will affect lightness or luminance values. Then correction should be applied and the uncertainty of the correction estimated. The uncertainty values will generally scale uniformly across all wavelengths, that is, the estimated uncertainties are fully correlated, and Equations (A2.18) and (A2.19) are used to calculate the contribution to the variances and covariances of the tristimulus values, respectively. Correlations are generally positive, although mixed signs (and hence the need for signed uncertainties) may occur for some forms of nonlinearity.

Offsets in the Spectral Values

In the presence of additive offsets b_i in the test channel and b_i^{Ref} in the reference channel, the transfer ratio at the ith wavelength becomes

$$t_i = \frac{S_i + b_i}{S_i^{\mathrm{Ref}} + b_i^{\mathrm{Ref}}} \tag{A2.47}$$

The signals are corrected so that the residual offsets are zero but have a nonzero uncertainty. For offsets in the test channel

$$u_s(S_i) = u_s(b_i) \tag{A2.48}$$

and in the reference channel

$$u_s(S_i) = \frac{S_i}{S_i^{\mathrm{Ref}}} u_s(b_i^{\mathrm{Ref}}) \tag{A2.49}$$

Note that offsets in the sample and test channels have different effects. Offsets are usually specified as a fraction of the maximum signal for the channel for source measurements or for single-beam reflectance measurements. Double-beam spectrophotometers used for reflectance measurements generally do not report the individual signals for the test and reference channels, but only their combined effect. Background levels, taken with the beam blocked, are reported directly as a zero reflectance; these levels can be analyzed for random and systematic offset uncertainty components, applied via Equation (A2.48) only.

Random Offset Components

Electronic noise in the background-corrected signals is random. Electronic noise is present in both the zero measurement and the signal measurement itself, and the noise in the corrected signal is $\sqrt{2}$ times that of the zero signal. Equations (A2.14) and (A2.15) are used to calculate the contribution to the variances and covariances of the tristimulus values, respectively.

Systematic Offset Components

Incorrect estimates of background through light reflected from a baffle or drift in the electronic offset between spectral scans leads to offsets that may vary with wavelength but scale across all wavelengths. They are correlated between wavelengths, and that correlation is positive, but generally uncorrelated between signal and reference channels. Estimated uncertainties for offsets are propagated to the spectral values through Equations (A2.48) and (A2.48) and then to the tristimulus values through Equations (A2.18) and (A2.19).

Wavelength Errors

Again, these have different effects in the reference and sample channels. If Δ is the offset in wavelength, for the sample channel we have

$$u_s(S_i) = \frac{\partial S_i}{\partial \lambda} u(\Delta) \tag{A2.50}$$

For the reference channel we have

$$u_s(S_i) = -\frac{S_i}{S_i^{\mathrm{Ref}}} \frac{\partial S_i^{\mathrm{Ref}}}{\partial \lambda} u(\Delta) \tag{A2.51}$$

The derivatives are calculated numerically. The method of combining the wavelength uncertainties in the two channels depends on the scanning process. Some transfers are made concurrently in wavelength, where the wavelength value is set and then both sample and reference signals are recorded. Other transfers are made sequentially, where the wavelength range is stepped for one channel and then stepped again for the other. For concurrent scanning, the wavelength setting is completely correlated between the sample and reference channels. Hence

$$u_s(S_i) = \left(\frac{\partial S_i}{\partial \lambda} - \frac{S_i}{S_i^{\mathrm{Ref}}} \frac{\partial S_i^{\mathrm{Ref}}}{\partial \lambda} \right) u(\Delta) \tag{A2.52}$$

For sequential scanning, the wavelength setting is random between the two channels, and it follows that

$$u_s^2(S_i) = \sqrt{\left(\frac{\partial S_i}{\partial \lambda}\right)^2 + \left(\frac{S_i}{S_{\mathrm{Ref},i}}\frac{\partial S_{\mathrm{Ref},i}}{\partial \lambda}\right)^2} u(\Delta) \tag{A2.53}$$

In the sequential mode of scanning, differences in exact wavelength setting also mean that the monochromator throughput may be different for the test and reference wavelengths, and we should strictly include a term for this effect where the throughput (and source irradiance in the case of reflectance measurements) varies significantly. Note that we have ignored the effect of wavelength uncertainty through the integration weights of Equation (A2.2); systematic effects here cancel, and random effects are negligible when averaged over a moderate number of input spectral values.

Random Wavelength Offsets

These arise from the precision of the calibration of the wavelength setting function, that is, from the accuracy of determining line centers or from mechanical effects such as friction or scanning accuracy of a mechanical system. The uncertainty in measured spectral irradiance is chosen according to the scanning method. The values at different wavelengths are uncorrelated; Equations (A2.14) and (A2.15) are used to calculate the contribution to the variances and covariances of the tristimulus values, respectively.

Systematic Wavelength Offsets

These can arise if the spectral lamp(s) used for calibrating the wavelength scale have different alignment to the broad-spectrum lamp used for measurement. They also arise from incorrect identification of calibration wavelengths, or through assuming an incorrect function, such as a linear step, for the scanning/setting mechanism. These systematic effects are fully correlated between the sample and reference measurements, and Equation (A2.52) applies for the uncertainty at a particular wavelength. The uncertainties of different wavelengths are fully correlated. Hence, we use Equations (A2.18) and (A2.19) to calculate the variances and covariances of the tristimulus values.

DETERMINING MEASUREMENT COMPONENTS

Some sources of uncertainty, both random and correlated between wavelengths, have been listed above as examples. It is not possible to cover all types of spectral instruments and possible sources of uncertainty covering source and reflectance measurements here. Part of the technical training and expertise in any field of measurement, including colorimetry, is in identifying components that can affect a measurement and adopting techniques to minimize errors. Instead, this section is intended to provide some general guidance on related topics. The GUM itself provides guidance for estimating standard uncertainties by direct statistical means and from an estimate of range limits, for example. Many of the national institutes and calibration accreditation bodies have also written useful interpretive documents for the GUM.[9–11]

Background Offsets

It is usual to record background readings that are then subtracted from the test and reference readings. Usual practice is to record a complete background spectrum. This is true even for double-beam reflectance measurements, where it is now common to sample data during a shutter-closed period at each wavelength. We can record a number of repeat spectra and analyze them to estimate the noise that might be present at a given wavelength, for a single scan, by statistical techniques. Where the background level is constant (either over the complete spectrum or in sections), we can estimate the noise from the standard deviation of the background over a number of wavelengths (but note the caveat of the next section). This is the random background noise component, typically 0.1% of the spectral value itself for reasonable signal levels. In a stable system, the mean of this background is applied as a correction over the whole spectrum. The uncertainty of the systematic offset is then the standard deviation of this mean, a factor of $\sqrt{N}$ lower than the random level if the background noise is random, where N is the number of data points. Averaging over typically 80 points in a spectrum then means that the systematic offset component is of order 0.01% for a typical system.

The statistical analysis should be performed over a number of repeats of spectra recorded under the measurement conditions, that is, once the system is stable and for the scan conditions of the measurement, including averaging over a number of repeated single scans where such averaging is part of the measurement sequence.

Where the spectral values are close to zero, the addition of noise can lead to nonphysical negative values. These values should remain in any colorimetric calculations—removing negative values is equivalent to introducing an offset.

Noise versus Drift

Noise and drift in spectral measurements are closely related. One becomes the other depending on the timescale of a spectral scan. Short-term (compared to the scan time for the spectrum) random effects, apparent at each wavelength, are noise. Long-term random effects appear as noise between repeat spectra, that is, as a drift. Random slow noise over a number of repeat spectra will contribute to the standard deviation of measurements at a given wavelength, but these deviations are in fact correlated. Where repeated spectra are analyzed for standard deviation at a given wavelength, the covariance between different wavelengths should also be determined. This is given as

$$u(x_i, x_j) = \overline{x_i x_j} - \overline{x}_i \overline{x}_j \qquad \text{(A2.54)}$$

where the superimposed bar indicates the mean value. Plotting the correlation coefficients (with the diagonal variance terms replaced with an average over surrounding values) shows whether there is any systematic effect (drift) contributing to the variation at each wavelength.

Some drift components are not random. Thermal changes during a warm-up period can change gains or signal zeroes in a continuous manner. These effects are correlated between wavelengths, as are temperature corrections applied to a detector calibrated at one temperature and used at another.

Source Noise

Source noise contributes to uncertainties for both source colorimetry and reflectance colorimetry. This component can be analyzed as for background offsets (including correlations for slowly varying noise), by recording repeat spectra of reference reflectors or reference lamps, preferably not the main laboratory standards but the ones of similar characteristics. For thermal lamps, current fluctuations cause temperature changes and noise that depends on the wavelength. Slowly varying fluctuations similarly vary at different wavelength but are fully correlated between wavelengths.

Band-Limited Spectra

Some samples such as LED sources are known to be restricted to a limited wavelength range. Routine measurements and calculations use the whole visible range, thus adding only noise where the spectral value is zero. A more accurate calculation to provide lower relative uncertainties is obtained by limiting the calculation and uncertainty estimate to the known spectral range of the sample.

Wavelength Uncertainties

The wavelength scale of spectrometers used for source color measurements is calibrated using one or more atomic emission lamps. A mercury lamp is the most commonly used. Care must be taken to correctly identify the wavelength of emission lines, bearing in mind that some may come from the gas used to carry a discharge (typically argon), from atomic emission wavelength tables.[12] Some lines may be present in second order; these can be identified by using a cut-on glass filter. Only isolated lines, fully resolved and uncontaminated by nearby emissions, should be used for wavelength calibration. For a scanning monochromator, the resolution should be increased to calibrate the relative position of the scale. Where this is not possible, such as for a fiber-coupled array spectrometer, the slit function (properly determined about the centroid wavelength) should be fitted to the recorded line.[3] The measured line positions are fitted to a scanning function; the standard deviation of that fit is an estimate of the random wavelength positioning error. Note that in many cases only a few lines are available for calibration and a Student's t-multiplier should be applied to the fit.[1]

For a monochromator with fixed entrance optics and an array detector, the standard deviation of the mean of the fit (standard deviation divided by $\sqrt{N}$, where N is the number of calibration lines) is an estimate of the systematic offset that might apply to the wavelength scale (provided the fit function properly represents the wavelength at a particular position in the array). The illumination method for the calibration lamp should be the same as for the source to be measured—typically this means that a diffuser is used at the entrance of the monochromator. For a scanning monochromator where sources may be measured at lower resolution (wider slits to improve throughput) than used for calibration of the wavelength scanning function, a wavelength offset of some fraction of the bandwidth may apply, caused by a nonuniform illumination of the grating. This can be estimated by repeat positioning of the source or by checking some of the calibration line positions for the reduced-resolution setting with a diffuser in the input optics.

The wavelength scale of double-beam spectrophotometers used for reflectance measurements is usually calibrated with rare earth glasses[13,14] (transmittance), solutions[15] (transmittance), or material embedded in a surface[16] (reflectance). Wavelengths of features in these materials are calibrated for particular instrument resolutions. Various algorithms may be used to identify locations of the spectral features.[15]

Nonlinearity

One component often requiring correction is the linearity of the detection system. This component is fully correlated at all wavelengths. The correction and its uncertainty are given as relative scaling factors that are dependent on the measured value. For single-beam source measurements, linearity is determined by comparing signals of superimposed beams and is usually a property of the detector itself. Linearity of single-beam transmittance or reflectance instruments and double-beam instruments is determined by recording signals from artifacts of known values, neutral density filters for transmittance, and gray tiles for reflectance.

Sphere-based reflectance measurements are inherently nonlinear, as the system response includes the sphere gain. This is dependent on the average reflectance of the sphere surface, in turn dependent on the reflectance of the surface being measured and its relative area compared to that of the complete sphere. Also for a double-beam instrument where internal measurements may not be made available to the user, applying an incorrect value of the zero recorded during the measurement cycle (e.g., recording data before the system has come to equilibrium after the closing of a shutter) can affect signals low in value and appear as a nonlinearity.

Corrections

Complex modification of the measured spectral data such as bandwidth correction, smoothing, interpolation, and others will alter the uncertainties and generally introduce correlations between the altered values. These correlations then need to be taken into account if the altered data are used to calculate color values. It is much simpler to calculate the uncertainties due to the measurement process alone using the measured values for which the uncertainties, including correlations, are well established, and then to apply the correction only to refine the calculated color values. The correction process can be modeled to determine its uncertainty. Care must be taken with smoothing of data, as this can reduce the effective resolution of the spectrum.

A simple example of data modification is the interpolation of a relatively smooth spectrum recorded at a relatively large data interval. Integrals of sparse data multiplied with the colour-matching functions assume that the functions are linear within the wavelength interval. This assumption leads to errors when the colour-matching functions are changing significantly within the wavelength step. Hence, the data are interpolated to a finer wavelength grid so that a more accurate integral can be calculated. The uncertainty in that integral is properly estimated by propagating the uncertainties of the measured data through the integral formula, at the original wavelength interval.[8]

CONCLUSION

The tristimulus values, and hence their uncertainties, vary strongly throughout color space; it is not possible to provide accurate color uncertainties as a single value applicable over the whole of color space. The estimation of the uncertainty in color quantities from spectral measurements follows a relatively simple path. Each component contributing to the uncertainty of the spectral measurement is identified and that component uncertainty, classed as either systematic or independent at the different measurement wavelengths, is then propagated to the uncertainties and correlations of the tristimulus values, using simple sum expressions. Uncertainties and correlations of the desired color values are propagated from those of the tristimulus values by simple matrix multiplication.

Spectral measurements are made as a transfer from a reference artifact. Uncertainties in the reference artifact need to be fully specified, including correlations. These can be specified directly, often able to be averaged over the range of visible wavelengths, or indirectly by providing the systematic uncertainties (preferably by component) separately from the combined random uncertainties or total uncertainty at each wavelength.

Representative examples of uncertainties in color values for surfaces and sources, calculated using the above principles, are available elsewhere.[17,18]

REFERENCES

1. International Organisation for Standardisation (1993), *Guide to the Expression of Uncertainty in Measurement*, Geneva.
2. Stearns EI, Stearns RE (1988) An example of a method for correcting radiance data for bandpass error. *Color Res. Appl.*, **13**, 257–259.

3. Gardner JL (2006) Bandwidth correction for LED chromaticity. *Color Res. Appl.*, **31,** 374–380.
4. CIE (2004) *Colorimetry*, 3rd ed., Publication 15:2004, CIE Central Bureau, Vienna.
5. ISO 10526/CIE S005/E (1999) Joint ISO/CIE standard: CIE standard illuminants for colorimetry.
6. CIE Draft Standard DS 014-2.2/E:2004 *Colorimetry - Part 2: CIE Standard Illuminants.*
7. Gardner JL ((2006) Uncertainty estimates in radiometry, in *Experimental Methods in the Physical Sciences* Vol. **41**, (Eds. A. Parr, R. Datla, and J. Gardner), Elsevier, pp. 291–325.
8. Gardner JL (2003) Uncertainties in interpolated spectral data, *J. Res. Natl. Inst. Stand. Technol.* **108**, 69–78.
9. Taylor BN, Kuyatt CE (2000) *Guidelines for Evaluating and Expressing the Uncertainty of NIST Measurement Results*, National Institute for Standard Technology, Technical Note TN 1297, available from www.nist.gov.
10. Cook RR (2002) *Assessment of Uncertainties of Measurement for Calibration and Testing Laboratories*, available from http://www.nata.asn.au/
11. Bell S (1999) *A Beginners Guide to Uncertainty in Measurement. NPL Measurement Good Practice Guide No. 11*, National Physical Laboratory, UK, available from www.npl.co.uk
12. Tables of emission lines can be found at http://physics.nist.gov/PhysRefData/Handbook/index.html
13. Vandenbelt JM (1961) Holmium oxide filter for checking the wavelength scale of recording spectrophotometers. *J. Opt. Soc. Am.*, **51**, 802–803.
14. McNeirney J, Slavin W (1961) A wavelength standard for ultraviolet-visible-near infrared spectrophotometry. *Appl. Opt.*, **1**, 365–367.
15. Travis JC, Zwinkels JC, Mercader F, Ruiz A, Early EA, Smith MV, Noel M, Maley M, Kramer GW, Eckerle KL, Duewer DL (2002) An international evaluation of holmium oxide solution reference materials for wavelength calibration in molecular absorption spectrometry. *Anal. Chem..*, **74**, 3408–3415.
16. Weidner VR, Barnes PY, Eckerle KL (1986) A wavelength standard for the near infrared based on the reflectance of rare-earth oxides. *J. Res. Natl. Bur. Stand.*, **91**, 243–253.
17. Early EA, Nadal ME (2004) Uncertainty analysis for reflectance colorimetry. *Color Res. Appl.*, **29**, 205–216.
18. Gardner JL (2005) *Uncertainties in Color Measurements*. National Measurement Institute Australia Technical Report NMI TR8, available from www.measurement.gov.au

APPENDIX 3

USE OF CIE COLORIMETRY IN THE PULP, PAPER, AND TEXTILE INDUSTRIES

ROBERT HIRSCHLER[a] and JOANNE ZWINKELS[b]
[a]*SENAI/CETIQT Colour Institute, Rua Dr. Manuel Cotrim, 195, Rio de Janeiro, RJ, Brazil 20961-040.*
[b]*National Research Council of Canada, Institute for National Measurement Standards, 1200 Montreal Road, Ottawa, Ontario, Canada KIA OR6.*

INTRODUCTION

In this appendix, we would like to show in two examples the use of CIE colorimetry in different industries, where the applications were made not by the CIE but some sister organizations. Dr. Joanne Zwinkels will review the application of CIE colorimetry in the pulp and paper industry, and Dr. Robert Hirschler will show examples of CIE colorimetry in the textile industry. For fundamental colorimetric terms and equations, check the main part of the book.

PULP AND PAPER APPLICATIONS

Introduction

The pulp and paper industry has been one of the primary driving forces and beneficiaries of the recent developments in CIE colorimetry. Although this industry sector desires to have accurate colorimetry and to follow the recommendations and findings of the CIE, the more important economic issues have been obtaining reliable, repeatable, and reproducible measurements.

Colorimetry: Understanding the CIE System, Edited by János Schanda

CIE Colorimetry is widely used in the specification, measurement, and control of optical properties in pulp and paper goods. The optical properties of brightness, whiteness, opacity, and glossiness are used in their grading and, therefore, are important in establishing their commercial value. Color is also measured, particularly for newspaper, where the color is referred to as the *shade* of the paper. The economic impact of these optical measurements to the paper industry was highlighted in a 2002 Metrotrade presentation at the National Conference of Standards Laboratories where it was estimated that the cost due to extra bleaching in increasing paper whiteness by 1 unit, for example, from 79 to 80, was 4 USD/ton which amounted to 100M USD per year for Canadian producers alone.[1]

Since 1937, the paper industry has been using the blue reflectance factor as the primary indicator of optical quality of pulp after bleaching.[2] This quantity was referred to as brightness* and different brightness scales were used in different countries, depending upon the measuring instrument. For example, in the early 1960s, in Canada and the United States, the brightness was measured in terms of TAPPI or the so-called G.E. Brightness, using a General Electric instrument with a bidirectional 45°/0° geometry, whereas in Europe, the brightness was measured in terms of Elrepho brightness using a Carl-Zeiss Elrepho reflectance meter with an integrating sphere.[3]

Commencing in the late 1960s, more and more Canadian mills shifted to the use of an integrating sphere and the measurement of European and Canadian newsprint was standardized with the measurement geometry of the Elrepho, that is, d/0 and the shade of the newsprint was defined in terms of the CIE 1931 2° standard observer functions and expressed as Y-value, dominant wavelength, and excitation purity. Limits for the color differences were defined in the CIE 1964 $U^*V^*W^*$ system, and the Elrepho reflectometer with its three colorimetric filters was adopted for the measurement of the so-called Elrepho tristimulus values, R_X, R_Y, and R_Z as the primary data for the calculation of shade parameters. The luminosity or Y-value was measured using a gelatine blue filter, and the brightness was measured using a precision optical filter with effective wavelength of 457 nm, which were located in the No. 8 and 10 positions, respectively, of the Elrepho instrument. Weiss determined experimentally an approximate interrelationship between these measured Elrepho tristimulus values and the CIE tristimulus values.[4] These equations were later determined more rigorously by calculating the constants by inverting the equations relating the CIE tristimulus values and the three-filter Elrepho response functions using the colorimetric data from CIE Publication No. 15 for the CIE Illuminant C and the 1931 2° Observer. The constants calculated using this CIE method were then published in international standards for the paper industry, as[5]

$$R_X = (X - 0.16707 \times Z)/0.78319$$
$$R_Y = Y$$
$$R_Z = Z/1.18225$$

*The pulp and paper industry uses some CIE defined terms in a somewhat different meaning; for example, in this appendix the word "brightness" is used to describe a specially defined radiometric quantity, and is not the perceptual correlate of luminance, as used in standard CIE terminology.

The Zeiss Elrepho reflectometer became the "de facto" reference instrument for the paper industry worldwide for traceable measurements of the optical properties of paper, pulp, and board. This instrument was a filter-based reflectometer with diffuse illumination and normal viewing geometry and a large gloss trap that excluded the specular component of glossy papers. The optical components of the instrument, that is, its combination of incandescent lamp, integrating sphere lining, filters, and photocells, were chosen to produce approximately the spectral characteristics of the CIE color-matching curves for CIE Illuminant C. At that time, no system for absolute colorimetric calibration was available, and the paper industry assumed that the Elrepho instrument was reasonably accurate.[6] At that time, it was also not appreciated that a problem with a filter instrument, such as the Elrepho, is that it did not detect the presence of metamerism.

In 1971, the International Organization for Standardization (ISO) adopted the brightness scale established by the Elrepho reflectometer and standardized this quantity as ISO brightness. Since then, ISO brightness has been the key specification in any commercial agreement on the purchase of bleached and semibleached market pulp.

In the early 1980s, new Elrepho-type instruments were introduced into the market that claimed to be more accurate than the original Zeiss Elrepho. These included the Elrepho 2000, the auto-Elrepho, and the Technidyne Micro TB-1C. However, these instruments were found to be in poor agreement with the Zeiss Elrepho.[6]

When Zeiss discontinued the production of the Elrepho reflectometer, there was an urgent need for standardized procedures to provide reliable paper colorimetry, and this is when the paper industry became increasingly reliant on the benefits of CIE colorimetry.

Beneficiaries of CIE Colorimetry

CIE Illuminant C and CIE Standard Geometry d/0

The historical attachment of the paper industry to the Zeiss Elrepho instrument has been steadfast. The needs to ensure good interinstrument agreement and to link the ISO brightness measurements to the Zeiss Elrepho were the driving forces for ISO to develop a series of international standards. The ISO technical committee TC6 on paper, board, and pulps was tasked with standardization including terminology, sampling procedures, test methods, product quality specifications, and the establishment and maintenance of appropriate calibration systems.

The first of these optical property standards were ISO 2469, ISO 2470, and ISO 2471 that dealt with standardizing measurements of diffuse reflectance factors, ISO brightness, and opacity (paper backing), respectively.[7–9] In developing these standards, ISO TC6 looked to the CIE for guidance on recommended practices. These three ISO standards, issued in 1973, made specific reference to CIE recommendations on measurement geometry and illumination conditions. However, the ISO standard 2469 introduced further rigorous specifications and restrictions on the geometry of the measuring instrument including sphere diameter, gloss trap diameter, thickness of the test piece aperture, and half-angle subtended by the gloss trap from the center of the test piece apertures, and associated tolerances. The net effect was that the ISO specification does not strictly comply with CIE recommendations. For example, the recommended measurement geometry in ISO 2469 specifies that the sum total of the areas of the apertures in the sphere is not to exceed 10% of the area of the sphere. At first glance, this appears to follow the CIE recommendation.[10] However, the CIE standard specifies 10% of the internal reflecting area of the sphere. This subtle distinction is important because the ISO standard also specifies a large gloss trap with a black annulus surrounding the measurement port area that reduces the effective internal reflecting area to below 90%.

Thus, an instrument that complies with CIE geometric specifications is not necessarily compliant for use in the paper industry. This departure from CIE recommendations has been hotly debated in the standards community, but within ISO it is maintained that after having made the decision to adopt a certain procedure, that is, conformance with the geometry of the obsolete Elrepho, the critical requirement is for measurement reproducibility.

To further complicate standardization of paper colorimetry, a new generation of Xenon flash instruments with array detection became available in the 1980s and marketed for paper applications. At the same time, the use of fluorescent whitening agents (FWAs) was becoming more widespread in the manufacture of paper. These materials absorb UV radiation and emit in the violet–blue range from 380 nm to 480 nm. Problems arose with these new flash instruments for paper samples containing FWAs because the Xenon source is more UV-rich than the incandescent lamp of the old Elrepho, so that measurements with these two different types of instruments did not agree. This resulted in an urgent need to agree on an standard international method of measuring ISO brightness for fluorescent paper standards and to what level the UV content of the instrument should be adjusted.

In the mid-1990s, the ISO TC6/WG3 approved a recommendation that linked the ISO brightness on the Zeiss Elrepho to a C-illuminant adjustment based on a CIE whiteness (see Chapter 3) (C/2°) measurement.[8] In the revision of ISO 2470 in 1999, a procedure was specified for adjusting the spectral conditions of the measuring reflectometer so that the UV content corresponded to that of the CIE illuminant C. It also gave a description for the use of abridged spectrophotometers, where the instrument's spectral conditions are to be achieved using an adjustable filter with a cutoff wavelength of 395 nm. The filter shall be adjusted so that the UV content of the illumination corresponds to that of the CIE illuminant C using a reference standard with an assigned CIE whiteness (C/2°) value. An important consequence of this revision is that the ISO working group introduced the concept of indoor whiteness and ISO brightness, which were linked together and to the *CIE Illuminant C*.

It was later discovered that an instrument calibrated to CIE whiteness used slightly different settings of the UV-adjustment filter than a calibration of the instrument to ISO brightness. This was also the case for calibration to CIE tristimulus values, X, Y, or Z where only one reflectance variable was found to be in calibration. Although ISO TC6/WG3 recognized this limitation, it still felt that the simplicity of the one point adjustment outweighed the increased precision of quantity-specific calibrations. Jordan later carried out a detailed investigation of this effect and showed that a one-point calibration was inadequate because it presumed that there was only one fluorescent moiety present in these fluorescently whitened papers.[11] This study of a large number of fluorescent white papers showed a composite nature of the excitation spectrum suggesting that the emission band is due to a multiplet of fluorescent moieties.

Other CIE Standard Illuminants and Standardized Light Sources

In the past 5 years, the paper industry has moved towards accepting CIE recommendations of Standard Illuminant D65 in colorimetric specifications. The earlier reluctance to adopting this illumination condition was because it excites about twice as much fluorescence as typical indoor illumination under which paper is normally handled and printed. However, ISO TC6 has decided that the advantages of adopting this CIE recommendation outweighs this practical limitation.

Currently, there are ISO pulp and paper standards for Color, D65/10°, CIE whiteness (D65/10°) and D65 brightness (see Table A3.1). While the paper industry has been reliant on CIE Illuminant C for many of its international standards, the graphics art industry has largely standardized on CIE Illuminant D50/2° conditions. In view of the close relationship between the pulp and paper industry and the graphic arts industry, ISO TC6 has also recently developed a standard for Color, D50/2° (see Table A3.1).

ISO TC6 has also looked to the CIE for recommendations on standardizing other light sources. For example, the ISO Standard 14358 - 2 on method of exposure of

TABLE A3.1 International standards for pulp and paper that are underpinned by CIE colorimetry standards

ISO 2469	Paper, board, and pulps—Measurement of diffuse radiance factor (diffuse reflectance factor)
ISO 2470	Paper, board, and pulps—Measurement of diffuse blue reflectance factor (ISO brightness); later superseded by:
ISO 2471	Paper and board—Determination of opacity (paper backing)—Diffuse reflectance method
ISO 5631	Paper and board—Determination of color (C/2°)—Diffuse reflectance method
ISO 11475	Paper and board—Determination of CIE whiteness, D65/10° (outdoor daylight)
ISO 11476	Paper and board—Determination of CIE whiteness, C/2° (indoor lighting conditions)
ISO 16692	Paper and board—Determination of color (D65/10°)—Diffuse reflectance method
ISO 16693	Paper and board—Measurement of D65 brightness (Diffuse blue reflectance factor under UV(D65) conditions)
ISO 20313	Paper and board—Determination of color (D50/10°)—Diffuse reflectance method
ISO 22891	Paper—Determination of transmittance by diffuse reflectance measurement
ISO 14358-2	Paper and board—Method of exposure to a laboratory light source, Part 2—Xenon arc

paper and board to a laboratory light source relies on CIE Publication No. 85:89 as a normative reference for specifying the quality and intensity of solar radiation.[12]

CIE Color Spaces

As previously mentioned, the pulp and paper industry specified the shade of their products for many years solely in terms of dominant wavelength and excitation purity.

In the 1970s, the Technical Association of the Pulp and Paper Industry (TAPPI) introduced the use of CIE color scales in a proposed new method T 527 for measuring the color of paper and paperboard in CIE *Y*, *x*, *y* using the CIE 1931 Standard Observer as an alternate method to the luminance factor, dominant wavelength and excitation purity, and the method employing the Hunter Lab color scale.[13]

In the late 1980s, there was interest in transferring the CIELAB system to a single color space. The main argument in favor of this change was that the CIELAB system was found to be in better agreement with the visual assessments of color differences and this gave a stronger foundation for establishing tolerances on these colorimetric parameters.[6] In this way, the customer's ΔE tolerance could be converted to the accuracy required by the paper producer in establishing their target tristimulus values.

CIE Reference Standards

At the time that ISO standardized the measurement of brightness to ISO brightness based on the d/0 geometry of the Elrepho reflectometer, it also recommended that the reference standard be changed from magnesium oxide to the "perfect reflecting diffuser (PRD)" as recommended in 1959 by the CIE Expert Committee on Colorimetry.[14] This was an ideal uniform diffuser having a reflectance of unity. In 1969, the CIE formally recommended the PRD as the primary standard for reflectance measurements of opaque surfaces (see Chapter 3). Although there was no physical standard that corresponded exactly to the PRD, the characteristics of suitable white transfer standards, such as pressed barium sulfate powder, could be measured relative to the PRD by standardizing laboratories having developed methods for measuring absolute reflection values. The uncertainty in preparing smoked magnesium oxide samples was about 1%, so this adoption of the perfect diffuser had immediate practical benefits to the paper industry in improving measurement accuracy. A further consequence of this change in reflectance scale was a lowering of the ISO brightness by about 1.5%.[14]

The need for improved reliability of these optical property measurements also motivated the adoption of a hierarchical international calibration system by ISO. This system was also first introduced in the set of standards issued in 1973. The CIE-recommended ultimate reference of the PRD was defined as an ISO reference standard of Level 1 (IR1), and reference standards of Levels 2 and 3 were proposed whereby only certain laboratories that were equipped for absolute reflectance factor measurements were appointed by ISO/TC6 as "Standardizing Laboratories." These laboratories would issue ISO reference standards of Level 2 (IR2) to certain

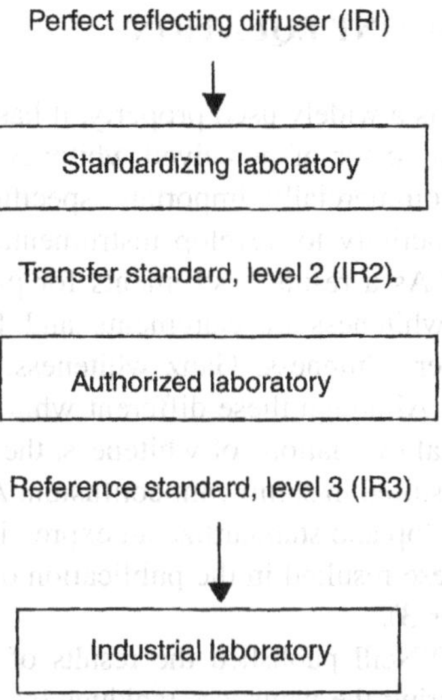

FIGURE A3.1 The ISO hierarchy of calibration of optical standards.

"Authorized Laboratories" for calibrating their reference instruments. These authorized laboratories were also appointed by ISO/TC6, which could then issue ISO reference standards of Level 3 (IR3) on demand to industrial laboratories. This hierarchical calibration system is illustrated schematically in Figure A3.1. The adoption of these ISO international standards based on CIE Colorimetry along with an international system of calibration has proved to be quite satisfactory and has provided a level of international agreement that was previously not available.

The ISO TC6 working group on optical properties, WG3, is entrusted with the task of developing test methods for the measurement of optical properties of paper, pulp, and board. This Working Group has established a liaison with CIE and relies on CIE recommendations and standards to underpin the credibility of these standard test methods. The current ISO TC6 standards for optical property measurement are listed in Table A3.1.

The membership of ISO TC6 WG3 includes the ISO standardizing laboratories that are also active members of the CIE. This situation has also helped to advance CIE and paper colorimetry. For example, for many years, the paper industry used Japanese opal glasses as an ISO IR 2 standard. However, in 2003, Zwinkels published the results of an investigation of the photoluminescent effect in opal glasses used as diffuse reflectance standards.[15] It was shown that the fluorescent effect in the Japanese opal glass, Everwhite, was significant and its use as a reflectance standard would produce erroneous colorimetric results, particularly for calibrating color-measuring instruments that used unfiltered xenon sources that would excite this fluorescence. As a consequence of this work, the paper industry has discontinued the use of this material as an IR2 standard, and it motivated a further investigation into the influence of the illuminant on paper colorimetry.[11]

CIE WHITENESS AND TINT EQUATIONS

Although ISO brightness is a widely used property, it has been recognized that it is not necessarily a good indicator of perceived whiteness. As the attribute of perceived whiteness is a commercially important specification in paper, this has resulted in a significant activity to develop instrumental whiteness measurement methods and formulas.[16] As a result, instruments for paper colorimetry provided a plethora of different whiteness measurements and formulas. These included Stensby whiteness, Hunter whiteness, Ganz whiteness, Taube whiteness, Berger whiteness, among others. Although these different whiteness methods were somewhat correlated with visual evaluations of whiteness, the existence of so many different whiteness scales resulted in significant confusion. As a consequence, the CIE launched an effort to develop and standardize an expression for whiteness based on psychometric studies. These resulted in the publication of the CIE whiteness equation[17,18] (see also Chapter 3).

In 1991, Jordan and O'Neill published the results of an extensive study of the whiteness of paper comparing the whiteness rankings by colorimetric measurements and visual ranking under indoor viewing conditions.[19] It was shown that the CIE whiteness equation using colorimetric data measured with an instrument using a tungsten–halogen lamp gave excellent correspondence. The CIE whiteness equation is now exclusively recommended in the International Standards for paper colorimetry.[20]

The shade of a paper is an attribute of the visual perception of color, which is an undesirable property in white papers. As in the case of whiteness, for many years the paper industry used a large number of equations for tint.

The need to have a single recommended a whiteness formula and a tint formula motivated the development of CIE whiteness and CIE tint index equations. Ganz recommended the whiteness formula[16]

$$W = Y - 800 \cdot (x - x_{10}) - 1700 \cdot (y - y_{10})$$

as a standard whiteness formula of neutral hue preference to be used with CIE standard illuminant D65 and both the CIE 1931 2° and CIE 1964 10° observers. This formula was adopted by the CIE as the recommended CIE whiteness equation (see Chapter 3). ISO has developed a new series of international standards that refers to CIE Publication 15.2 and recommends the use of the CIE whiteness equations exclusively. In 1999, ISO TC6 broke new ground with the publication of ISO 11475:1999 Paper and board—Determination of CIE Whiteness, D65/10° (outdoor daylight), which was the first international standard for the determination of CIE Whiteness.

In the same way, there previously existed a number of different tint formulas. Because the tint readings from different equations were not comparable, this made quality assurance of paper goods difficult and motivated the CIE to develop standardized procedures and terminology. Two standard tint formulas were proposed for standard illuminant D65, one each for the CIE 1931, 2° standard observer

$$T = -1000 \cdot (x - x_0) + 700 \cdot (y - y_0)$$

and one for the CIE 1964 10° standard observer

$$T = -900 \cdot (x - x_0) + 800 \cdot (y - y_0)$$

These tint formulas were based on lines of constant tint running approximately parallel to a dominant wavelength of 470 nm. These CIE tint formulas have since been further refined to be based on lines of constant tint running parallel to a dominant wavelength of 466 nm and are found in CIE Publication 15.3.[21]

Harmonized Terminology

For many years, the vocabulary for color terms used in various ISO standards for the paper industry was ambiguous and lacked a rigorous logic, particularly when describing color effects with fluorescently whitened paper. In recognizing this problem, Dr. A. Bristow, the convenor of ISO TC6/WG3, in 2003, looked to the CIE for guidance in developing a logical sequence of definitions. He stated that "we are on safe ground if we keep rigidly to CIE terminology, we do not invent our own. Indeed our liaison with CIE requires that we do so."[22]

This new approach of harmonizing terminology with CIE has impacted the development of new ISO standards. In the early development of the ISO Standard on the Determination of transmittance (ISO/CD22891), the CIE was consulted on terminology, and the quantity previously referred to as luminous reflectance factor was corrected to luminance factor (C/2) to correspond more closely to CIE-recommended terminology; the measured quantity here is the CIE *Y* tristimulus value determined for CIE Illuminant C and CIE 2° Observer conditions.

DRIVING FORCE IN THE DEVELOPMENT OF CIE COLORIMETRY

Establishment of New CIE Technical Committees

The CIE and the pulp and paper industry through the auspices of ISO have worked cooperatively for many years to facilitate the implementation of their objectives, and CIE colorimetry is currently the basis of all optical property standards prepared and published by ISO. In order to facilitate this interaction, the CIE and ISO TC6 have established formal liaisons that report on standards development in areas of mutual interest. ISO recognizes that the CIE has expertise in matters related to the science, technology, and standardization in the fields of light and lighting. CIE, in turn, tries to incorporate the relevant technical recommendations of ISO TC6 and other partner organizations, in establishing their technical committees and standards development. ISO TC6, as we have seen, relies on the CIE to provide reference data that are accurate and reliable and to provide traceability for their international standards.

In recent years, the demands of the pulp and paper industry have influenced the establishment of new CIE technical committees and the development of several CIE colorimetry standards. When the revision of CIE Publication 15.2 was being drafted by CIE TC 1-48, it caused great concern in the paper standardizing community. Of particular concern was that an earlier draft of this publication deprecated the use of CIE Illuminant C and did not explicitly refer to the d/0 geometry, which was the standard illuminant and measurement geometry adopted in several ISO paper standards. ISO TC6 wanted to ensure that the CIE continued to provide normative references for these measurement conditions.

In 2003, the Chairman of ISO TC6 visited the CIE Central Bureau to meet the General Secretary and the Technical Manager in order to highlight the importance of ISO dependence on CIE colorimetry standards. He discussed the problematic wording proposed in the draft revision of CIE Publication 15 and made the case for CIE providing continued traceability for six very important ISO/TC6 standards to the worldwide pulp and paper industry that specify CIE Illuminant C conditions.

These issues raised by ISO TC6 were discussed at some length at the CIE Division 1 meetings in San Diego and the outcomes were received positively by the pulp and paper industry. Notably, the text of CIE Publication 15:2004 was changed and now only states "Illuminant C does not have the status of a CIE standard but its relative spectral power distribution, tristimulus values and chromaticity coordinates are given in Table T.1 and Table T.3, as many practical measurement instruments and calculations still use this illuminant."[21] An explicit recommendation for the d/o geometry was also given. There was considerable discussion about whether to distinguish daylight applications as being either "indoor daylight" or "outdoor daylight." There was no clear consensus on this distinction but, as a consequence of this discussion, a new CIE technical committee on Indoor Daylight was established.

Practical Simulator of Illuminant D65

In 1963, the CIE adopted the Standard D Illuminants, whose spectral power distributions were a more accurate representation of daylight than those of Standard Illuminants B and C, which had been introduced in 1931. However, the D illuminants were only specified by a tabular set of data, whereas the earlier recommendation of illuminants B and C had also included a description of practical sources to simulate these illuminant conditions. The need for standard sources to represent these D illuminants, particularly D65, has grown in importance with the increasing use of optical brightening agents in the manufacture of various whitened goods (Figure A3.2).

In 1990, the Chairman of the TAPPI Optical Properties Committee wrote to the US CIE Division 1 member requesting the formation of a new CIE Technical Committee to provide a recommendation for a standard source for whiteness measurement.[23] Coincidentally, this lack of a CIE recommendation for realizing standard sources was also highlighted by R. Hunt in a 1992 Letter to the Editor in Color Research and Application.[24] This motivated the CIE to set up a reportership (R1-09 Practical daylight sources) and to subsequently establish a technical committee (TC 1-44) whose terms of reference were to recommend practical daylight sources for colorimetry. This proved to be a daunting task, and the terms of

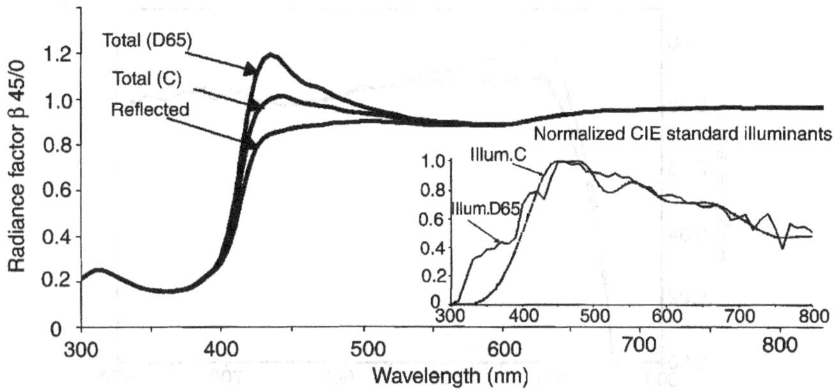

FIGURE A3.2 Influence of CIE illuminant on the total spectral radiance factors of a typical fluorescently whitened paper standard; the corresponding reflected component, which is independent of CIE illuminant, is also shown.

reference were modified to the measurements of commercially available daylight sources and daylighting booths and recommendations on acceptable systems. Thus, more than 40 years have passed since the D illuminants were introduced, and the CIE has not developed a standardized procedure for simulating D65. In the absence of such a recommendation, industry has developed its own industry-specific procedures for calibrating instrument spectral conditions to conform to the shape of D65. In the case of the paper industry, the present ISO standards recommend a practice of adjusting the net UV content by attenuating the Xenon lamp with a filter at 395 nm in order to adjust the ratio of UV light to visible light in the illumination to match CIE Illuminant D65 or C. The underlying premise for this procedure to provide reliable colorimetry is that the sample has only one fluorescent moiety so that the ratio of its emission to excitation spectrum remains constant.

However, it was shown by Jordan that fluorescently brightened paper has significant excitation below 300 nm so that short-wavelength UV light from a Xenon lamp excites fluorescence beyond the reach of the daylight illuminants.[11] It was also found that a variety of fine white papers differed in their relative sensitivity to short-wavelength (280 nm) and long-wavelength (370 nm) UV light so that the net emission spectrum depended upon the relative spectral amounts of UV available and not just the total UV amount which is the presumption with a one point UV adjustment using a calibrated D65 fluorescent standard. In order to perform reliable and accurate paper colorimetry of fluorescent papers, it is necessary that the source accurately simulates the relative spectral distribution of D65 and not just provides an equivalent total UV content.

This study emphasized the need to have an instrument source that accurately simulates the relative spectral distribution of D65 throughout the entire excitation and emission bands of the fluorescent species. This, in turn, has stimulated a renewed effort in CIE to provide a recommendation for a practical simulation of D65 illuminant (Figure A3.3).

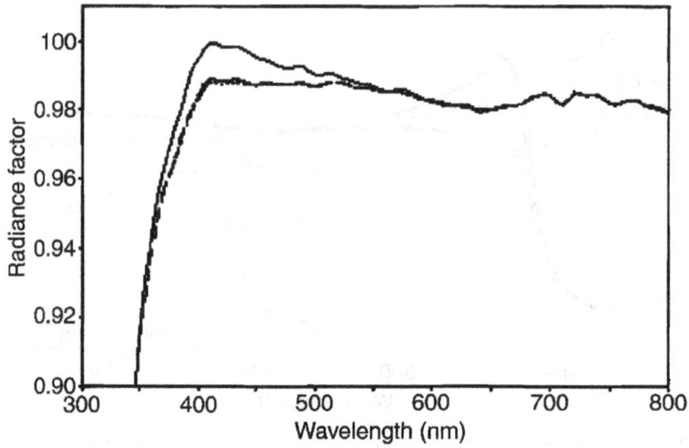

FIGURE A3.3 Reflected (dashed) and total (solid) 45/0 radiance factor of white Japanese opal reflectance standard for D65 and equienergy spectral illumination. This reference standard was used for many years as an ISO Level 2 (IR 2) standard for paper colorimetry.

FUTURE NEEDS

An area in which the paper industry is currently seeking guidance is the development of a standard procedure to correct for differences in measurement geometry between the ISO standardizing laboratory and the ISO authorized laboratories for the calibration of ISO Level 2 fluorescent standards. The CIE-recommended reference geometry for these measurements is a 45/0 geometry and the Standardizing Laboratories have developed reference instruments that conform to this CIE recommendation. However, the paper industry has standardized on a d/0 geometry, and a standardized procedure is needed for how to apply this geometric correction. This issue of geometric correction is also important to the textile industry.

CONCLUSION

It can be seen that the pulp and paper industry has greatly benefited from and contributed to CIE Colorimetry. It is expected that this interdependence and close cooperation will continue to grow as more laboratories seek certification or accreditation according to the requirements of ISO/IEC 17025. The international standards for optical property measurements that are used by the pulp and paper industry worldwide rely on CIE recommendations and traceability for measurement geometries, illuminants, and colorimetric calculation procedures. The CIE needs to ensure that it continues to provide the necessary normative references to support this important industry sector.

TEXTILE APPLICATIONS

Introduction

The textile industry is one of the most color-conscious ones; here color control has always been a prime concern. The application of CIE Colorimetry in the textile industry started not long after the publication of the system, and among its first advocates we find Hardy,[25] Judd[26] and somewhat later Davidson and Godlove[27] and Stearns.[28]

The main areas where CIE Colorimetry has been successfully practiced in the textile industry are the following:

- characterization of colorants in CIELAB space;
- determination of the standard depth of color (based on CIELAB coordinates);
- quality control and pass/fail determination based on the CIELAB color difference formula and newer formulas derived from CIELAB;
- shade sorting and tapering using CIELAB space and color difference formulas;
- fastness evaluation (CIELAB color differences);
- determination of whiteness and yellowness using formulas based on tristimulus values and chromaticity coordinates including the CIE-recommended whiteness and tint formulas;
- Recipe formulation (based on a combination of spectral and tristimulus-matching techniques).

CIELAB Color Space and Its Derivations

The textile industry had made many attempts of using instrumental color measurement in an industrial scale long before CIELAB became the officially recommended color space and color difference equation. In those days the MacAdam ellipses, the Simon–Goodwin charts,[29] and the FMC-2 formula[30] were more popular in the United States and the Adams–Nickerson (ANLAB)[31] space in Europe (particularly in the United Kingdom and Germany). Many of the concepts developed in ANLAB space have been transferred directly into CIELAB after 1976.

Characterization of the Buildup of Colorants and of Colorant Combinations
The analysis of the behavior of a series of textile dyeing or prints on a color diagram or in color space gives valuable information on the colorimetric characteristics of colorants (dyes and pigments), and these charts are considered excellent visual aids in the selection of the range of colorants to be selected for a new application. In its simplest form, the analysis consists of plotting the color coordinates of the series of dyeings or prints on an a^*–b^* and an L^*–C^* diagram (see Figure A3.4).

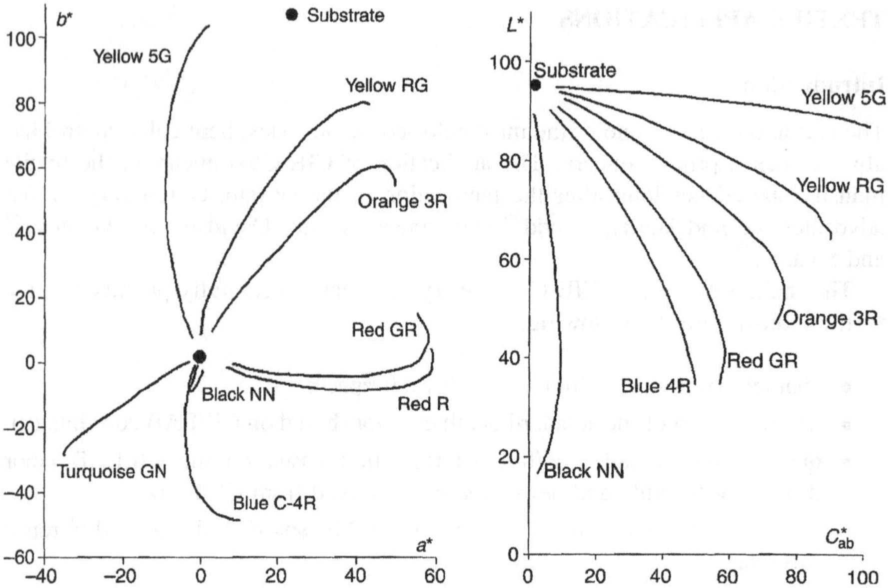

FIGURE A3.4 Series of textiles dyeings with increasing concentration on an a^*–b^* (left) and an L^*–C^* (right) diagram. On the a^*–b^* plane, the lines connecting the individual points turn back towards the neutral points with increasing concentration. On the L^*–C^* diagram, lightness decreases with increasing concentration while chroma increases until a "saturation point" after which it also starts to decrease.

Standard Depth (SD)

AATCC[32] defines *depth* as a "departure of a colored object from white and frequently associated with either concentration or efficiency of a colorant." Standard depth (SD) scales are a series of colored samples of different hue, chroma, and lightness that are considered *visually* to have the same depth. This concept was created in the 1920s by the major German and Swiss dye manufacturers,[33] and the scales in 1/25, 1/12, 1/6, 1/3, 1/1, and 2/1 SD have been used since the 1960s practically in their present form. These scales are used primarily in the determination of the fastness properties of colorants (which depend very much on the SD level in which they are tested).

There have been a number of attempts to define SD by instrumental methods, but there is no universally accepted conversion formula based on the CIE tristimulus values, and therefore current national and international standards, including ISO 105-A01,[34] recommend the use of visual methods based on a range of 18 dyeings (of different hues) considered as SD 1/1. Double depths and fractional depths (2/1 resp. 1/3, 1/6, etc.) provide supplementary indices. ISO/TC38/SC1 requested in 1976 a colorimetric method of determining 1/1 SD, and a formula was suggested in 1985 by Christ[35] that is far from being perfect but, in lieu of anything better, it was adopted in 1995 as ISO 105-A06.[36] It is considered as an alternative to, but not as a replacement for the visual method implied in ISO 105-A01.

Color Difference Evaluation

The dream of every quality controller in the textile (as well as in any other) industry is to have one single color difference number representing the commercially acceptable tolerance limit. Some of the pioneering work leading up to the "improvements" on ANLAB and later CIELAB was done in the textile industry, starting with the much-referenced work of Davidson and Friede[37] in 1953. In the 1970 s, two UK companies started to work on the modifications of the then available color difference formulas, and both tried to arrive at SNSP (single number shade passing) by local modifications of the CIE $L^*a^*b^*$ space. The UK chain Marks and Spencer has never published the M&S formula, but the results of J&P Coats with the JPC79 color difference formula have been extremely well received, and after some modifications it was published in 1984 as the CMC(l:c) formula by the Color Measurement Committee of the Society of Dyers and Colorists (see also Chapter 4).

The CMC formula has been used with great success in the textile industry, and although adopted in national and international standards (AATCC,[38] ISO 105-J03[39]) it has never achieved "official" CIE recognition, its eventual replacement by the CIEDE2000 formula is still being studied by the AATCC and the SDC.[40] Aspland and Shanbhag[41] compared the characteristics of CIEDE2000 and CMC; industrial studies by Gay and Hirschler[42] showed the CIEDE2000 formula to perform better than most previous formulas; nearly (but not quite) as well as CMC (see Figure A3.5).

Further investigations are urged by many researchers (e.g., Kuehni[43]) claiming that the precision of the currently available formulas may not be acceptable for applications in the textile industry.

Shade Sorting, Tapering

Textile coloration (dyeing and printing) has long been considered an art rather then science. With the advent of sophisticated methods of process control the uniformity,

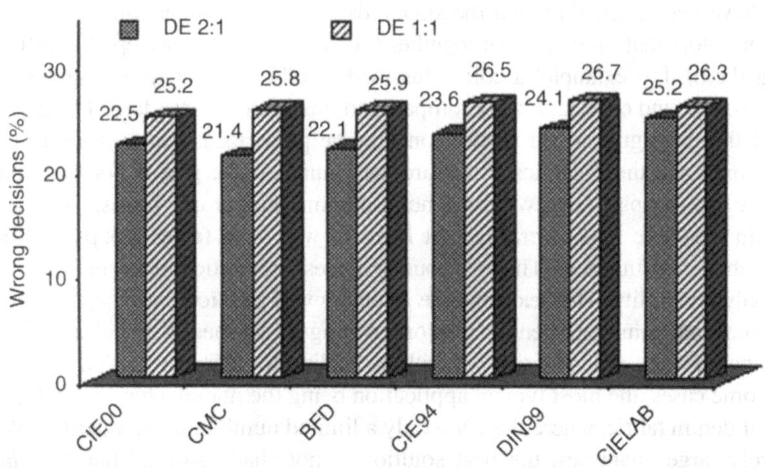

FIGURE A3.5 Wrong decisions (%) of instrumental evaluations for different color difference formulas as compared to the "verdict" of a panel of visual observers.

repeatability, and reproducibility have improved significantly, but it still cannot be guaranteed that the packages of yarn or pieces of fabric dyed or printed in one lot be always of "exactly" the same color (i.e., the color difference between the individual packages or pieces be less than the commercially acceptable). This is even more so when lot-to-lot differences are considered. Increasing the level of process control increases production costs and the textile industry has to accept the fact that a smaller or larger percentage of the production shows unacceptably large color differences as compared to the respective "standard" of each particular color.

Luckily, however, extremely small color differences between the production and the "standard" are only very rarely required; it is normally only in the garment industry where very close tolerances are required between the textile fabric pieces going into one piece of garment or one set (e.g., jacket and trousers of a suit). In these cases the solution may be "shade sorting," originally suggested by Simon[44] and widely known as the "555" method or one of its modifications known as "sequencing" and "clustering."[45] These methods are based on the idea that if the pieces of a production lot (or of many production lots) are grouped according to their color coordinates, a number of groups may be formed with very small color differences within the lot, but greater than minimum among the lots. The obvious advantage of this grouping is that the user (normally the garment manufacturer) will get textile pieces which may be sewn together and not show perceptible (or, at least, unacceptable) differences, while most of the production can be processed as first quality and not as second or third (due to the out-of-tolerance color).

The original "555 shade sorting" published in 1961 used a three-dimensional grid divided into lightness, chroma, and hue directions in the UCC Color Difference System[46]; today of course CIELAB ΔL^*, ΔC^*, and ΔH^*, or the equivalents in CMC or CIEDE2000, are generally used, although some people prefer to build the grid along the L^*, a^*, b^* axes.

The advantage of the 555 shade sorting (with a fixed grid structure) is that once the boxes have been formed around the standards, the same designation will always be used for colors that may be sewn together (i.e., those showing acceptable difference among them). For example, a color classified as 645 will always be lighter, duller (less chroma), and of equal hue as compared to the reference standard. The disadvantage of the fixed grid is that production may be pulverized, that is, sorted into too many boxes and the "identical" lots are very small. If the grid is not fixed but we optimize the groups so that we have the minimum number of "boxes," we can still maintain the same tight tolerance; but here we will have fewer groups with much more fabric within each. This of course makes production cheaper, but it is a once-only possibility: The next time the grouping will position the groups differently, we cannot maintain consistency from one sorting to another. Aspland et al.[47] compared the different shade-sorting algorithms leading to different results.

In some cases, the most typical application being the manufacturing of blue jeans made of denim fabric, where there are only a limited number of "standard" colors in relatively large quantities, the best solution is not shade sorting, but *tapering or sequencing*. In this technique, there are no groups or clusters (although successful tapering may be preceded by a preliminary sorting process); the pieces to be sewn together are arranged in such an order that the difference between neighboring pieces be minimized by some colorimetric-mathematical criterion.[45]

Fastness Evaluation

Color fastness (colorfastness) of textiles—or textile fastness in short—is a fundamental technical characteristic of dyed and printed textile materials: it shows how resistant the given substrate—dyed or printed by a given method using a particular set of dyes/pigments at given concentration levels—is to different treatments simulating the usual tear and wear of the product. The AATCC Technical Manual[48] lists 28 colorfastness test methods and procedures from *Colorfastness to acids and alkalis* through *Dry-cleaning, Laundering, Light, Perspiration etc.* to *Colorfastness to water spotting*. The common in all these methods (all different in their physical-chemical treatment of the textile samples) is the use of a Gray Scale for visually evaluating changes in color of textiles resulting from colorfastness tests, or the alternative method of instrumental assessment of the change of color.

In some of the tests, the change in color of the test sample is evaluated using a Gray Scale, in some others also the staining of an adjacent piece of fabric with a different Gray Scale. The AATCC Evaluation Procedure 1 (Gray Scale for Color Change)[49] describes the pair of reference chips for Colorfastness grade 5 as "neutral grey in color and having a Y tristimulus value of 12 ± 1. The color difference of the pair is $0.0 + 0.2$." The other grade pairs are described by CIELAB color difference units from 0.8 (Grade 4–5) to 13.6 (Grade 1).

AATCC EP2 (Gray Scale for Staining)[50] uses similar methods of description, but the reference chips here are white (Y not less than 85) and the color difference values for the same grades are much larger, 2.2 CIELAB units for Grade 4–5 up to 34.1 for Grade 1. It is emphasized in the EP that these values assigned to the Gray Scale grades serve only for the instrumental measurement and confirmation that a Gray Scale is within tolerance, and are *not* to be used for assigning a Gray Scale based on instrumental measurement of two specimens.

AATCC EP7[51] (which is the same as ISO 105-A05) describes the *Instrumental Assessment of the Change in Color of a Test Specimen* using a special color difference formula converting CIELAB coordinates into Gray Scale ratings.

The ISO formula for assessing staining[52] has not found its way into the AATCC Technical Manual. Moreover, the British Standard Institution TCI/81, Color Fastness and Color Measurement Committee, initiated large-scale interlaboratory trials whose results were recently published.[53] The authors of this study showed that the present ISO standard for assessing staining fastness gave a poor performance, whereas the formula for assessing change in color gave reasonable correlation with the available visual data. New formulas are proposed for both situations.

Determination of Whiteness

Whiteness indices are fairly widely used in the textile industry. For nonfluorescent specimens (which is the case of untreated textile substrates), a range of indices have successfully been used, such as the Berger,[54] Hunter,[55] Stensby,[56] and more recently the Ganz[57] and the CIE[58] formulas (see also Chapter 3). The complications start when the specimens are fluorescent, their whiteness is determined by instrumental measurement, and it is expected that the measurement results give reasonable correlation with visual assessments. The number one unsolved technical problem in textile colorimetry today is probably that of the determination of the degree of whiteness of fluorescent (optically brightened) textiles, and the basic reasons for this are the following:

1. Visual assessments are generally performed under nonstandard illumination conditions (acceptable D65 simulators for visual inspection are extremely scarce in the textile industry), and thus the results are not reliable.
2. Even under the best-controlled conditions, the concept of "white" or "whiter" is subjective, observers within the same organization show significant disagreement in ranking samples according to whiteness.
3. There are very few color-measuring reflectance spectrophotometers in industry, which are adequate—let alone ideal—for the measurement of fluorescent samples.

For the instrumental evaluation of fluorescent whites, the ideal instrument would be a bispectral spectrophotometer (currently not available commercially); the next best would be a 45/0 reflectance spectrophotometer with UV calibration capability (currently available only from one supplier). The vast majority of textile companies and laboratories use sphere instruments with UV calibration for the measurement of fluorescent whites.

AATCC and ISO[59] standardized on the CIE recommended whiteness and tint formulas accepting its known restrictions "to samples that are called "white" commercially, that do not differ much in color and fluorescence, and that are measured on the same instrument at nearly the same time." Should there be a necessity to lift these restrictions, the UV-adjustment of the measuring instrument may be performed as described by Griesser.[60]

A detailed study[61] based on the measurements of a number of commercially available color-measuring spectrophotometers, a bispectral instrument, and a large number of visual evaluations under very strict (spectroradiometric) control of the illumination confirmed that it is really necessary to adjust the UV radiation of the sample illumination when fluorescent samples are to be measured on different instruments or at different times, a basic fact not always so obvious for the industrial user. It is also necessary to adjust the instrument-specific parameters when using the Ganz–Griesser whiteness formula. If for some reason this is not possible it is then more appropriate to use the CIE whiteness formula than to use the standard parameters for the Ganz–Griesser whiteness formula.

The final conclusion of the study was that the interinstrument agreement of instrumental whiteness evaluation may not yet be as good as desirable, but compared to the reproducibility of visual evaluation, the application of CIE colorimetry represents significant improvement.

Recipe Formulation

If there is one area in the textile industry where the application of colorimetry has long been an unqualified success, it is that of recipe formulation, also known as "instrumental match prediction" or "computerized color matching."

The first algorithm was already published in 1944 by Park and Stearns,[62] but it was necessary to have access first to the special analog[63] and then to the general-purpose digital computers[64] to introduce instrumental match prediction in an industrial scale. The most popular algorithm, which served for decades as the basis of commercial color-matching software packages, was published by Allen.[65] The novelty of his approach was the determination of a good starting recipe (spectrally), which then could be iterated until it gave the expected result (in terms of colorimetric values).

For nonfluorescent substrates and colorants, the colorimetric calculations suggested by Allen are fully satisfactory. Where problems arise, they can in most cases be traced back to one or more of the following factors:

- poor quality calibration dyeings,
- problems in the repeatability of laboratory and production processes,
- problems in the reproducibility from laboratory to bulk,
- fluorescence of the substrate and/or one or more of the dyestuffs.

Simon[66] suggested a method for recipe prediction with fluorescent dyes already in 1972, and several new methods have been suggested since then, but the question is not satisfactorily resolved at the industrial application level.

As for the other sources of inaccuracies listed above, the solution lies in the textile technology more than in colorimetry or even computation, but nevertheless a number of highly sophisticated methods based on neural networks have been put forward[67,68] to improve the accuracy of the computerized colorant formulation in the textile industry.

Future Needs

There are quite a few areas in the application of CIE Colorimetry in the textile industry where CIE might help to answer open questions whether by forming Technical Committees or by encouraging further research within sister organizations.

1. *Instrumental determination of standard depth*: Results of recent research (such as published by Hawkyard and Kelly[69] and Hawkyard and Haque[70]) should be evaluated and compared to established methods[36] to arrive at a usable and generally accepted depth formula based on CIELAB coordinates. This is one of the fields where forming a TC may help in arriving at a new, CIE-endorsed formula.

2. *Color difference evaluation*: There are two CIE TCs working in this field (TC1-55 Uniform Colour Space for Industrial Colour Difference Evaluation and TC1-63 Validity of the Range of CIEDE2000) but their activities do not seem to cover the question whether CIEDE2000 is adequate and recommended for pass/fail evaluations in the textile industry. As the textile industry has been using instrumental color quality control—and building up important databases—with significant success using CIELAB-based formulas, (but not yet CIEDE2000) this question needs to be answered urgently.
3. *Fastness evaluation*: CIE has not been active so far in the area of this application of CIE Colorimetry, but the formation of a TC may help in developing new, more directly applicable methods, in cooperation with other organizations such as ISO, AATCC, and BSI.
4. *Determination of whiteness*: In this area future needs are not related to the development or acceptance of the relevant formulas (AATCC and ISO have already standardized the CIE-recommended whiteness and tint formulas) but to the recommendation of the most adequate light sources (daylight simulators) and measurement geometry. TC1-44 (Practical Daylight Sources for Colorimetry) is preparing a Technical Report on the best available daylight simulator technologies for visual evaluation and for instrumental measurement. CIE—probably within a new TC—will have to revise the recommendations regarding the geometry for the measurement of fluorescent samples: the current recommendation of bidirectional geometry is implemented by only a very small fraction of the color-measuring spectrophotometers with UV-calibration capability available in the textile industry. The question of the UV calibration would also probably have to be revised.

Conclusion

CIE Colorimetry has been embraced with enthusiasm by the textile industry. Color-measuring instruments and recipe prediction systems have become commonplace and are put to good use in hundreds of textile mills and laboratories worldwide. In many applications current CIE recommendations are used without any modification; in some areas further cooperation between CIE, ISO, BSI (and other standardizing organizations), and professional bodies related to the textile industry (AATCC, SDC and others) is needed to achieve the acceptance and widespread application of more of the textile-specific methods.

REFERENCES

1. Howarth, P (2002) Metrotrade Task 2.2 "Optical measurements in the paper industry," in: *Presentation at the National Conference of Standards Laboratories*, San Diego, August 2002.

2. Hunter RS, Harold RW (Eds.), (1987) *The Measurement of Appearance*, J. Wiley.
3. Budde W, Chapman S (1975), Measurement of brightness and opacity according to ISO standards. *Transactions*, **2**, 1–4.
4. Weiss GR, Bruce AM (1982) Luminosity "Y" vs 457 nm brightness interrelationship, *Pulp Paper Canada*, **83**, T50–T57.
5. SCAN-G.5:94 (1994).
6. Bristow JA (1994) The calibration of instruments for the measurement of paper whiteness, *Color Res. Appl.*, **19**, 475–483.
7. International Standard ISO 2469:1994 *Paper, board and pulps—Measurement of diffuse reflectance factor.*
8. International Standard ISO 2470:1999 *Paper and board—Measurement of diffuse blue reflectance factor (ISO brightness).*
9. International Standard ISO 2471:1998 *Paper and board—Determination of opacity (paper backing)—Diffuse reflectance method.*
10. CIE (1971) Colorimetry Publication 15
11. Jordan B, Zwinkels J, McGarry PM (2003) The influence of the illuminant on the luminescent radiance factor spectrum of a reference fluorescent paper, in: *Proceedings of TAGA*, Montreal, QC, April 2003, pp. 420–434.
12. CIE (1989) Solar Spectral Irradiance, Publication 85.
13. TAPPI Standard (1972) Color of paper and paperboard in CIE Y,x,y, or Y, dominant wavelength and excitation purity, proposed new suggested method T527, *Tappi*, **55**, 1125–1129.
14. Budde W (1960) Standards of Reflectance, *J. Opt. Soc. Am.*, **50**, 217–220.
15. Zwinkels J, Gauthier F (2003) Investigation of photoluminescent effect in opal glasses used as diffuse reflectance standards, *Proc. SPIE*, **4826**, 70.
16. Ganz E (1979) Whiteness formulas: a selection. *Appl. Opt.*, **16**, 1073–1076.
17. CIE (1986) *Colorimetry.* Publication 15.2, 2nd ed.
18. Brockes A (1982) The evaluation of whiteness, *CIE J.*, **1**, 38–39.
19. Jordan BD, O'Neill MA (1991) The whiteness of paper—Colorimetry and visual ranking. *Tappi J.*, **74**, 93–101.
20. Bristow JA (1988) Measuring and reporting newsprint shade, in: *Proceedings of IFRA Symposium*, April 12–13, Copenhagen.
21. CIE (2004) *Colorimetry*, Publication 15.3:2004.
22. Bristow JA (2005) Response to Revised Draft D_{65} Colour Methods, ISO TC6/WG3:N614.
23. Popson J (1990) Personal correspondence to US CIE Division 1.
24. Hunt RW (1992) Letter to the Editor: Standard sources to represent daylight, *Color Res. Appl.*, **17**, 293.
25. Hardy AC (1937) The applicability of the spectrophotometer to the solution of color problems in the textile industry, *Am. Dyest. Rep.*, **26**, 67–71.
26. Judd DB (1939) Specification of uniform color tolerances for textiles, *Textile Res. J.*, **9**, 253–308.
27. Davidson HR, Godlove IH (1950) Applications of the automatic tristimulus integrator to textile mill practice, *Am. Dyest. Rep.*, **39**, 78–84.

28. Stearns EI (1951) What's new in spectrophotometry: Progress of spectrophotometry in the textile industry to 1951, *Am. Dyest. Rep.*, **40**, 563–574.
29. Simon FT, Goodwin WJ (1958) Rapid graphical computation of small color differences, *Am. Dyest. Rep.*, **47**, 105–112.
30. Chickering KD (1971) FMC color difference formulas: clarification concerning usage, *J. Opt. Soc. Am.*, **61**, 118–122.
31. McLaren K (1970) The Adams–Nickerson colour-difference formula, *J. Soc. Dyers Colour.*, **86**, 354–366.
32. AATCC (2005) *Standard depth scales for depth determination*, Evaluation Procedure 4, AATCC Technical Manual/2005, p. 383.
33. Rabe P (1956) The standard depth of shade in relation to the assessment of color-fastness of dyes on textiles, in: *Proc. Perkin Centennial 1856-1956*, AATCC: New York, 25–30.
34. ISO 105-A01 (1994) *Textiles—Tests for color fastness: General principles of testing.*
35. Christ HA (1985) Eine farbmetrische Formel zur Bestimmung der Standardfarbtiefe, *Textilveredlung*, **20**, 241–242.
36. ISO 105-A06 (1995) *Textiles—Tests for colour fastness: Instrumental determination of 1/1 standard depth of color.*
37. Davidson HR, Friede E (1953) The size of acceptable color tolerances, *J. Opt. Soc. Am.*, **43**, 581–589.
38. AATCC (2005) Test Method 173-1998 CMC: Calculation of small color differences for acceptability. *AATCC Technical Manual/2005*, 311–313.
39. ISO 105-J03 1995 *Textiles—Tests for color fastness—Part J03: Calculation of color differences.*
40. Hinks D (2005) *Optimizing color control throughout the supply chain*, National Textile Center Annual Report: November. http://www.ntcresearch.org/projectapp/index.cfm?project=C04-NS11
41. Aspland R, Shanbhag P (2004) Comparison of color difference equations for textiles: CMC(2:1) and CIEDE2000, *AATCC Rev.*, **4**, 26–30.
42. Gay J, Hirschler R (2003) Field trials for CIEDE2000—Correlation of visual and instrumental pass/fail decisions in industry, in: *Proceedings of the 25th Session of the CIE, San Diego, June 25–July 2*, D1 pp. 38–41.
43. Kuehni RG (2003) Colour difference formulas: accurate enough for objective colour quality control of textiles? *Color. Technol.*, **119**, 164–169.
44. Simon FT (1961) Small color differences computation and control., *Die Farbe*, **10**, 225–234.
45. Willis, RF (2000) Modern color sorting methods: sequencing and clustering, in: *AATCC Color Science Symposium*, Greenville, SC, March 22–23.
46. Simon FT, Goodwin WJ (1958) Rapid graphical computation of small color differences, *Amer. Dyest. Rep.*, **47**, 105–112.
47. Aspland JR, Balasaygun, KD, Jarvis JP, Whitaker TH (2000) Alternative mathematical approaches to shade sorting, *Color Res. Appl.*, **25**, 368–375.
48. AATCC (2005) *Technical Manual* **80.**
49. AATCC (2005) EP1: Gray scale for Color change, *AATCC Technical Manual*, **80**, 379–380 (technically equivalent to ISO 105-A02).

50. AATCC (2005) EP2: Gray scale for staining, *AATCC Technical Manual*, **80**, 381–382 (technically equivalent to ISO 105-A03).
51. AATCC (2005) EP7: Instrumental assessment of the change in color of a test specimen. *AATCC Technical Manual*, **80**, 392-393 (technically equivalent to ISO 105-A05).
52. ISO (1989) 105-A04: *Textiles—Tests for Colour Fastness, Part A04: Method for instrumental assessment of degree of staining of adjacent fabrics.*
53. Cui G, Luo MR et al. Grading textile fastness, *Coloration Technol.*
    Part I: Using a digital camera system, **119** (2003) 212–218
    Part II: Development of a new staining formula, **119** (2003) 219-224
    Part III: Development of a new fastness formula for assessing the change of colour, **120** (2004) 226–230
    Part IV: An interlaboratory trial using DigiEye systems **120** (2004) 231–235.
54. Berger A (1959) Weissgradformeln und ihre praktische Bedeutung, *Farbe*, **8**, 187–202.
55. Hunter R (1958) Description and measurement of white surface, *J. Opt. Soc. Am.*, **48**, 597–605.
56. Stensby P (1973) Questions in regard to whiteness evaluation, *J. Color Appear.*, **2**, 39–42.
57. Ganz E (1972) Whiteness measurement, *J. Color Appear.*, **1**, 33–41.
58. CIE Technical Report (2004) *Colorimetry*, 3rd ed. Publication 15:2004, CIE Central Bureau, Vienna.
59. AATCC (2005) Test Method 110-2000: Whiteness of textiles, *AATCC Technical Manual*, **80**, 167-168 (technically equivalent to ISO 105–J02).
60. Griesser R (1994) Assessment of whiteness and tint of fluorescent substrates with good interinstrument correlation, *Color Res. Appl.*, **19**, 446–460.
61. Gay JK, Melo CC, Hirschler R (2004) Instrumental whiteness evaluation: Practical results of inter-instrument agreement tests, in: *Proceedings of the AIC 2004 "Color and Paints"* Interim Meeting of the International Color Association, Porto Alegre, Brazil, November 3–5, 2004 Downloaded on March 14, 2006 from http://www.fadu.uba.ar/sitios/sicyt/color/aic2004.htm.
62. Park RH, Stearns EI (1944) Spectrophotometric formulation, *J. Opt. Soc. Am.*, **34**, 112–113.
63. Davidson HR, Hemmendinger H (1958) A colorant mixture computer, *J. Opt. Soc. Am.*, **48**, 281–287.
64. Alderson JV, Atherton E, Derbyshire AN (1961) Modern physical techniques in colour formulation, *J. Soc. Dyers Col.*, **77**, 657–669.
65. Allen E (1966) Basic equations used in computer color matching, *J. Opt. Soc. Am.*, **56**, 1256–1259.
66. Simon FT (1972) The two-mode method for measurement and formulation with fluorescent colorants, *J. Color Appear.*, **1**, (4) 5–11.
67. Westland S et al (1991) An intelligent approach to computer recipe prediction, *J. Soc. Dyers Colour.*, **107**, 235- 237.
68. Bishop JM, Bushnell MJ, Westland S (1991) Application of neural networks to computer recipe prediction, *Color Res. Appl.*, **16**, 3–9.
69. Hawkyard CJ, Kelly M (2000) A new approach to the assessment of standard depth, *J. Soc. Dyers Colour.*, **116**, 339–344.
70. Hawkyard CJ, Haque A (2005) Depth assessment for mixture shades, *Color. Technol.*, **121**, 86–88.

APPENDIX 4

LIST OF CIE PUBLICATIONS

RECOMMENDATIONS

17.4 International lighting vocabulary, 4th ed. (Joint publication IEC/CIE), 1987.
23 International recommendations for motorway lighting, 1973.
39.2 Recommendations for surface colours for visual signalling, 2nd ed., 1983.

STANDARDS

CIE S004-2001 Colours of light signals, 2001.
ISO 16508/CIE S006 Road traffic light—200 mm roundel signals photometric properties, 1999.
ISO 17166/CIE S007 Erythema reference action spectrum and standard erythema dose, 1998.
ISO 8995/CIE S008-2001 Lighting of indoor work places, 2001.
CIE S009:2002 Photobiological safety of lamps and lamp systems, 2002.
ISO 23539/CIE S010:2005 Photometry—The CIE system of physical photometry, 2005.
ISO 15469/CIE S011:2003 Spatial distribution of daylight—CIE standard general sky, 2003.
ISO 23603/CIE S012:2005 Standard method of assessing the spectral quality of daylight simulators for visual appraisal and measurement of colour, 2005.

CIE S013:2003 International standard global UV index, 2003.
CIE S 014-1:2006 Colorimetry—Part 1: CIE Standard Colorimetric Observers, 2006.
CIE S 014-2: 2006 Colorimetry—Part 2: CIE Standard Illuminants, 2006.
CIE S 015:2005 Lighting of outdoor work places, 2005.
CIE S 016:2005 Lighting of outdoor work places—Lighting requirements for safety and security, 2005.
DS 018.2:2007 Standard file format for luminaire photometric data, 2007.
DS 019.2:2005 Photocarcinogenesis action spectrum (non-melanoma skin), 2005.
CIE S 020:2007 Emergency lighting, 2007.

TECHNICAL COMMITTEE REPORTS

1 Guide lines for minimising urban sky glow near astronomical observatories (Joint publication IAU/CIE), 1980.
13.3 Method of measuring and specifying colour rendering of light sources, 1995.
15 Colorimetry, 3rd ed., 2004.
16 Daylight, 1972.
18.2 The basis of physical photometry, 2nd ed., 1983.
19.21 An analytic model for describing the influence of lighting parameters upon visual performance, 2nd ed., Vol. 1: Technical foundations, 1981.
19.22 An analytic model for describing the influence of lighting parameters upon visual performance, 2nd ed., Vol. 2: Summary and application guidelines, 1981.
23 International recommendations for motorlighting, 1973.
31 Glare and uniformity in road lighting installations, 1976.
32 Lighting in situations requiring special treatment (in road lighting), 1977.
33 Depreciation of installation and their maintenance (in road lighting), 1977.
34 Road lighting lantern and installation data: photometrics, classification and performance, 1977.
38 Radiometric and photometric characteristics of materials and their measurement, 1977.
40 Calculations for interior lighting: Basic method, 1978.
41 Light as a true visual quantity: Principles of measurement, 1978.
42 Lighting for tennis, 1978.
43 Photometry of floodlights, 1979.
44 Absolute methods for reflection measurements, 1979.
45 Lighting for ice sports, 1979.
46 A review of publications on properties and reflection values of material reflection standards, 1979.
47 Road lighting for wet conditions, 1979.

48 Light signals for road traffic control, 1980.

49 Guide on the emergency lighting of building interiors, 1981.

51.2 A method for assessing the quality of daylight simulators for colorimetry, 1999.

52 Calculations for interior lighting: Applied method, 1982.

53 Methods of characterising the performance of radiometers and photometers, 1982.

54.2 Retroreflection: Definition and measurement, 2001.

55 Discomfort glare in the interior working environment, 1983.

57 Lighting for football, 1983.

58 Lighting for sports halls, 1983.

59 Polarisation: Definitions and nomenclature, instrument polarisation, 1984.

60 Vision and the visual display unit work station, 1984.

61 Tunnel entrance lighting: A survey of fundamentals for determining the luminance in the threshold zone, 1984.

62 Lighting for swimming pools, 1984.

63 The spectroradiometric measurement of light sources, 1984.

64 Determination of the spectral responsivity of optical radiation detectors, 1984.

65 Electrically calibrated thermal detectors of optical radiation (absolute radiometers), 1985.

66 Road surfaces and lighting (joint technical report CIE/PIARC), 1984.

67 Guide for the photometric specification and measurement of sports lighting installations, 1986.

69 Methods of characterising illuminance meters and luminance meters: Performance, characteristics and specifications, 1987.

70 The measurement of absolute luminous intensity distributions, 1987.

72 Guide to the properties and uses of retroreflectors at night, 1987.

73 Visual aspects of road markings (joint technical report CIE/PIARC; French translation: Aspects visuels des marquages routiers is available from PIARC), 1988.

74 Roadsigns, 1988.

75 Spectral luminous efficiency functions based upon brightness matching for monochromatic point sources, 2° and 10° fields, 1988.

76 Intercomparison on measurement of (total) spectral radiance factor of luminescent specimens, 1988.

77 Electric light sources: State of the art—1987, 1988.

78 Brightness–luminance relations: Classified bibliography, 1988.

79 A guide for the design of road traffic lights, 1988.

80 Special metamerism index: Change in observer, 1989.

81 Mesopic photometry: History, special problems and practical solutions, 1989.

82 CIE History 1913–1988, 1990.

83 Guide for the lighting of sports events for colour television and film systems, 1989.

84 Measurement of luminous flux, 1989.
85 Solar spectral irradiance, 1989.
86 CIE 1988 2° spectral luminous efficiency function for photopic vision, 1990.
87 Colorimetry of self-luminous displays—A bibliography, 1990.
88 Guide for the lighting of road tunnels and underpasses, 2nd ed., 2004.
89 Technical Collection 1990:
 89/1 Results of a CIE detector response intercomparison
 89/2 Photobiological effects of sunlamps
 89/3 On the deterioration of exhibited museum objects by optical radiation
 89/4 Guide for the measurement of underground mine lighting.
90 Sunscreen testing (UV.B), 1991.
93 Road lighting as an accident countermeasure, 1992.
94 Guide for floodlighting, 1993.
95 Contrast and visibility, 1992.
96 Electric light sources—State of the art, 1992.
97 Maintenance of indoor electric lighting systems, 2nd ed., 2005.
98 Personal dosimetry of UV radiation, 1992.
99 Lighting education (1983–1989), 1992.
100 Fundamentals of the visual task of night driving, 1992.
101 Parametric effects in colour-difference evaluation, 1993.
102 Recommended file format for electronic transfer of luminaire photometric data, 1993.
103 Technical Collection 1993:
 103/1 Colour appearance analysis
 103/2 Industrial lighting and safety at work
 103/3 Reference action spectra for ultraviolet induced erythema and pigmentation of different human skin types
 103/4 Biologically effective emissions and hazard potential of desktop luminaires incorporating tungsten halogen lamps
 103/5 The economics of interior lighting maintenance
 103/6 Clarification of maintained illuminance and associated terms.
104 Daytime running lights (DRL), 1993.
105 Spectroradiometry of pulsed optical radiation sources, 1993.
106 CIE Collection in Photobiology and Photochemistry, 1993:
 106/1 Determining ultraviolet action spectra
 106/2 Photokeratitis
 106/3 Photoconjunctivitis
 106/4 A reference action spectrum for ultraviolet induced erythema in human skin
 106/5 Photobiological effects in plant growth
 106/6 Malignant melanoma and fluorescent lighting
 106/7 On the quantification of environmental exposures: limitations of the concept of risk-to-benefit ratio
 106/8 Terminology for photosynthetically active radiation for plants.

107 Review of the official recommendations of the CIE for the colours of signal lights, 1994.
108 Guide to recommended practice of daylight measurement, 1994.
109 A method of predicting corresponding colours under different chromatic and illuminance adaptation, 1994.
110 Spatial distribution of daylight—Luminance distributions of various reference skies, 1994.
111 Variable message signs, 1994.
112 Glare evaluation system for use within outdoor sports and area lighting, 1994.
113 Maintained nighttime visibility of retroreflective road signs, 1995.
114 CIE Collection in photometry and radiometry, 1994:
- 114/1 Survey of reference materials for testing the performance of spectrophotometers and colorimeters
- 114/2 International intercomparison on transmittance measurement—Report of results and conclusions
- 114/3 Intercomparison of luminous flux measurements on HPMV lamps
- 114/4 Distribution temperature and ratio temperature
- 114/5 Terminology relating to non-selective detectors
- 114/6 Photometry of thermally sensitive lamps.

115 Recommendations for the lighting of roads for motor and pedestrian traffic, 1995.
116 Industrial colour-difference evaluation, 1995.
117 Discomfort glare in interior lighting, 1995.
118 CIE Collection in colour and vision, 1995:
- 118/1 Evaluation of the attribute of appearance called gloss
- 118/2 Models of heterochromatic brightness matching
- 118/3 Brightness-luminance relations
- 118/4 CIE guidelines for co-ordinated research on evaluation of colour appearance models for reflection print and self-luminous display image comparisons
- 118/5 Testing colour appearance models: Guidelines for co-ordinated research
- 118/6 Report on color difference literature
- 118/7 CIE guidelines for co-ordinated future work on industrial colour-difference evaluation.

121 Photometry and goniophotometry of luminaires, 1996.
122 The relationship between digital and colorimetric data for computer-controlled CRT displays, 1996.
123 Low Vision—Lighting needs for the partially sighted, 1997.
124 CIE Collection in Colour and Vision, 1997:
- 124/1 CIE TC 1-31 Report: Colour notations and colour order systems
- 124/2 CIE TC 1-18 Chairman(s Report: On the course of the disability glare function and its attribution to components of ocular scatter

- 124/3 Next step in industrial colour difference evaluation, Report on a colour difference research meeting.

125 Standard erythemal dose—A review, 1997.
126 Guidelines for minimizing sky glow, 1997.
127 Measurement of LEDs, 1997.
128 Guide to the lighting for open-cast mines, 1998.
129 Guide for lighting exterior work areas, 1998.
130 Practical methods for the measurement of reflectance and transmittance, 1998.
132 Design methods for lighting of roads, 1999.
134 CIE Collection in Photobiology and Photochemistry, 1999.

- 134/1 CIE TC 6-26 Report: Standardization of the terms UV-A1, UV-A2, and UV-B
- 134/2 CIE TC 6-30 Report: UV protection of the eye
- 134/3 CIE TC 6-38 Report: Recommendation on photobiological safety of lamps. A review of standards

135 CIE Collection 1999: Vision and colour, physical measurement of light and radiation.

- 135/1 Disability glare
- 135/2 Colour rendering, closing remarks
- 135/3 Virtual metamers for assessing the quality of simulators of CIE illuminant D50 (Supplement 1-1999 to CIE 51-1981)
- 135/4 Some recent developments in colour-difference evaluation
- 135/5 Visual adaptation to complex luminance distribution
- 135/6 45°/0° Spectral reflectance factors of pressed polytetrafluoroethylene (PTFE) power (Reprint of NIST Technical Note 1413)

136 Guide to the lighting of urban areas, 2000.
137 The conspicuity of traffic signs in complex background, 2000.
138 CIE Collection 2000: Photobiology and Photochemistry.

- 138/1 Blue-light photochemical retinal hazard
- 138/2 Action spectrum for photocarcino-genesis (non-melanoma skin cancers)
- 138/3 Standardized protocols for photocarcino-genesis safety testing
- 138/4 A proposed global UV index.

139 The influence of daylight and artificial light on diurnal and seasonal variations in humans. A bibliography, 2001.
140 Road lighting calculations, 2000.
141 Testing of supplementary systems of photometry, 2001.
142 Improvement to industrial colour-difference evaluation, 2001.
143 International recommendations for colour vision requirements for transport, 2001.
144 Road surface and road marking reflection characteristics, 2001.
145 The correlation of models for vision and visual performance, 2002.
146/147 Collection on Glare, 2002:

- 146 CIE equations for disability glare
- 147 Glare from small, large and complex sources.

148 Action spectroscopy of skin with tunable lasers, 2002.
149 The use of tungsten filament lamps as secondary standard sources, 2002.
150 Guide on the limitation of the effects of obtrusive light from outdoor lighting installations, 2003.
151 Spectral weighting of solar ultraviolet radiation, 2003.
153 Report on an intercomparison of measurements of the luminous flux of high pressure sodium lamps, 2003.
154 The maintenance of outdoor lighting systems, 2003.
155 Ultraviolet air disinfection, 2003.
156 Guidelines for the evaluation of gamut mapping algorithms, 2004.
157 Control of damage to museum objects by optical radiation, 2004.
158 Ocular lighting effects on human physiology and behaviour, 2004.
159 A colour appearance model for colour management systems: CIECAM02, 2004.
160 A review of chromatic adaptation transforms, 2004.
161 Lighting design methods for obstructed interiors, 2004.
162 Chromatic adaptation under mixed illumination condition when comparing softcopy and hardcopy images, 2004.
163 The effects of fluorescence in the characterization of imaging media, 2004.
164 Hollow light guide technology and applications, 2005.
165 CIE 10 degree photopic photometric observer, 2005.
166 Cognitive colour, 2005.
167 Recommended practice for tabulating spectral data for use in colour computations, 2005.
168 Criteria for the evaluation of extended-gamut colour encodings, 2005.
169 Practical design guidelines for the lighting of sport events for colour television and filming, 2005.
170-1 Fundamental chromaticity diagram with physiological axes—Part 1, 2006.
171 Test cases to assess the accuracy of lighting computer programs, 2006.
172 UV protection and clothing, 2006.
173 Tubular daylight guidance systems, 2006.
174 Action spectrum for the production of previtamin D_3 in human skin, 2006.
175 A framework for the measurement of visual appearance, 2006.
176 Geometric tolerances for colour measurements, 2006.
177 Colour rendering of white LED light sources, 2007.
179 Methods for characterizing tristimulus colorimeters for measuring the colour of light, 2007.

PROCEEDINGS OF THE SESSIONS

1921 Paris
1924 Geneva
1927 Bellagio
1928 Saranac
1931 Cambridge

1935		Berlin
1939		Scheweningen
1948		Paris
1951		Stockholm
1955		Zurich
1959	4–7	Bruxelles (Vol. A, B, C, D)
1963	11	Vienna (Vol. A, B, C, D)
1967	14	Washington (Vol. A, B)
1971	21	Barcelona (Vol. A, B, C)
1975	36	London
1979	50	Kyoto
1983	56	Amsterdam
1987	71	Venice, Vol. 1–2
1991	91	Melbourne, Vol. 1–2
1995	119-120	New Delhi, Vol. 1–2
1999	133	Warsaw, Vol. 1–2
2003	152	San Diego, Vol. 1–2

DISCS AND OTHER PUBLICATIONS

D001 Disc version of CIE Colorimetric Data (S001 and S002 Tables), 1988.

D002 Disc version of CIE Colorimetric and Colour Rendering Data (Publ. 13.2 and 15 Tables), 2004.

D005 A method for assessing the quality of D65 daylight simulators for colorimetry (based on CIE 51-1981) 1994.

D006 Automatic quality control of daylight measurement—Software for IDMP stations (computer program to CIE 108-1994), 1994.

D007 A computer program implementing the "Method of predicting corresponding colours under different chromatic and illuminance adaptation" (described in CIE 109-1994), 1994.

D008 Computer program to calculate CRIs (according to CIE 13.3-1995), 1995.

SPECIAL PUBLICATIONS

x005 Proceedings of the CIE Seminar '92 on Computer programs for light and lighting.

x006 Japan CIE Session at PRAKASH 91.

x007 Proceedings of the CIE Symposium '93 on Advanced Colorimetry.

x008 Urban sky glow—A worry for astronomy (Proceedings of a Symposium of CIE TC 4-21), 1994.

x009 Proceedings of the CIE Symposium '94 on Advances in Photometry.

x010 Proceedings of the CIE Expert Symposium '96 Colour Standards for Image Technology.

x011 Special volume, 23rd Session, New Delhi '95, Late papers.
x012 NPL—CIE-UK Visual Scales Conference.
x013 Proceedings of the CIE LED Symposium '97 on Standard Methods for Specifying and Measuring LED Characteristics, 1998.
x014 Proceedings of the CIE Expert Symposium '97 on Colour Standards for Imaging Technology, 1998.
x015 Proceedings of the First CIE Symposium on Lighting Quality, 1998.
x016 Proceedings of the CIE/ICNIRP Conference on Measurements of Optical Radiation Hazards, 1998.
x017 Special volume, 24th Session, Warsaw '99, Late papers, 2000.
x018 Proceedings of the CIE Symposium '99 "75 Years of CIE Photometry," 2000.
x019 Proceedings of three CIE workshops on Criteria for Road Lighting, 2001.
x020 Proceedings of the CIE Symposium 2001 "Uncertainty Evaluation, Methods for Analysis of Uncertainties in Optical Radiation Measurement," 2001.
x021 Proceedings of the CIE Expert Symposium 2000 "Extended range colour spaces," 2001.
x022 Proceedings of the 2nd CIE Expert Symposium on LED measurement, "Standard methods for specifying and measuring LED and LED cluster characteristics," 2001.
x023 Proceedings of two CIE Workshops on photometric measurement systems for road lighting installations, 2002.
x024 Proceedings of the CIE/ARUP Symposium on Visual Environment, 2002.
x025 Proceedings of the CIE Symposium 2002 "Temporal and spatial aspects of light and colour perception and measurement," 2003.
x026 Proceedings of the CIE Symposium '04. LED Sources: Physical Measurement and Visual and Photobiological Assessment, 2005.
x027 Proceedings of the CIE Symposium '04. Light and Health: non-visual effects, 2004
x028 Proceedings of the CIE Symposium '05. Vision and Lighting in Mesopic Conditions, 2005.
x029 Proceedings of the 2nd CIE Expert Symposium on Measurement Uncertainty, 2006.
x030 Proceedings of the ISCC/CIE Expert Symposium '06 "75 Years of the CIE Standard Colorimetric Observer", 2006.
x031 Proceedings of the 2nd CIE Expert Symposium on Lighting and Health, 2006.
x032 Proceedings of the CIE Symposium on Visual Appearance, 2007.

CIE PUBLICATIONS ON CD-ROM

A CD-ROM with all current CIE Technical Reports and Standards is available from IHS, Information Handling Services, 15 Inverness Way East, M/S B203 Englewood, CO 80112-5776 USA.

CIE-Journal Vol. 1–Vol. 8 1982–1989. **CIE NEWS** No. 1–No. 78 1986–2006.

For latest information on CIE publications see the CIE home page on the World Wide Web: **http://www.cie.co.at/**

All CIE Publications can be ordered online at: **http://www.techstreet.com/cgi-bin/joint.cgi/cie**

You can reach this site also via a link from the CIE central Web site: **http://www.cie.co.at**

Please use the CIE online store for all your orders.

GLOSSARY

Action spectrum	Relative spectral effectiveness of optical radiation, for a specified actinic phenomenon, in a specified system.
Additive color mixing	Stimulation that combines on the retina the actions of various colour stimuli in such a manner that they cannot be perceived individually.
Additivity	Asserted property of lights A, B, C, and D such that, if A matches C and B matches D, the then A + B matches C + D, where "+" stands for additive mixing of spectral power distributions.
Anti-prime (AP)	Pertaining to three wavelengths (497 nm, 579 nm and 653 nm) identified by W. A. Thornton as having particularly bad properties as narrow-band primary wavelengths for color matching.
Brightness	Attribute of a visual sensation according to which an area appears to emit more or less light.
Chroma	Chromaticness, colourfulness, of an area judged as a proportion of the brightness of a similarly illuminated area that appears white or highly transmitting.
Chromatic adaptation	Adaptation by stimuli in which the dominant effect is that of different relative spectral distributions.

Colorimetry: Understanding the CIE System, Edited by János Schanda

Chromaticity — Property of a colour stimulus defined by its chromaticity coordinates, or by its dominant or complementary wavelength and purity taken together.

Chromaticity diagram — A plane diagram in which points specified by chromaticity coordinates represent the chromaticities of colour stimuli.

Chromaticness — See colorfulness

Color appearance model — A model describing colour appearance, built from descriptors of colour stimuli.

Color matching function — The tristimulus values of monochromatic stimuli of equal radiant power. In the CIE colorimetric systems the $\bar{x}(\lambda)$, $\bar{y}(\lambda)$, $\bar{z}(\lambda)$ functions (in the CIE 1931 standard colorimetric system) or the $\bar{x}_{10}(\lambda)$, $\bar{y}_{10}(\lambda)$, $\bar{z}_{10}(\lambda)$ functions (in the CIE 1964 standard colorimetric system).

Color perception — Characteristic of visual perception that can be described by attributes of hue, brightness (or lightness) and colourfulness (or saturation or chroma).

Color stimulus — Visible radiation entering the eye and producing a sensation of colour, either chromatic or achromatic.

Color temperature — The temperature of a Planckian radiator whose radiation has the same chromaticity as that of a given stimulus.

Colorfulness — Attribute of a visual sensation according to which the perceived colour of an area appears to be more or less chromatic.

Contrast — In the perceptual sense: Assessment of the difference in appearance of two or more parts of a field seen simultaneously or successively (hence: brightness contrast, lightness contrast, colour contrast, simultaneous contrast, successive contrast, etc.).

In the physical sense: Quantity intended to correlate with the perceived brightness contrast, usually defined by one of a number of formulae which involve the luminances of the stimuli considered: for example by the proportional variation in contrast near the luminance threshold, or by the ratio of luminances for much higher luminances.

Contrast sensitivity (S_c) — Reciprocal of the least perceptible (physical) contrast, usually expressed as $L/\Delta L$, where L is the average luminance and ΔL is the luminance difference threshold.

Note. The value of S_c depends on a number of factors including the luminance the viewing conditions and the state of adaptation.

Coplanarity	The state of being entirely contained in a (two-dimensional) plane.
Covering theory	Parametric extension of a theory that is under experimental test. The theory-under-test will be confirmed if the covering theory best fits the experimental data for covering-theory parameters that are close to the values corresponding to the theory-under-test.
Diffuse transmittance	Ratio of the diffusely transmitted part of the (whole) transmitted flux, to the incident flux.
Fluorescence	The emission of optical radiation (light) when a substance is exposed to any type of electromagnetic radiation, where the emitted radiation generally appears within 10 nanoseconds after the excitation. This is due to an "allowed" transition generally from an excited singlet state to a ground singlet state.
Forward-matrix method	Method of predicting the weightings of three new primaries needed to match a test light, whereby the (presumed known) 3×3 matrix of new-primary weightings to match each old primary is multiplied by the vector of old-primary weightings needed to match the same test light.
Fovea	Central part of the retina, thin and depressed, which contains almost exclusively cones and forming the site of most distinct vision.
Fundamental	A linear combination of color-matching functions that is held to represent one of the basic responsivities of color (i.e., cone spectral responsivities weighted by the transmittance of the intra-ocular media).
Helmholtz-Kohlrausch Effect	The Helmholtz-Kohlrausch effect refers to change in brightness of perceived colour produced by increasing the purity of a colour stimulus while keeping its luminance constant within the range of photopic vision.
Helson-Judd effect	The Helson-Judd effect refers to tendency, in a coloured illumination, for light colours to be tinged with the hue of the illuminant, and for dark colours to be tinged with the complementary hue.
Hue	Attribute of a visual sensation according to which an area appears to be similar to one of the perceived colours red, yellow, green and blue, or to a combination of two of them.
Hue angle	It, ranged from 0 to 360 degree in the Redness-Greenness (a), and Yellowness-Blueness (b) plane,

	is based on the concept of equal perceived difference.
Hue composition	It describes colour appearances in terms of four unitary hues, ranged from 0 (pure red), 100 (pure yellow), 200 (pure green), 300 (pure blue), and back to pure red at 400.
Hunt Effect	An object or image illuminate by a higher luminance would appear more colourful than when illuminated by a lower luminance. This is known as Hunt Effect.
Inverse-matrix method	Method of predicting the weightings of three new primaries needed to match a test light, whereby the 3×3 matrix of old-primary weightings that match each new primary is inverted and multiplied by the vector of old-primary weightings needed to match the same test light.
Lightness	The brightness of an area judged relative to the brightness of a similarly illuminated area that appears to be white or highly transmitting.
Lightness contrast effect	A colour appears lighter against a dark background than against a light background.
Luminance	Quantity defined by the formula $L_v = \frac{d^2\Phi_v}{dA \cdot \cos\theta \cdot d\Omega}$, where $d\Phi_v$ is the luminous flux transmitted by an elementary beam passing through the given point and propagating in the solid angle $d\Omega$ containing the given direction; dA is the area of a section of that beam containing the given point; θ is the angle between the normal to that section and the direction of the beam.
Macula lutea	Layer of photostable pigment covering parts of the retina in the foveal region.
Maximum saturation method	Color-matching method whereby a superposition of two of the primaries is adjusted to match a superposition of the third primary and the test light. The chromaticity of the light in each matching field is on the boundary of the primary (e.g., RGB) triangle, hence the name "maximum saturation."
Maxwell method	Color-matching method whereby the chromaticity of each matching field is constrained to a particular white (in the middle of the primary or RGB triangle).
Maxwell spot	Visual manifestation of the yellow pigmentation (called macular pigmentation) in the central four degrees of visual field in the human retina. Color matches within the Maxwell spot may not remain matches when viewed outside the spot, and vice versa.

Mesopic vision	Vision by the normal eye intermediate between the photopic and scotopic vision.
Photopic vision	Vision by the normal eye in which cones are the principle active photoreceptors. Photopic vision normally occurs when the eye is adapted to levels of luminance of at least 10 candelas per square metre.
Photopigment	organic pigment found in visual cone or rod photoreceptors, whose light-incurred bleaching triggers a visual response.
photopigment-depletion theory (pigment-bleaching hypothesis)	Theory of color-matching that takes into account the wavelength-dependent change in light absorption that occurs when photopigments are bleached by light during the visual process.
primary	One of three lights that are used as color-matching proxies for a test light in a color match.
Prime-color (PC)	Pertaining to three wavelengths (452 nm, 533 nm and 607 nm) identified by W. A. Thornton that are good narrowband-primary wavelengths for color matching due to their subtending [approximately] a maximum tristimulus volume per watt.
proportionality	Asserted property of lights A and B such that, if A matches B, then *k*A matches *k*B for any positive scale factor *k*.
Reflectance factor	Ratio of the radiant or luminous flux reflected in the directions delimited by the given cone of that reflected in the same directions by a perfect reflecting diffuser identically irradiated or illuminated.
Saturation	Chromaticness, colourfulness, of an area judged in proportion to its brightness.
Scotopic vision	Vision by the normal eye in which rods are the principle active photoreceptors. Scotopic vision normally occurs when the eye is adapted to levels of luminance of less than some thousandths of a candela per square metre.
Spectral distribution	Quotient of the radiant or luminous or photon quantity $dX(\lambda)$ contained in an elementary range of $d\lambda$ of wavelength at the wavelength λ by that range: $X_\lambda = \frac{dX(\lambda)}{d\lambda}$
Spectral luminous efficiency function	Ratio of the radiant flux at wavelength λ_m to that at wavelength λ such that both radiations produce equally intense luminous sensations under specified photometric conditions and λ_m is chosen so that the maximum value if this ratio is equal to 1.

Spectrograph — Instrument for measuring radiometric quantities in narrow wavelength intervals over a given spectral region.

Spectro(radio)meter — Instrument for measuring spectrally resolved radiometric quantities in one shot over a given spectral region (e.g. by exposing a photographic plate or irradiating a CCD (linear) matrix.

Spectrophotometer — Instrument for measuring the ratio of two values of a radiometric quantity at the same wavelength.

Spectroreflectometer — Instrument for measuring the ratio of two irradiance values at the same wavelength.

Standard colorimetric observer (CIE 1931) — Ideal observer whose colour-matching properties correspond to the CIE colour-matching functions $\bar{x}(\lambda)$, $\bar{y}(\lambda)$, $\bar{z}(\lambda)$ adopted by the CIE in 1931

Stevens Effect — A surface colour illuminated by a higher luminance would appear brighter than the same surface illuminated by a lower luminance.

Surface color — Colour perceived as belonging to a surface from which the light appears to be diffusely reflected or radiated.

Symmetric-matching experiment — Color-matching experiment in which all variables except the matching lights (e.g., state of adaptation and surround of the test field) are constrained to be the same on both sides of the match.

Symmetry — Asserted property of lights A and B such that, if A matches B then B matches A.

transformability of primaries — Effectiveness of the linear-algebra prediction of new-primary weightings needed to match a test light, as compared with actual settings obtained by experiment.

Transformation of primaries — Use of linear algebra to predict the weightings of three new primaries needed to match a test light, given certain information about the new primaries and the vector of old-primary weightings needed to match the same test light. (See forward-matrix and inverse-matrix methods.)

Transitivity — Asserted property of lights A, B, and C such that, if A matches B and B matches C, then A matches C.

Troland (td) — Unit used to express a quantity proportional to retinal illuminance produced by a light stimulus. When the eye is viewing a surface of uniform luminance, the number of trolands is equal to the product of the area in square millimetres of the limiting pupil, natural or artificial, by the luminance of the surface in candelas per square metre.

trichromatic units	Units of the primary lights comprising a trichromatic match, developed by W. D. Wright. Sometimes called "T-units".
Tristimulus colorimeter	Instrument for measuring the tristimulus values of a colour stimulus.
Tristimulus value	Amounts of the three reference colour stimuli, in a given trichromatic system, required to match the colour of the stimulus considered.
Visible radiation	Any optical radiation capable of causing a visual sensation directly. In colorimetry one usual takes the wavelength limits of visible radiation as 380 nm and 780 nm.
White point	An achromatic reference stimulus in a chromaticity diagram that corresponds to the stimulus producing in an image area the perception of white.

INDEX

Terms

Terms with a letter G as page number refer to definitions to be found in the Glossary. This technique is usually used if more than one reference page number is in the list. Otherwise the page number, where the definition of the term can be found, is in Italic.

Colorimetry: Understanding the CIE System, Edited by János Schanda

Printed in the USA/Agawam, MA
September 16, 2026

910661.028